THE CYTOKINE *FactsBook*

Second Edition

Other books in the FactsBook Series:

Robin Callard and Andy Gearing
The Cytokine FactsBook

Steve Watson and Steve Arkinstall
The G-Protein Linked Receptor FactsBook

Shirley Ayad, Ray-Boot Handford, Martin J. Humphries, Karl E. Adler and C. Adrian Shuttleworth
The Extracellular Matrix FactsBook, 2nd edn

Grahame Hardie and Steven Hanks
The Protein Kinase FactsBook
The Protein Kinase FactsBook CD-Rom

Edward C. Conley
The Ion Channel FactsBook
I: Extracellular Ligand-Gated Channels

Edward C. Conley
The Ion Channel FactsBook
II: Intracellular Ligand-Gated Channels

Edward C. Conley and William J. Brammar
The Ion Channel FactsBook
IV: Voltage-Gated Channels

Kris Vaddi, Margaret Keller and Robert Newton
The Chemokine FactsBook

Marion E. Reid and CHristine Lomas-Francis
The Blood Group Antigen FactsBook

A. Neil Barclay, Marion H. Brown, S.K. Alex Law, Andrew J. McKnight, Michael G. Tomlinson and P. Anton van der Merwe
The Leucocyte Antigen FactsBook, 2nd edn

Robin Hesketh
The Oncogene and Tumour Suppressor Gene FactsBook, 2nd edn

Jeffrey K. Griffith and Clare E. Sansom
The Transporter FactsBook

Tak W. Mak, Josef Penninger, John Rader, Janet Rossant and Mary Saunders
The Gene Knockout FactsBook

Bernard J. Morley and Mark J. Walport
The Complement FactsBook

Steven G.E. Marsh, Peter Parham and Linda Barber
The HLA FactsBook

Hans G. Drexler
The Leukemia-Lymphoma Cell Line FactsBook

Clare M. Isacke and Michael A. Horton
The Adhesion Molecule FactsBook, 2nd edn

Marie-Paule Lefranc and Gérard Lefranc
The Immunoglobulin FactsBook

Marie-Paule Lefranc and Gérard Lefranc
The T Cell Receptor FactsBook

Vincent Laudet and Hinrich Gronemeyer
The Nuclear Receptor FactsBook

THE CYTOKINE *FactsBook*

Second Edition

Katherine A. Fitzgerald
Department of Biochemistry & Biotechnology Institute, Trinity College, Dublin, Ireland

Luke A.J. O'Neill
Department of Biochemistry & Biotechnology Institute, Trinity College, Dublin, Ireland

Andy J.H. Gearing
Biocomm International, Melbourne, Australia

Robin E. Callard
Institute of Child Health, University College London, London, UK

ACADEMIC PRESS

A Harcourt Science and Technology Company

San diego San Francisco New York Boston
London Sydney Tokyo

This book is printed on acid-free paper.

Academic Press
A Harcourt Science and Technology Company
Harcourt Place, 32 Jamestown Road, London NW1 7BY, UK
http://www.academicpress.com

Academic Press
A Harcourt Science and Technology Company
525 B Street, Suite 1900, San Diego, California 92101-4495, USA
http://www.academicpress.com

ISBN 0-12-155142-3

Library of Congress Catalog Number: 2001092474

A catalogue record for this book is available from the British Library

INSTRUCTIONS ON GAINING ACCESS TO THE ON-LINE VERSION OF THE CYTOKINE FACTSBOOK, SECOND EDITION

Access for a limited period to an on-line version of The Cytokine FactsBook, Second Edition is included in the purchase price of the print edition.

The on-line version has been uniquely and persistently identified by the Digital Object Identifier (DOI)

doi: 10.1006/bkcf.2001

By following the link

http://dx.doi.org/10.1006/bkcf.2001

from any Web Browser, buyers of the Cytokine FactsBook, Second edition will find instructions on how to register for access.

If you have any problems with accessing the on-line version, e-mail tscap@hbtechsupport.com

Typeset by Bibliocraft, Dundee, Scotland
Transferred to Digital Printing in 2008
01 02 03 04 05 06 07 BC 9 8 7 6 5 4 3 2 1

Contents

Preface

The updating of the *Cytokine FactsBook* was started by Robin Callard in 1998. Given the rate at which new cytokines were being discovered and characterized, the work was then shared out, with Kate Fitzgerald, Luke O'Neill and Andy Gearing contributing. Kate Fitzgerald in particular did much of the work on novel cytokines. There are many people who have helped with advice and information during the writing of this book. In particular, we would like to mention Ewan Robson (Dept. of Pathology, University of Cambridge) and Eleanor Dunn (Trinity College Dublin) for help with database searches and structural models respectively. We also wish to thank Tessa Picknett, Margaret MacDonald and Emma Parkinson for their patience and hard work in getting the manuscript into press.

R.E.C. also wishes to acknowledge funding from Action Research, the Leukaemia Research Fund, the Medical Research Council, and the Wellcome Trust.

A.J.H.G. wishes to thank Kate Owen for cheery help, British Biotech for employment, Academic Press for very nearly understanding I had a day job (sorry Tessa!) and finally Frances, Jamie and Catherine.

K.F. wishes to acknowledge funding from Enterprise Ireland and the European Union and members (both past and present) of the Inflammation Group, Trinity College Dublin (in particular Ewan Robson and Eleanor Dunne) for much appreciated help.

L.O.N. wishes to acknowledge funding from the Health Research Board, Enterprise Ireland, Bioresearch Ireland, the European Union and finally Margaret, Stevie and Sam.

There will undoubtedly be some omissions and errors in this volume although we hope they will be infrequent. We would greatly appreciate being informed of any inaccuracies by writing to the Editor, Cytokine FactsBook, Academic Press, Harcourt Place, 32 Jamestown Road, London, NW1 7BY, UK, so that these can be rectified in future editions.

Katherine Fitzgerald

Luke O'Neill

Andy Gearing

Robin Callard

Abbreviations

CSF	Colony-stimulating factor
DAG	1,2-Diacylglycerol
EBV	Epstein–Barr virus
FADD	Fas-associated death domain
FNIII	Fibronectin type III domain
GAG	Glycosaminoglycan
GAP	GTPase-activating protein
GF	Growth factor
GPI	Glycosyl-phosphatidylinositol
IFN	Interferon
Ig	Immunoglobulin
IL	Interleukin
IP_3	Inositol 1,4,5-trisphosphate
IRAK	IL-1 receptor associated kinase
JAK	Janus kinase
LAK	Lymphokine-activated killer
LPS	Lipopolysaccharide
LRR	Leucine-rich region
LTR	Long terminal repeat
M_r	Molecular ratio
NK	Natural killer
ORF	Open reading frame
PGE_2	Prostaglandin E_2
PHA	Phytohaemagglutinin
PI	Phosphatidylinositol
PIP_2	Phosphatidylinositol bisphosphate
PKC	Protein kinase C
PLC	Phospholipase C
PLD	Phospholipase D
PTK	Protein tyrosine kinase
SCID	Severe combined immunodeficiency
STAT	Signal transducer of activated transcription
TCR	T-cell receptor
TIR	Toll/IL-1 receptor
TLR	Toll-like receptor
TRADD	TNF receptor-associated death domain
TRAF	TNF receptor-associated factor
4PS	IL-4 induced phosphotyrosine substrate

Abbreviations for all the cytokines are not included here as the abbreviation and full name appears at the beginning of each entry.

'Now what I want is, Facts. Teach these boys and girls nothing but Facts. Facts alone are wanted in life. Plant nothing else, and root out everything else. You can only form the minds of reasoning animals upon Facts: nothing else will ever be of any service to them.'

From Hard Times, by Charles Dickens

Section I

THE INTRODUCTORY CHAPTERS

1 Introduction

AIMS OF THE BOOK

The main aim of this book is to provide a compendium of human and murine cytokines and their receptors. It is an updated version of the original *Cytokine FactsBook* published in 1994. The number of cytokines covered in the 1994 edition was 50. This version provides an update on these cytokines and an additional 51 cytokines are covered. The information provided is confined largely to physicochemical properties and includes amino acid sequences. The biological properties are not treated in detail but are described briefly, as are the major signal transduction pathways activated by each cytokine. There are also introductory chapters on the nature of cytokines and cytokine families, the cytokine network, cytokine receptor superfamilies and chemokines.

WHAT IS A CYTOKINE?

Cytokines are soluble proteins or glycoproteins produced by leukocytes, and in many cases other cell types, which act as chemical communicators between cells, but not as effector molecules in their own right. They have many roles, but a unifying feature of most cytokines is that they are regulators of host defence against pathogens and/or the inflammatory response. Most are secreted, but some can be expressed on the cell membrane, and others are held in reservoirs in the extracellular matrix. Cytokines bind to specific receptors on the surface of target cells that are coupled to intracellular signal transduction and second messenger pathways. Most cytokines are growth and/or differentiation factors and they generally act on cells within the haematopoietic system. There are four key features displayed by most cytokines:

1 Pleiotropy – Most cytokines have more than one action, for example IL-6 will provoke hepatocytes to produce acute phase proteins and it is also a growth factor for B cells.
2 Redundancy – Most cytokines have biological effects also observed in another cytokine, for example both IL-2 and IL-15 promote T cell proliferation.
3 Potency – Most cytokines act in the nanomolar to femtomolar range.
4 Action as part of a network or cascade – Most cytokines are part of a cascade of cytokines released in succession, often act in synergy, and are often counter-regulated by inhibitory cytokines or soluble receptors.

Most of these features can be explained in terms of receptors and signal transduction. Cytokines are pleiotropic because their receptors are expressed on multiple cell types and the signalling pathways activated will increase gene expression specific for that cell type. Similarities in the predicted amino acid sequences within the cytosolic domain of cytokine receptors lead to similar signals being generated by different cytokines, hence the redundancy. Cytokines are potent because of the very high affinity which cytokine receptors have for their ligands: most cytokine receptors have a negligible 'off' rate.

The number of proteins which may be classified as cytokines will become apparent when the full human genome sequence is annotated. At this stage it is possible to predict the number of homologues for certain cytokines – for example there are at least 15 proteins that are similar in sequence to TNF, while at least five proteins encode homologues of IL-1. Cytokines are likely to account for a significant percentage of the thousands of secreted proteins made by cells. Attempts to annotate and describe the precise function of all cytokines represents a huge challenge but will be necessary for a full understanding of how the immune system in particular functions in health and disease.

Most of the molecules covered in this book fall easily within these definitions of cytokines, but some do not. Erythropoietin (Epo) is not produced by leukocytes, but does act on haematopoietic precursors to generate red blood cells, and its receptor belongs to the type I (haematopoietin) receptor superfamily. Nerve growth factor (NGF), neurotrophin-3 (NT-3), and brain-derived neurotrophic factor (BDNF) are all members of the same family of cytokines, which are produced and act predominantly in the nervous system, but NGF also affects B cells, and its low-affinity receptor is related to the tumour necrosis factor receptor (TNFR). Not all soluble peptide mediators are considered to be cytokines (e.g. insulin) and these exceptions have not been included. In the end, the decision whether to include a molecule as a cytokine or not must be somewhat arbitrary. If there are any omissions which offend, please let the Editor know and we will try to include them in the next edition. Information in this book is provided only for human and murine (or in some cases rat) cytokines that have been cloned, and for which there is a reasonable body of biological information. Where the receptors have been cloned, they are also included.

CYTOKINE FAMILIES

It should become clear from reading the entries in this book that cytokine nomenclature owes little to any systematic relationships between molecules. This is a reflection of the different historical approaches to naming new cytokines which were based either on cell of origin or initial defining bioassay. These systems have created anomalies such as tumour necrosis factor, originally defined as causing necrosis of solid tumours, but which is now thought to be primarily an immunomodulatory and proinflammatory cytokine, and which has proven ineffective as an anticancer agent in several clinical trials. The interleukin nomenclature, which merely assigns a sequential number to new factors, is a rational system, but it has not been universally applied to new factors. A consensus on the grouping of cytokines into families has settled on classifying cytokines based on the structure of their receptors[1]. One somewhat surprising finding has been that although there are many cytokines, receptors for these cytokines fall into a relatively small number of families, with highly conserved structural features[1–6]. As stated above, this provides an explanation for the redundancy in biological effects among cytokines, with many different cytokines affecting T-cell function for example. Redundancy can now be explained in terms of similar signal transduction pathways being triggered by highly similar receptors.

There are six major families of cytokine receptors (each family being defined by similar sequences in their cytosolic domains) and most of the cytokines with known receptors described in this book fall into one of the six families (Table 1). Within each family there are homologies in receptor sequence both intracellularly and extracellularly but there are also differences, particularly extracellularly where additional domains can be found. These subtleties are discussed in more detail in Chapter 4. Broadly speaking, the functions of cytokines fitting into each family are also conserved and these are also listed in Table 1. Because cytokines are so pleiotropic, however, it is difficult to classify them according to function. Because the receptors show homologies extracellularly, the structures of the cytokines themselves are also conserved, with subtle differences allowing cytokines to discriminate between receptors. The structural features of cytokines are therefore also listed in Table 1 and solid models of representatives of each of the families, some as a complex with their receptor, are shown in Figure 1.

Receptors for three cytokines which do not fit into these families have been left out. The TGFβ receptor has a serine/threonine kinase domain and is the only cell activator to have a

Table 1.1 *Cytokine families grouped according to receptor structure*

Family	Key receptor feature	Members	Shared function	Cytokine structure
Haematopoietin receptors (type I receptors)	Cytosolic box 1/2	IL-6R, G-CSFR, gp130, IL-12R	T/B-cell activation	4 α-helical bundles
	WSXWS sequence extracellularly	LIFR, IL-2Rβ, IL-2Rγ, IL-4R, IL-3Rα, IL-7Rα, IL-9R, GM-CSFR, EpoR, prolactin R, GH-R, Tpo-R; β-chain for IL-3R, GM-CSFR and IL-5R	Haemato-poiesis	
Interferon receptors (type II receptors)	Cytosolic box 1/2 Fibronectin domains extracellularly	IFNα/βR, IFN-Rα/β, IL-10R	Antiviral (not IL-10)	4 α-helical bundles
TNF receptors	Cytosolic death domain Four Cys-rich regions extracellularly	p55 TNFR, p75 TNRR (no death domain), LTβR, NGFR, CD40, CD30, CD27, 4-1BB, OX40, TRAMP (DR3), TRAILR (DR4)	Proinflam-matory	Jelly roll motif
IL-1/Toll-like receptors	Cytosolic Toll/ IL-1R (TIR) domain Ig domains (IL-1R subgroup) or leucine-rich repeats (TLR subgroup) extracellularly	IL-1RI, IL-1RII (no TIR), IL-1RAcP IL-18R α/β-chains, T1/ST2, IL-1Rrp2, TIGGIR-1, IL- 1RAPL, SIGIRR TLR1-TLR10	Proinflam-matory	β-trefoil
Tyrosine kinase receptors	Cytosolic tyrosine kinase domain	M-CSFR, EGFR, TGF, IGFs, FGFs	Growth factors	β-sheet
Chemokine receptors	7 transmembrane spanning regions	IL-8, MCPs, RANTES, Eotaxin	Chemotactic	Triple-stranded antiparal-lel β-sheet in Greek key motif

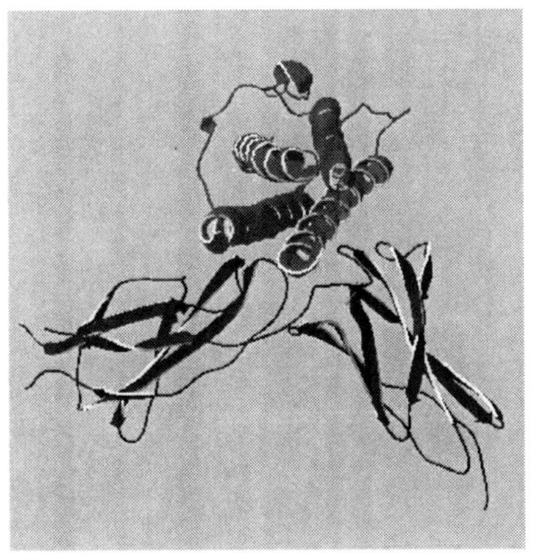

Figure 1-1a

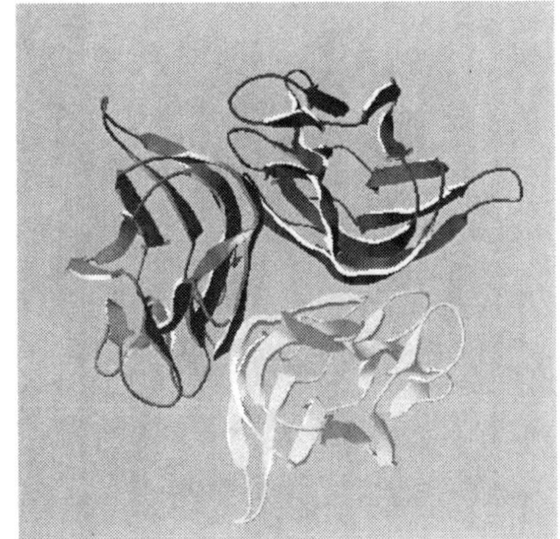

Figure 1-1c

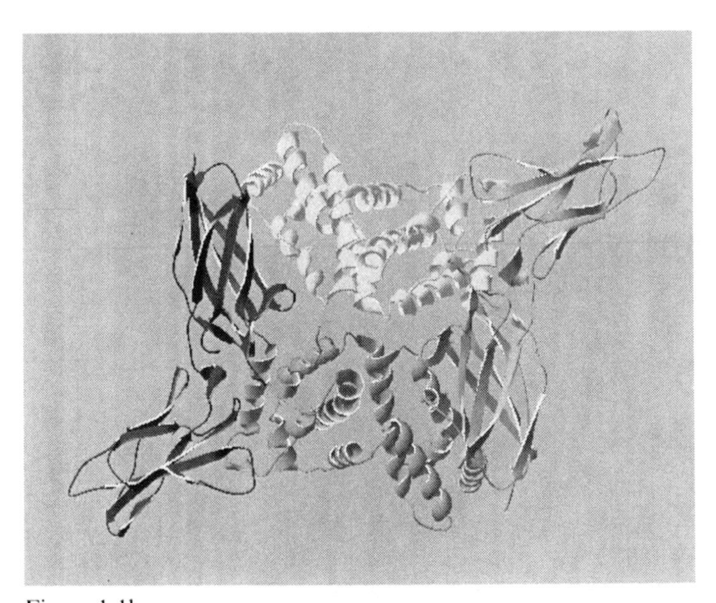

Figure 1-1b

Figure 1.1 *Crystal structures of cytokines and cytokine/receptor complexes. All structures are depicted in ribbon format. Coordinates were downloaded from the Protein Data Bank and viewed using the Swiss-PDB Viewer v3.51. (a) Granulocyte colony-stimulating factor with receptor (PDB I.D. 1CD9). (b) Interferon γ with receptor (PDB I.D. 1FYH). (c) TNF: individual monomers (PDB I.D. 2TNF). (d) Interleukin-1β with Type I IL-1 receptor (PDB I.D. 1ITB). (e) Basic fibroblast growth factor (PDB I.D. 1BFF). (f) IL-8 (PDB I.D. 3IL-8).*

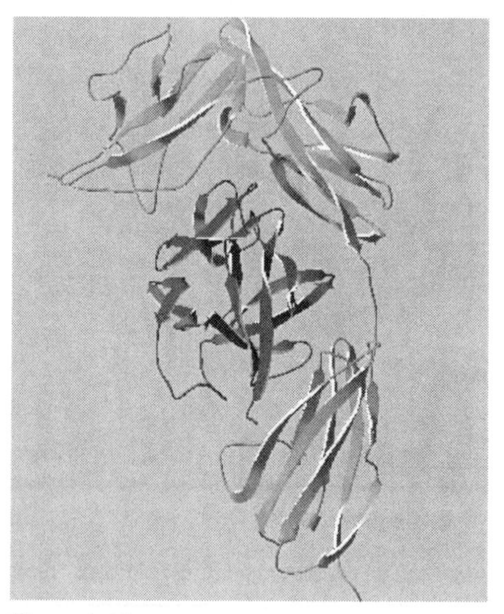

Figure 1-1d

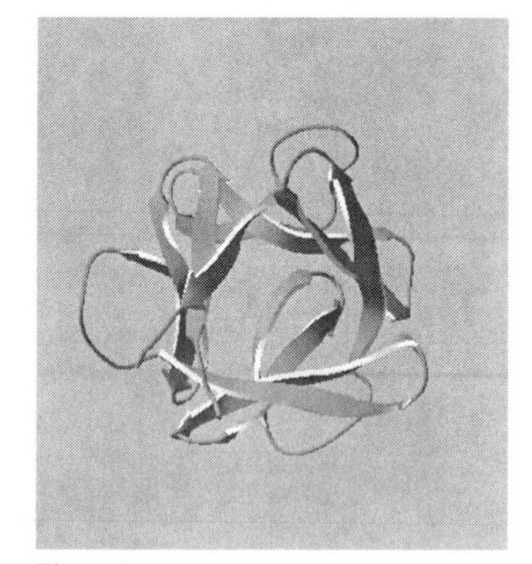

Figure 1-1e

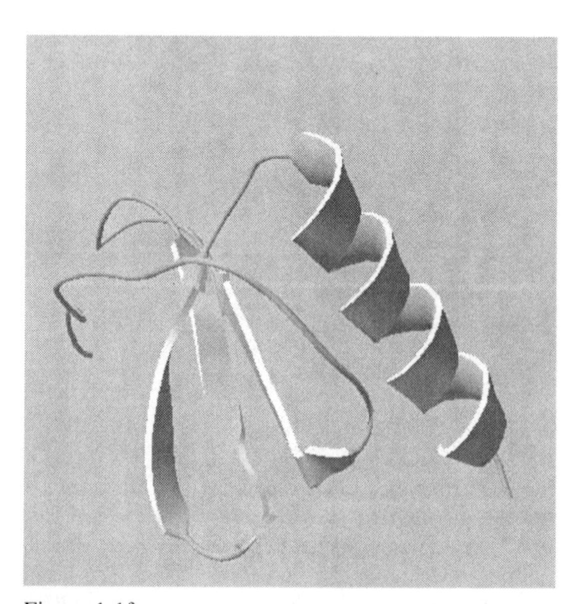

Figure 1-1f

receptor with such a domain so far described. The α-chains of the receptor complexes for IL-2 and IL-15 contain complement control protein domains involved in ligand binding.

Overall, it seems unlikely that cytokine nomenclature will reflect the family relationships described here, either in terms of receptor families or structures of cytokines. General principles on the functioning of cytokines within families have emerged, however, and provide a framework from which to approach the complexities inherent in the study of cytokine function.

References

[1] Taniguchi, T. (1995) Science 268, 251–255.
[2] Stroud, R.M. et al. (2000) In The Cytokine Reference, Academic Press, London, pp. 21–34.
[3] Bazan, J.F. (1990) Proc. Natl Acad. Sci. USA 87, 6934–6938.
[4] Ihle, J.N. et al. (1995) Annu. Rev. Immunol. 13, 369–398.
[5] Aggarwal, B.B. et al. (2000) In The Cytokine Reference, Academic Press, New York, pp. 1619–1631.
[6] O'Neill, L.A.J. and Dinarello, C.A. (2000) Immunol. Today 21, 206–209.

2 Organization of the data

Cytokine entries include the following information:

Other names

Most cytokines have more than one name. This section lists names which appear in the literature and can be directly related to the protein under discussion. We have used the interleukin nomenclature whenever this has been assigned unequivocally. All other cytokines are entered under their most commonly accepted name. Alternative names are also listed.

THE MOLECULE

At the beginning of each entry is a brief description of the molecule and its main biological properties.

Crossreactivity

The degree of amino acid sequence homology between human and mouse cytokines is given when known, together with cross-species reactivity. In some cases, comparisons with other species are also given.

Sources

A list of cell types known to produce each cytokine is included.

Bioassays

Bioassays for each cytokine are described in brief. For the most part, these have been taken from methods used by the National Institute for Biological Standards, South Mimms, UK as described in refs 1 and 2.

Physicochemical properties of the cytokines

This table includes basic physicochemical information on human and mouse cytokines. The number of amino acids and predicted molecular weight for the mature proteins are calculated after removal of the signal peptide and propeptides where relevant. In some cases the cleavage point of the signal peptide has been determined by sequencing and in others from computer prediction. In cases where the position of the signal peptide is unclear, the predicted molecular weight for the unprocessed precursor is given. Potential *N*-linked glycosylation sites are identified by the consensus sequence AsnXSer/Thr except when X is Pro or for the sequence AsnXSer/ThrPro which is not usually glycosylated. The number of potential sites in the table is for the extracellular portion of the molecule only.

3-D structure

Information on the tertiary structure of each cytokine is taken from the SwissProt database, original papers or from Macromolecular Structures 19911993 published by Current Biology[3]. It includes data derived from X-ray or NMR structures, or predictions based on molecular models. In some cases because of the high degree of homology in amino acid sequence, the structures are predicted to resemble that of previously solved ones.

Gene structure and chromosomal localization

The chromosomal localization for the human cytokines is taken from original papers and/or the Human Gene Mapping (HGM 11)[4]. Mouse mapping data are taken from ref. 5 or original papers in some cases. The gene structure is taken from original papers with the number of exons and introns indicated where known.

Amino acid sequences

Human and mouse amino acid sequences are given for each cytokine and receptor where known. In some cases, the murine sequence is not available and the rat sequence is given instead. The sequences for most entries were taken directly from the SwissProt Protein Sequence Database, which is a database of protein sequences produced collaboratively by Amos Bairoch (University of Geneva) and the EMBL Data Library. The data in SwissProt are derived from translations of DNA sequences from the EMBL Nucleotide Sequence Database, adapted from the Protein Identification Resource (PIR) collection, extracted from the literature or directly submitted by researchers. SwissProt contains high-quality annotation, is nonredundant and is cross-referenced to several other databases, notably the EMBL, PROSITE pattern database and PDB. In cases where a SwissProt entry does not exist for a given protein, a TrEMBL entry is given. TrEMBL is a protein sequence database supplementing the SwissProt Data Bank. TrEMBL contains the translations of all coding sequences (CDS) present in the EMBL database that are not yet integrated in SwissProt. TrEMBL can be considered as a preliminary section of SwissProt. In certain cases, cDNA nucleic acid sequences were obtained from EMBL and/or Genbank and translated using the translation tool from the Expasy Molecular Biology Server (http://www.expasy.ch/tools/#translate).

In all cases, the accession number is listed with the sequence. In all sequences, the single-letter amino acid code is used (Table 1). The numbering starts with the N-terminal amino acid after removal of the signal peptide. If the N-terminus has not been unequivocally assigned, the signal sequence is predicted according to consensus rules[6] and numbered to −1. Propeptides removed during post-translational modifications are shown in italics. The transmembrane portions of the sequences for cytokine receptors and cytokine precursor proteins (in cases where the precursor is a transmembrane protein) are underlined. Potential *N*-linked glycosylation sites marked by N in bold type are predicted by the presence of sequences AsnXSer or AsnXThr with the exceptions AsnProSer/Thr which are not normally glycosylated and AsnXSer/ThrPro which are often not glycosylated[7–9]. *O*-Linked glycosylation occurs at Ser or Thr residues. Although there is no clear-cut sequence motif that invariably indicates *O*-linked glycosylation, it usually occurs where there is a preponderance of Ser, Thr and Pro. Sequence motifs of particular interest are annotated under the sequence.

THE RECEPTORS

A brief description of the cytokine receptors with comments on important features is given in this section. The criteria used for defining cytokine receptor superfamilies are given in Chapter 4. The symbols used to represent the various domains, glycosylation and membrane attachment are given in Figure 1.

Table 1 *Single-letter amino acid codes*

	Amino acid	Code	
Small hydrophilic	Serine	Ser	S
	Threonine	Thr	T
	Proline	Pro	P
	Alanine	Ala	A
	Glycine	Gly	G
Acid, acid amide	Asparagine	Asn	N
Hydrophilic	Aspartic	Asp	D
	Glutamine	Gln	Q
	Glutamic	Glu	E
Basic	Histidine	His	H
	Arginine	Arg	R
	Lysine	Lys	K
Small hydrophobic	Methionine	Met	M
	Isoleucine	Ile	I
	Leucine	Leu	L
	Valine	Val	V
Aromatic	Phenylalanine	Phe	F
	Tyrosine	Tyr	Y
	Tryptophan	Trp	W
Sulfhydryl	Cysteine	Cys	C

Distribution

The tissue distribution of the receptors has been determined in some cases by ligand (cytokine) binding studies. Otherwise, it is assumed from mRNA expression or biological response.

Physicochemical properties of the receptors

This table includes the number of amino acids in both the precursor and the mature processed protein, predicted and expressed molecular ratio (M_r), and the number of *N*-linked glycosylation sites in the extracellular portion of the mature protein. The affinity of the receptor for its cytokine is also given. These data have been taken from binding studies with the natural receptor or the cloned receptor expressed on cells transfected with receptor cDNA.

Signal transduction

This section describes in brief what is known about the intracellular signal transduction pathways coupled to the cytokine receptors.

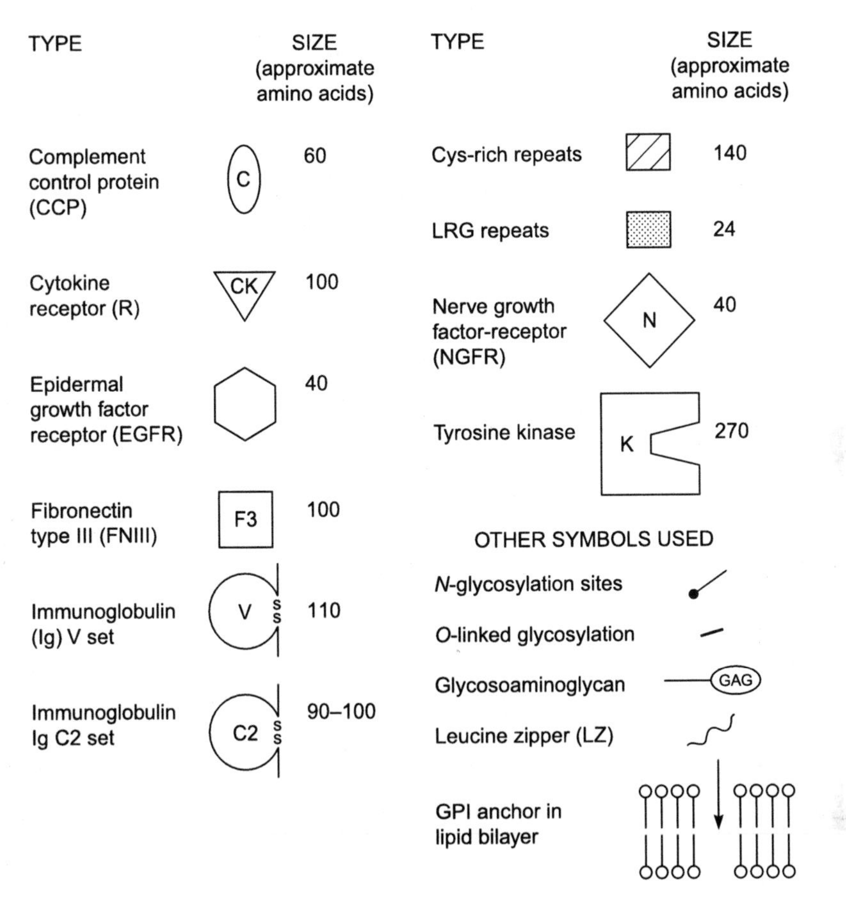

Figure 1 *Models for domains and repeats found in leukocyte membrane proteins. These models are used in the diagrams drawn for the entries in Section II.*

Chromosomal location

The chromosomal location of the receptors is taken from original papers and/or HGM 11[4,5].

Amino acid sequences of cytokine receptors

Sequence data for human and mouse receptors are given as described above for the cytokines. Transmembrane domains are underlined and short sequences of particular interest, such as the WSXWS motif in fibronectin type III (FNIII) domains, are annotated under the sequence.

References

1. Thorpe, R. et al. (1992) Blood Rev. 6, 133–148.
2. Wadhwa, M. et al. (1992) In Cytokines: A Practical Approach, Balkwill, F.R. ed., IRL Press, Oxford, pp. 309–330.
3. Hendrickson, W.A. and Wuthrich, K. eds (1991, 1992, 1993) Macromolecular Structures, Current Biology, London.

[4] Human Gene Mapping 11 (HGM 11) (1991) Cytogenet. Cell Genet. 58, 1–2200.
[5] Copeland, N.G. et al. (1993) Science 262, 57–66.
[6] von Heijne, G. (1983) Eur. J. Biochem. 133, 17–21.
[7] Bause, E. and Hettkamp, H. (1979) FEBS Lett. 108, 341–344.
[8] Kornfeld, R. and Kornfeld, S. (1985) Annu. Rev. Biochem. 54, 631–664.
[9] Gavel, Y. and von Heijne, G. (1990) Protein Eng. 3, 433–442.
[10] Barclay, N.A. et al. (1993) The Leucocyte Antigen FactsBook, Academic Press, London.

3 The cytokine network

INTERACTIONS BETWEEN CYTOKINES

Individually, cytokines are potent molecules which, *in vitro*, can cause changes in cell proliferation, differentiation and movement at nanomolar to picomolar concentrations. Injection of cytokines into animals and humans either systemically or locally can also have profound effects on leukocyte migration and function, haematopoietic cell numbers, temperature regulation, acute phase responses, tissue remodelling and cell survival. The individual entries in this book describe the specific properties of each cytokine, but delineation of the mechanisms by which cytokines cause these effects is complicated by the tendency of cytokines to affect the expression of other cytokines and/or their receptors. In addition, it is clear that there are no circumstances *in vivo* in which cytokines are produced individually. Rather they are produced together with other cytokines in patterns characteristic of the particular stimulus or disease.

The potency of cytokines, and the potential for amplification and damage which excessive cytokine production carries, has resulted in elaborate controls on cytokine production and action. The current view of cytokine biology reflecting these concepts is of a network of positive and negative cytokines, and cytokine inhibitors and inducers, which combine to give an overall biological or clinical response[1–6].

From experiments using cytokine proteins as agonists, or by blocking cytokine production or action with drugs or monoclonal antibodies, the contributions which many individual cytokines make to particular aspects of immunity, inflammation and haematopoiesis have been delineated. In addition, the use of transgenic mice which overexpress certain cytokines, or mice deficient in cytokines, their receptors or signalling proteins activated by them, have allowed investigators to determine more precisely the role of many individual cytokines[7]. What emerges is a highly complex network of interacting cytokines, and for the first time investigators are attempting to provide complex mathemathical models to predict cytokine function. Apparent redundancy in the action of cytokines and the production of inhibitory cytokines or soluble receptors contribute to a chaotic network of interactions[8].

Cytokines act in cascades and networks

The effect of a cytokine *in vivo* will depend on the immediate cytokine environment. This is best illustrated in the context of inflammation; at an inflamed site, many different cytokines and their inhibitors can be detected. Apparent redundancy in the actions of cytokines would imply that blocking a given cytokine would have only a marginal effect. This is not the case, however, as is best illustrated by the profound anti-inflammatory effect of anti-TNF strategies, which currently comprise neutralizing antibodies or soluble receptors[9]. At the same time, anticytokine measures will be induced, such as increased production of the IL-1 receptor antagonist (IL-1Ra) and the inhibitory cytokine IL-10. It is the balance of activating and inhibitory cytokines that will determine outcome[6]. Recently, it has been suggested that certain polymorphisms in the promoter regions of cytokine genes might lead to imbalances in cytokine networks. For example, individuals with a polymorphism in the gene for IL-1β, which gives rise to the overproduction of IL-1 in response to *Helicobacter pylori*, show a higher incidence of gastric cancer because of extensive gastritis[10]. It is likely that many more imbalances in the cytokine network will be described because of polymorphisms, and will be found to be important in the pathogenesis of many diseases.

Th1/Th2 polarization

A particularly informative paradigm with regard to cytokine networks has been the description of Th1/Th2 cytokine patterns. Work by Mossman and colleagues identified two groups of cytokines produced by different T helper (Th) lymphocytes which favour either cell-mediated and inflammatory immunity (Th1) or antibody-mediated humoral immunity (Th2)[11–13]. The Th2 cytokines, such as IL-4, IL-5, IL-6, IL-10 and IL-13, generally promote B-cell function, specifically favouring IgE and IgA responses. A major role here appears to be in host defence against gastrointestinal nematodes. The Th1 cytokines, such as IFNγ and IL-2, can also have some antibody-promoting effects, but favour an IgG2a response. The main role of Th1 cells is actually to clear intracellular bacteria by promoting cellular immunity. These two cytokine groups are also antagonistic in that IFNγ inhibits Th2 cytokine production, whereas IL-10, IL-4 and IL-13 inhibit Th1 cytokine production, probably via effects on antigen-presenting cells. The initiating factors which predispose towards Th1 or Th2 responses are not fully understood. Likely determinants include pathogen-derived molecules, the nature of the peptide being presented by antigen-presenting cells and the nature of costimulatory molecules, all in the context of genetic background[14].

Along with their roles in host defence, polarized Th1 and Th2 responses can also contribute to the pathogenesis of immune-mediated diseases. Th1 responses are involved in organ-specific autoimmune diseases while Th2 responses are implicated in allergic disorders in genetically susceptible individuals. A key question has therefore concerned the molecular basis for polarization. Both Th1 and Th2 cells are derived from a common precursor termed Th0. The issue of polarization therefore lies at the heart of cellular differentiation, and discoveries in this area are proving informative to other areas of biology. The cytokine IL-12, which is produced by antigen-presenting cells, promotes differentiation into Th1 cells, whilst IL-4 produced by Th0 cells promotes their differentiation into Th2 cells[14]. CD28 is expressed on the surface of Th1 and Th2 cells and is an important costimulus to promote activation of both cell types, while a homologue of CD28, termed ICOS, may selectively promote activation of Th2 cells[15–18]. Both are activated by different molecules on the surface of antigen-presenting cells, with B7 activating CD28 and B7h activating ICOS[19]. Different populations of dendritic cells may be involved in this process. In humans, lymphoid dendritic cells promote Th2 cell development while myeloid cells activate Th1 cells[20].

Transcriptional control is critical since it is the specific transcription of distinct genes that governs polarization. The transcription factors STAT6, GATA3, c-Maf, JunB and NFATc drive Th2 cytokines, while STAT4, T-bet and NFκB are implicated in Th1 cytokines[14–16]. Specific signals that activate these transcription factors have been studied, with Jun N-terminal kinase and p38 MAP kinase being essential for Th1 cell function[15]. Most recently, it has become clear that some transcription factors, such as T-bet, alter chromatin structure, allowing a Th cell to express specific genes[16]. Detailed molecular understanding of these processes may provide opportunities for therapeutic intervention. Inhibitors of specific cell surface proteins, such as T1/ST2 or the chemokine receptor CCR3 on Th2 cells, or IL-18α or CCR5 on Th1 cells may also provide possible means to control specific populations. Table 1 summarizes current knowledge on molecules specific for Th1 and Th2 cells[14–19].

REGULATION OF CYTOKINE PRODUCTION AND ACTION

Gene expression

There is some evidence for the constitutive expression of haematopoietic cytokines such as M-CSF, G-CSF, SCF, IL-6 and Epo, which are necessary to maintain steady state

Table 3.1 *Th1 and Th2 cells*

	Th1	Th2
Roles	Inflammation and organ-specific autoimmunity	Humoral immunity and allergy
Cytokines produced	IL-2, IFN-γ, IL-18, LT	IL-4, IL-5, IL-9. IL-10, IL-13
Activators	IL-12, IL-18, CD28	IL-4, ICOS
Inhibitors	IL-10, IL-4	IFN-γ
Cell surface molecules	LAG-3, IL-12R-β2, CCR5, CXCR3	ICOS, CD30, CD62, CCR3,4,8, CXCR4
Transcripiton factors	T-bet, STAT-4, NF- κB	STAT-6, GATA-3, c-Maf, JunB, NFATc
Signals	JNK, p38 MAPK	

haematopoiesis. Moreover, several cytokines are presynthesized, and stored either in cytoplasmic granules, e.g. GM-CSF, TGFβ, PF-4, PDGF[21]; as membrane proteins, e.g. TNFα, IL-1β, EGF, TGFα[22]; or complexed with cell surface binding proteins or extracellular matrix, e.g. TGFβ, MIP-1β, IL-8[23,24]. These pools of cytokine protein are available for rapid release in response to stimulation. Most cytokines, however, are not constitutively expressed in adult animals, but are rapidly produced in response to stimulation. The stimuli for gene expression are well characterized, particularly for the haematopoietic cells. In general, infectious agents such as bacteria, viruses, fungi and parasites, as well as mechanical injury and toxic stimuli are potent cytokine inducers.

In addition to classical antigens, infectious agents also express many nonspecific cytokine-inducing molecules, e.g. endotoxins[25]. It has recently emerged that Toll-like receptors (TLRs), which are members of the IL-1 receptor superfamily, are receptors for such pathogen-derived molecules. TLR-4 appears to be the receptor for LPS, TLR-2 is responsive to products from gram-positive bacteria (e.g. peptidoglycan)[26] and TLR-9 responds to CpG DNA from bacteria[27]. The so-called Toll/IL-1R (TIR) domain in the cytosolic portions of these receptors signals cytokine induction. TLRs may therefore be the key receptors important for adjuvancy in the immune response. Two signals activated are the transcription factor NFκB and p38 MAP kinase. NFκB is critical for the induction of many cytokine genes, including IL-2, IL-6, IL-8, MCP-1, TNF and GM-CSF. The role of p38 MAP kinase is to stabilize cytokine mRNAs, which are unstable because of multiple AU repeats in their 3′ ends. How p38 stabilizes these mRNAs is not known. Inhibition of both signals will block the production of many cytokines.

For many cells, cytokines are themselves potent cytokine inducers. Some, such as IL-1, TNF and IFNγ, are particularly potent inducers of cytokine gene expression, and are referred to as proinflammatory cytokines[28]. Others, like IL-1 and TNF, are even capable of stimulating their own production[29], again acting via signals such as NFκB and p38 MAP kinase. As stated above, specific transcription factors play a key role in the induction of specific cytokines from Th1 and Th2 cells.

In addition to the many positive stimuli, several mediators act to limit or prevent cytokine gene expression, or to limit cytokine action. The classical inhibitors of cytokine gene expression are the glucocorticoid hormones and the synthetic steroids which are widely used as immunosuppressants and anti-inflammatory drugs[30]. They are thought to act as part of an intracellular glucocorticoid receptor complex which binds to glucocorticoid-response elements present in the IL-1, IL-2, IL-3, IL-6, IL-8 and IFNγ genes. Many glucocorticoid-sensitive genes do not have glucocorticoid-response elements however, and the steroid/receptor complex may instead compete with activated transcription factors such as NFκB, for basal transcription factors such as p300/CBP, which are required for induced gene expression. Agents such as cyclosporin A, FK506 and rapamycin also act via cytokine pathways[31–33]. Prostaglandins are also known to inhibit the production of cytokines. As described above, many cytokines act to inhibit cytokine production. The antagonistic Th1 and Th2 cytokines are the best-studied examples of this. Other examples include TGFβ, which is a broad-spectrum inhibitor of cytokine production, and IL-10, which is a potent inhibitor of TNF and IL-1 production by monocytes[34,35]. The mechanism of action of IL-10 is still not fully understood.

Processing

Control of cytokine function can also be achieved by regulating the processing of precursor forms. Many cytokines are initially produced as biologically active integral membrane proteins which need to be proteolytically cleaved to release the active molecule. Examples in this category include EGF, TGFα, IL-1β, IL-1α, TNFα[22]. Alternatively, cytokines such as TGFβ are produced as secreted but biologically inactive precursors, which are enzymatically processed to the active forms[36]. Progress has been made in the identification of enzymes involved in processing. IL-1β-converting enzyme is the founder of the caspase family of cysteine proteases, and is now termed caspase-1. It is required for processing of both pro-IL-1β and pro-IL-18. TNFα-converting enzyme (TACE) is the processing enzyme for TNF, cleaving the membrane-bound form of TNF. It is a member of the ADAMs family of metalloproteinsases[37].

Sequestration

Some growth factors such as TGFβ, FGF, LIF and IL-1 are sequestered on extracellular matrix in connective tissues, skin and bone[23]. This serves as a sink of active cytokine that can be rapidly released when the matrix is broken down during injury or tissue repair. Other cytokines, such as GM-CSF, IL-3 and SCF, are localized to stromal cell layers in bone marrow where they stimulate haematopoiesis[38]. Others, including MIP-1β, and IL-8, bind to endothelial cells at sites of inflammation where they promote leukocyte extravasation[24]. In general, these cytokines are sequestered onto glycosaminoglycans such as heparin, decorin or CD44-like molecules.

Soluble binding proteins

Several binding proteins for cytokines are found in blood and tissue fluids. Some of these are secreted forms of the specific cell membrane receptors such as for TNF, IL-1, IL-2 and IL-6, whereas others are less specific such as α_2-macroglobulin[39–41]. These binding proteins may serve as passive carriers of the cytokines, either extending their half-lives or promoting their excretion, or they may act as circulating inhibitors, limiting systemic effects of the cytokines. The soluble form of the IL-6 receptor is unusual in that it complexes with IL-6 to form a biologically active molecule which binds directly to the IL-6 receptor signalling chain[42].

Receptor antagonists

To date the only naturally occurring cytokine receptor antagonist that has been identified is IL-1Ra. It is structurally related to IL-1α and IL-1β and binds to the IL-1 receptors, but does not cause signal transduction, thereby acting as a specific receptor antagonist[43–45].

Receptor modulation

Control of cytokine function can also be achieved by modulating receptor number through controlling gene expression, internalization or receptor shedding. Modulation of receptor affinity or function can also occur by control of receptor phosphorylation, or by competition for shared receptor chains or signal transduction molecules[39,46].

Viruses and the cytokine network

A major area of interest over the past 5 years has been the mechanisms used by viruses to avoid elimination by the immune system. Viruses use an array of strategies here, including inhibition of the humoral response, inhibition of apoptosis, modulation of MHC function and evasion of cytotoxic T lymphocytes[47]. A key mechanism has concerned subversion of the cytokine network. Viruses achieve this in a number of ways. Poxviruses can make a range of soluble cytokine-binding proteins, including those for TNF, IL-1, interferons, IL-18 and chemokines. In some cases these are hijacked from the host genome but in other cases (e.g. chemokine-binding proteins) the virus may have coevolved the gene with the host. Other viruses which make various cytokine-binding proteins are Herpes virus 8, Cytomegalovirus and Shope sarcoma virus. Some viruses make homologues of cytokines (termed virokines) which the virus uses either as an antagonist (e.g. viral MIPs from Herpes virus 8) or an agonist (viral IL-10 from Epstein–Barr virus). Viral proteins can also inhibit cytokine signalling, There are many examples of viral proteins that can inhibit interferon signalling, including E1A from adenovirus which blocks STAT1 function and Tat from HIV, which blocks PKR activity (a key signal for responses to viral double-stranded RNA). Two vaccinia proteins, A46R and A52R, block IL-1, IL-18 and TLR-4 signalling[48]. African swine fever virus makes a homologue of the NFκB inhibitor I-κB. Poxviruses make an inhibitor of caspase-1 termed CrmA, which blocks IL-1β production.

Viruses have therefore acquired a range of strategies to manipulate the cytokine network, attesting to its importance for antiviral immunity. Viruses are in fact teaching us the key components of immunity, and in particular, which cytokines are critical for antiviral host defense. Table 2 lists the major strategies used by particular viruses to manipulate the cytokine network.

Table 3.2 *Viruses and the cytokine network*

Strategy	Used by	Function
Cytokine-binding proteins		
vIFNα/βR	Vaccinia virus (B18R)	Sequester IFNα/β
vIFNγR	Vaccinia virus (B8R)	Sequester IFNγ
vIL-18BP	Molluscum contagiosum virus	Sequester IL-18
vIL-1RII	Vaccinia virus (B15R)	Sequester IL-1β
vCKBP	Myxoma virus (M-T7)	Sequester chemokines
vCKR	Human herpes virus (ORF-74)	Sequester chemokines
Virokines		
vCK	Human herpes virus (vMIP-II)	Chemokine antagonist
	Murine cytomegalovirus (MCK-1)	Chemokine agonist
vGF	Vaccinia virus (C11R)	EGF homologue: replication
vIL-10	Epstein–Barr virus (BCRF-1)	IL-10 homologue: inhibit Th1 cells
vIL-6	Human herpes virus (K2)	IL-6 homologue: angiogenic
Interference with signal transduction		
STAT inhibition	Adenovirus (E1A)	Blocks IFN signalling
PKR inhibition	HIV (Tat)	Blocks dsRNA-dependent PKR
Inhibition of TNF signalling	Adenovirus (E3)	Prevent TNF cytotoxicity
I-κB homologue	African swine fever virus (A238L)	Inhibition of NFκB
Inhibition of Toll/IL-1R signalling	Vaccinia (A46R, A52R)	Inhibition of NFκB
Inhibition of cytokine production		
Caspase-1 inhibition	Cow pox virus (CrmA)	Inhibits processing of IL-1β and IL-18

Note: For additional examples of viral proteins which use these strategies see ref. 47.

References

[1] Balkwill, F. and Burke, F. (1989) Immunol. Today 10, 299–304.
[2] Wong, G.C. and Clark, S.C. (1988) Immunol. Today 9, 137.
[3] Paul, W.E. (1989) Cell 57, 521–524.
[4] Chatenoud, L. (1992) Eur. Cytokine Netw. 3, 509–513.
[5] Aria, K. et al. (1992) J. Dermatol. 19, 575–583.
[6] Feldmann, M. and Brennan, F.M. (2000) In The Cytokine Reference, Academic Press, London, pp. 35–51.
[7] Casciari, J.J. et al. (1996) Cancer Chemother. Biol. Response Modif. 16, 315–346.
[8] Callard, R. et al. (1999) Immunity 11, 507–513.
[9] Maini, R.N. and Taylor, P.C. (2000) Annu. Rev. Med. 51, 207–229.
[10] El-Omar, E.M. et al. (2000) Nature 404, 398–402.
[11] Mosmann, T.R. and Coffman, R.L. (1989) Annu. Rev. Immunol. 7, 145–173.
[12] Mosmann, T.R. et al. (1986) J. Immunol. 136, 2348–2357.
[13] Moller, G. (1991) Immunol. Rev. 123, 2–229.
[14] O'Garra, A. and Arai, N. (2000) Trends Cell Biol. 10, 542–550.
[15] Dong, C. and Flavell, R.A. (2000) Science's STKE: www.stke.org/cgi/content/full/OC_sigtrans;2000/49/pe1
[16] Glimcher, L.H. and Murphy, K.M. (2000) Genes Dev. 14, 1693–1711.
[17] McAdam, A.J. et al. (2000) J. Immunol. 165, 5035–5040.
[18] Dong, C et al. (2001) Nature 409, 97–101.
[19] Ling V. et al. (2000) J. Immunol. 164, 1653–1657.
[20] Hartgers, F.C. et al. (2000) Immunol. Today 21, 542–545.
[21] Jyung, R.W. and Mustoe, T.A. (1993) In Clinical Applications of Cytokines, Oppenheim, J.J. et al. eds, Oxford University Press, Oxford.
[22] Massague, J. and Pandiella, A. (1993) Annu. Rev. Biochem. 62, 515–541.
[23] Noble, N.A. et al. (1992) Prog. Growth Factor Res. 4, 369–382.
[24] Tanaka, Y. et al. (1993) Immunol. Today 14, 111–115.
[25] Sturk, A. et al. eds (1991) Bacterial Endotoxins: Cytokine Mediators and New Therapies for Sepsis, Wiley-Liss, New York.
[26] O'Neill, L.A.J. and Dinarello, C.A. (2000) Immunol. Today 21, 206–209.
[27] Hemmi, H. et al. (2000) Nature 408, 740–745.
[28] Aria, K. et al. (1990) Annu. Rev. Biochem. 59, 783–836.
[29] Spriggs, D.R. et al. (1990) Cancer Res. 50, 7101–7107.
[30] Almalwi, W.Y. et al. (1990) Prog. Leuk. Biol. 10A, 321–326.
[31] Elliot, J.F. et al. (1984) Science 226, 1439–1441.
[32] Bierer, B.E. et al. (1990) Proc. Natl Acad. Sci. USA 87, 9231–9235.
[33] Henderson, D.J. et al. (1992) Immunology 73, 316–321.
[34] Sporn, M.B. and Roberts, A.B. (1990) Peptide Growth Factors and Their Receptors, Springer-Verlag, Berlin.
[35] Howard, M. et al. (1993) J. Exp. Med. 177, 1205–1208.
[36] Harper, J.G. et al. (1993) Prog. Growth Factor Res. 4, 321–335.
[37] Gearing, A.J. et al. (1994) Nature 370, 555–557.
[38] Gordon, M.Y. et al. (1987) Nature 326, 403–405.
[39] Cosman, D. (1993) Cytokine 5, 95–106.
[40] Van Zee, K.J. et al. (1992) Proc. Natl Acad. Sci. USA 89, 4845–4849.
[41] James, K. (1990) Immunol. Today 11, 163–166.
[42] Taga, T. et al. (1989) Cell 58, 573–581.
[43] Carter, D.B. et al. (1990) Nature 344, 633–638.

[44] Hannum, C.H. et al. (1990) Nature 343, 336–340.
[45] Eisenberg, S.P. et al. (1990) Nature 343, 341–346.
[46] Ullrich, A. and Schlessinger, J. (1990) Cell 61, 203–212.
[47] Alcami, A. and Koszinowski, U.H. (2000) Immunol. Today 21, 447–455.
[48] Bowie, A.G. et al. (2000) Proc. Natl Acad. Sci. USA 97, 10162–10167.

4 Cytokine Receptor Superfamilies

INTRODUCTION

The receptors for many cytokines have now been cloned and analysis of their primary structures has enabled many of them to be grouped into superfamilies based on common homology regions. There is no agreed nomenclature for most of the superfamilies. In order to achieve conformity within the FactsBook series, we have used the same nomenclature as Barclay et al.[1] based on the most commonly used names. The main superfamilies recognized today are the haematopoietic receptor superfamily (also termed type I cytokine receptors), the interferon receptor superfamily (also termed type II cytokine receptors), the TNF receptor superfamily, the IL-1/Toll-like receptor superfamily, the tyrosine kinase receptor superfamily and the chemokine receptor superfamily. The term 'superfamily' was first used by Dayhoff et al.[2] for proteins with amino acid sequence homology of 50% or less and 'family' for those with homology of more than 50%. In fact, homologies within superfamilies are often 15–25% and it can be difficult to establish an evolutionary relationship rather than a chance similarity until structural information is obtained. Conserved amino acids in superfamilies are often clustered in domains or repeats, but even then the homology may be low with the conserved amino acids clustered in small sequences throughout the 40–110 residues that make up superfamily domains and repeats. It is also important to note that many cytokine receptors have combinations of different domains or repeats.

A domain is a sequence or segment of a protein which is likely to form a discrete structural unit. Three criteria are used to identify a domain. (1) Tertiary structure may be used to define a domain. Domains established by their tertiary structure include Ig (C1 and C2), complement control protein, fibronectin type III (FNIII) and the haematopoietic cytokine receptor domain. (2) Superfamily segments that exist as the sole extracellular sequence, or as a sequence contiguous with hinge-like regions containing a high content of Ala, Gly, Pro, Ser and Thr residues may be considered as domains. (3) Superfamily segments coded for by single exons which can be readily spliced to form a new gene with an open reading frame may also be considered as domains.

Superfamily segments may not always be independent structural units (domains), in which case the term 'repeat' may be used. For example, in the TNF receptor superfamily, blocks of three or four repeats are found without intervening sequences between the repeats. The pattern of exons does not correspond to the repeats, and in this superfamily it seems that a precursor structure evolved by gene duplication of a primordial repeat. Additional members of the superfamily have evolved by duplication and divergence of this repeat.

The term 'motif' describes a smaller sequence pattern which probably does not form a folded structural unit. An example is the Trp–Ser–X–Trp–Ser (WSXWS) motif in the FNIII domain found in most members of the haematopoeitic receptor superfamily.

Apart from extracellular domains, receptor families also often have conserved cytosolic domains which are involved in signal transduction. Haematopoietic cytokine and interferon receptors have a box 1 motif that is required for signalling. Some members of the TNF receptor superfamily have a death domain. Members of the Toll/IL-1 receptor family have a Toll/IL-1R (TIR) domain which forms a discrete structure, while growth factors have tyrosine kinase domains.

THE HAEMATOPOIETIC RECEPTOR SUPERFAMILY

This is the largest cytokine receptor superfamily by far and includes IL-2R β- and γ-chains, IL-4R, IL-3R α- and β-chains, IL-5R α- and β-chains, IL-6R, gp130, IL-9R, IL-12R,

G-CSFR, GM-CSFR, CNTFR, LIFR, EpoR, PRLR and GHR. The IL-7R is also usually included in this family but the case for its inclusion is weak. It has a very low ALIGN score[2] and few of the conserved amino acids in the domain[1].

Domain structure of the haematopoietic receptor superfamily

The extracellular regions of the superfamily all contain combinations of haematopoietic cytokine, FNIII and in some cases C2 Ig domains. The haematopoietic domain has a length of about 100 amino acids with a characteristic Cys–X–Trp motif and three other conserved Cys residues. The FNIII domain was first identified in an extracellular matrix protein, but it is also common in membrane molecules in the nervous system which often have a characteristic Ig domain[3]. It includes the Trp–Ser–X–Trp–Ser (WSXWS) motif required for ligand binding and signal transduction[4]. The extracellular binding domain of the growth hormone receptor was the first to be co-crystallized with its ligand (GH) and the three-dimensional structure solved at a resolution of 2.8 Å by X-ray crystallography[5,6] and NMR[7]. The receptor has two domains (haematopoietic receptor and FNIII) of about 100 residues each, consisting of two antiparallel β-sheets of four and three strands with a similar folding pattern to the Ig C2 set domain, and the domains of the PapD chaperone protein (Figure 1). Interestingly, the WSXWS motif is located away from the binding surfaces. However, this motif is highly conserved and is essential for ligand binding and receptor-mediated signal transduction[4]. Two GHR molecules crystallized with a single GH ligand had identical surfaces of the two GHR subunits binding to nonidentical surfaces on the GH ligand. The areas of contact between the two GHR molecules and GH were of unequal size (1230 $Å^2$ and 900 $Å^2$) and the area of contact between the two GHR chains was 500 $Å^2$. This structure is compatible with ligand-induced receptor dimerization.

Structures of the haematopoeitic cytokine receptor complexes

For some members of the cytokine receptor superfamily, transfection of receptor cDNA results in high-affinity ligand binding similar to that observed with responsive cells (e.g. EpoR, G-CSFR and GHR), suggesting that only one receptor subunit is required for high-affinity binding. Ligand-induced receptor homodimerization has been shown for GH/GHR and G-CSF/G-CSFR interactions, and may occur with other members of this group. For example, a mutation in the extracellular domain of the EpoR has been shown to result in homodimerization and signal transduction without ligand binding.

In contrast to this group, receptor-binding studies have revealed the existence of more than one binding affinity for several members of the cytokine receptor superfamily. Typically, these are high (10–100 pM) and low (1–10 nM) affinity binding sites. For these receptors, additional subunits have been identified which are required for high-affinity receptor expression. These additional subunits are sometimes referred to as affinity converters or converter chains. Moreover, some of these subunits are shared by more than one cytokine receptor giving rise to heterodimeric and in some cases heterotrimeric structures. The elucidation of these complex multisubunit receptors allows for an understanding of the functional crossreactivity and redundancy of some cytokines. These are grouped in the following discussion according to the identity of the shared subunit(s).

GM-CSFR β-chain users

IL-3, IL-5 and GM-CSF all have unique and specific low-affinity receptors with similar structures including short intracytoplasmic segments which are unable to transduce a signal.

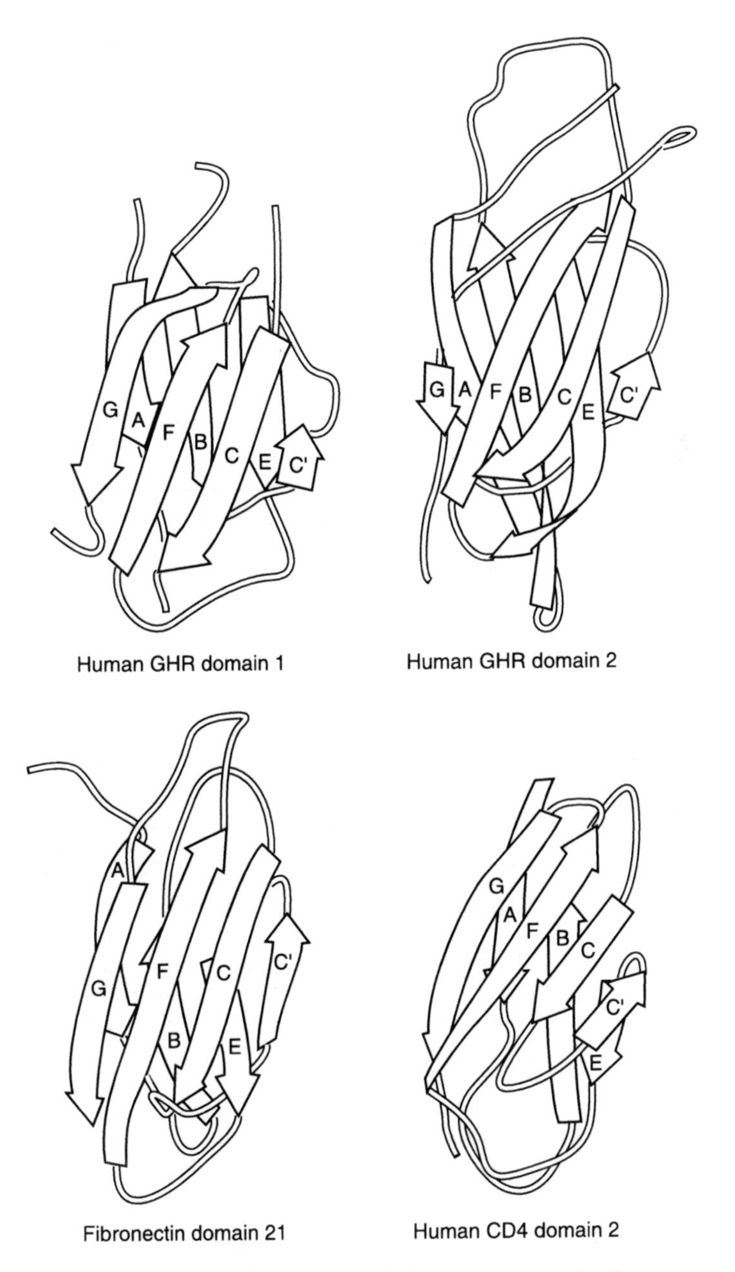

Figure 1 *The folding patterns of the haematopoietic receptor superfamily.*

In humans, a second β-chain (KH97) common to all three receptors converts the low-affinity receptors to high-affinity receptors but does not itself bind ligand[8,9]. The KH97 β-subunit has a much longer intracytoplasmic portion, and is able to transduce a signal. In the mouse, there are two β-subunits, AIC2A and AIC2B[10–13]. Of these, AIC2B is similar to the human

KH97 and associates with all three low-affinity receptors to form high-affinity receptors, but does not itself bind ligand. On the other hand, AIC2A binds IL-3 with low affinity and can associate with the IL-3 receptor α-chain to form a high affinity IL-3 receptor, but it does not associate with either the GM-CSF or IL-5 receptor α-chains[9,13,14].

IL-6R β-chain (gp130) users

Another group of receptors (IL-6R, CNTFR, LIFR, OSMR, IL-11R and CT-1R) share a common signalling subunit known as gp130 which has some structural homology with the G-CSFR. The IL-6R binds IL-6 with low affinity but is itself unable to signal. This IL-6/IL-6R complex binds two gp130 molecules to form a gp130 disulfide-linked homodimer[15]. Dimerization results in tyrosine phosphorylation of gp130 and is required for signal transduction. The high-affinity CNTFR complex has a similar structure[16]. Binding of CNTF to the CNTFR results in association with gp130 and the low-affinity receptor for LIF (LIFR) to form a gp130/LIFR heterodimeric structure. Like the gp130 homodimer, formation of the gp130/LIFR heterodimer results in tyrosine phosphorylation and is required for signal transduction. The LIFR/gp130 heterodimer is also a high-affinity receptor for LIF, OSM, CNTF and CT-1. LIF binds to the LIFR with low affinity but does not bind to gp130, whereas OSM binds to gp130 with low affinity, but not to the LIFR. The LIFR/gp130 heterodimer is required for high-affinity binding to both ligands, and for signal transduction. There is, however, some evidence for different signalling pathways activated by LIF and OSM[17], and a specific receptor for OSM has been cloned and characterized[18]. gp130 may also form part of the IL-11 receptor[19].

IL-2R γ-chain users

The IL-2 receptor is also a complex of three chains[20]. The IL-2R α-chain (Tac, CD25 or p55) is not a member of the haematopoietic receptor superfamily, but has two domains with homology to the complement control protein. It binds IL-2 with low affinity (K_d 1.4×10^{-8} M), but does not signal. The other two components are the IL-2R β-chain (p75), and the IL-2R γ-chain (p64), both of which are members of the superfamily[21,22]. The IL-2R β-chain binds IL-2 with intermediate affinity (K_d 1.2×10^{-7} M), but the γ-chain (p64) does not bind IL-2. Receptor complexes consisting of α/γ or β/γ heterodimers bind IL-2 with an affinity of about 10^{-9} M. The high-affinity receptor complex is an α/β/γ heterotrimer. It has an equilibrium K_d of 1.3×10^{-11} M and a dissociation half-life of 50 minutes[22].

Both β/γ and α/β/γ complexes are able to mediate signal transduction. The β-chain is also a component of the receptor complexes for IL-15, IL-9 and IL-4. At least two distinct cytoplasmic regions of the IL-2 β-chain are involved in IL-2-mediated cellular signalling. A serine-rich region is critical for induction of c-*myc* and cell proliferation (signal 1), whereas an acidic region is required for physical association with *src*-like protein tyrosine kinase (PTK) $p56^{lck}$, activation of $p21^{ras}$, and induction of c-*fos* and c-*jun* (signal 2)[20]. The IL-2R γ-chain has an SH2-like homology domain which may be involved in binding to phosphotyrosine residues on other signalling proteins. Mutations in the IL-2R γ-chain have been shown to cause X-linked severe combined immunodeficiency (X-SCID) in humans[23]. This is a lethal disease characterized by absent or greatly reduced T cells and severely depressed cell-mediated and humoral immunity. The IL-2R γ-chain is also a functional component of the IL-4R[24,25], the IL-7R[26] and the IL-15 receptor, and there is some evidence that it is also part of the IL-9 and IL-13 receptor complexes. It is therefore known as the common γ-chain (γc).

Table 4.1 *Cytokine signalling via Jaks and STATs*

Jak/STAT	Activated by	Phenotype of null mice
STAT1	IFNα/β, IFNγ	No response to IFNs. Impaired innate immunity
STAT2	IFNα/β	Impaired antiviral response
STAT3	IL-2, IL-7, IL-9, IL-15	Fetal lethal (fetal growth blunted)
	IL-6, IL-11, OSM, CNTF, LIF, CT-1	
	Leptin	
	GH	
	IL-10	
STAT4	IL-12	Impaired Th1 development
STAT5a	IL-2, IL-7, IL-9, IL-15	Impaired breast development
	IL-3, IL-5, GM-CSF	
	GH	
	Prolactin	
	Epo, Tpo	
STAT5b	IL-2, IL-7, IL-9, IL-15	Impaired breast development
	IL-3, IL-5, GM-CSF	Dwarfism
	GH	Defective T-cell proliferation
	Prolactin	
	Epo, Tpo	
STAT6	IL-4, IL-13	Defective Th2 development and antibody class switch
Jak1	IL-2, IL-7, IL-9, IL-15	Perinatal lethal, defective innate immunity
	IL-4, IL-13	
	IL-6, IL-11, OSM, CNTF, LIF, CT-1	
	IL-12	
	IFNα/β, IFNγ	
Jak2	IL-13	Fetal lethal, haematopoietic failure
	IL-6, IL-11, OSM, CNTF, LIF, CT-1	
	IL-12	
	GH	
	Prolactin	
	Epo, Tpo	
	IFNγ	
Jak3	IL-2, IL-7, IL-9, IL-15	Severe combined immunodeficiency (similar to X-linked SCID)
	IL-4	

Signal transduction by haematopoietic cytokine receptors

Haematopoietic receptors and also the interferon family of receptors signal via the activation of tyrosine kinases termed Janus kinases (Jaks), which activate transcription factors termed signal transducers of activated transcription (STATs)[27,28]. This process has become one of the best characterized mechanisms whereby cytokines induce signals, and the pathways elucidated provide a direct link between receptors and changes in gene expression. Jaks associate constitutively with the cytosolic domains of receptors. Upon receptor dimerization (induced by ligand binding), Jaks transphosphorylate each other on tyrosine and also phosphorylate tyrosine residues on receptors. STATs are then recruited via their SH2 domains. The STATs themselves then become phosphorylated and dimerize via reciprocal SH2 domain/phosphotyrosine interactions. They are then able to translocate to the nucleus, bind consensus elements on target genes, increasing their expression. Table 1 lists the particular Jaks and STATs activated by each cytokine, and the phenotypes of corresponding null mice. Specificity comes from the particular set of Jaks and STATs activated. A number of proteins which negatively regulate Jak signalling have also been found, and are termed suppressers of cytokine signalling (SOCS). They bind Jaks and inactivate them, preventing signal transduction. They are induced by the cytokines which activate Jaks and act as negative feedback inhibitors[29].

Soluble haematopoietic cytokine receptors

Alternatively spliced mRNA giving rise to secreted receptors lacking the transmembrane region and the cytoplasmic proximal charged residues that anchor the protein in the membrane have been described for the IL-4R[30], IL-5R[31], IL-7R[32], IL-9R[33], GM-CSFR[34], G-CSFR[35], GHR[36] and LIFR[37]. Soluble receptors can also be generated by proteolysis of cell surface receptors, e.g. GHR, and by phospholipase action on the GPI-linked CNTFR. The *in vivo* function of soluble molecules is not established. They may act as antagonists by competitive ligand binding, but the affinity of cytokines for soluble receptors is generally low. sIL-4R has been shown to inhibit IL-4-mediated responses[38]. Alternatively, soluble receptors could act as transport proteins to carry cytokines to sites where they are required for biological activity or even to sites where they would have no biological activity for removal from the body. More interestingly, soluble receptors complexed with cytokine may act as agonists. Such a role has been described for IL-6R/IL-6 complexes which bind to gp130 and transduce a signal[39]. Similarly CNTF/CNTFR complexes bind to LIFR/gp130 heterodimers[40], and the IL-12 p35/p340 complex binds to the IL-12 receptor[41,42].

THE INTERFERON RECEPTOR SUPERFAMILY

The interferon receptor superfamily, otherwise known as type II cytokine receptors, includes the IFNα/β receptor, IFNγ receptor, IL-10 receptor and tissue factor[43–46]. The members of this family are single spanning transmembrane glycoproteins characterized by either one (IFNγR and IL-10R) or two (IFNα/βR) homologous extracellular regions of about 200 amino acids, each of which has two FNIII domains. The first of these FNIII domains has two conserved tryptophans and a pair of conserved cysteines, whereas the second has a unique disulfide loop formed from the second pair of conserved cysteines, but no WSXWS motif characteristic of type I cytokine receptors. Signal transduction by the IFN receptors involves phosphorylation and activation of Jak and STATs as described above.

THE TNF RECEPTOR SUPERFAMILY

The TNF receptor superfamily includes both cytokine receptors (TNFR I (p55), TNFR II (p75) and NGFR) and several other leukocyte surface glycoproteins, including CD27, CD30, CD40, OX40, 4-1BB and Fas antigen. The members of this superfamily are characterized by three or four cysteine-rich repeats of about 40 amino acids in the extracellular part of the molecule[47]. These repeats contain four or six conserved cysteine residues in a pattern characteristic of the superfamily, but distinct from cysteine-rich repeats in other molecules such as low-density lipoprotein. Several members of this superfamily including the NGFR and TNFR II (p75) also have a hinge-like region characterized by a lack of cysteine residues and a high proportion of Ser, Thr and Pro, suggesting that it may be glycosylated with *O*-linked sugars. The gene structure shows that the repeats do not correspond to particular exons, suggesting that the receptors did not arise by duplication of exons encoding single repeats. A primordial gene encoding four repeats may have arisen by unequal crossing-over. Other members of the TNF receptor superfamily would then have evolved by gene duplication and divergence.

One unusual feature of the TNF receptor superfamily is that some of the receptors bind more than one ligand. Moreover, unlike most other cytokines, the ligands are dimeric or trimeric. Both TNFα and TNFβ bind TNFR I and TNFR II with high affinity even though these two cytokines are only 31% identical in amino acid sequence (see the TNF entry page 474). Similarly, NGF, NT-3 and BDNF all bind the NGFR.

Our understanding of signal transduction by TNF has seen great progress. Signal transduction by the type I TNFR leads in some transformed cell types to cell death (by apoptosis and/or necrosis), but in most cell types TNF increases the expression of a large number of proinflammatory genes. Signalling occurs by a series of protein–protein interactions which lead to activation of caspase-8 or caspase-2 in the case of the apoptotic pathway, or protein kinases in the case of inflammatory gene expression. The death domain of the type I TNFR mediates signalling and recruits the adapter protein TNF receptor-associated death domain (TRADD). For apoptosis signalling, TRADD recruits Fas-associated death domain (FADD), which via its death effector domain recruits and activates caspase-8[48]. This triggers an apoptotic cascade. TRADD can also recruit a protein kinase termed RIP (receptor interacting protein), which in turn recruits another adapter termed RAIDD[47–51]. This has a death effector domain which recuits caspase-2, again promoting apoptosis. RIP can also couple via an as yet unknown kinase to the I-κB kinase (IKK) complex, leading to the activation of the central proinflammatory transcription factor NFκB[49]. This process can also be activated by another adapter, Traf-2, which activates the IKK complex, although again the linking kinase is not known[50]. Traf-2 and RIP are also implicated in the activation of Jun N-terminal kinase and p38 MAP kinase by TNF[51]. These kinases act to stabilize mRNAs induced by TNF which contain AU repeats in their 3′ untranslated regions[52]. The role of the type II TNFR in signalling is not fully understood. Because it lacks a death domain its signalling will differ from the type I TNFR[53] and it may have a role in passing TNF to the type I TNF receptor.

IL-1/TOLL-LIKE RECEPTORS

The type I IL-1 receptor is the founder member of the IL-1 receptor/Toll-like receptor superfamily[54,55]. All members have a conserved cytsolic domain, termed the Toll/IL-1 receptor (TIR) domain. The family splits into two subfamilies based on extracellular homologies. IL-1RI belongs to one subfamily and is defined by Ig domains extracellularly. The Ig superfamily is the largest known superfamily of proteins with about 40% of known leukocyte membrane polypeptides containing one or more Ig domains. Cytokine receptors with Ig domains in their extracellular sequences include IL-1R and IL-18R. Other cytokine receptors with Ig domains, but without TIR domains are IL-6R, FGFR, PDGFR, M-CSFR and SCFR (c-*kit*). The Ig domains are characterized by a structural unit of about 100 amino acids with a distinct folding pattern known as the Ig-fold[56], consisting of a sandwich of two β-sheets, each made up from antiparallel β-strands of about 5–10 amino acids with a conserved disulfide bond between the two sheets in most but not all domains. The tertiary structure has been solved by X-ray crystallography for Ig V and Ig C domains, β_2-microglobulin, the α3 domain of MHC class I, CD4 domains 1 and 2, CD8 and the type I IL-1R. Ig and TCR V regions have common sequence patterns and are known as the V set. The Ig, TCR and MHC antigen C type domains all share the same sequence patterns and are known as the C1 set. A third set of domains with similar structure to C domains but with some sequence homology with the V set are known as the C2 set. The CD4 domain 2 is a C2-set sequence as are most of the Ig domains in the cytokine receptors. For a detailed description of the immunoglobulin domain sets see ref. 1.

The second subfamily in the IL-1 receptor/Toll-like receptor superfamily have leucine-rich repeats extracellularly, and include the Toll-like receptors, which are essential for recogntion of microbial products. For example TLR-4 recognizes LPS[57] and TLR-9 recognizes bacterial CpG DNA[58]. TLRs are therefore critical for innate immune responses.

The TIR domain, which is found in all superfamily members, consists of around 200 amino acids, with three particular boxes being highly conserved[54,55]. Box 2 forms a loop in the structure of the TIR domain which is required for signalling[59]. Signalling by the TIR domain is analogous to TNF signalling in that it involves a series of protein–protein interactions which lead to the activation of protein kinases. Proteins involved include the adapter MyD88, which has a TIR domain and is recruited to the receptor complex; IL-1 receptor-associated kinase (IRAK)[60] and IRAK-2[61], and Traf-6 which leads to the activation of the I-κB kinase complex, promoting activation of the key transcription factor NFκB. Many of the proinflammatory genes induced by IL-1 are NFκB-regulated. These proteins are involved in signalling by IL-1, IL-18 and TLRs.

For IL-1 and IL-18 these receptors complex with highly homologous receptor accessory proteins[55]. These do not bind ligand but contain TIR domains. Similarly TLRs dimerize, with TLR-2 associating with either TLR-1 or TLR-6, and TLR-4 homodimerizing[62].

THE PROTEIN TYROSINE KINASE RECEPTOR SUPERFAMILY

Receptor tyrosine kinases all have a large glycosylated extracellular ligand-binding domain, a single transmembrane domain and an intracellular tyrosine kinase catalytic domain[63,64]. This superfamily of receptors can be divided into four subclasses. Subclass I receptors are monomeric with two extracellular cysteine-rich repeat sequences and include the EGFR. Disulfide-linked heterotetrameric (α2β2) structures with similar cysteine-rich sequences are found in subclass II receptors such as insulin receptor and insulin growth factor 1 receptor (IGF-1R). Subclass III (PDGFR, CSF-1R, SCFR) and IV (FGFR, flg, bek) receptors have five and three extracellular Ig domains respectively. The tyrosine kinase domain of these two subclasses is interrupted by a hydrophilic insertion sequence of varying length. Another group is the trkA, trkB and trkC receptors for the NGF family of neurotrophins, including NGF, BDNF, NT-3 and NT-4[65,66]. Signal transduction by these receptors is mediated by ligand-induced oligomerization[67]. Oligomerization may be induced by monovalent ligands such as EGF, inducing conformational changes in the receptor, or bivalent ligands such as PDGF and CSF-1 that mediate oligomerization of nearby receptors. A significant feature of signal transduction by these receptors is ligand-induced autophosphorylation. The resultant phosphotyrosine residues on the cytoplasmic domains of these receptors bind to SH2 domains of other proteins involved in the signalling cascade.

The tyrosine kinase domain contains 11 highly conserved subdomains[63]. One of these contains the Gly–X–Gly–X–X–Gly–X(15–20)–Lys consensus sequence which forms part of the binding site for ATP. Another subdomain contains an invariant Lys which appears to be involved in the phosphotransfer reaction, and a third contains the Pro–Ile/Val–Lys/Arg–Trp–Thr/Met–Ala–Pro–Glu sequence characteristic of tyrosine kinases.

CHEMOKINE RECEPTOR SUPERFAMILY

Chemokine receptors have seven transmembrane domains, characterisitc of G protein-coupled receptors[68]. The superfamily also includes viral chemokine receptors US28 from cytomegalovirus and ECRF3 from Herpes saimari virus which may bind specific chemokines and activate signal transduction pathways[69]. The Duffy blood group antigen is similarly related to this family, and binds IL-8 and related chemokines[70].

The receptors have a characteristic structure of a relatively short acidic extracellular N-terminal sequence followed by seven transmembrane spanning domains with three extracellular and three intracellular loops. There are conserved cysteine residues in the N-terminal sequence and in the third extracellular loop which form a disulfide bond required for ligand binding. A second disulfide bond is probably formed between conserved cysteine residues in the first and second extracellular loop. The receptor ligands are highly cationic (pI >8.5) proteins which are as large or larger than the projected extracellular domains of their receptors and it is likely that contact points of ligand and receptor includes membrane or intracellular portions (reviewed in ref. 67).

The receptors are coupled to heterotrimeric GTP-binding proteins which induce PIP_2 hydrolysis and activate kinases, phosphatases and ion channels. The α-subunit is pertussis toxin sensitive (presumably $G_i\alpha2$ or $G_i\alpha3$).

References

1 Barclay, A.N. et al. (1993) The Leucocyte Antigen FactsBook, Academic Press, London.
2 Dayhoff, M.O. et al. (1983) Meth. Enzymol. 91, 524–545.
3 Patthy, L. (1990) Cell 61, 13–14.
4 Miyazaki, T. et al. (1991) EMBO J. 10, 3191–3197.
5 de Vos, A.M. et al. (1992) Science 255, 306–312.
6 Cunningham, B.C. et al. (1991) Science 254, 821–825.
7 Baron, M. et al. (1992) Biochemistry 31, 2068–2073 .
8 Hayashida, K. et al. (1990) Proc. Natl Acad. Sci. USA 87, 9655–9659.
9 Lai, C.F. et al. (1995) Ann. N.Y. Acad. Sci. 762, 189–205.
10 Kitamura, T. et al. (1991) Proc. Natl Acad. Sci. USA 88, 5082–5086.
11 Gorman, D.M. et al. (1990) Proc. Natl Acad. Sci. USA 87, 5459–5463.
12 Takaki, S. et al. (1991) EMBO J. 10, 2833–2838.

[13] Hara, T. and Miyajima, A. (1992) EMBO J. 11, 1875–1884.
[14] Cosman, D. (1993) Cytokine 5, 95–106.
[15] Murakami, M. et al. (1993) Science 260, 1808–1810.
[16] Davis, S. et al. (1993) Science 260, 1805–1808.
[17] Gearing, D.P. (1993) Adv. Immunol. 53, 31–58.
[18] Mosley, B. et al (1996) J. Biol. Chem. 271, 32635–32640
[19] Yin, T. et al. (1993) J. Immunol. 151, 2555–2561.
[20] Minami, Y. et al. (1993) Annu. Rev. Immunol. 11, 245–267.
[21] Takeshita, T. et al. (1992) Science 257, 379–382.
[22] Taniguchi, T. and Minami, Y. (1993) Cell 73, 5–8.
[23] Noguchi, M. et al. (1993) Cell 73, 147–157.
[24] Russell, S.M. et al. (1993) Science 262, 1880–1883.
[25] Kondo, M. et al. (1993) Science 262, 1874–1877.
[26] Noguchi, M. et al. (1993) Science 262, 1877–1880.
[27] Ihle, J.N. (1996) Cell 84, 331–334
[28] Leonard, W.J. and O'Shea, J.J. (1998) Annu. Rev. Immunol. 16, 293–322
[29] Nicola, N.A. and Greenhaugh, C.J. (2000) Exp. Hematol. 28, 1105–1112
[30] Mosley, B. et al. (1989) Cell 59, 335–348.
[31] Murata, Y. et al. (1992) J. Exp. Med. 175, 341–351.
[32] Goodwin, R.G. et al. (1990) Cell 60, 941–951.
[33] Renauld, J.C. et al. (1992) Proc. Natl Acad. Sci. USA 89, 5690–5694.
[34] Raines, M.A. et al. (1991) Proc. Natl Acad. Sci. USA 88, 8203–8207.
[35] Fukunaga, R. et al. (1990) Proc. Natl Acad. Sci. USA 87, 8702–8706.
[36] Sadeghi, H. et al. (1990) Mol. Endocrinol. 4, 1799–1805.
[37] Gearing, D.P. et al. (1991) EMBO J. 10, 2839–2848.
[38] Fanslow, W.C. et al. (1991) J. Immunol. 147, 535–540.
[39] Taga, T. et al. (1989) Cell 58, 573–581.
[40] Davis, S. et al. (1993) Science 259, 1736–1739.
[41] Gearing, D.P. and Cosman, D. (1991) Cell 66, 9–10.
[42] Chizzonite, R. et al. (1992) J. Immunol. 148, 3117–3124.
[43] Aguet, M. et al. (1988) Cell 55, 273–280.
[44] Bazan, J.F. (1990) Proc. Natl Acad. Sci. USA 87, 6934–6938.
[45] Lutfalla, G. et al. (1992) J. Biol. Chem. 267, 2802–2809.
[46] Ho, A.S.Y. et al. (1993) Proc. Natl Acad. Sci. USA 90, 11267–11271.
[47] Mallett, S. and Barclay, A.N. (1991) Immunol. Today 12, 220–223.
[48] Hsu, H. et al. (1996) Cell 84, 299–308.
[49] Stanger, B.Z. et al. (1995) Cell 81, 513–523.
[50] Rothe, M. et al. (1995) Science 269, 1424–1427.
[51] Wallach, D. et al. (1999) Annu. Rev. Immunol. 17, 331–367.
[52] Ridley, S.H. et al (1997) J Immunol 158, 3165–3173
[53] Tartaglia, L.A. and Goeddel, D.V. (1992) Immunol. Today 13, 151–153.
[54] O'Neill, L.A.J. and Dinarello, C.A. (2000) Immunol. Today 21, 206–209.
[55] O'Neill, L.A.J. (2000) Science's STKE http://www.stke.org/cgi/content/full/OC_sigtrans;2000/44/re1
[56] Amzel, L.M. and Poljak, R. J. (1979) Annu. Rev. Biochem. 48, 961–967.
[57] Poltorak, A. et al. (2000) Science 282, 2085–2088.
[58] Hemmi H. et al. (2000) Nature 408, 740–745.
[59] Xu, Y. et al. (2000) Nature 408, 111–115.
[60] Cao et al. (1996) Nature 383, 443–446.

[61] Muzio, M et al. (1997) Science 278, 1612–1615.
[62] Ozinsky, A. et al. (2000) Proc. Natl Acad. Sci. USA 97, 13766–13771.
[63] Hanks, S.K. et al. (1988) Science 241, 42–52.
[64] Ullrich, A. and Schlessinger, J. (1990) Cell 61, 203–212.
[65] Bradshaw, R.A. et al. (1993) Trends Biochem. Sci. 18, 48–52.
[66] Schneider, R. and Schweiger, M. (1991) Oncogene 6, 1807–1811.
[67] Schlessinger, J. (1988) Trends Biochem. Sci. 13, 443–447.
[68] Gerard, C. and Gerard, N.P. (1994) Curr. Opin. Immunol. 6, 140–145.
[69] Ahuja, S.K. and Murphy, P.M. (1993) J. Biol. Chem. 268, 20691–20694.
[70] Neote, K. et al. (1993) J. Biol. Chem. 268, 12247–12249.

Section II

THE CYTOKINES AND THEIR RECEPTORS

Other names

Lymphocyte-activating factor (LAF), endogenous pyrogen (EP), leukocyte endogenous mediator (LEM), mononuclear cell factor (MCF), catabolin.

THE MOLECULES

Interleukin 1 (IL-1) has a very wide range of biological activities on many different target cell types including B cells, T cells and monocytes[1–3]. *In vivo*, it induces hypotension, fever, weight loss, neutrophilia and acute phase response. IL-1α and IL-1β are distinct molecular forms of IL-1 derived from two different genes. IL-1α is mostly cell associated and IL-1β is mostly secreted. Although the amino acid sequence homology between the α and β forms is only about 20%, these molecules bind to the same receptor and have very similar if not identical biological properties. An IL-1 receptor antagonist (IL-1Ra) has been described which is structurally related to IL-1β and binds to the IL-1 receptor[4]. Intracellular forms of human IL-1Ra have also been identified that are splice variants of IL-1Ra[5]. The antagonist is made by the same cells that secrete IL-1 and may be an important physiological regulator. A cysteine protease (converting enzyme) which releases mature IL-1β has also been cloned and is termed caspase-1, since it is the founder member of the caspase family of cysteine proteases[6,7]. A cowpox virus-derived inhibitor (CRMA) of the IL-1-converting enzyme has been shown to inhibit the host inflammatory response[8].

Crossreactivity

There is 62% amino acid sequence homology between human and mouse IL-1α and 68% for IL-1β. Both forms crossreact between humans and mice. There is 77% sequence homology between human and mouse IL-1Ra.

Sources

A wide variety of cells secrete IL-1, including monocytes, tissue macrophages, Langerhans cells, dendritic cells, T lymphocytes, B lymphocytes, natural killer (NK) cells, large granular lymphocytes (LGL), vascular endothelium and smooth muscle, fibroblasts, thymic epithelia, astrocytes, microglia, glioma cells, keratinocytes and chondrocytes.

Bioassays

Activation of murine thymocytes or murine T cell lines. PGE_2 induction in fibroblasts using an IL-1-neutralizing antibody as control. *In vivo* (rabbit) pyrogen assay.

Physicochemical properties of IL-1α and IL-1β

Property	IL-1α		IL-1β	
	Human	Mouse	Human	Mouse
pI	5	5	7	7
Amino acids – precursor	271	270	269	269
– mature[a]	159	156	153	159
M_r (K) – predicted	18.0	18.0	17.4	17.4
– expressed	17.5	17.4	17.3	17.5
Potential *N*-linked glycosylation sites[b]	2	3	1	2
Disulfide bonds	0	0	0	0

[a] After proteolytic removal of propeptide.
[b] IL-1 is not normally glycosylated.

3-D structure

The structure of IL-1α has been determined at a resolution of 2.7 Å by X-ray crystallography and IL-1β at lower resolution by NMR spectroscopy. Both forms of IL-1 are stable tetrahedral globular proteins formed by an antiparallel six-stranded barrel closed at one end by a six-stranded β-sheet to form a bowl-like structure.

Gene structure[11–14]

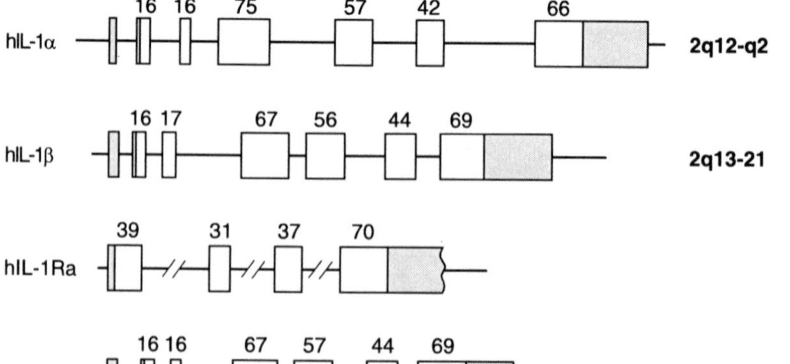

The gene for IL-1α is on human chromosome 2q12–q21, and contains six coding exons of 16, 16, 75, 57, 42 and 66 amino acids. Human IL-1β is on chromosome 2q13–q21, with six exons of 16, 17, 67, 56, 44 and 66 amino acids. Human IL-1Ra is on chromosome 2q and contains four exons. Mouse IL-1α, IL-1β and IL-1Ra are all on chromosome 2 and have similar structures to the human genes.

Amino acid sequence for human IL-1α[15]

Accession code: SwissProt P01583

1 *MAKVPDMFED LKNCYSENEE DSSSIDHLSL NQKSFYHVSY GPLHEGCMDQ*

51 *SVSLSISETS KTSKLTFKES MVVVATNGKV LKKRRLSLSQ SITDDDLEAI*

101 *A*N*DSEEEIIK PR*SAPFSFLS NVKYNFMRII KYEFILNDAL **N**QSIIRANDQ

151 YLTAAALHNL DEAVKFDMGA YKSSKDDAKI TVILRISKTQ LYVTAQDEDQ

201 PVLLKEMPEI PKTITGSETN LLFFWETHGT KNYFTSVAHP NLFIATKQDY

251 WVCLAGGPPS ITDFQILENQ A

Propeptide 1–112 (in italics) is removed to form the mature protein. Conflicting sequence A→S at position 114.

Amino acid sequence for human IL-1β[15]

Accession code: SwissProt P01584

1 *MAEVPKLASE MMAYYSGNED DLFFEADGPK QMKCSFQDLD LCPLDGGIQL*

51 *RISDHHYSKG FRQAASVVVA MDKLRKMLVP CPQTFQENDL STFFPFIFEE*

101 *EPIFFDTWDN EAYVHD*APVR SL**N**CTLRDSQ QKSLVMSGPY ELKALHLQGQ

151 DMEQQVVFSM SFVQGEESND KIPVALGLKE KNLYLSCVLK DDKPTLQLES

201 VDPKNYPKKK MEKRFVFNKI EINNKLEFES AQFPNWYIST SQAENMPVFL

251 GGTKGGQDIT DFTMQFVSS

Propeptide 1–116 (in italics) is removed to form the mature protein. Conflicting sequence K→E at position 6, D→H at position 20, E→Q at position 111, G→A at position 177, and R→P at position 214.

Amino acid sequence for human IL-1 receptor antagonist (IL-1Ra)[4,16]

Accession code: SwissProt P18510

-7 MA LETIC

-25 MEICRGLRSH LITLLLFLFH SETIC

1 RPSGRKSSKM QAFRIWDVNQ KTFYLRNNQL VAGYLQGPNV NLEEKIDVVP

51 IEPHALFLGI HGGKMCLSCV KSGDETRLQL EAV**N**ITDLSE NRKQDKRFAF

101 IRSDSGPTTS FESAACPGWF LCTAMEADQP VSLTNMPDEG VMVTKFYFQE

151 DE

A second intracellular form of the IL-1 receptor antagonist has been reported[5] with a short seven-amino-acid N-terminal sequence MALETIC instead of the longer signal sequence of the secreted form.

Amino acid sequence for mouse IL-1α[17]

Accession code: SwissProt P01582

1 *MAKVPDLFED LKNCYSENED YSSAIDHLSL NQKSFYDASY GSLHETCTDQ*
51 *FVSLRTSETS KMS***N***FTFKES RVTVSATSSN GKILKKRRLS FSETFTEDDL*
101 *QSITHDLEET IQPR*SAPYTY QSDLRYKLMK LVRQKFVM**N**D SL**N**QTIYQDV
151 DKHYLSTTWL NDLQQEVKFD MYAYSSGGDD SKYPVTLKIS DSQLFVSAQG
201 EDQPVLLKEL PETPKLITGS ETDLIFFWKS INSKNYFTSA AYPELFIATK
251 EQSRVHLARG LPSMTDFQIS

Propeptide 1–114 (in italics) is removed to form the mature protein.

Amino acid sequence for mouse IL-1β[18]

Accession code: SwissProt: P10749

1 *MATVPELNCE MPPFDSDEND LFFEVDGPQK MKGCFQTFDL GCPDESIQLQ*
51 *ISQQHI***N***KSF RQAVSLIVAV EKLWQLPVSF PWTFQDEDMS TFFSFIFEEE*
101 *PILCDSWDDD DNLLVCD*VPI RQLHYRLRDE QQKSLVLSDP YELKALHLNG
151 QNINQQVIFS MSFVQGEPSN DKIPVALGLK GKNLYLSCVM KDGTPTLQLE
201 SVDPKQYPKK KMEKRFVFNK IEVKSKVEFE SAEFPNWYIS TSQAEHKPVF
251 LG**N**NSGQDII DFTMESVSS

Propeptide 1–117 (in italics) is removed to form the mature protein.

Amino acid sequence for mouse IL-1 receptor antagonist (IL-1Ra)[19,20]

Accession code: SwissProt P25085

-26 MEICWGPYSH LISLLLILLF HSEAAC
1 RPSGKRPCKM QAFRIWDTNQ KTFYLRNNQL IAGYLQGPNI KLEEKIDMVP
51 IDLHSVFLGI HGGKLCLSCA KSGDDIKLQL EEV**N**ITDLSK NKEEDKRFTF
101 IRSEKGPTTS FESAACPGWF LCTTLEADRP VSLTNTPEEP LIVTKFYFQE
151 DQ

THE IL-1 RECEPTORS

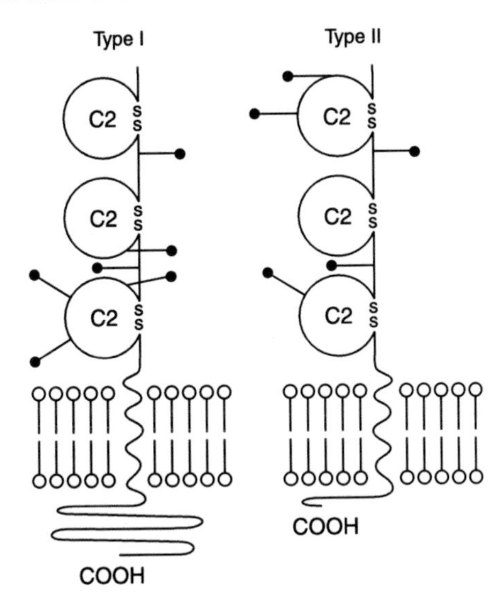

There are two IL-1 receptors. The type I receptor (CDw121a) is a transmembrane glycoprotein with an M_r of 80 000. It is a member of the Ig superfamily with three Ig-SF C2 set domains. The cytoplasmic domain is highly conserved between human and mouse with 78% sequence identity and it is essential for signal transduction. The receptor binds IL-1α, IL-1β and IL-1Ra mature proteins. In transfection experiments, human IL-1 receptor cDNA resulted in expression of low- and high-affinity receptors with dissociation constants similar to those found on the original T-cell clone ($K_d \sim 10^{-9}$ and $\sim 10^{-11}$ M). The type II receptor (CDw121b) is a glycoprotein with an M_r of 60 000 with three Ig-SF C2 set domains in its extracellular region, a single transmembrane segment, and a short (29-amino-acid) cytoplasmic domain. The type II receptor can bind IL-1α, IL-1β and IL-1Ra. An open reading frame in the vaccinia virus genome has significant homology with the type II IL-1R[21]. There is also significant homology between the transmembrane region of type II IL-1R and the membrane-spanning segment of the Epstein–Barr virus protein BHRF1[21]. This suggests an interaction with a common second subunit. A soluble receptor found in normal human serum and secreted by the human B cell line RAJI which binds preferentially to IL-1β has also been described[22]. Signalling by the type I IL-1 receptor requires an accessory protein, termed IL-1R AcP[23,24]. This does not bind IL-1, but forms a dimer with IL-1RI when liganded. Signalling is mediated via a cytosolic domain on both proteins, termed the Toll/IL-1 receptor domain (see Chapter 4). This occurs in a wide range of receptors, including those for IL-18, Toll-like receptors (which bind microbial products) and orphan receptors including T1/ST2 and IL-1Rrp2[25].

Distribution

The type I IL-1 receptor is widely distributed on many cell types including T cells, B cells, monocytes, NK cells, basophils, neutrophils, eosinophils, dendritic cells, fibroblasts, endothelial cells, vascular endothelial cells and neural cells. The type II receptor is also expressed on a number of different tissues including T cells, B cells, monocytes and keratinocytes.

Physicochemical properties of the IL-1 receptors

	IL-1α		IL-1β	
Property	Human	Mouse	Human	Mouse
Amino acids – precursor	569	576	398	410
– mature[a]	552	557	385	397
M_r (K) – predicted	63.5	64.6	44.0	44.3
– expressed	80	80–90	60–68	60
Potential *N*-linked glycosylation sites[b]	6	7	5	4
Affinity K_d (M)[c]	$\sim 10^{-9}$	$\sim 10^{-9}$	$\sim 10^{-9}$	$\sim 10^{-9}$

[a] After removal of predicted signal peptide.
[b] There are five conserved *N*-linked glycosylation sites in human and murine IL-1 receptors.
[c] The affinity of IL-1α, IL-1β and IL-1Ra binding to type I and type II receptors is described in detail in ref. 21.

Signal transduction

IL-1 signalling is analogous to TNF signalling in that it involves a series of protein-protein interactions which lead to the activation of protein kinases. The pathways activated lead to a profound increase in the expression of proinflammatory genes. For a detailed review on IL-1 signalling, see refs 25 and 26. As stated above, IL-1 signalling involves the binding of IL-1 to IL-1RI, which then recruits IL-1R AcP. Both of these proteins have a TIR domain, which recruits the adapter protein MyD88, which itself possesses a TIR domain[27]. MyD88 then recruits a kinase termed IL-1 receptor-associated kinase (IRAK)[28]. A second kinase, IRAK-2, is also part of the complex[29]. Another protein, Tollip, acts to promote IRAK recruitment to the complex[30]. IRAK then recruits Traf-6 which leads to the activation of the I-κB kinase complex, promoting activation of the key transcription factor NFκB. Many of the proinflammatory genes induced by IL-1 are NFκB-regulated. In addition, IL-1 activates Jun N-terminal kinase and p38 MAP kinase via Traf-6. These are involved in stabilizing mRNAs which contain AU repeats in their 3' untranslated regions. A role for low-molecular-weight G proteins in IL-1 signalling has also been suggested, with Rac participating in the pathway to enhanced transactivation by NFκB[31] and a Ras-like G protein on the pathway to p38 MAP kinase[32].

Chromosomal location

Human type I IL-1R maps to 2q12 and the type II IL-1R to 2q12-q22. The mouse type I and type II receptors both map to the centromere-proximal region of chromosome 1[33].

Amino acid sequence for human type I IL-1 receptor[34]

Accession code: SwissProt P14778

```
-17  MKVLLRLICF IALLISS
  1  LEADKCKERE EKIILVSSAN EIDVRPCPLN PNEHKGTITW YKDDSKTPVS
 51  TEQASRIHQH KEKLWFVPAK VEDSGHYYCV VRNSSYCLRI KISAKFVENE
101  PNLCYNAQAI FKQKLPVAGD GGLVCPYMEF FKNENNELPK LQWYKDCKPL
151  LLDNIHFSGV KDRLIVMNVA EKHRGNYTCH ASYTYLGKQY PITRVIEFIT
201  LEENKPTRPV IVSPANETME VDLGSQIQLI CNVTGQLSDI AYWKWNGSVI
251  DEDDPVLGED YYSVENPANK RRSTLITVLN ISEIESRFYK HPFTCFAKNT
301  HGIDAAYIQL IYPVTNFQKH MIGICVTLTV IIVCSVFIYK IFKIDIVLWY
351  RDSCYDFLPI KASDGKTYDA YILYPKTVGE GSTSDCDIFV FKVLPEVLEK
401  QCGYKLFIYG RDDYVGEDIV EVINENVKKS RRLIIILVRE TSGFSWLGGS
451  SEEQIAMYNA LVQDGIKVVL LELEKIQDYE KMPESIKFIK QKHGAIRWSG
501  DFTQGPQSAK TRFWKNVRYH MPVQRRSPSS KHQLLSPATK EKLQREAHVP
551  LG
```

Ig-SF C2 set domains at residues 20–86, 118–186 and 224–302. Disulfide bonds between Cys27–79, 125–179 and 231–295.

Amino acid sequence for human type II IL-1 receptor[21]

Accession code: SwissProt P27930

```
-13  MLRLYVLVMG VSA
  1  FTLQPAAHTG AARSCRFRGR HYKREFRLEG EPVALRCPQV PYWLWASVSP
 51  RINLTWHKND SARTVPGEEE TRMWAQDGAL WLLPALQEDS GTYVCTTRNA
101  SYCDKMSIEL RVFENTDAFL PFISYPQILT LSTSGVLVCP DLSEFTRDKT
151  DVKIQWYKDS LLLDKDNEKF LSVRGTTHLL VHDVALEDAG YYRCVLTFAH
201  EGQQYNITRS IELRIKKKKE ETIPVIISPL KTISASLGSR LTIPCKVFLG
251  TGTPLTTMLW WTANDTHIES AYPGGRVTEG PRQEYSENNE NYIEVPLIFD
301  PVTREDLHMD FKCVVHNTLS FQTLRTTVKE ASSTFSWGIV LAPLSLAFLV
351  LGGIWMHRRC KHRTGKADGL TVLWPHHQDF QSYPK
```

Ig-SF C2 set domains at residues 30–102, 132–201 and 238–320. Disulfide bonds between Cys37–95, 139–194 and 245–313.

Amino acid sequence for mouse type I IL-1 receptor[35]

Accession code: SwissProt P13504

```
-19  MENMKVLLGL ICLMVPLLS
  1  LEIDVCTEYP NQIVLFLSVN EIDIRKCPLT PNKMHGDTII WYKNDSKTPI
 51  SADRDSRIHQ QNEHLWFVPA KVEDSGYYYC IVRNSTYCLK TKVTVTVLEN
101  DPGLCYSTQA TFPQRLHIAG DGSLVCPYVS YFKDENNELP EVQWYKNCKP
151  LLLDNVSFFG VKDKLLVRNV AEEHRGDYIC RMSYTFRGKQ YPVTRVIQFI
201  TIDENKRDRP VILSPRNETI EADPGSMIQL ICNVTGQFSD LVYWKWNGSE
251  IEWNDPFLAE DYQFVEHPST KRKYTLITTL NISEVKSQFY RYPFICVVKN
301  TNIFESAHVQ LIYPVPDFKN YLIGGFIILT ATIVCCVCIY KVFKVDIVLW
351  YRDSCSGFLP SKASDGKTYD AYILYPKTLG EGSFSDLDTF VFKLLPEVLE
401  GQFGYKLFIY GRDDYVGEDT IEVTNENVKK SRRLIIILVR DMGGFSWLGQ
451  SSEEQIAIYN ALIQEGIKIV LLELEKIQDY EKMPDSIQFI KQKHGVICWS
501  GDFQERPQSA KTRFWKNLRY QMPAQRRSPL SKHRLLTLDP VRDTKEKLPA
551  ATHLPLG
```

Ig-SF C2 set domains at residues 20–87, 119–187 and 225–303. Disulfide bonds between Cys27–80, 126–180 and 232–296. Thr537 is phosphorylated by PKC.

Amino acid sequence for mouse type II IL-1 receptor[21]

Accession code: SwissProt P27931

```
-13  MFILLVLVTG VSA
  1  FTTPTVVHTG KVSESPITSE KPTVHGDNCQ FRGREFKSEL RLEGEPVVLR
 51  CPLAPHSDIS SSSHSFLTWS KLDSSQLIPR DEPRMWVKGN ILWILPAVQQ
101  DSGTYICTFR NASHCEQMSV ELKVFKNTEA SLPHVSYLQI SALSTTGLLV
151  CPDLKEFISS NADGKIQWYK GAILLDKGNK EFLSAGDPTR LLISNTSMDD
201  AGYYRCVMTF TYNGQEYNIT RNIELRVKGT TTEPIPVIIS PLETIPASLG
251  SRLIVPCKVF LGTGTSSNTI VWWLANSTFI SAAYPRGRVT EGLHHQYSEN
301  DENYVEVSLI FDPVTREDLH TDFKCVASNP RSSQSLHTTV KEVSSTFSWS
351  IALAPLSLII LVVGAIWMRR RCKRRAGKTY GLTKLRTDNQ DFPSSPN
```

Ig-SF C2 set domains at residues 44–114, 144–213 and 250–332. Disulfide bonds between Cys51–107, 151–206 and 257–325.

References

[1] Dinarello, C.A. (1989) Adv. Immunol. 44, 153–205.

[2] Fuhlbrigge, R.C. et al. (1989) In The Year in Immunology 1988: Immunoregulatory Cytokines and Cell Growth, Vol. 5, Cruse, J.M. and Lewis, R.E. eds, Karger, Basle, pp. 21–37.

[3] di Giovine, F.S. and Duff, G.W. (1990) Immunol. Today 11, 13–20.

[4] Eisenberg, S.P. et al. (1990) Nature 343, 341–346.

[5] Arend, W.P. (1993) Interleukin-1 receptor antagonist. Adv. Immunol. 54, 167–227.

[6] Cerretti, D.P. et al. (1992) Science 256, 97–100.

[7] Thornberry, N.A. et al. (1992) Nature 356, 768–774.

[8] Ray, C.A. et al. (1992) Cell 69, 597–604.

[9] Graves, B.J. et al. (1990) Biochemistry 29, 2679–2684.
[10] Priestle, J.P. et al. (1988) EMBO J. 7, 339–349.
[11] Lennard, A. et al. (1992) Cytokine 4, 83–89.
[12] Bensi, G. et al. (1987) Gene 52, 95–101.
[13] Telford, J.L. et al. (1986) Nucleic Acids Res. 14, 9955–9963.
[14] Furutani, Y. et al. (1986) Nucleic Acids Res. 14, 3167–3179.
[15] March, C.J. et al. (1985) Nature 315, 641–647.
[16] Carter, D.B. et al. (1990) Nature 344, 633–638.
[17] Lomedico, P.T. et al. (1984) Nature 312, 458–462.
[18] Gray, P.W. et al. (1986) J. Immunol. 137, 3644–3648.
[19] Matsushime, H. et al. (1991) Blood 78, 616–623.
[20] Zahedi, K. et al. (1991) J. Immunol. 146, 4228–4233.
[21] McMahan, C.J. et al. (1991) EMBO J. 10, 2821–2832.
[22] Symons, J.A. et al. (1991) J. Exp. Med. 174, 1251–1254.
[23] Greenfeeder, S.A. et al. (1995) J. Biol. Chem. 271, 13757–13765.
[24] Wesche, H. et al. (1997) J. Biol. Chem. 272, 7727–7731.
[25] O Neill, L.A.J. and Dinarello, C.A. (2000) Immunol. Today 21, 206–209.
[26] O Neill, L.A.J. (2000) Science's Signal Transduction Knowledge Environment http://www.stke.org/cgi/content/full/OC_sigtrans;2000/44/re1
[27] Wesche, H. et al. (1997) Immunity 7, 837–847.
[28] Cao, Z. et al. (1996) Nature 383, 443–446.
[29] Muzio, M. et al. (1997) Science 278, 1612–1615.
[30] Burns, K. et al. (2000) Nature Cell Biol. 2, 346–351.
[31] Jeffries, C.A. and O Neill, L.A.J. (2000) J. Biol. Chem. 275, 3114–3120.
[32] Palsson, E. et al. (2000) J. Biol. Chem. 275, 7818–7825.
[33] Copeland, N.G. et al. (1991) Genomics 9, 44–50.
[34] Sims, J.E. et al. (1989) Proc. Natl Acad. Sci. USA 86, 8946–8950.
[35] Curtis, B.M. et al. (1990) J. Immunol. 144, 1295–1303.

Other names
T-cell growth factor (TCGF).

THE MOLECULE

Interleukin 2 (IL-2) is a T cell-derived cytokine which was first described as a T-cell growth factor (TCGF). It is now known to stimulate growth and differentiation of T cells, B cells, NK cells, LAK cells, monocytes, macrophages and oligodendrocytes[1–3].

Crossreactivity
There is about 60% homology between human and mouse IL-2. Human IL-2 is active on mouse lymphocytes, but mouse IL-2 is not active on human lymphocytes.

Sources
T cells.

Bioassays
Proliferation of activated T cells or IL-2-dependent T-cell lines. Proliferation of B cells costimulated with anti-IgM.

Physicochemical properties of IL-2

Property	Human	Mouse
pI	8.2	?
Amino acids – precursor	153	169
– mature[a]	133	149
M_r (K) – predicted	15.4	17.2
– expressed	15–20	15–30
Potential *N*-linked glycosylation sites[b]	0	0
Disulfide bonds[c]	1	1

[a] After removal of signal peptide.
[b] *O*-Linked glycosylation of the threonine residue at position 3 in human and mouse IL-2 is responsible for size and charge heterogeneity of the mature protein. Glycosylation is not required for biological activity.
[c] Disulfide bonds between Cys58 and 105 (human) and 72 and 120 (mouse).

3-D structure
The IL-2 structure has been determined at a resolution of 3 Å[4]. It is a globular protein composed of six α-helical regions (A-E) with a disulfide bridge between Cys58 and Cys105 (human) following a bent loop between α-helices A and B. α-Helices C, D, E and F form an apparent antiparallel α-helical bundle. An alternative folding based on the structures of GM-CSF and IL-4 has been proposed[5]. A site on the B α-helix is thought to bind to the p55 chain of the receptor[6].

Gene structure[7–9]

Scale

Exons 50 aa

Translated

Untranslated

Introns 1Kb

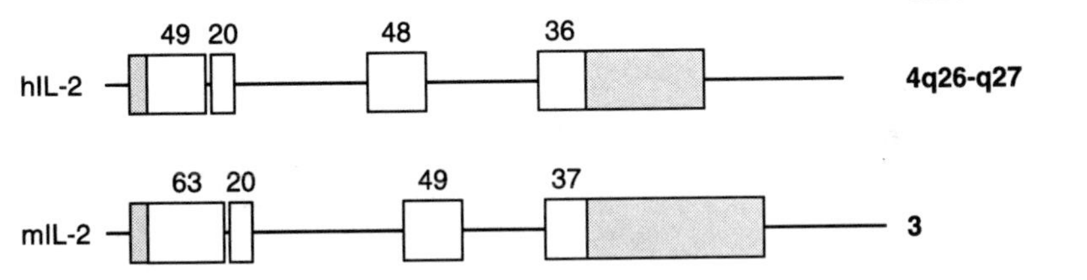

The genes for IL-2 are on human chromosome 4q26–q27, spanning four exons of 49, 20, 48 and 36 amino acids. The gene for mouse IL-2 is on mouse chromosome 3 also spanning four exons of 63, 20, 49 and 37 amino acids.

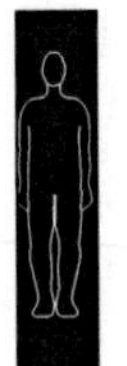

Amino acid sequence for human IL-2[10–12]

Accession code: SwissProt P01585

```
-20  MYRMQLLSCI ALSLALVTNS
  1  APTSSSTKKT QLQLEHLLLD LQMILNGINN YKNPKLTRML TFKFYMPKKA
 51  TELKHLQCLE EELKPLEEVL NLAQSKNFHL RPRDLISNIN VIVLELKGSE
101  TTFMCEYADE TATIVEFLNR WITFCQSIIS TLT
```

Amino acid sequence for mouse IL-2[13,14]

Accession code: SwissProt P04351

```
-20  MYSMQLASCV TLTLVLLVNS
  1  APTSSSTSSS TAEAQQQQQQ QQQQQQHLEQ LLMDLQELLS RMENYRNLKL
 51  PRMLTFKFYL PKQATELKDL QCLEDELGPL RHVLDLTQSK SFQLEDAENF
101  ISNIRVTVVK LKGSDNTFEC QFDDESATVV DFLRRWIAFC QSIISTSPQ
```

Mouse IL-2 has an insertion of 12 repeat glutamine residues (15–26) which is not found in human IL-2.

THE IL-2 RECEPTOR

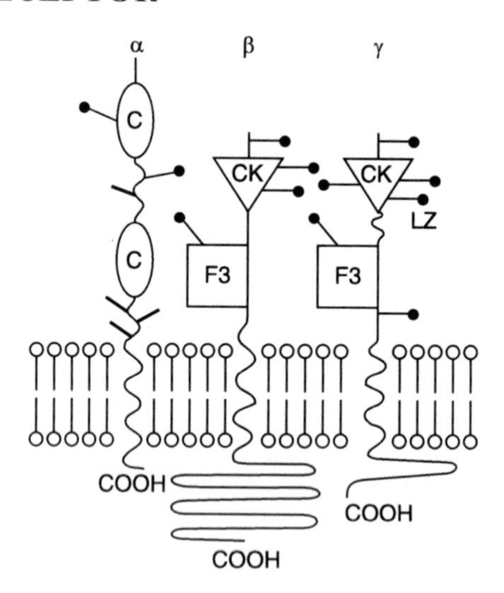

The IL-2 receptor is a complex of three distinct polypeptide chains[15]. The α-chain (Tac, p55 or CD25) binds IL-2 with low affinity (K_d 1.43 × 10^{-8} M) and a short dissociation half-life of 1.7 s. A longer p75 subunit (β-chain) CD122 binds IL-2 with an affinity of 1.23 × 10^{-7} M (K_d). The third p64 subunit (γ-chain) does not bind IL-2. Intermediate affinity (K_d 10^{-9} M) IL-2R complexes are formed from α or β/γ heterodimers. The high-affinity receptor complex is an α/β/γ heterotrimer. It has an equilibrium K_d of 1.33 × 10^{-11} M and a half-life of 50 min[16]. The IL-2R β- and γ-chains are both members of the cytokine receptor superfamily. In addition, the γ-chain has a leucine zipper motif between the CK and FNIII domains, and an SH2-like domain at the proximal end of the cytoplasmic region. The α-chain is not a member of the CKR-SF, but it has two domains in its extracellular region with homology to the complement control protein (CCP). The γ-chain is also a functional component of the IL-4R, IL-7R and IL-15R, and possibly IL-9R and IL-13R and is now known as the common γ (γc) chain.

Mutations in the IL-2R γ-chain are responsible for X-linked severe combined immunodeficiency (X-SCID) in humans[17]. This is a lethal disease characterized by absent or greatly reduced T cells, and severely depressed cell-mediated and humoral immunity.

Distribution

T cells, B cells, NK cells, monocytes and macrophages.

Physicochemical properties of the IL-2 receptor

Property	α-Chain		β-Chain		γ-Chain	
	Human	Mouse	Human	Mouse	Human	Mouse
Amino acids – precursor	272	268	551	539	369	369
– mature[a]	251	247	525	513	347	348
M_r (K) – predicted	28.4	28.4	59.4	57.7	39.9	39.8
– expressed	55	50–60	70–75	70–75	64	?
Potential *N*-linked glycosylation sites	2	4	4	6	6	7
Affinity K_d (M)[b]	10^{-8}	10^{-8}	10^{-7}	10^{-7}	none	none

[a] After removal of predicted signal peptide.
[b] The high-affinity receptor (K_d 10^{-11} M) is a complex of the p55 (α), p75 (β) and p64 (γ) chains. α/β and β/γ receptor complexes have intermediate affinities of 10^{-10} M and 10^{-9} M respectively[15].

Signal transduction

At least two distinct cytoplasmic regions of the IL-2R β-chain are involved in IL-2-mediated cellular signalling. A serine-rich region is critical for induction of c-*myc* and cell proliferation (signal 1) whereas an acidic region is required for physical association with *src*-like protein tyrosine kinase (PTK) p56^{lck}, for activation of p21^{ras}, and induction of c-*fos* and c-*jun* (signal 2)[15]. Binding of IL-2 to its receptor induces tyrosine phosphorylation of the β-chain[18]. The IL-2R γ-chain has an SH2-like homology domain which may be involved in binding to phosphotyrosine residues on other signalling proteins. There is conflicting evidence for phosphatidylinositol hydrolysis and elevation of cAMP[19,20].

Chromosomal location

The human IL-2R α-chain is located on 10p15–p14, the β-chain on 22q11.2–q12 (a region associated with some lymphoid tumours), and the γ-chain on Xq13. The mouse β-chain is on chromosome 15, and the γ-chain is on the X chromosome.

Amino acid sequence for human IL-2 receptor p55 (α-chain)[21–23]

Accession code: SwissProt P01589

```
-21  MDSYLLMWGL LTFIMVPGCQ A
  1  ELCDDDPPEI PHATFKAMAY KEGTMLNCEC KRGFRRIKSG SLYMLCTGNS
 51  SHSSWDNQCQ CTSSATRNTT KQVTPQPEEQ KERKTTEMQS PMQPVDQASL
101  PGHCREPPPW ENEATERIYH FVVGQMVYYQ CVQGYRALHR GPAESVCKMT
151  HGKTRWTQPQ LICTGEMETS QFPGEEKPQA SPEGRPESET SCLVTTTDFQ
201  IQTEMAATME TSIFTTEYQV AVAGCVFLLI SVLLLSGLTW QRRQRKSRRT
251  I
```

Amino acid sequence for human IL-2 receptor p75 (β-chain)[24]

Accession code: SwissProt P14784

```
-26  MAAPALSWRL PLLILLLPLA TSWASA
  1  AVNGTSQFTC FYNSRANISC VWSQDGALQD TSCQVHAWPD RRRWNQTCEL
 51  LPVSQASWAC NLILGAPDSQ KLTTVDIVTL RVLCREGVRW RVMAIQDFKP
101  FENLRLMAPI SLQVVHVETH RCNISWEISQ ASHYFERHLE FEARTLSPGH
151  TWEEAPLLTL KQKQEWICLE TLTPDTQYEF QVRVKPLQGE FTTWSPWSQP
201  LAFRTKPAAL GKDTIPWLGH LLVGLSGAFG FIILVYLLIN CRNTGPWLKK
251  VLKCNTPDPS KFFSQLSSEH GGDVQKWLSS PFPSSSFSPG GLAPEISPLE
301  VLERDKVTQL LLQQDKVPEP ASLSSNHSLT SCFTNQGYFF FHLPDALEIE
351  ACQVYFTYDP YSEEDPDEGV AGAPTGSSPQ PLQPLSGEDD AYCTFPSRDD
401  LLLFSPSLLG GPSPPSTAPG GSGAGEERMP PSLQERVPRD WDPQPLGPPT
451  PGVPDLVDFQ PPPELVLREA GEEVPDAGPR EGVSFPWSRP PGQGEFRALN
501  ARLPLNTDAY LSLQELQGQD PTHLV
```

Amino acid sequence for human common γ-chain[25]

Accession code: SwissProt P31785

```
-22  MLKPSLPFTS LLFLQLPLLG VG
  1  LNTTILTPNG NEDTTADFFL TTMPTDSLSV STLPLPEVQC FVFNVEYMNC
 51  TWNSSSEPQP TNLTLHYWYK NSDNDKVQKC SHYLFSEEIT SGCQLQKKEI
101  HLYQTFVVQL QDPREPRRQA TQMLKLQNLV IPWAPENLTL HKLSESQLEL
151  NWNNRFLNHC LEHLVQYRTD WDHSWTEQSV DYRHKFSLPS VDGQKRYTFR
201  VRSRFNPLCG SAQHWSEWSH PIHWGSNTSK ENPFLFALEA VVISVGSMGL
251  IISLLCVYFW LERTMPRIPT LKNLEDLVTE YHGNFSAWSG VSKGLAESLQ
301  PDYSERLCLV SEIPPKGGAL GEGPGASPCN QHSPYWAPPC YTLKPET
```

Leucine zipper motif between residues 143 and 164. SH2-like domain between residues 266 and 299.

Amino acid sequence for mouse IL-2 receptor p55 (α-chain)[26,27]

Accession code: SwissProt P01590

```
-21  MEPRLLMLGF LSLTIVPSCR A
  1  ELCLYDPPEV PNATFKALSY KNGTILNCEC KRGFRRLKEL VYMRCLGNSW
 51  SSNCQCTSNS HDKSRKQVTA QLEHQKEQQT TTDMQKPTQS MHQENLTGHC
101  REPPPWKHED SKRIYHFVEG QSVHYECIPG YKALQRGPAI SICKMKCGKT
151  GWTQPQLTCV DEREHHRFLA SEESQGSRNS SPESETSCPI TTTDFPQPTE
201  TTAMTETFVL TMEYKVAVAS CLFLLISILL LSGLTWQHRW RKSRRTI
```

Conflicting sequence[27] M→T at position −15, T→A at position 97, E→V at position 119, F→L at position 195, and E→V at position 206.

Amino acid sequence for mouse IL-2 receptor p75 (β-chain)[28]

Accession code: SwissProt P16297

```
-26  MATIALPWSL SLYVFLLLLA TPWASA
  1  AVKNCSHLEC FYNSRANVSC MWSHEEALNV TTCHVHAKSN LRHWNKTCEL
 51  TLVRQASWAC NLILGSFPES QSLTSVDLLD INVVCWEEKG WRRVKTCDFH
101  PFDNLRLVAP HSLQVLHIDT QRCNISWKVS QVSHYIEPYL EFEARRRLLG
151  HSWEDASVLS LKQRQQWLFL EMLIPSTSYE VQVRVKAQRN NTGTWSPWSQ
201  PLTFRTRPAD PMKEILPMSW LRYLLLVLGC FSGFFSCVYI LVKCRYLGPW
251  LKTVLKCHIP DPSEFFSQLS SQHGGDLQKW LSSPVPLSFF SPSGPAPEIS
301  PLEVLDGDSK AVQLLLLQKD SAPLPSPSGH SQASCFTNQG YFFFHLPNAL
351  EIESCQVYFT YDPCVEEEVE EDGSRLPEGS PHPPLLPLAG EQDDYCAFPP
401  RDDLLLFSPS LSTPNTAYGG SRAPEERSPL SLHEGLPSLA SRDLMGLQRP
451  LERMPEGDGE GLSANSSGEQ ASVPEGNLHG QDQDRGQGPI LTLNTDAYLS
501  LQELQAQDSV HLI
```

Amino acid sequence for mouse common γ-chain[29]

Accession code: SwissProt P34902

```
-21  MLKLLLSPRS FLVLQLLLLR A
  1  GWSSKVLMSS ANEDIKADLI LTSTAPEHLS APTLPLPEVQ CFVFNIEYMN
 51  CTWNSSSEPQ ATNLTLHYRY KVSDNNTFQE CSHYLFSKEI TSGCQIQKED
101  IQLYQTFVVQ LQDPQKPQRR AVQKLNLQNL VIPRAPENLT LSNLSESQLE
151  LRWKSRHIKE RCLQYLVQYR SNRDRSWTEL IVNHEPRFSL PSVDELKRYT
201  FRVRSRYNPI CGSSQQWSKW SQPVHWGSHT VEENPSLFAL EAVLIPVGTM
251  GLIITLIFVY CWLERMPPIP PIKNLEDLVT EYQGNFSAWS GVSKGLTESL
301  QPDYSERFCH VSEIPPKGGA LGEGPGGSPC SLHSPYWPPP CYSLKPEA
```


References

[1] Smith, K.A. (1984) Annu. Rev. Immunol. 2, 319–333.
[2] Smith, K.A. (1988) Science 240, 1169–1176.
[3] Kuziel, W.A. and Greene, W.C. (1991) In The Cytokine Handbook, Thomson, A. ed., Academic Press, London, pp. 83–102.
[4] Brandhuber, B.J. et al. (1987) Science 238, 1707–1709.
[5] Bazan, J.F. (1992) Science 257, 410–413.
[6] Sauve, K. et al. (1991) Proc. Natl Acad. Sci. USA 88, 4636–4640.
[7] Holbrook, N.J. et al. (1984) Proc. Natl Acad. Sci. USA 81, 1634–1638.
[8] Fujita, T. et al. (1983) Proc. Natl Acad. Sci. USA 80, 7437–7441.
[9] Fuse, A. et al. (1984) Nucleic Acids Res. 12, 9323–9331.
[10] Taniguchi, T. et al. (1983) Nature 302, 305–310.
[11] Maeda, S. et al. (1983) Biochem. Biophys. Res. Commun. 115, 1040–1047.
[12] Devos, R. et al. (1983) Nucleic Acids Res. 11, 4307–4323.
[13] Yokota, T. et al. (1985) Proc. Natl Acad. Sci. USA 82, 68–72.
[14] Kashima, N. et al. (1985) Nature 313, 402–404.
[15] Minami, Y. et al. (1993) Annu. Rev. Immunol. 11, 245–267.
[16] Taniguchi, T. and Minami, Y. (1993) Cell 73, 5–8.
[17] Noguchi, M. et al. (1993) Cell 73, 147–157.
[18] Asao, H. et al. (1990) J. Exp. Med. 171, 637–644.

[19] Rigley, K.P. and Harnett, M. (1990) In Cytokines and B Lymphocytes, Callard, R. ed., Academic Press, London, pp. 39–63.
[20] Tigges, M.A. et al. (1989) Science 243, 781–786.
[21] Leonard, W.J. et al. (1984) Nature 311, 626–631.
[22] Cosman, D. et al. (1984) Nature 312, 768–771.
[23] Nikaido, T. et al. (1984) Nature 311, 631–635.
[24] Hatakeyama, M. et al. (1989) Science 244, 551–556.
[25] Takeshita, T. et al. (1992) Science 257, 379–382.
[26] Miller, J. et al. (1985) J. Immunol. 134, 4212–4217.
[27] Shimuzu, A. et al. (1985) Nucleic Acids Res. 13, 1505–1516.
[28] Kono, T. et al. (1990) Proc. Natl Acad. Sci. USA 87, 1806–1810.
[29] Cao, X. et al. (1993) Proc. Natl Acad. Sci. USA 90, 8464–8468.

Other names

Mast cell growth factor (MCGF), multi-colony stimulating factor (multi-CSF), eosinophil-CSF (E-CSF), haematopoietic cell growth factor (HCGF), burst-promoting activity (BPA), P-cell stimulating factor activity, thy-1 inducing factor, WEHI-3 growth factor.

THE MOLECULE

Interleukin 3 (IL-3) is a haematopoietic growth factor which stimulates colony formation of erythroid, megakaryocyte, neutrophil, eosinophil, basophil, mast cell and monocytic lineages[1]. IL-3 may also stimulate multipotent progenitor cells, but it is more likely to be important in committing progenitor cells to a differentiation pathway rather than self-renewal of primitive stem cells. Many of the activities of IL-3 are enhanced or depend upon costimulation with other cytokines. IL-3 does not stimulate lymphocyte colony formation, but it is a growth factor for B lymphocytes and it activates monocytes, suggesting that it may have an additional immunoregulatory role. IL-3 has been used clinically to expand haematopoietic precursors after bone marrow transplantation, aplastic anaemia and chemotherapy[2].

Crossreactivity

Amino acid sequence homology between mouse and human IL-3 is only 29% and there is no cross-species reactivity.

Sources

Activated T cells, mast cells, eosinophils.

Bioassays

Proliferation of TF-1 (human erythroleukaemia), MO7e (human megakaryoblastic leukaemia) or AML-193 (acute myeloid leukaemia) cell lines. Stimulation of erythroid, granulocyte and macrophage colony formation in bone marrow colony assay.

Physicochemical properties of IL-3

Property	Human	Mouse
pI	4–8	4–8
Amino acids – precursor	152	166
– mature[a]	133	140
M_r (K) – predicted	15.1	15.7
– expressed	14–30	28
Potential *N*-linked glycosylation sites	2	4[b]
Disulfide bonds	1	2

[a] After removal of predicted signal peptide.
[b] Glycosylation only at positions 16 and 86 (see sequence). Glycosylation is not required for biological activity.

3-D structure

Similar to IL-4, GM-CSF and M-CSF.

Gene structure[3]

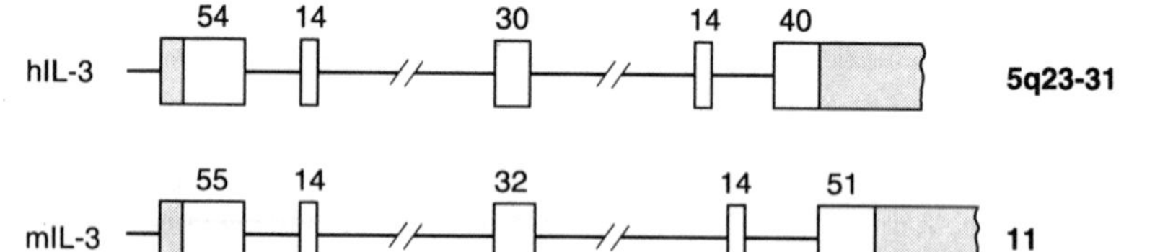

IL-3 is located on human chromosome 5q23 (five exons of 54, 14, 30, 14 and 40 amino acids) and mouse chromosome 11 (five exons of 55, 14, 32, 14 and 51 amino acids).

Amino acid sequence for human IL-3[4,5]

Accession code: SwissProt P08700

```
-19  MSRLPVLLLL QLLVRPGLQ
  1  APMTQTTPLK TSWVNCSNMI DEIITHLKQP PLPLLDFNNL NGEDQDILME
 51  NNLRRPNLEA FNRAVKSLQN ASAIESILKN LLPCLPLATA APTRHPIHIK
101  DGDWNEFRRK LTFYLKTLEN AQAQQTTLSL AIF
```

Conflicting sequence[5] P→S at position 8. Disulfide bond between Cys16–84.

Amino acid sequence for mouse IL-3[6,7]

Accession code: SwissProt P01586

```
-26  MVLASSTTSI HTMLLLLLML FHLGLQ
  1  ASISGRDTHR LTRTLNCSSI VKEIIGKLPE PELKTDDEGP SLRNKSFRRV
 51  NLSKFVESQG EVDPEDRYVI KSNLQKLNCC LPTSANDSAL PGVFIRDLDD
101  FRKKLRFYMV HLNDLETVLT SRPPQPASGS VSPNRGTVEC
```

Disulfide bonds between Cys17–80 and 79–140.

THE IL-3 RECEPTOR

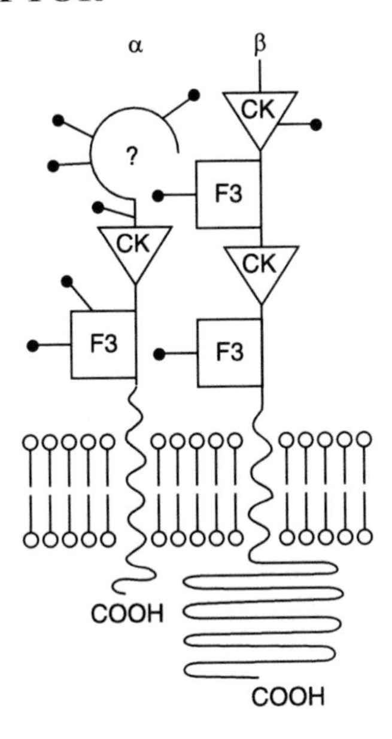

The high-affinity IL-3 receptor is formed by association of a low-affinity IL-3-binding α-subunit (IL-3Rα) (CD123) with a second β-subunit[8]. In humans, the β-subunit (KH97) is common to the IL-3, IL-5 and GM-CSF receptors, but does not itself bind any of these cytokines[9]. In mice, there are two β-subunits, AIC2A and AIC2B[10]. Of these, AIC2B is similar to KH97 and is shared with IL-5 and GM-CSF receptors, but does not itself bind cytokine[11,12]. In contrast, AIC2A binds IL-3 and associates only with the IL-3R α-chain. The IL-3R α-subunit is a member of the CKR-SF. It has an N-terminal region of about 100 amino acids with homology to a similar sequence in the α-subunits of the IL-5 and GM-CSF receptors, followed by a CKR domain and an FNIII domain that contains the WSXWS motif. It has a short cytoplasmic domain, and is unable to signal. The β-subunit has two homologous segments in the extracellular region, each with a CKR domain followed by an FNIII domain. It has a much longer cytoplasmic domain with a Ser/Pro-rich region in common with the IL-2R β-chain, the IL-4R and the EpoR.

Distribution

Haematopoietic progenitor cells derived from pluripotent stem cells, monocytes, pre-B cells and B cells[13].

Physicochemical properties of the IL-3 receptor

Property	α-Chain Human	α-Chain Mouse	β-Chain Human KH97	β-Chain Mouse AIC2A	β-Chain Mouse AIC2B
Amino acids – precursor	378	396	897	878	896
– mature[a]	360	380	881	856	874
M_r (K) – predicted	41.3	41.5	95.7	94.7	96.6
– expressed	70	60–70	120	110–120	120–140
Potential *N*-linked glycosylation sites	6	5	3	2	3
Affinity K_d (M)[b]	10^{-7}	53×10^{-8}	none	10^{-8}	none

[a] After removal of predicted signal peptide.
[b] High-affinity receptor is a complex of α- and β-chains. Human IL-3R α/β (KH97) K_d 10^{-10} M[14], mouse α/β (AIC2A) K_d 33×10^{-10} M, mouse α/β (AIC2B) K_d 43×10^{-10} M[15]. KH97 and AIC2B also associate with IL-5Rα and GM-CSFRα subunits.

Signal transduction

IL-3 binding to its receptor results in rapid tyrosine and serine/threonine phosphorylation of a number of cellular proteins[16–18], including the IL-3R β-subunit itself[19]. The cloned IL-3 receptor has no consensus sequence for tyrosine kinase, indicating that the receptor and associated tyrosine kinase are separate molecules. There is also some evidence for PKC translocation. Signal transduction requires the presence of the β-subunit.

Chromosomal location

The human IL-3 receptor α-chain is located in the pseudo-autosomal region of the sex chromosomes at Yp13.3 and Xp22.3. The β-chain is on chromosome 22. AIC2A and AIC2B are both on mouse chromosome 15.

Amino acid sequence for human IL-3 receptor α-chain[14]

Accession code: SwissProt P26951

```
-18  MVLLWLTLLL IALPCLLQ
  1  TKEDPNPPIT NLRMKAKAQQ LTWDLNRNVT DIECVKDADY SMPAVNNSYC
 51  QFGAISLCEV TNYTVRVANP PFSTWILFPE NSGKPWAGAE NLTCWIHDVD
101  FLSCSWAVGP GAPADVQYDL YLNVANRRQQ YECLHYKTDA QGTRIGCRFD
151  DISRLSSGSQ SSHILVRGRS AAFGIPCTDK FVVFSQIEIL TPPNMTAKCN
201  KTHSFMHWKM RSHFNRKFRY ELQIQKRMQP VITEQVRDRT SFQLLNPGTY
251  TVQIRARERV YEFLSAWSTP QRFECDQEEG ANTRAWRTSL LIALGTLLAL
301  VCVFVICRRY LVMQRLFPRI PHMKDPIGDS FQNDKLVVWE AGKAGLEECL
351  VTEVQVVQKT
```

Amino acid sequence for human IL-3 receptor β-chain (KH97)[9]

Accession code: SwissProt P32927

-16 MVLAQGLLSM ALLALC
1 WERSLAGAEE TIPLQTLRCY NDYTSHITCR WADTQDAQRL V**N**VTLIRRVN
51 EDLLEPVSCD LSDDMPWSAC PHPRCVPRRC VIPCQSFVVT DVDYFSFQPD
101 RPLGTRLTVT LTQHVQPPEP RDLQISTDQD HFLLTWSVAL GSPQSHWLSP
151 GDLEFEVVYK RLQDSWEDAA ILLS**N**TSQAT LGPEHLMPSS TYVARVRTRL
201 APGSRLSGRP SKWSPEVCWD SQPGDEAQPQ NLECFFDGAA VLSCSWEVRK
251 EVASSVSFGL FYKPSPDAGE EECSPVLREG LGSLHTRHHC QIPVPDPATH
301 GQYIVSVQPR RAEKHIKSSV NIQMAPPSL**N** VTKDGDSYSL RWETMKMRYE
351 HIDHTFEIQY RKDTATWKDS KTETLQNAHS MALPALEPST RYWARVRVRT
401 SRTGYNGIWS EWSEARSWDT ESVLPMWVLA LIVIFLTTAV LLALRFCGIY
451 GYRLRRKWEE KIPNPSKSHL FQNGSAELWP PGSMSAFTSG SPPHQGPWGS
501 RFPELEGVFP VGFGDSEVSP LTIEDPKHVC DPPSGPDTTP AASDLPTEQP
551 PSPQPGPPAA SHTPEKQASS FDFNGPYLGP PHSRSLPDIL GQPEPPQEGG
601 SQKSPPPGSL EYLCLPAGGQ VQLVPLAQAM GPGQAVEVER RPSQGAAGSP
651 SLESGGGPAP PALGPRVGGQ DQKDSPVAIP MSSGDTEDPG VASGYVSSAD
701 LVFTPNSGAS SVSLVPSLGL PSDQTPSLCP GLASGPPGAP GPVKSGFEGY
751 VELPPIEGRS PRSPRNNPVP PEAKSPVLNP GERPADVSPT SPQPEGLLVL
801 QQVGDYCFLP GLGPGPLSLR SKPSSPGPGP EIKNLDQAFQ VKKPPGQAVP
851 QVPVIQLFKA LKQQDYLSLP PWEVNKPGEV C

Amino acid sequence for mouse IL-3 receptor α-chain[15]

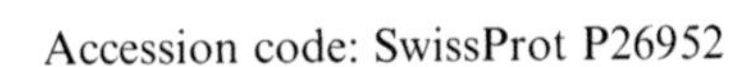

Accession code: SwissProt P26952

-16 MAANLWLILG LLASHS
1 SDLAAVREAP PTAVTTPIQN LHIDPAHYTL SWDPAPGADI TTGAFCRKGR
51 DIFVWADPGL ARCSFQSLSL CHVT**N**FTVFL GKDRAVAGSI QFPPDDDGDH
101 EAAAQDLRCW VHEGQLSCQW ERGPKATGDV HYRMFWRDVR LGPAHNRECP
151 HYHSLDVNTA GPAPHGGHEG CTLDLDTVLG STPNSPDLVP QVTITV**N**GSG
201 RAGPVPCMDN TVDLQRAEVL APPTLTVEC**N** GSEAHARWVA RNRFHHGLLG
251 YTLQV**N**QSSR SEPQEY**N**VSI PHFWVPNAGA ISFRVKSRSE VYPRKLSSWS
301 EAWGLVCPPE VMPVKTALVT SVATVLGAGL VAAGLLLWWR KSLLYRLCPP
351 IPRLRLPLAG EMVVWEPALE DCEVTPVTDA

Amino acid sequence for mouse IL-3 receptor (AIC2A) β-chain[20]

Accession code: SwissProt P26954

-22 MDQQMALTWG LCYMALVALC WG
1 HEVTEEEETV PLKTLECYND YTNRIICSWA DTEDAQGLI**N** MTLLYHQLDK
51 IQSVSCELSE KLMWSECPSS HRCVPRRCVI PYTRFSNGDN DYYSFQPDRD
101 LGIQLMVPLA QHVQPPPPKD IHISPSGDHF LLEWSVSLGD SQVSWLSSKD
151 IEFEVAYKRL QDSWEDASSL HTSNFQVNLE PKLFLPNSIY AARVRTRLSA
201 GSSLSGRPSR WSPEVHWDSQ PGDKAQPQNL QCFFDGIQSL HCSWEVWTQT
251 TGSVSFGLFY RPSPAAPEEK CSPVVKEPQA SVYTRYRCSL PVPEPSAHSQ
301 YTVSVKHLEQ GKFIMSYYHI QMEPPIL**N**QT KNRDSYSLHW ETQKIPKYID
351 HTFQVQYKKK SESWKDSKTE NLGRVNSMDL PQLEPDTSYC ARVRVKPISD
401 YDGIWSEWSN EYTWTTDWVM PTLWIVLILV FLIFTLLLAL HFGRVYGYRT
451 YRKWKEKIPN PSKSLLFQDG GKGLWPPGSM AAFATKNPAL QGPQSRLLAE
501 QQGVSYEHLE DNNVSPLTIE DPNIIRDPPS RPDTTPAASS ESTEQLPNVQ
551 VEGPIPSSRP RKQLPSFDFN GPYLGPPQSH SLPDLPGQLG SPQVGGSLKP
601 ALPGSLEYMC LPPGGQVQLV PLSQVMGQGQ AMDVQCGSSL ETTGSPSVEP
651 KENPPVELSV EKQEARDNPM TLPISSGGPE GSMMASDYVT PGDPVLTLPT
701 GPLSTSLGPS LGLPSAQSPS LCLKLPRVPS GSPALGPPGF EDYVELPPSV
751 SQAATSPPGH PAPPVASSPT VIPGEPREEV GPASPHPEGL LVLQQVGDYC
801 FLPGLGPGSL SPHSKPPSPS LCSETEDLDQ DLSVKKFPYQ PLPQAPAIQF
851 FKSLKY

Amino acid sequence for mouse IL-3 receptor AIC2B[10]

Accession code: SwissProt P26955

-22 MDQQMALTWG LCYMALVALC WG
1 HGVTEAEETV PLKTLQCYND YTNHIICSWA DTEDAQGLI**N** MTLYHQLEKK
51 QPVSCELSEK LMWSECPSSH RCVPRRCVIP YTRFSITNED YYSFRPDSDL
101 GIQLMVPLAQ NVQPPLPK**N**V SISSSEDRFL LEWSVSLGDA QVSWLSSKDI
151 EFEVAYKRLQ DSWEDAYSLH TSKFQVNFEP KLFLPNSIYA PRVRTRLYPG
201 SSLSGRPSRW SPEAHWDSQP GDKAQPQNLQ CFFDGIQSLH CSWEVWTQTT
251 GSVSFGLFYR PSPVAPEEKC SPVVKEPPGA SVYTRYHCSL PVPEPSAHSQ
301 YTVSVKHLEQ GKFIMSYNHI QMEPPTL**N**LT KNRDSYSLHW ETQKMAYSFI
351 EHTFQVQYKK KSDSWEDSKT ENLDRAHSMD LSQLEPDTSY CARVRVKPIS
401 NYDGIWSKWS EEYTWKTDWV MPTLWIVLIL VFLILTLLLI LRFGCVSVYR
451 TYRKWKEKIP NPSKSLLFQD GGKGLWPPGS MAAFATKNPA LQGPQSRLLA
501 EQQGESYAHL EDNNVSPLTI EDPNIIRVPP SGPDTTPAAS SESTEQLPNV
551 QVEGPTPNRP RKQLPSFDFN GPYLGPPQSH SLPDLPDQLG SPQVGGSLKP
601 ALPGSLEYMC LAPGGQVQLV PLSQVMGQGQ AMDVQCGSSL ETSGSPSVEP
651 KENPPVELSM EEQEARDNPV TLPISSGGPE GSMMASDYVT PGDPVLTLPT
701 GPLSTSLGPS LGLPSAQSPS LCLKLPRVPS GSPALGPPGF EDYVELPPSV
751 SQAAKSPPGH PAPPVASSPT VIPGEPREEV GPASPHPEGL LVLQQVGDYC
801 FLPGLGPGSL SPHSKPPSPS LCSETEDLVQ DLSVKKFPYQ PMPQAPAIQF
851 FKSLKHQDYL SLPPWDNSQS GKVC

References

[1] Ihle, J.N. (1991) In Peptide Growth Factors and their Receptors I, Sporn, M.B. and Roberts, A.B. eds, Springer-Verlag, New York, pp. 541–575.
[2] Ihle, J.N. (1992) Chem. Immunol. 51, 65–106.
[3] Miyatake, S. et al. (1985) Proc. Natl Acad. Sci. USA 82, 316–320.
[4] Otsuka, T. et al. (1988) J. Immunol. 140, 2288–2295.
[5] Yang, Y-C. et al. (1986) Cell 47, 3–10.
[6] Yokota, T. et al. (1984) Proc. Natl Acad. Sci. USA 81, 1070–1074.
[7] Fung, M.C. et al. (1984) Nature 307, 233–237.
[8] Cosman, D. (1994) Cytokine 5, 95–106.
[9] Hayashida, K. et al. (1990) Proc. Natl Acad. Sci. USA 87, 9655–9659.
[10] Gorman, D.M. et al. (1990) Proc. Natl Acad. Sci. USA 87, 5459–5463.
[11] Kitamura, T. et al. (1991) Proc. Natl Acad. Sci. USA 88, 5082–5086.
[12] Takaki, S. et al. (1991) EMBO J. 10, 2833–2838.
[13] Park, L.S. et al. (1989) J. Biol. Chem. 264, 5420–5427.
[14] Kitamura, T. et al. (1991) Cell 66, 1165–1174.
[15] Hara, T. and Miyajima, A. (1992) EMBO J. 11, 1875–1884.
[16] Isfort, R. et al. (1988) J. Biol. Chem. 263, 19203–19209.
[17] Sorensen, P.H. et al. (1989) Blood 73, 406–418.
[18] Murata, Y. et al. (1990) Biochem. Biophys. Res. Commun. 173, 1102–1108.
[19] Sorensen, P. et al. (1989) J. Biol. Chem. 264, 19253–19258.
[20] Itoh, N. et al. (1990) Science 247, 324–327.

Other names

B cell-stimulating factor 1 (BSF-1).

THE MOLECULE

Interleukin 4 (IL-4) is a pleiotropic cytokine derived from T cells and mast cells with multiple biological effects on B cells, T cells and many nonlymphoid cells including monocytes, endothelial cells and fibroblasts. It also induces secretion of IgG1 and IgE by mouse B cells and IgG4 and IgE by human B cells. The IL-4-dependent production of IgE and possibly IgG1 and IgG4 is due to IL-4-induced isotype switching[1–3]. IL-4 appears to share this property with IL-13.

Crossreactivity

Two regions of human IL-4 (amino acids 1–90 and 129–149) share approximately 50% sequence homology with the corresponding regions of mouse IL-4. In contrast, the region from amino acid positions 91–128 shares very little homology with the corresponding region of mouse IL-4. There is no cross-species reactivity between human and mouse IL-4.

Sources

Mast cells, T cells, some mouse B-cell lymphomas, bone marrow stromal cells.

Bioassays

Human: Proliferation of PHA T-cell blasts in the presence of blocking anti-IL-2 or anti-IL-2R antibody; proliferation of MO7 cell line; increased expression of CD23 or surface IgM on human tonsillar B cells.

Mouse: Proliferation of CTLL in the presence of anti-IL-2 or anti-IL-2R antibody. Increased expression of MHC class II on murine B cells.

Physicochemical properties of IL-4

Property	Human	Mouse
pI	10.5	6.5
Amino acids – precursor	153	140
– mature[a]	129	120
M_r (K) – predicted	15.0	13.6
– expressed	15–19	15–19
Potential *N*-linked glycosylation sites	2[b]	3
Disulfide bonds	3	3

[a] After removal of signal peptide.
[b] Asn38 is glycosylated.

3-D structure

The three-dimensional structure of IL-4 has been determined by NMR[4] and X-ray crystallography[5]. It has a compact globular structure with a predominantly hydrophobic core. A four α-helix bundle with the helices arranged in a left-handed antiparallel bundle with two overhand connections containing a two-stranded antiparallel β-sheet make up most of the molecule. The structure is similar to that of GM-CSF, M-CSF and IL-3.

Gene structure[6,7]

The gene for IL-4 is on human chromosome 5q31 (four exons of 45, 16, 59 and 33 amino acids) and mouse chromosome 11 (four exons of 44, 16, 51 and 29 amino acids).

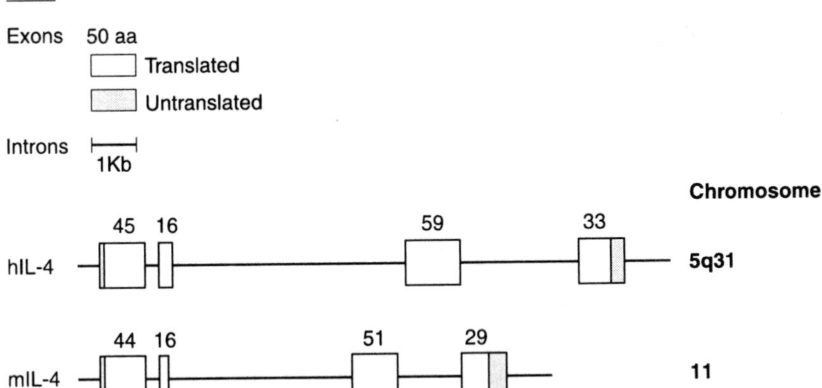

Amino acid sequence for human IL-4[8]

Accession code: SwissProt P05112

```
-24   MGLTSQLLPP LFFLLACAGN FVHG
  1   HKCDITLQEI IKTLNSLTEQ KTLCTELTVT DIFAASKNTT EKETFCRAAT
 51   VLRQFYSHHE KDTRCLGATA QQFHRHKQLI RFLKRLDRNL WGLAGLNSCP
101   VKEANQSTLE NFLERLKTIM REKYSKCSS
```

Disulfide bonds between Cys3–127, 24–65 and 46–99. Asn38 is glycosylated.

Amino acid sequence for mouse IL-4[9,10]

Accession code: SwissProt P07750

```
-20   MGLNPQLVVI LLFFLECTRS
  1   HIHGCDKNHL REIIGILNEV TGEGTPCTEM DVPNVLTATK NTTESELVCR
 51   ASKVLRIFYL KHGKTPCLKK NSSVLMELQR LFRAFRCLDS SISCTMNESK
101   STSLKDFLES LKSIMQMDYS
```

Disulfide bonds between Cys5–87, 27–67 and 49–94.

THE IL-4 RECEPTOR

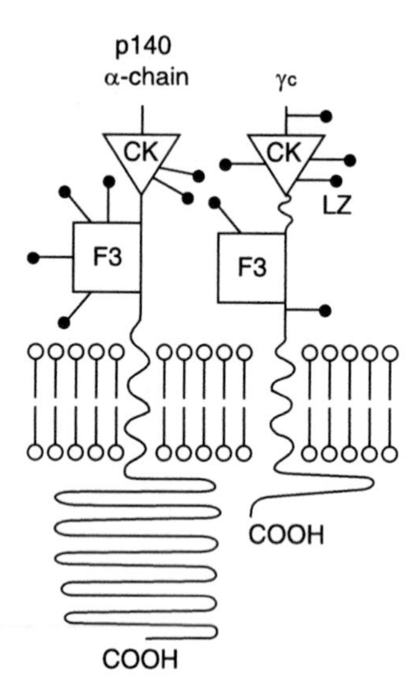

The IL-4 receptor is a complex consisting at least of two chains: a high-affinity IL-4-binding (p140, α-chain) (CD 124) chain and the IL-2R γ-chain, also known as the common γ-chain (γc). The high-affinity (K_d 10^{-10} M) IL-4-binding chain belongs to the cytokine receptor superfamily. It has a CKR domain and an FNIII domain containing the WSXWS motif in the extracellular region[11]. The cytoplasmic domain contains Ser/Pro-rich regions similar to those present in the IL-2R and GM-CSFR β-chains. A soluble form of the high-affinity mouse IL-4 receptor is produced by alternative mRNA splicing[12]. The soluble form of the mouse IL-4R and a recombinant extracellular domain of the human IL-4R are both potent IL-4 antagonists[13]. The IL-2R γ-chain is a functional component of the IL-4R and augments IL-4 binding affinity[14,15]. A second low-affinity (K_d 10^{-8} M) IL-4 receptor has also been identified and partially sequenced, but not yet cloned[16]. The high- and low-affinity receptors appear to be coupled to different signal transduction pathways[17].

Distribution

High-affinity receptors for IL-4 are present in low numbers on pre-B cells, resting B cells and resting T cells; and in high numbers after activation. They are also present on haematopoietic progenitor cells, mast cells, macrophages, myeloid cells, endothelial cells, epithelial cells, fibroblasts, muscle cells, neuroblasts, brain stroma and bone marrow stroma[18–21].

Physicochemical properties of the IL-4 receptor

Property	Human	Mouse
Amino acids – precursor	825	810
– mature[a]	800	785 (205)[b]
M_r (K) – predicted	87	85.1 (23.9)[b]
– expressed	140	138–145 (32–41)[b]
Potential *N*-linked glycosylation sites	6	5
Affinity K_d (M)	10^{-10}	10^{-10}

[a] After removal of predicted signal peptide.
[b] Secreted form of murine IL-4R in parenthesis.

Signal transduction

IL-4 binding to the high-affinity receptor results in tyrosine phosphorylation of cellular proteins including 4PS, which associates with the p85 subunit of PI 3-kinase after cytokine stimulation[22,23], and translocation of PKC[24]. Activation of a tyrosine phosphatase and dephosphorylation of an 80 000 molecular weight protein has also been reported[25]. Another signal transduction pathway for human IL-4 has been shown to involve breakdown of phosphatidylinositol bisphosphate (PIP_2) releasing IP_3 and intracellular calcium followed after a lag period of 10–15 min by a rise in intracellular cAMP[26]. This pathway may be coupled to a low-affinity receptor[16,17].

Chromosomal location

The high-affinity human IL-4 receptor is on chromosome 16p12.1–p11.2. The mouse IL-4 receptor is on chromosome 7.

Amino acid sequence for human IL-4 receptor[27]

Accession code: SwissProt P24394

```
-25  MGWLCSGLLF PVSCLVLLQV ASSGN
  1  MKVLQEPTCV SDYMSISTCE WKMNGPTNCS TELRLLYQLV FLLSEAHTCI
 51  PENNGGAGCV CHLLMDDVVS ADNYTLDLWA GQQLLWKGSF KPSEHVKPRA
101  PGNLTVHTNV SDTLLLTWSN PYPPDNYLYN HLTYAVNIWS ENDPADFRIY
151  NVTYLEPSLR IAASTLKSGI SYRARVRAWA QCYNTTWSEW SPSTKWHNSY
201  REPFEQHLLL GVSVSCIVIL AVSLLCYVSI TKIKKEWWDQ IPNPARSRLV
251  AIIIQDAQGS QWEKRSRGQE PAKCPHWKNC LTKLLPCFLE HNMKRDEDPH
301  KAAKEMPFQG SGKSAWCPVE ISKTVLWPES ISVVRCVELF EAPVECEEEE
351  EVEEEKGSFC ASPESSRDDF QEGREGIVAR LTESLFLDLL GEENGGFCQQ
401  DMGESCLLPP SGSTSAHMPW DEFPSAGPKE APPWGKEQPL HLEPSPPASP
451  TQSPDNLTCT ETPLVIAGNP AYRSFSNSLS QSPCPRELGP DPLLARHLEE
501  VEPEMPCVPQ LSEPTTVPQP EPETWEQILR RNVLQHGAAA APVSAPTSGY
551  QEFVHAVEQG GTQASAVVGL GPPGEAGYKA FSSLLASSAV SPEKCGFGAS
601  SGEEGYKPFQ DLIPGCPGDP APVPVPLFTF GLDREPPRSP QSSHLPSSSP
651  EHLGLEPGEK VEDMPKPPLP QEQATDPLVD SLGSGIVYSA LTCHLCGHLK
701  QCHGQEDGGQ TPVMASPCCG CCCGDRSSPP TTPLRAPDPS PGGVPLEASL
751  CPASLAPSGI SEKSKSSSSF HPAPGNAQSS SQTPKIVNFV SVGPTYMRVS
```

Amino acid sequence for mouse IL-4 receptor[12]

Accession code: SwissProt P16382

-25 MGRLCTKFLT SVGCLILLLV TGSGS

1 IKVLGEPTCF SDYIRTSTCE WFLDSAVDCS SQLCLHYRLM FFEFSE**N**LTC

51 IPRNSASTVC VCHMEMNRPV QSDRYQMELW AEHRQLWQGS FSPSGNVKPL

101 APD**N**LTLHT**N** VSDEWLLTWN NLYPSNNLLY KDLISMV**N**IS REDNPAEFIV

151 Y**N**VTYKEPRL SFPINILMSG VYYTARVRVR SQILTGTWSE WSPSITWYNH

201 FQLPLIQRLP LGVTISCLCI PLFCLFCYFS ITKIKKIWWD QIPTPARSPL

251 VAIIIQDAQV PLWDKQTRSQ ESTKYPHWKT CLDKLLPCLL KHRVKKKTDF

301 PKAAPTKSLQ SPGKAGWCPM EVSRTVLWPE NVSVSVVRCM ELFEAPVQNV

351 EEEEDEIVKE DLSMSPENSG GCGFQESQAD IMARLTENLF SDLLEAENGG

401 LGQSALAESC SPLPSGSGQA SVSWACLPMG PSEEATCQVT EQPSHPGPLS

451 GSPAQSAPTL ACTQVPLVLA DNPAYRSFSD CCSPAPNPGE LAPEQQQADH

501 LEEEEPPSPA DPHSSGPPMQ PVESWEQILH MSVLQHGAAA GSTPAPAGGY

551 QEFVQAVKQG AAQDPGVPGV RPSGDPGYKA FSSLLSSNGI RGDTAAAGTD

601 DGHGGYKPFQ NPVPNQSPSS VPLFTFGLDT ELSPSPLNSD PPKSPPECLG

651 LELGLKGGDW VKAPPPADQV PKPFGDDLGF GIVYSSLTCH LCGHLKQHHS

701 QEEGGQSPIV ASPGCGCCYD DRSPSLGSLS GALESCPEGI PPEANLMSAP

751 KTPSNLSGEG KGPGHSPVPS QTTEVPVGAL GIAVS

Alternative mRNA splicing gives rise to three forms of the mouse IL-4 receptor. One is full-length as shown, a second lacks the cytoplasmic domain (amino acids 233–285), and a third secreted form terminates with the sequence PSNENL which replaces HFQLPL at position 200–205 due to a 114-bp insertion at nucleotide number 598[12].

References

1. Ohara, J-I. (1989) In The Year in Immunology 1988: Immunoregulatory Cytokines and Cell Growth, Vol. 5, Cruse, J.M. and Lewis, R.E. eds, Karger, Basle, pp. 126–159.
2. Callard, R.E. (1991) Br. J. Haematol. 78, 293–299.
3. Paul, W.E. (1991) Blood 77, 1859–1870.
4. Powers, R. et al. (1992) Science 256, 1673–1677.
5. Walter, M.R. et al. (1992) J. Biol. Chem. 267, 20371–20376.
6. Arai, N. et al. (1989) J. Immunol. 142, 274–282.
7. Otsuka, T. et al. (1987) Nucleic Acids Res. 15, 333–344.
8. Yokota, T. et al. (1986) Proc. Natl Acad. Sci. USA 83, 5894–5898.
9. Noma, Y. et al. (1986) Nature 319, 640–646.
10. Lee, F. et al. (1986) Proc. Natl Acad. Sci. USA. 83, 2061–2065.
11. Beckmann, M.P. et al. (1992) Chem. Immunol. 51, 107–134.
12. Mosley, B. et al. (1989) Cell 59, 335–348.
13. Garrone, P. et al. (1991) Eur. J. Immunol. 21, 1365–1369.
14. Russell, S.M. et al. (1993) Science 262, 1880–1883.
15. Kondo, M. et al. (1993) Science 262, 1874–1877.
16. Fanslow, W.C. et al. (1993) Blood 81, 2998–3005.
17. Rigley, K.P. et al. (1991) Int. Immunol. 3, 197–203.
18. Lowenthal, J.W. et al. (1988) J. Immunol. 140, 456–464.

[19] Ohara, J. and Paul, W.E. (1987) Nature 325, 537–540.
[20] Park, L.S. et al. (1987) Proc. Natl Acad. Sci. USA 84, 1669–1673.
[21] Park, L.S. et al. (1987) J. Exp. Med. 166, 476–488.
[22] Wang, L.-M. et al. (1992) EMBO J. 11, 4899–4908.
[23] Wang, L.-M. et al. (1993) Proc. Natl Acad. Sci. USA 90, 4032–4036.
[24] Harnett, M.M. et al. (1991) J. Immunol. 147, 3831–3836.
[25] Mire-Sluis, A.R. and Thorpe, R. (1991) J. Biol. Chem. 266, 18113–18118.
[26] Finney, M. et al. (1990) Eur. J. Immunol. 20, 151–156.
[27] Idzerda, R.L. et al. (1990) J. Exp. Med. 171, 861–873.

Other names

Eosinophil differentiation factor (EDF), eosinophil colony-stimulating factor (E-CSF), B cell growth factor II (BCGFII), B cell differentiation factor for IgM (BCDFμ), IgA enhancing factor, T cell-replacing factor (TRF).

THE MOLECULE

Interleukin 5 (IL-5) is a T cell-derived glycoprotein which stimulates eosinophil colony formation and is an eosinophil differentiation factor in humans and mice. It is also a growth and differentiation factor for mouse but not human B cells[1-3].

Crossreactivity

There is 71% homology between mouse and human IL-5 and significant cross-reactivity in functional assays.

Sources

Mast cells, T cells and eosinophils.

Bioassays

Human: Eosinophil differentiation; proliferation of TF1 cell line.

Mouse: Eosinophil differentiation; proliferation of BCL1 or B13 B-cell lines.

Physicochemical properties of IL-5

Property	Human	Mouse
pI (calculated)	7	7.8
Amino acids – precursor	134	133
– mature[a]	115	113
M_r (K) – predicted	13.1	13.1
– expressed[b]	45	40–50
Potential *N*-linked glycosylation sites	2	3
Disulfide bonds[c]	2	2

[a] After removal of predicted signal peptide.
[b] Homodimer.
[c] Interchain.

3-D structure

IL-5 is an antiparallel disulfide-linked homodimer. The monomer is biologically inactive. The structure of the dimer has been determined at a resolution of 2.4 Å[4]. It has a novel two-domain structure with each domain showing significant structural homology to the cytokine fold in GM-CSF, M-CSF, IL-2, IL-4 and growth hormone. The IL-5 structure is made up of two left-handed bundles of four α-helices with two short β-sheets on opposite sides of the molecule. The C-terminal strand and helix of one chain of the dimer together with three helices and one strand at the N-terminal end of the other chain make up the bundle of four helices and a β-sheet. This dimeric structure of IL-5 is unique. A 3-D image and PDB file are available from SwissProt P05113.

Gene structure[5–7]

Scale

Exons

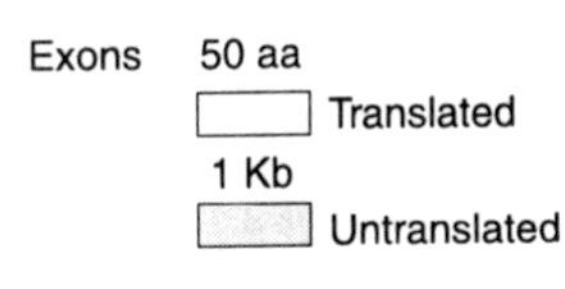

Introns

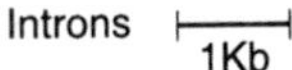

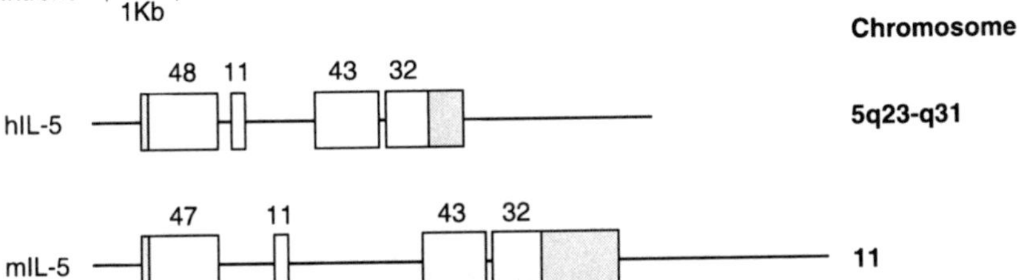

The gene for IL-5 is on human chromosome 5q23–q31 (four exons of 48, 11, 43 and 32 amino acids) and mouse IL-5 chromosome 11 (4 exons of 47, 11, 43 and 32 amino acids).

Amino acid sequence for human IL-5[5,8,9]

Accession code: SwissProt P05113

-19 MRMLLHLSLL ALGAAYVYA
1 IPTEIPTSAL VKETLALLST HRTLLIA**N**ET LRIPVPVHKN HQLCTEEIFQ
51 GIGTLESQTV QGGTVERLFK **N**LSLIKKYID GQKKKCGEER RRVNQFLDYL
101 QEFLGVMNTE WIIES

IL-5 is produced as an antiparallel homodimer formed by two interchain disulfide bonds between Cys44 and 86 of each chain.

Amino acid sequence for mouse IL-5[10]

Accession code: SwissProt P04401

-20 MRRMLLHLSV LTLSCVWATA
1 MEIPMSTVVK ETLTQLSAHR ALLTS**N**ETMR LPVPTHKNHQ LCIGEIFQGL
51 DILK**N**QTVRG GTVEMLFQ**N**L SLIKKYIDRQ KEKCGEERRR TRQFLDYLQE
101 FLGVMSTEWA MEG

IL-5 is produced as an antiparallel homodimer formed by two interchain disulfide bonds between Cys42 and 84 of each chain.

THE IL-5 RECEPTOR

The IL-5 receptor consists of a low-affinity (K_d 10^{-9} M) binding α-chain (CD125) and a nonbinding β-chain shared with the IL-3R and the GM-CSFR[11–14]. Both chains belong to the cytokine receptor superfamily. The extracellular region of the α-chain consists of an N-terminal region of about 100 amino acids with homology to a similar sequence in the α-chains of the IL-3 and GM-CSF receptors, followed by a CKR domain and an FNIII domain that contains the WSXWS motif. The human β-chain (KH97) and mouse β-chain (AIC2B) also belong to the CKR-SF. Both are described in the IL-3 entry. Soluble forms of the human and mouse IL-5R are produced by alternative mRNA splicing[13,15–17].

Distribution

In humans the IL-5 receptor is expressed on eosinophils and basophils. In mice it is expressed on eosinophils and B cells.

Physicochemical properties of the IL-5 receptor

Property	Human	Mouse
Amino acids – precursor	420	415
– mature[a]	400	398
M_r (K) – predicted	45.6	45.3
– expressed	60	60
Potential *N*-linked glycosylation sites	6	4
Affinity K_d (M)[b]	53×10^{-10}	53×10^{-9}

[a] After removal of predicted signal peptide.
[b] High-affinity receptor formed by association of α-chain with human (KH97) and mouse (AIC2B) β-chains has K_d of 53×10^{-11} or 1.53×10^{-10} M respectively. The amino acid sequences and properties of the β-chains are given in the entry for IL-3.

Signal transduction

Activation of Jak2, STAT3 and MAP kinase.

Chromosomal location

The IL-5R is on human chromosome 3p26[18,19] and mouse chromosome 6[16].

Amino acid sequence for human IL-5 receptor α-chain[3–16]

Accession code: SwissProt Q01344

```
-20  MIIVAHVLLI LLGATEILQA
  1  DLLPDEKISL LPPVNFTIKV TGLAQVLLQW KPNPDQEQRN VNLEYQVKIN
 51  APKEDDYETR ITESKCVTIL HKGFSASVRT ILQNDHSLLA SSWASAELHA
101  PPGSPGTSVV NLTCTTNTTE DNYSRLRSYQ VSLHCTWLVG TDAPEDTQYF
151  LYYRYGSWTE ECQEYSKDTL GRNIACWFPR TFILSKGRDW LAVLVNGSSK
201  HSAIRPFDQL FALHAIDQIN PPLNVTAEIE GTRLSIQWEK PVSAFPIHCF
251  DYEVKIHNTR NGYLQIEKLM TNAFISIIDD LSKYDVQVRA AVSSMCREAG
301  LWSEWSQPIY VGNDEHKPLR EWFVIVIMAT ICFILLILSL ICKICHLWIK
351  LFPPIPAPKS NIKDLFVTTN YEKAGSSETE IEVICYIEKP GVETLEDSVF
```

Two truncated soluble (extracellular) variants of the human IL-5 receptor have also been cloned[3–16]. Both differ from the membrane-anchored isoform by a sequence switch at nucleotide position 1243. In humans, the major soluble isoform is encoded by a specific exon.

Amino acid sequence for human IL-5 receptor β-chain

Accession code: SwissProt P32927

See Amino acid sequence for the human IL-3 receptor β-chain (KH97) under IL-3 entry (page **51**).

Amino acid sequence for mouse IL-5 receptor α-chain[5]

Accession code: SwissProt P21183

```
-17  MVPVLLILVG ALATLQA
  1  DLLNHKKFLL LPPVNFTIKA TGLAQVLLHW DPNPDQEQRH VDLEYHVKIN
 51  APQEDEYDTR KTESKCVTPL HEGFAASVRT ILKSSHTTLA SSWVSAELKA
101  PPGSPGTSVT NLTCTTHTVV SSHTHLRPYQ VSLRCTWLVG KDAPEDTQYF
151  LYYRFGVLTE KCQEYSRDAL NRNTACWFPR TFINSKGFEQ LAVHINGSSK
201  RAAIKPFDQL FSPLAIDQVN PPRNVTVEIE SNSLYIQWEK PLSAFPDHCF
251  NYELKIYNTK NGHIQKEKLI ANKFISKIDD VSTYSIQVRA AVSSPCRMPG
301  RWGEWSQPIY VGKERKSLVE WHLIVLPTAA CFVLLIFSLI CRVCHLWTRL
351  FPPVPAPKSN IKDLPVVTEY EKPSNETKIE VVHCVEEVGF EVMGNSTF
```

Alternative soluble forms of the receptor with no transmembrane region due to deletions in nucleotides 986–1164 and 986–1079 have been identified. The amino acid sequences of these two forms from residue 312 are:

```
312 ETFE
312 GVIYGPGCFH RFRPQRVTSK ISLWLLNMRN LRMKPKLKLY IVWKRLDLKS
    WEIPRFDGIL PF
```

Amino acid sequence for mouse IL-5 receptor β-chain

Accession code: SwissProt P26955

See Amino acid sequence for the mouse IL-3 receptor (AIC2A) β-chain AIC2B under IL-3 entry (page **51**) .

References

[1] Sanderson, C.J. et al. (1988) Immunol. Rev. 102, 29–50.
[2] McKenzie, A.N. and Sanderson, C.J. (1992) Chem. Immunol. 51, 135–152.
[3] Takatsu, K. (1992) Curr. Opin. Immunol. 4, 299–306.
[4] Milburn, M.V. et al. (1993) Nature 363, 172–176.
[5] Campbell, H.D. et al. (1987) Proc. Natl Acad. Sci. USA 84, 6629–6633.
[6] Tanabe, T. et al. (1987) J. Biol. Chem. 262, 16580–16584.
[7] Campbell, H.D. et al. (1988) Eur. J. Biochem. 174, 345–352.
[8] Azuma, C. et al. (1986) Nucleic Acids Res. 14, 9149–9158.
[9] Yokota, T. et al. (1987) Proc. Natl Acad. Sci. USA 84, 7388–7392.

[10] Kinashi, T. et al. (1986) Nature 324, 70–73.
[11] Miyajima, A. (1992) Int. J. Cell Cloning 10, 126–134.
[12] Kitamura, T. et al. (1991) Cell 66, 1165–1174.
[13] Tavernier, J. et al. (1991) Cell 66, 1175–1184.
[14] Mita, S. et al. (1991) Int. Immunol. 3, 665–672.
[15] Takaki, S. et al. (1990) EMBO J. 9, 4367–4374.
[16] Gough, N.M. and Raker, S. (1992) Genomics 12, 855–856.
[17] Murata, Y. et al. (1992) J. Exp. Med. 175, 341–351.
[18] Tuypens, T. et al. (1992) Eur. Cytokine Netw. 3, 451–459.
[19] Isobe, M. et al. (1992) Genomics 14, 755–758.

Other names

Interferon-β2 (IFNβ2), 26-kDa protein, B cell-stimulatory factor 2 (BSF-2), hybridoma/plasmacytoma growth factor (HPGF or IL-HP1), hepatocyte-stimulating factor (HSF), monocyte granulocyte inducer type 2 (MGI-2), cytotoxic T cell-differentiation factor and thrombopoietin.

THE MOLECULE

Interleukin 6 (IL-6) is a multifunctional cytokine secreted by both lymphoid and nonlymphoid cells which regulates B and T cell function, haematopoiesis and acute phase reactions[1–3].

Crossreactivity

There is 42% homology between mouse and human IL-6. Human IL-6 is functional on mouse cells but mouse IL-6 has no activity on human cells.

Sources

IL-6 is made by lymphoid cells (T cells, B cells), and many nonlymphoid cells, including macrophages, bone marrow stromal cells, fibroblasts, keratinocytes, mesangium cells, astrocytes and endothelial cells.

Bioassays

Proliferation by IL-6-dependent B9 cell line. Increased Ig secretion by CESS or other EBV-transformed human lymphoblastoid B cell lines.

Physicochemical properties of IL-6

Property	Human	Mouse
pI (calculated)	6.2	6.5
Amino acids – precursor	212	211
– mature[a]	183[b]	187
M_r (K) – predicted	20.8	21.7
– expressed	26	22-29
Potential *N*-linked glycosylation sites	2	0[c]
Disulfide bonds[d]	2	2[d]

[a] After removal of predicted signal peptide.
[b] N-terminal amino acids of human IL-6 derived from a T-cell line, an osteosarcoma cell line and a liposarcoma cell line are Pro, Ala and Val respectively, indicating heterogeneity in the signal peptide cleavage site.

3-D structure

IL-6 has a four antiparallel α-helical structure similar to IL-11, LIF, OSM and GM-CSF. A 3-D image and PDB file are available from SwissProt P05231.

Gene structure[4,5]

Scale

Exons 50 aa

Translated

Untranslated

Introns 1Kb

Chromosome

hIL-6: 6 1/3, 63 2/3, 38, 49, 55 — 7p21p14

mIL-6: 6 1/3, 61 2/3, 38, 50, 55 — 5

The gene for IL-6 is on human chromosome 7p21p14 (five exons of $6^{1/3}$, $63^{2/3}$, 38, 49 and 55 amino acids) or mouse chromosome 5 (five exons of $6^{1/3}$, $61^{2/3}$, 38, 50 and 55 amino acids).

Amino acid sequence for human IL-6[6,7]

Accession code: SwissProt P05231

```
-29  MNSFSTSAFG PVAFSLGLLL VLPAAFPAP
  1  VPPGEDSKDV AAPHRQPLTS SERIDKQIRY ILDGISALRK ETCNKSNMCE
 51  SSKEALAENN LNLPKMAEKD GCFQSGFNEE TCLVKIITGL LEFEVYLEYL
101  QNRFESSEEQ ARAVQMSTKV LIQFLQKKAK NLDAITTPDP TTNASLLTKL
151  QAQNQWLQDM TTHLILRSFK EFLQSSLRAL RQM
```

The N-terminal amino acid of IL-6 from different sources has been reported as Ala, Pro and Val, indicating some heterogeneity in the signal peptide cleavage site.

Amino acid sequence for mouse IL-6[8]

Accession code: SwissProt P08505

```
-24  MKFLSARDFH PVAFLGLMLV TTTA
  1  FPTSQVRRGD FTEDTTPNRP VYTTSQVGGL ITHVLWEIVE MRKELCNGNS
 51  DCMNNDDALA ENNLKLPEIQ RNDGCYQTGY NQEICLLKIS SGLLEYHSYL
101  EYMKNNLKDN KKDKARVLQR DTETLIHIFN QEVKDLHKIV LPTPISNALL
151  TDKLESQKEW LRTKTIQFIL KSLEEFLKVT LRSTRQT
```

THE IL-6 RECEPTOR

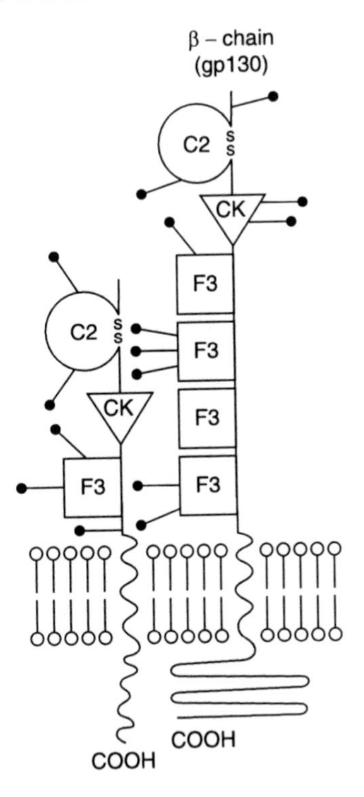

High-affinity (K_d 10^{-11} M) IL-6 receptors are formed by the noncovalent association of two subunits[3,9]. The IL-6R α-chain (CD126) binds IL-6 with low affinity (K_d 10^{-9} M) but does not signal. The extracellular region consists of an Ig-SF C2 set domain at the N-terminus followed by a CKR-SF domain and an FNIII domain which includes the WSXWS motif. The β-chain (gp130) (CD130) does not itself bind IL-6, but associates with the α-chain/IL-6 complex and is responsible for signal transduction[9]. The extracellular domain of gp130 consists of an Ig-SF C2 set domain at the N-terminus followed by a CKR-SF domain and four FNIII domains, only the first of which contains the WSXWS motif. gp130 is also an oncostatin M receptor and an affinity converter (β-chain) for the LIF and CNTF receptors[3,10]. A mouse variant of the IL-6 receptor has been cloned in which the intracellular domain has been replaced by part of a long terminal repeat (LTR) sequence from the intracisternal A particle (IAP) gene[11]. Cells transfected with this variant bind IL-6 and respond in proliferation assays.

Distribution

The IL-6 receptor is expressed on activated but not resting B cells, plasma cells, T cells, monocytes and many other nonlymphoid cells including epithelial cells, fibroblasts, hepatocytes and neural cells.

Physicochemical properties of the IL-6 receptor

Property	IL-6R α-chain		gp130	
	Human	Mouse	Human	Mouse
Amino acids – precursor	468	460	918	917
– mature[a]	449	441	896	895
M_r (K) – predicted	49.9	48.6	101	100
– expressed	80	80	130	130
Potential *N*-linked glycosylation sites	5	3	10	9
Affinity K_d (M)[b]	10^{-9}	2.53×10^{-9}	none	none

[a] After removal of predicted signal peptide.
[b] The high-affinity (K_d 10^{-11} M) receptor is formed by association of the IL-6R α-chain with gp130.

Signal transduction

The signal transducing molecule in the IL-6 receptor is the gp130 chain[3]. IL-6 binds to the IL-6R (α-chain) and the IL-6/IL-6R complex then binds two gp130 molecules to form a gp130 disulfide-linked homodimer[12]. Dimerization results in tyrosine phosphorylation of gp130. IL-6 has also been shown to induce phosphorylation of the CD40 B cell surface antigen[13]. A motif of about 60 amino acids proximal to the transmembrane domain in the cytoplasmic domain of gp130 is required for signal transduction. Two short segments within this motif are highly conserved among many cytokine receptor signal transducers and may be important for association with a tyrosine kinase.

Chromosomal location

The IL-6R α-chain is on human chromosome 1. In humans, gp130 has two distinct chromosomal loci on chromosomes 5 and 17. The presence of two distinct gp130 gene sequences is restricted to primates and is not found in other vertebrates[14].

Amino acid sequence for human IL-6 receptor α-chain[15]

Accession code: SwissProt P08887

```
-19  MLAVGCALLA ALLAAPGAA
  1  LAPRRCPAQE VARGVLTSLP GDSVTLTCPG VEPEDNATVH WVLRKPAAGS
 51  HPSRWAGMGR RLLLRSVQLH DSGNYSCYRA GRPAGTVHLL VDVPPEEPQL
101  SCFRKSPLSN VVCEWGPRST PSLTTKAVLL VRKFQNSPAE DFQEPCQYSQ
151  ESQKFSCQLA VPEGDSSFYI VSMCVASSVG SKFSKTQTFQ GCGILQPDPP
201  ANITVTAVAR NPRWLSVTWQ DPHSWNSSFY RLRFELRYRA ERSKTFTTWM
251  VKDLQHHCVI HDAWSGLRHV VQLRAQEEFG QGEWSEWSPE AMGTPWTESR
301  SPPAENEVST PMQALTTNKD DDNILFRDSA NATSLPVQDS SSVPLPTFLV
351  AGGSLAFGTL LCIAIVLRFK KTWKLRALKE GKTSMHPPYS LGQLVPERPR
401  PTPVLVPLIS PPVSPSSLGS DNTSSHNRPD ARDPRSPYDI SNTDYFFPR
```

Amino acids 24–81 form an Ig-like domain with potential disulfide bonds between Cys28 and Cys77.

Amino acid sequence for human IL-6 receptor β-chain (gp130)[16]

Accession code: SwissProt P40189

```
-22  MLTLQTWVVQ ALFIFLTTES TG
  1  ELLDPCGYIS PESPVVQLHS NFTAVCVLKE KCMDYFHVNA NYIVWKTNHF
 51  TIPKEQYTII NRTASSVTFT DIASLNIQLT CNILTFGQLE QNVYGITIIS
101  GLPPEKPKNL SCIVNEGKKM RCEWDGGRET HLETNFTLKS EWATHKFADC
151  KAKRDTPTSC TVDYSTVYFV NIEVWVEAEN ALGKVTSDHI NFDPVYKVKP
201  NPPHNLSVIN SEELSSILKL TWTNPSIKSV IILKYNIQYR TKDASTWSQI
251  PPEDTASTRS SFTVQDLKPF TEYVFRIRCM KEDGKGYWSD WSEEASGITY
301  EDRPSKAPSF WYKIDPSHTQ GYRTVQLVWK TLPPFEANGK ILDYEVTLTR
351  WKSHLQNYTV NATKLTVNLT NDRYLATLTV RNLVGKSDAA VLTIPACDFQ
401  ATHPVMDLKA FPKDNMLWVE WTTPRESVKK YILEWCVLSD KAPCITDWQQ
451  EDGTVHRTYL RGNLAESKCY LITVTPVYAD GPGSPESIKA YLKQAPPSKG
501  PTVRTKKVGK NEAVLEWDQL PVDVQNGFIR NYTIFYRTII GNETAVNVDS
551  SHTEYTLSSL TSDTLYMVRM AAYTDEGGKD GPEFTFTTPK FAQGEIEAIV
601  VPVCLAFLLT TLLGVLFCFN KRDLIKKHIW PNVPDPSKSH IAQWSPHTPP
651  RHNFNSKDQM YSDGNFTDVS VVEIEANDKK PFPEDLKSLD LFKKEKINTE
701  GHSSGIGGSS CMSSSRPSIS SSDENESSQN TSSTVQYSTV VHSGYRHQVP
751  SVQVFSRSES TQPLLDSEER PEDLQLVDHV DGGDGILPRQ QYFKQNCSQH
801  ESSPDISHFE RSKQVSSVNE EDFVRLKQQI SDHISQSCGS GQMKMFQEVS
851  AADAFGPGTE GQVERFETVG MEAATDEGMP KSYLPQTVRQ GGYMPQ
```

Amino acid sequence for mouse IL-6 receptor α-chain[11]

Accession code: SwissProt P22272

```
-19  MLTVGCTLLV ALLAAPAVA
  1  LVLGSCRALE VANGTVTSLP GATVTLICPG KEAAGNVTIH WVYSGSQNRE
 51  WTTTGNTLVL RDVQLSDTGD YLCSLNDHLV GTVPLLVDVP PEEPKLSCFR
101  KNPLVNAICE WRPSSTPSPT TKAVLFAKKI NTTNGKSDFQ VPCQYSQQLK
151  SFSCQVEILE GDKVYHIVSL CVANSVGSKS SHNEAFHSLK MVQPDPPANL
201  VVSAIPGRPR WLKVSWQHPE TWDPSYYLLQ FQLRYRPVWS KEFTVLLLPV
251  AQYQCVIHDA LRGVKHVVQV RGKEELDLGQ WSEWSPEVTG TPWIAEPRTT
301  PAGILWNPTQ VSVEDSANHE DQYESSTEAT SVLAPVQESS SMSLPTFLVA
351  GGSLAFGLLL CVFIILRLKQ KWKSEAEKES KTTSPPPPPY SLGPLKPTFL
401  LVPLLTPHSS GSDNTVNHSC LGVRDAQSPY DNSNRDYLFP R
```

Amino acids 24–77 form an Ig-like domain with potential disulfide bonds between Cys28 and Cys73. Conflicting sequence A→R at position 355.

Amino acid sequence for mouse IL-6 receptor β-chain (gp130)[17]

Accession code: SwissProt Q00560

```
-22  MSAPRIWLAQ ALLFFLTTES IG
  1  QLLEPCGYIY PEFPVVQRGS NFTAICVLKE ACLQHYYVNA SYIVWKTNHA
 51  AVPREQVTVI NRTTSSVTFT DVVLPSVQLT CNILSFGQIE QNVYGVTMLS
101  GFPPDKPTNL TCIVNEGKNM LCQWDPGRET YLETNYTLKS EWATEKFPDC
151  QSKHGTSCMV SYMPTYYVNI EVWVEAENAL GKVSSESINF DPVDKVKPTP
201  PYNLSVTNSE ELSSILKLSW VSSGLGGLLD LKSDIQYRTK DASTWIQVPL
251  EDTMSPRTSF TVQDLKPFTE YVFRIRSIKD SGKGYWSDWS EEASGTTYED
301  RPSRPPSFWY KTNPSHGQEY RSVRLIWKAL PLSEANGKIL DYEVILTQSK
351  SVSQTYTVTG TELTVNLTND RYVASLAARN KVGKSAAAVL TIPSPHVTAA
401  YSVVNLKAFP KDNLLWVEWT PPPKPVSKYI LEWCVLSENA PCVEDWQQED
451  ATVNRTHLRG RLLESKCYQI TVTPVFATGP GGSESLKAYL KQAAPARGPT
501  VRTKKVGKNE AVLAWDQIPV DDQNGFIRNY SISYRTSVGK EMVVHVDSSH
551  TEYTLSSLSS DTLYMVRMAA YTDEGGKDGP EFTFTTPKFA QGEIEAIVVP
601  VCLAFLLTTL LGVLFCFNKR DLIKKHIWPN VPDPSKSHIA QWSPHTPPRH
651  NFNSKDQMYS DGNFTDVSVV EIEANNKKPC PDDLKSVDLF KKEKVSTEGH
701  SSGIGGSSCM SSSRPSISSN EENESAQSTA STVEYSTVVH SGYRHQVPSV
751  QVFSRSESTQ PLLDSEERPE DLQLVDSVDG GDEILPRQPY FKQNCSQPEA
801  CPEISHFERS NQVLSGNEED FVRLKQQQVS DHISQPYGSE QRRLFQEGST
851  ADALGTGADG QMERFESVGM ETTIDEEIPK SYLPQTVRQG GYMPQ
```

References

[1] Hirano, T. (1991) In The Cytokine Handbook, Thomson, A.W. ed., Academic Press, London, pp. 169–190.
[2] Kishimoto, T. (1989) Blood 74, 1–10.
[3] Kishimoto, T. et al. (1992) Science 258, 593–597.
[4] Yasukawa, K. et al. (1987) EMBO J. 6, 2939–2945.
[5] Tanabe, O. et al. (1988) J. Immunol. 141, 3875–3881.
[6] Hirano, T. et al. (1986) Nature 324, 73–76.
[7] May, L.T. et al. (1986) Proc. Natl Acad. Sci. USA 83, 8957–8961.
[8] van Snick, J. et al. (1988) Eur. J. Immunol. 18, 193–197.
[9] Taga, T. et al. (1989) Cell 58, 573–581.
[10] Gearing, D.P. et al. (1992) Science 255, 1434–1437.
[11] Sugita, T. et al. (1990) J. Exp. Med. 171, 2001–2009.
[12] Murakami, M. et al. (1993) Science 260, 1808–1810.
[13] Clark, E.A. and Shu, G. (1990) J. Immunol. 145, 1400–1406.
[14] Kidd, V.J. et al. (1992) Somat. Cell Mol. Genet. 18, 477–483.
[15] Yamasaki, K. et al. (1988) Science 241, 825–828.
[16] Hibi, M. et al. (1990) Cell 63, 1149–1157.
[17] Saito, M. et al. (1992) J. Immunol. 148, 4066–4071.

Other names

Lymphopoietin 1 (LP-1), pre-B cell growth factor.

THE MOLECULE

Interleukin 7 (IL-7) is a stromal cell-derived growth factor for progenitor B cells and T cells. The main population in the thymus responsive to IL-7 is CD4 – CD8 – . IL-7 also stimulates proliferation and differentiation of mature T cells[1,2].

Crossreactivity

There is 60% homology between human and mouse IL-7 and significant cross-species reactivity.

Sources

Bone marrow, thymic stromal cells and spleen cells.

Bioassays

Proliferation of B-cell precursors, 2bx murine pre-B cell line or mitogen-stimulated T cells.

Physicochemical properties of IL-7

Property	Human	Mouse
pI	9	8.7
Amino acids – precursor	177	154
– mature[a]	152	129
M_r (K) – predicted	17.4	14.9
– expressed	20–28	25
Potential *N*-linked glycosylation sites	3	2
Disulfide bonds[b]	3	3

[a] After removal of predicted signal peptide.
[b] All six Cys residues probably form disulfide bonds.

3-D structure

IL-7 has a four antiparallel α-helical structure. A 3-D image and PDB file are available from SwissProt P13232.

Gene structure[3]

Scale

Exons 50 aa
Translated
1 Kb
Untranslated

Introns
1Kb

~15kb
Chromosome

hIL-7: $3^{1/3}$, $45^{2/3}$, 27, 44, 18, 39 — 8p12-q13

mIL-7: $3^{1/3}$, $44^{2/3}$, 27, 44, 35 — ?

The human IL-7 gene is on chromosome 8q12–q13 with six coding exons of $3^{1/3}$, $45^{2/3}$, 27, 44, 18 and 39 amino acids. It spans at least 33 kb. Intron 2 has not been fully sequenced and the 5′ initiation site has not been identified. The 18 amino acids coded for by exon 5 are not found in mouse IL-7. The mouse IL-7 gene has five exons (one less than the human gene) and spans 56 kb or more. The 3′ end has not been cloned, and the position and size of exon 5 have not been established. Sizes of introns 1, 2 and 4 are not known. Chromosomal location is not known.

Amino acid sequence for human IL-7[4]

Accession code: SwissProt P13232

-25 MFHVSFRYIF GLPPLILVLL PVASS
1 DCDIEGKDGK QYESVLMVSI DQLLDSMKEI GSNCLNNEFN FFKRHICDAN
51 KEGMFLFRAA RKLRQFLKM**N** STGDFDLHLL KVSEGTTILL **N**CTGQVKGRK
101 PAALGEAQPT KSLEE**N**KSLK EQKKLNDLCF LKRLLQEIKT CWNKILMGTK
151 EH

The 18 amino acids from positions 96–113 are absent in murine IL-7 and are coded for by an extra exon in the human IL-7 gene.

Amino acid sequence for mouse IL-7[5]

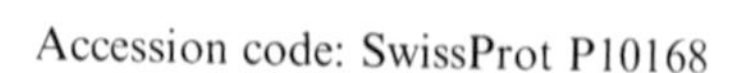

Accession code: SwissProt P10168

-25 MFHVSFRYIF GIPPLILVLL PVTSS
1 ECHIKDKEGK AYESVLMISI DELDKMTGTD SNCPNNEPNF FRKHVCDDTK
51 EAAFLNRAAR KLKQFLKM**N**I SEEFNVHLLT VSQGTQTLV**N** CTSKEEKNVK
101 EQKKNDACFL KRLLREIKTC WNKILKGSI

THE IL-7 RECEPTOR

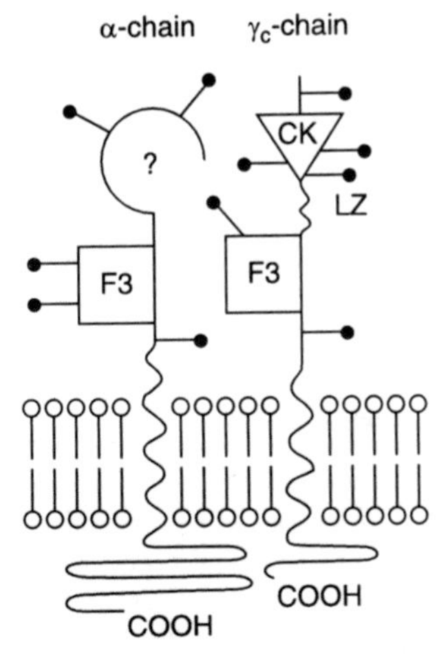

The IL-7 receptor is a complex consisting at least of an IL-7-binding chain (CD127) and the IL-2R γ-chain (γc-chain). The extracellular domain of the IL-7R-binding chain consists of an N-terminal region of about 100 amino acids with no clear sequence homology to other proteins, followed by an FNIII domain containing the WSXWS motif. Overall homology with the CKR-SF is poor, however, and it is unclear whether it should be included as a member of the CKR superfamily[6]. A soluble form of the IL-7R is produced by alternative splicing of the IL-7R gene[7]. The IL-2R γ-chain is a functional component of the IL-7R and augments IL-7 binding[8]. This may explain the two classes of IL-7-binding sites which have been described on human and murine cells with high ($K_d \sim 10^{-10}$ M) and low ($K_d \sim 10^{-8}$ M) affinity[9]. Similar high- and low-affinity binding sites have also been described on cells transfected with IL-7R cDNA[7]. A low-affinity IL-7 receptor has also been identified but not yet cloned[10].

Distribution

Thymocytes, T- and B-cell progenitors, mature T cells, monocytes, and some lymphoid and myeloid cell lines.

Physicochemical properties of the IL-7 receptor

Property	Human	Mouse
Amino acids – precursor	459	459
– mature[a]	439	439
M_r (K) – predicted	49.5	49.6
– expressed	68	68
Potential *N*-linked glycosylation sites	5	3
Affinity K_d (M) – high	23×10^{-10}	23×10^{-10}
– low	10^{-8}	10^{-8}

[a] After removal of predicted signal peptide.

Signal transduction

IL-7 stimulates tyrosine phosphorylation of cellular proteins and tyrosine kinase-dependent activation of PI-specific phospholipase C with hydrolysis of PIP_2 and release of IP_3[11]. IL-7R activates PI-3' kinase[12] and has been shown to associate with and activate p59^{fyn}[13]. Also activates Jak1 and Jak3 in mouse T cells, and STAT1 and STAT5 in B cells.

Chromosomal location

The IL-7R gene is on human chromosome 5p13[14], and mouse chromosome 15.

Amino acid sequence for human IL-7 receptor[7]

Accession code: SwissProt P16871

-20 MTILGTTFGM VFSLLQVVSG
1 ESGYAQNGDL EDAELDDYSF SCYSQLEV**N**G SQHSLTCAFE DPDV**N**TTNLE
51 FEICGALVEV KCLNFRKLQE IYFIETKKFL LIGKSNICVK VGEKSLTCKK
101 IDLTTIVKPE APFDLSVIYR EGANDFVVTF **N**TSHLQKKYV KVLMHDVAYR
151 QEKDENKWTH V**N**LSSTKLTL LQRKLQPAAM YEIKVRSIPD HYFKGFWSEW
201 SPSYYFRTPE I**N**NSSGEMDP ILLTISILSF FSVALLVILA CVLWKKRIKP
251 IVWPSLPDHK KTLEHLCKKP RKNLNVSFNP ESFLDCQIHR VDDIQARDEV
301 EGFLQDTFPQ QLEESEKQRL GGDVQSPNCP SEDVVVTPES FGRDSSLTCL
351 AGNVSACDAP ILSSSRSLDC RESGKNGPHV YQDLLLSLGT TNSTLPPPFS
401 LQSGILTLNP VAQGQPILTS LGSNQEEAYV TMSSFYQNQ

Fibronectin type III domain 108–204. Thr residue at position 262 is a PKC phosphorylation site. A soluble form of the human IL-7 receptor resulting from a 93-bp deletion terminates with the sequence

GLSLSYGPVS PIIRRLWNIF VRNQEKI

Amino acid sequence for mouse IL-7 receptor[7]

Accession code: SwissProt P16872

-20 MMALGRAFAI VFCLIQAVSG
1 ESGNAQDGDL EDADADDHSF WCHSQLEVDG SQHLLTCAF**N** DSDINTANLE
51 FQICGALLRV KCLTLNKLQD IYFIKTSEFL LIGSSNICVK LGQK**N**LTCKN
101 MAINTIVKAE APSDLKVVYR KEANDFLVTF NAPHLKKKYL KKVKHDVAYR
151 PARGES**N**WTH VSLFHTRTTI PQRKLRPKAM YEIKVRSIPH NDYFKGFWSE
201 WSPSSTFETP EPKNQGGWDP VLPSVTILSL FSVFLLVILA HVLWKKRIKP
251 VVWPSLPDHK KTLEQLCKKP KTSLNVSFIP EIFLDCQIHE VKGVEARDEV
301 EIFLPNDLPA QPEELETQGH RAAVHSANRS PETSVSPPET VRRESPLRCL
351 ARNLSTCNAP PLLSSRSPDY RDGDRNRPPV YQDLLPNSGN TNVPVPVPQP
401 LPFQSGILIP FSQRQPISTS SVLNQEEAYV TMSSFYQNK

Fibronectin type III domain 108–205. Thr residue at position 262 is a PKC phosphorylation site.

References

[1] Goodwin, R.G. and Namen, A.E. (1991) In The Cytokine Handbook, Thomson, A.W. ed., Academic Press, London, pp. 191–200.
[2] Henney, C.S. (1989) Immunol. Today 10, 170–173.
[3] Lupton, S.D. et al. (1990) J. Immunol. 144, 3592–3601.
[4] Goodwin, R.G. et al. (1989) Proc. Natl Acad. Sci. USA 86, 302–306.
[5] Namen, A.E. et al. (1988) Nature 333, 571–573.
[6] Barclay, A.N. et al. (1993) The Leucocyte Antigen FactsBook, Academic Press, London.
[7] Goodwin, R.G. et al. (1990) Cell 60, 941–951.
[8] Noguchi, M. et al. (1993) Science 262, 1877–1880.
[9] Park, L.S. et al. (1990) J. Exp. Med. 171, 1073–1089.
[10] Armitage, R.J. et al. (1992) Blood 79, 1738–1745.
[11] Uckun, F.M. et al. (1991) Proc. Natl Acad. Sci. USA 88, 3589–3593.
[12] Dadi, H.K. et al. (1993) Biochem. Biophys. Res. Commun. 192, 459–464.
[13] Venkitaraman, A.R. and Cowling, R.J. (1992) Proc. Natl Acad. Sci. USA 89, 12083–12087.
[14] Lynch, M. et al. (1992) Hum. Genet. 89, 566–568.

Other names

Neutrophil-attractant/activating protein (NAP-1), monocyte-derived neutrophil-activating peptide (MONAP), monocyte-derived neutrophil chemotactic factor (MDNCF), neutrophil-activating factor (NAF), leukocyte adhesion inhibitor (LAI), granulocyte chemotactic protein (GCP).

THE MOLECULE

Interleukin 8 (IL-8) is an inflammatory chemokine produced by many cell types, which mainly functions as a neutrophil chemoattractant and activating factor[1-3]. It also attracts basophils and a subpopulation of lymphocytes. IL-8 is a potent angiogenic factor[4].

Crossreactivity

There is no obvious murine homologue of human IL-8, although murine MIP-2 can compete with IL-8 for binding to its receptors on neutrophils[3]. IL-8 exhibits limited reactivity on neutrophils from other species[5].

Sources

IL-8 is secreted by multiple cell types, including monocytes, lymphocytes, granulocytes, fibroblasts, endothelial cells, bronchial epithelial cells, keratinocytes, hepatocytes, mesangial cells and chondrocytes.

Bioassays

IL-8 can be measured in neutrophil chemotaxis and activation assays[6,7].

Physicochemical properties of IL-8

Property	Human
pI	8.6
Amino acids – precursor	99
– mature[a,b]	77–72
M_r (K) – predicted	11.1
– expressed	6–8
Potential *N*-linked glycosylation sites	0
Disulfide bonds[c]	2

[a] IL-8 appears to exist as a dimer in solution.
[b] Several distinct N-termini are found in the natural protein[8].
[c] Disulfide bonds link residues 7–33 and 9–49.

3-D structure

X-ray crystallography and nmr studies have given compatible structures for IL-8[9,10]. The monomer unit consists of a triple-stranded antiparallel β-sheet in a Greek key, with a long C-terminal helix. The monomers are joined by hydrogen bonds from residues 25–29, 27–27 and 29–25 respectively.

Gene structure[11]

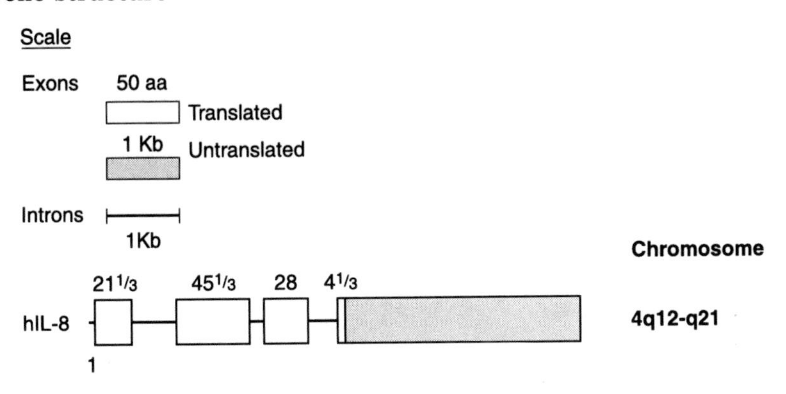

The gene for IL-8 is located on human chromosome 4q12–q21 and consists of four exons of $21^{1/3}$, $45^{1/3}$, 28 and $4^{1/3}$ amino acids.

Amino acid sequence for human IL-8[2]

Accession code: SwissProt P10145

```
-27  MTSKLAVALL AAFLISAALC EGAVLPR
  1  SAKELRCQCI KTYSKPFHPK FIKELRVIES GPHCANTEII VKLSDGRELC
 51  LDPKENWVQR VVEKFLKRAE NS
```

Conflicting sequence E→L at position 28.

THE IL-8 RECEPTORS (CXCR1 and CXCR2)

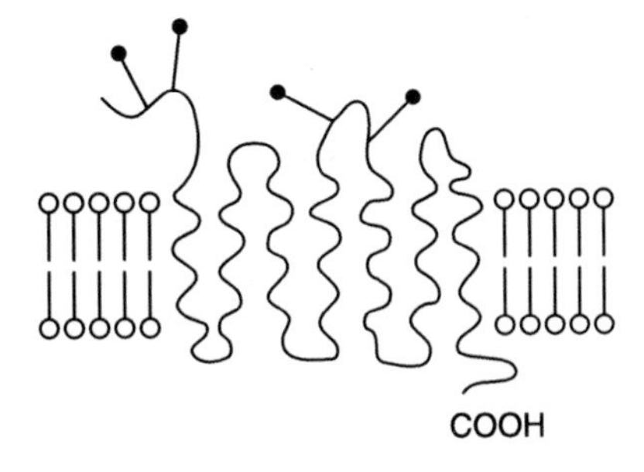

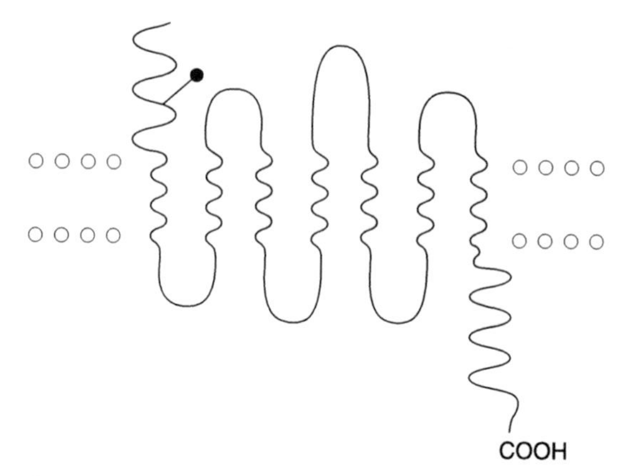

Two sequences for human IL-8 receptors have been published, CXCR1 (CDw128) and CXCR2[12,13]. Both are seven transmembrane spanning, G protein-linked receptors of the rhodopsin superfamily. CXCR1 has a predicted M_r of 40 000, and glycosylation at two potential *N*-linked sites would bring the weight up to that measured on neutrophils of 58 000–67 000. CXCR1 is 77% identical to CXCR2. CXCR2 has an unglycosylated molecular weight of 32 000.

CXCR1 selectively binds IL-8 with an EC_{50} of 4 nM compared with 40–65 nM for ENA-78, NAP-2 and GROα, β and γ, whereas CXCR2 receptor binds IL-8 with an EC_{50} of 4 nM and GRO/MGSA family members, ENA-78, GCP-2 and NAP-2 with comparable EC_{50} values of 1–11 nM[14,16]. A murine homologue of CXCR2 has been described that binds murine KC and MIP-2 with an EC_{50} of 5 nM but has poor affinity for human IL-8 (400 nm)[17].

Distribution

Neutrophils, basophils and lymphocytes have functional receptors. Saturable binding sites on monocytes, eosinophils, endothelial cells and erythrocytes have been demonstrated.

Physicochemical properties of the IL-8 receptors

Property	CXCR1	CXCR2	CXCR2
	Human	Human	Mouse
Amino acids – precursor	350	360	359
– mature	?	?	?
M_r (K) – predicted	40	41	40
– expressed	58–67	?	?
Potential *N*-linked glycosylation sites	2	1	1
Affinity K_d (M)	4×10^{-9} (EC_{50})	4×10^{-9} (EC_{50})	4×10^{-7} (EC_{50} for human IL-8)

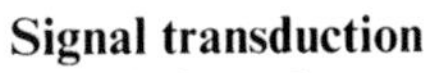

Signal transduction

Stimulation of neutrophils by IL-8 causes an immediate increase in intracellular Ca^{2+} which is pertussis toxin-sensitive. Inhibitors of PKC block neutrophil activation by IL-8.

Chromosomal location

The genes for both receptors are located on human chromosome 2q35[15,18].

Amino acid sequence for human CXCR1[12,13]

Accession code: SwissProt P25024

```
  1 MSNITDPQMW DFDDLNFTGM PPADEDYSPC MLETETLNKY VVIIAYALVF
 51 LLSLLGNSLV MLVILYSRVG RSVTDVYLLN LALADLLFAL TLPIWAASKV
101 NGWIFGTFLC KVVSLLKEVN FYSGILLLAC ISVDRYLAIV HATRTLTQKR
151 HLVKFVCLGC WGLSMNLSLP FFLFRQAYHP NNSSPVCYEV LGNDTAKWRM
201 VLRILPHTFG FIVPLFVMLF CYGFTLRTLF KAHMGQKHRA MRVIFAVVLI
251 FLLCWLPYNL VLLADTLMRT QVIQETCERR NNIGRALDAT EILGFLHSCL
301 NPIIYAFIGQ NFRHGFLKIL AMHGLVSKEF LARHRVTSYT SSSVNVSSNL
```

Disulfide bonds link Cys110–187.

Amino acid sequence for human CXCR2[12,13]

Accession code: SwissProt P25025

```
  1 MEDFNMESDS FEDFWKGEDL SNYSYSSTLP PFLLDAAPCE PESLEINKYF
 51 VVIIYALVFL LSLLGNSLVM LVILYSRVGR SVTDVYLLNL ALADLLFALT
101 LPIWAASKVN GWIFGTFLCK VVSLLKEVNF YSGILLLACI SVDRYLAIVH
151 ATRTLTQKRY LVKFICLSIW GLSLLLALPV LLFRRTVYSS NVSPACYEDM
201 GNNTANWRML LRILPQSFGF IVPLLIMLFC YGFTLRTLFK AHMGQKHRAM
251 RVIFAVVLIF LLCWLPYNLV LLADTLMRTQ VIQETCERRN HIDRALDATE
301 ILGILHSCLN PLIYAFIGQK FRHGLLKILA IHGLISKDSL PKDSRPSFVG
351 SSSGHTSTTL
```

Disulfide bonds link Cys119–196.

Amino acid sequence for mouse CXCR2[17,19]

Accession code: SwissProt P35343

```
  1 MGEFKVDKFN IEDFFSGDLD IFNYSSGMPS ILPDAVPCHS ENLEINSYAV
 51 VVIYVLVTLL SLVGNSLVML VILYNRSTCS VTDVYLLNLA IADLFFALTL
101 PVWAASKVNG WTFGSTLCKI FSYVKEVTFY SSVLLLACIS MDRYLAIVHA
151 TSTLIQKRHL VKFVCIAMWL LSVILALPIL ILRNPVKVNL STLVCYEDVG
201 NNTSRLRVVL RILPQTFGFL VPLLIMLFCY GFTLRTLFKA HMGQKHRAMR
251 VIFAVVLVFL LCWLPYNLVL FTDTLMRTKL IKETCERRDD IDKALNATEI
301 LGFLHSCLNP IIYAFIGQKF RHGLLKIMAT YGLVSKEFLA KEGRPSFVSS
351 SSANTSTTL
```

Disulfide bonds link Cys118–195.

References

[1] Matsushima, K. and Oppenheim J.J. (1989) Cytokine 1, 2–13.
[2] Matsushima, K. et al. (1988) J. Exp. Med. 167, 1883–1893.
[3] Oppenheim, J.J. et al. (1991) Annu. Rev. Immunol. 9, 617–648.
[4] Koch, A.E. et al. (1993) Science 258, 1798–1801.
[5] Rot, A. (1991) Cytokine 3, 21–27.
[6] Van Damme, J. and Conings, R. (1991) In Cytokines: A Practical Approach, Balkwill, F.A. ed., IRL Press, Oxford, pp. 187–196.
[7] Westwick, J. (1991) In Cytokines: A Practical Approach, Balkwill, F.A. ed., IRL Press, Oxford, pp. 197–204.
[8] Van Damme, J. et al. (1989) J. Exp. Med. 167, 1364–1376.
[9] Clore, G.M. and Gronenborn, A.M. (1991) J. Mol. Biol. 217, 611–620.
[10] Baldwin, E.T. et al. (1991) Proc. Natl Acad. Sci. USA 88, 502–506.
[11] Mukaida, N. et al. (1989) J. Immunol. 143, 1366–1371.
[12] Holmes, W.E. et al. (1991) Science 253, 1278–1280.
[13] Murphy, P.M. and Tiffany, H.L. (1991) Science 253, 1280–1283.
[14] Larosa, GJ et al. (1992) J. Biol.Chem. 25402–25406.
[15] Morris, S.W. et al. (1992) Genomics 14, 685–691.
[16] Ahuja, S.K and Murphy, PM (1996) J. Biol. Chem. 271, 20545–20550.
[17] Bozic, CR et al. (1994) J. Biol. Chem. 47: 29355–29358.
[18] Mollereau, C. et al. (1993) Genomics 16, 248–251.
[19] Ceretti, D.P. et al. (1993) Genomics 18, 410–413.

Other names

Human IL-9 has been known as P40. Mouse IL-9 has been known as P40, mast cell growth-enhancing activity and T cell growth factor III.

THE MOLECULE

Interleukin 9 (IL-9) is a cytokine which enhances the proliferation of T lymphocytes, mast cell lines, megakaryoblastic leukaemia cell lines and erythroid precursors[1–7].

Crossreactivity

There is about 55% homology between human and mouse IL-9[8]. Mouse IL-9 functions on human cells; human IL-9 is inactive on mouse cells[9].

Sources

IL-9 is produced by IL-2-activated Th2 lymphocytes and Hodgkin's lymphoma cells[10].

Bioassays

Mouse and human IL-9 can be assayed using the proliferation of PHA- plus IL-4-stimulated human lymphoblast lines[4].

Physicochemical properties of IL-9

Property	Human	Mouse
pI (calculated)	8.7	6.2–7.3 (9)
Amino acids – precursor	144	144
– mature	126	126
M_r (K) – predicted	14.1	14.0
– expressed	32–39	30–40
Potential *N*-linked glycosylation sites	4	4
Disulfide bonds	5	5

3-D structure

No information.

Gene structure[8–11]

Scale

Exons 50 aa

Translated

500 bp

Untranslated

Introns

500 bp

	Exons	Chromosome
hIL-9	38 12 11 44 39	5q31.1
mIL-9	38 12 11 44 39	13

The gene for IL-9 is located on human chromosome 5q31.1 (five exons of 38, 12, 11, 44 and 39 amino acids) and mouse chromosome 13 (five exons of 38, 12, 11, 44 and 39 amino acids).

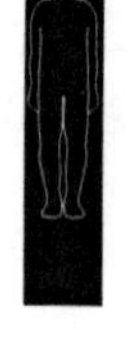

Amino acid sequence for human IL-9[1,8]

Accession code: SwissProt P15248

```
-18  MLLAMVLTSA LLLCSVAG
  1  QGCPTLAGIL DINFLINKMQ EDPASKCHCS ANVTSCLCLG IPSDNCTRPC
 51  FSERLSQMTN TTMQTRYPLI FSRVKKSVEV LKNNKCPYFS CEQPCNQTTA
101  GNALTFLKSL LEIFQKEKMR GMRGKI
```

Amino acid sequence for mouse IL-9[3]

Accession code: SwissProt P15247

```
-18  MLVTYILASV LLFSSVLG
  1  QRCSTTWGIR DTNYLIENLK DDPPSKCSCS GNVTSCLCLS VPTDDCTTPC
 51  YREGLLQLTN ATQKSRLLPV FHRVKRIVEV LKNITCPSFS CEKPCNQTMA
101  GNTLSFLKSL LGTFQKTEMQ RQKSRP
```

THE IL-9 RECEPTOR

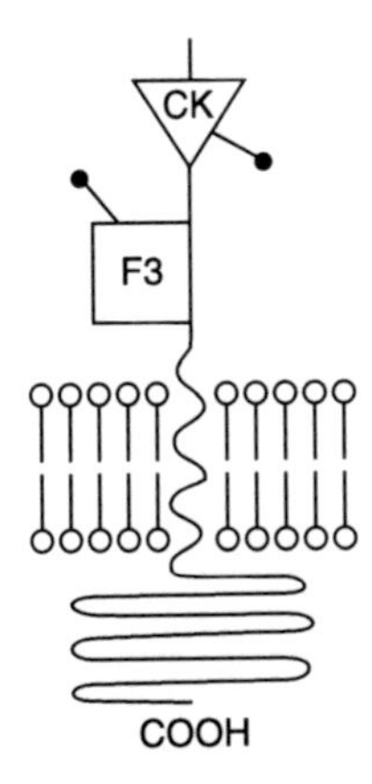

A single receptor type is detected on mouse T cell lines with a K_d of 6.73×10^{-11} M, and an M_r of 64 000, which is reduced to 54 000 after *N*-glycosidase F treatment[12]. The recombinant mouse receptor has a K_d of 1.93×10^{-10} M[13]. Both membrane-bound and soluble forms of the receptor exist[13]. The IL-2R γ-chain may associate with the IL-9R.

Distribution

IL-9 receptors are found on T helper clones, macrophages, some T-cell tumours and mast cell lines.

Physicochemical properties of the IL-9 receptor

Property	Human	Mouse
Amino acids – precursor	522	468
– mature	482	431
M_r (K) – predicted	57	52
– expressed	?	64
Potential *N*-linked glycosylation sites	2	2
Affinity K_d (M)	?	1.93×10^{-10}

Signal transduction

IL-9 causes tyrosine phosphorylation of four major proteins in MO7e cells, but not MAP kinase or Raf-1[14]. Activation of Jak1 and Jak3.

Chromosomal location

Human IL-9R is on subtelomeric pseudoautosomal region of chromosomes X and Y. It is expressed from both chromosomes and escapes X-inactivation.

Amino acid sequence for human IL-9 receptor[13]

Accession code: SwissProt Q01113

-40 MGLGRCIWEG WTLESEALRR DMGTWLLACI CICTCVCLGV
1 SVTGEGQGPR SRTFTCLTNN ILRIDCHWSA PELGQGSSPW LLFTSNQAPG
51 GTHKCILRGS ECTVVLPPEA VLVPSD**N**FTI TFHHCMSGRE QVSLVDPEYL
101 PRRHVKLDPP SDLQS**N**ISSG HCILTWSISP ALEPMTTLLS YELAFKKQEE
151 AWEQAQHRDH IVGVTWLILE AFELDPGFIH EARLRVQMAT LEDDVVEEER
201 YTGQ*WSEWS*Q PVCFQAPQRQ GPLIPPWGWP GNTLVAVSIF LLLTGPTYLL
251 FKLSPRVKRI FYQNVPSPAM FFQPLYSVHN GNFQTWMGAH RAGVLLSQDC
301 AGTPQGALEP CVQEATALLT CGPARPWKSV ALEEEQEGPG TRLPGNLSSE
351 DVLPAGCTEW RVQTLAYLPQ EDWAPTSLTR PAPPDSEGSR SSSSSSSSN
401 NNNYCALGCY GGWHLSALPG NTQSSGPIPA LACGLSCDHQ GLETQQGVAW
451 VLAGHCQRPG LHEDLQGMLL PSVLSKARSW TF

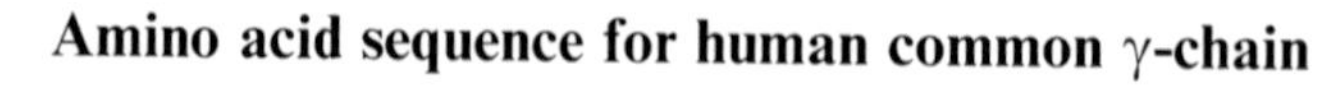

Amino acid sequence for human common γ-chain

Accession code: SwissProt P31785

See Amino acid sequence for human common γ-chain under the IL-2 entry (page **44**).

Amino acid sequence for mouse IL-9 receptor[13]

Accession code: SwissProt Q01114

-37 MALGRCIAEG WTLERVAVKQ VSWFLIYSWV CSGVCRG
1 VSVPEQGGGG QKAGAFTCLS NSIYRIDCHW SAPELGQESR AWLLFTSNQV
51 TEIKHKCTFW DSMCTLVLPK EEVFLPFD**N**F TITLHRCIMG QEQVSLVDSQ
101 YLPRRHIKLD PPSDLQS**N**VS SGRCVLTWGI NLALEPLITS LSYELAFKRQ
151 EEAWEARHKD RIVGVTWLIL EAVELNPGSI YEARLRVQMT LESYEDKTEG
201 EYYKSH*WSEW S*QPVSFPSPQ RRQGLLVPRW QWSASILVVV PIFLLLTGFV
251 HLLFKLSPRL KRIFYQNIPS PEAFFHPLYS VYHGDFQSWT GARRAGPQAR
301 QNGVSTSSAG SESSIWEAVA TLTYSPACPV QFACLKWEAT APGFPGLPGS
351 EHVLPAGCLE LEGQPSAYLP QEDWAPLGSA RPPPPDSDSG SSDYCMLDCC
401 EECHLSAFPG HTESPELTLA QPVALPVSSR A

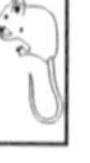

Amino acid sequence for mouse common γ-chain

Accession code: SwissProt P34902

See Amino acid sequence for mouse common γ-chain under the IL-2 entry (page **44**).

References

[1] Renauld, J.-C. et al. (1990) Cytokine 2, 9–12.
[2] Uyttenhove, C. et al. (1988) Proc. Natl Acad. Sci. USA 85, 6934–6938.
[3] Van Snick, J. et al. (1989) J. Exp. Med. 169, 363–368.
[4] Yang, Y. et al. (1989) Blood 74, 1880–1884.
[5] Yang, Y.C. (1992) Leuk. Lymphoma 8, 441–447.
[6] Moeller, J. (1990) J. Immunol. 144, 4231–4234.
[7] Williams, D.E. et al. (1990) Blood 76, 906–914.

[8] Renauld, J.-C. et al. (1990) J. Immunol. 144, 4235–4241.
[9] Birner, A. et al. (1992) Exp. Hematol. 20, 541–545.
[10] Gruss, H.J. et al. (1992) Cancer Res 52, 1026–1031.
[11] Kelleher, K. et al. (1991) Blood 77, 1436–1441.
[12] Druez, C. et al. (1990) J. Immunol. 145, 2494–2499.
[13] Renauld, J.C. et al. (1992) Proc. Natl Acad. Sci. USA 89, 5690–5694.
[14] Miyazawa, K. et al. (1992) Blood 80, 1685–1692.

Other names
Cytokine synthesis inhibitory factor (CSIF).

THE MOLECULE

Interleukin 10 (IL-10) is an acid-labile cytokine secreted by Th0 and Th2 subsets of CD4+ T lymphocytes that blocks activation of cytokine synthesis by Th1 T cells, activated monocytes and NK cells[1–3]. IL-10 also stimulates and/or enhances proliferation of B cells, thymocytes and mast cells[1,2,4], and it cooperates with TGFβ to stimulate IgA production by human B cells[5]. There is a high degree of homology (70%) between IL-10 and an open reading frame (BCRF1), in the EBV genome. The protein encoded by BCRF1 exhibits some of the activities of IL-10 and has been designated vIL-10.

Crossreactivity
There is 72% homology between human and mouse mature IL-10. Human IL-10 is active on mouse cells, but murine IL-10 is not active on human cells.

Sources
Th0 and Th2 subsets of CD4+ murine T cells, activated CD4+ and CD8+ human T cells, and murine Ly-1+ B cells, monocytes, macrophages, keratinocytes.

Bioassays
Inhibition of IFNγ synthesis by mitogen/antigen-activated Th1 clones or activated PBMCs. Proliferation of MC/9 mouse mast cell line.

Physicochemical properties of IL-10

Property	Human	Mouse
pI	8	8.1
Amino acids – precursor	178	178
– mature[a]	160	160
M_r (K) – predicted	18.6	18.8
– expressed	35–40	17–21
Potential *N*-linked glycosylation sites	1	2
Disulfide bonds[b]	2	2

[a] After removal of predicted signal peptide.
[b] Possible disulfide bonds.

3-D structure
Similar to IFNγ. A 3-D image and PDB file are available from SwissProt P22301.

Gene structure[6]

The gene for IL-10 is mouse chromosome 1 (five exons of 55, 20, 51, 22 and 30 amino acids). Human IL-10 is on syntenic region of chromosome 1.

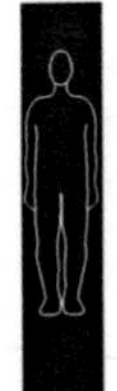

Amino acid sequence for human IL-10[7]

Accession code: SwissProt P22301

```
-18   MHSSALLCCL VLLTGVRA
  1   SPGQGTQSEN SCTHFPGNLP NMLRDLRDAF SRVKTFFQMK DQLDNLLLKE
 51   SLLEDFKGYL GCQALSEMIQ FYLEEVMPQA ENQDPDIKAH VNSLGENLKT
101   LRLRLRRCHR FLPCENKSKA VEQVKNAFNK LQEKGIYKAM SEFDIFINYI
151   EAYMTMKIRN
```

Amino acid sequence for mouse IL-10[8]

Accession code: SwissProt P18893

```
-18   MPGSALLCCL LLLTGMRI
  1   SRGQYSREDN NCTHFPVGQS HMLLELRTAF SQVKTFFQTK DQLDNILLTD
 51   SLMQDFKGYL GCQALSEMIQ FYLVEVMPQA EKHGPEIKEH LNSLGEKLKT
101   LRMRLRRCHR FLKCENKSKA VEQVKSDFNK LEDQGVYKAM NEFDIFINCI
151   EAYMMIKMKS
```

THE IL-10 RECEPTOR

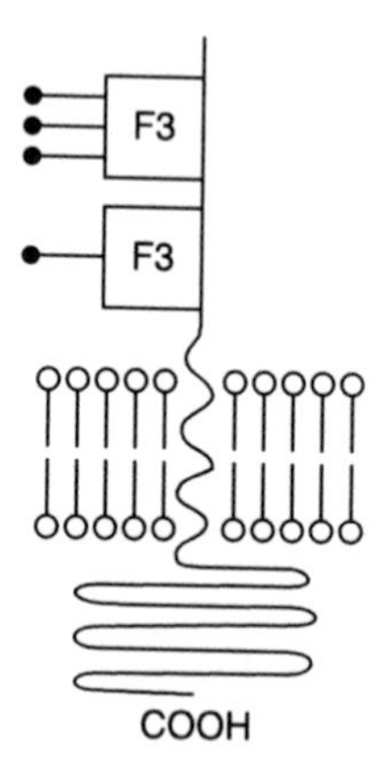

The IL-10 receptor belongs to the class II CKR family which also includes the IFNγ receptor, IFNα/β receptor and tissue factor[9–11]. The extracellular region of 220 amino acids consists of two homologous FNIII domains. The first of these has two conserved tryptophans and a pair of conserved cysteines, whereas the second has a unique disulfide loop formed from the second pair of conserved cysteines, but no WSXWS motif characteristic of class I cytokine receptors[12].

Distribution

IL-10R mRNA is expressed in B cells, thymocytes, the mast cell line MC/9 and the macrophage cell line IG.18LA. Human IL-10R mRNA is restricted mostly to haematopoietic cells and cell lines.

Physicochemical properties of the IL-10 receptor

Property	Human	Mouse
Amino acids – precursor	578	575
– mature[a]	557	559
M_r (K) – predicted	61	63
– expressed	90–110	110
Potential *N*-linked glycosylation sites	6	4
Affinity K_d (M)	23×10^{-10}	73×10^{-11}

[a] After removal of predicted signal peptide.

Signal transduction

Activation of Jak1 and Tyk2. Phosphorylation of STAT1a and STAT3.

Chromosomal location

The gene for the IL-10R is on human chromosome 11.

Amino acid sequence for human IL-10 receptor[13]

Accession code: SwissProt Q13651

-21 MLPCLVVLLA ALLSLRLGSD A
1 HGTELPSPPS VWFEAEFFHH ILHWTPIP**N**Q SESTCYEVAL LRYGIESWNS
51 IS**N**CSQTLSY DLTAVTLDLY HSNGYRARVR AVDGSRHS**N**W TVTNTRFSVD
101 EVTLTVGSVN LEIHNGFILG KIQLPRPKMA PA**N**DTYESIF SHFREYEIAI
151 RKVPG**N**FTFT HKKVKHE**N**FS LLTSGEVGEF CVQVKPSVAS RSNKGMWSKE
201 ECISLTRQYF TVTNVIIFFA FVLLLSGALA YCLALQLYVR RRKKLPSVLL
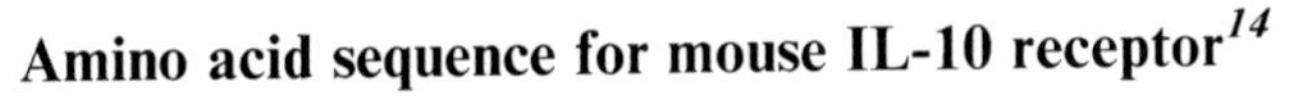
251 FKKPSPFIFI SQRPSPETQD TIHPLDEEAF LKVSPELKNL DLHGSTDSGF
301 GSTKPSLQTE EPQFLLPDPH PQADRTLGNG EPPVLGDSCS SGSSNSTDSG
351 ICLQEPSLSP STGPTWEQQV GSNSRGQDDS GIDLVQNSEG RAGDTQGGSA
401 LGHHSPPEPE VPGEEDPAAV AFQGYLRQTR CAEEKATKTG CLEEESPLTD
451 GLGPKFGRCL VDEAGLHPPA LAKGYLKQDP LEMTLASSGA PTGQWNQPTE
501 EWSLLALSSC SDLGISDWSF AHDLAPLGCV AAPGGLLGSF NSDLVTLPLI
551 SSLQSSE

Amino acid sequence for mouse IL-10 receptor[14]

Accession code: SwissProt Q61727

-16 MLSRLLPFLV TISSLS
1 LEFIAYGTEL PSPSYVWFEA RFFQHILHWK PIP**N**QSESTY YEVALKQYG**N**
51 STWNDIHICR KAQALSCDLT TFTLDLYHRS YGYRARVRAV DNSQYS**N**WTT
101 TETRFTVDEV ILTVDSVTLK AMDGIIYGTI HPPRPTITPA GDEYEQVFKD
151 LRVYKISIRK FSELK**N**ATKR VKQETFTLTV PIGVRKFCVK VLPRLESRIN
201 KAEWSEEQCL LITTEQYFTV TNLSILVISM LLFCGILVCL VLQWYIRHPG
251 KLPTVLVFKK PHDFFPANPL CPETPDAIHI VDLEVFPKVS LELRDSVLHG
301 STDSGFGSGK PSLQTEESQF LLPGSHPQIQ GTLGKEESPG LQATCGDNTD
351 SGICLQEPGL HSSMGPAWKQ QLGYTHQDQD DSDVNLVQNS PGQPKYTQDA
401 SALGHVCLLE PKAPEEKDQV MVTFQGYQKQ TRWKAEAAGP AECLDEEIPL
451 TDAFDPELGV HLQDDLAWPP PALAAGYLKQ ESQGMASAPP GTPSRQWNQL
501 TEEWSLLGVV SCEDLSIESW RFAHKLDPLD CGAAPGGLLD SLGSNLVTLP
551 LISSLQVEE

The Cys pairs 59–67 and 188–209, Trp residues 29 and 53, and Tyr residue 154 are conserved in the cytokine receptor class II superfamily.

References

[1] Moore, K.W. et al. (1993) Annu. Rev. Immunol. 11, 165–190.
[2] Malefyt, R.D.W. et al. (1992) Curr. Opin. Immunol. 4, 314–320.
[3] Howard, M. and O Garra, A. (1992) Immunol. Today 13, 198–200.
[4] Go, N.F. et al. (1990) J. Exp. Med. 172, 1625–1631.
[5] Defrance, T. et al. (1992) J. Exp. Med. 175, 671–682.
[6] Kim, J.M. et al. (1992) J. Immunol. 148, 3618–3623.
[7] Vieira, P. et al. (1991) Proc. Natl Acad. Sci. USA 88, 1172–1176.
[8] Moore, K.W. et al. (1990) Science 248, 1230–1234.
[9] Bazan, J.F. (1990) Proc. Natl Acad. Sci. USA 87, 6934–6938.
[10] Aguet, M. (1991) Br. J. Haematol. 79, 6–8.

[11] Lutfalla, G. et al. (1992) J. Biol. Chem. 267, 2802–2809.
[12] Bazan, J.F. (1990) Cell 61, 753–754.
[13] Liu, Y. et al. (1994) J. Immunol. 152, 1821–1829.
[14] Ho, A.S.Y. et al. (1993) Proc. Natl Acad. Sci. USA 90, 11267–11271.

Other names
Adipogenesis inhibitory factor[1].

THE MOLECULE

Interleukin 11 (IL-11) is a growth factor for plasmacytomas, haematopoietic multipotential and committed megakaryocytic and macrophage progenitor cells, and inhibits adipogenesis in mouse pre-adipocytes[1–5]. IL-11 is distantly related to IL-6, LIF and OSM[6].

Crossreactivity
Human IL-11 is active on mouse cells.

Sources
IL-1-stimulated fibroblasts, bone marrow stromal cell lines.

Bioassays
IL-11 can be assayed by its proliferative effects on some murine IL-6-dependent plasmacytoma lines such as T1165[7].

Physicochemical properties of IL-11

Property	Human	Mouse
pI (calculated)	11.2	11.2
Amino acids – precursor	199	199
– mature	178	178
M_r (K) – predicted	19.1	19.2
– expressed	23	?
Potential *N*-linked glycosylation sites	0	0
Disulfide bonds	0	0

3-D structure
No information.

Gene structure

The gene for IL-11 is located on human chromosome 19q13.3–13.4[6,8].

Amino acid sequence for human IL-11[2]

Accession code: SwissProt P20809

```
-21  MNCVCRLVLV VLSLWPDTAV A
  1  PGPPPGPPRV SPDPRAELDS TVLLTRSLLA DTRQLAAQLR DKFPADGDHN
 51  LDSLPTLAMS AGALGALQLP GVLTRLRADL LSYLRHVQWL RRAGGSSLKT
101  LEPELGTLQA RLDRLLRRLQ LLMSRLALPQ PPPDPPAPPL APPSSAWGGI
151  RAAHAILGGL HLTLDWAVRG LLLLKTRL
```

Amino acid sequence for mouse IL-11[9]

Accession code: SwissProt P47873

```
-21  MNCVCRLVLV VLSLWPDRVV A
  1  PGPPAGSPRV SSDPRADLDS AVLLTRSLLA DTRQLAAQMR DKFPADGDHS
 51  LDSLPTLAMS AGTLGSLQLP GVLTRLRVDL MSYLRHVQWL RRAGGPSLKT
101  LEPELGALQA RLERLLRRLQ LLMSRLALPQ AAPDQPVIPL GPPASAWGSI
151  RAAHAILGGL HLTLDWAVRG LLLLKTRL
```

THE IL-11 RECEPTOR

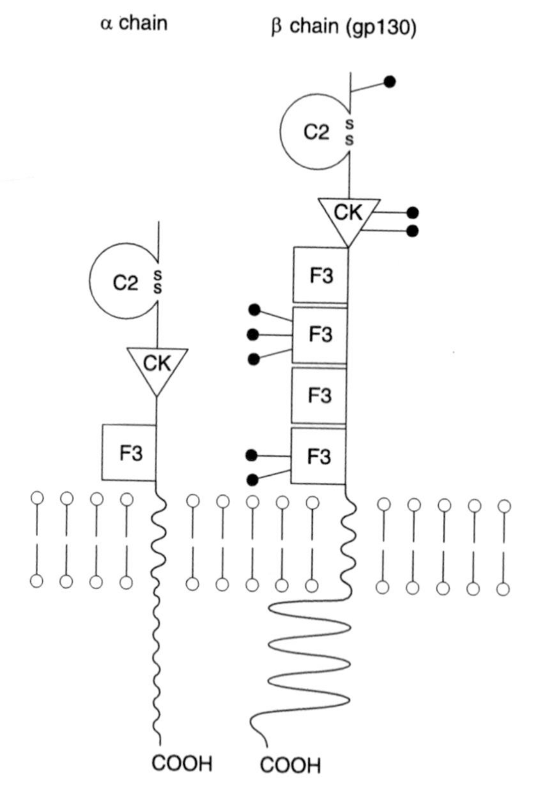

The IL-11 high-affinity binding receptor complex is composed of a low-affinity IL-11 ligand-binding α-chain[10] and the signal-transducing subunit, gp130, which is shared with other members of the IL-6 family such as IL-6, LIF, oncostatin M, CT-1 and CNTF[8]. IL-11 binding to IL-11Rα occurs with low affinity (K_d ~10 nM) and does not transduce an intracellular signal. In association with gp130, the affinity of the receptor complex for IL-11 increases significantly and converts the complex into a high-affinity signal transducing complex (K_d ~400–800 pM)[11]. The murine IL-11R α-chain is composed of an extracellular domain, a transmembrane domain and a cytoplasmic tail, while human IL-11R exists in two isoforms, one of which lacks a cytoplasmic domain. The IL-11R α-chain does not need to be membrane-bound to elicit a signal, as soluble forms of the IL-11R α-chain have been generated which are capable of activating gp130. A second murine IL-11Rα locus has been identified (IL-11Rα2) adjacent to the IL-11Rα1 described here, but there is no evidence of a second human locus.

Distribution

IL-11Rα is widely expressed.

Physicochemical properties of the IL-11 receptors

Property	IL-11Rα	
	Human	Mouse
Amino acids – precursor	422	432
– mature[a]	?	409
M_r (K) – predicted	45.22	46.65
– expressed	?	?
Potential *N*-linked glycosylation sites	?	2
Affinity K_d (M)	10×10^{-9}	10×10^{-9}
+gp130	$4–8 \times 10^{-10}$	$4–8 \times 10^{-10}$

[a] After removal of predicted signal peptide.

Signal transduction

Signal transduction by IL-11 is similar to other receptors which utilize gp130. It involves activation of Jak1 and Jak2, which in turn activate STAT3.

Chromosomal location

The gene for IL-11Rα is located on human chromosome 9p13 and mouse chromosome 4.

Amino acid sequence for human IL-11Rα

Accession code: SPTREMBLE Q16542

Amino acid sequence for mouse IL-11Rα

Accession code: SPTREMBLE Q64385

```
-23  MVLVAVATALV SSS
  1  SPCPQAWGPP GVQYGQPGRP VMLCCPGVSA GTPVSWFRDG DSRLLQGPDS
 51  GLGHRLVLAQ VDSPDEGTYV CQTLDGVSGG MVTLKLGFPP ARPEVSCQAV
101  DYENFSCTWS PGQVSGLPTR YLTSYRKKTL PGAESQRESP STGPWPCPQD
151  PLEASRCVVH GAEFWSEYRI NVTEVNPLGA STCLLDVRLQ SILRPDPPQG
201  LRVESVPGYP RRLHASWTYP ASWRRQPHFL LKFRLQYRPA QHPAWSTVEP
251  IGLEEVITDA VAGLPHAVRV SARDFLDAGT WSAWSPEAWG TPSTGPLQDE
301  IPDWSQGHGQ QLEAVVAQED SPAPARPSLQ PDPRPLDHRD PLEQVAVLAS
351  LGIFSCLGLA VGALALGLWL RLRRSGKDGP QKPGLLAPMI PVEKLPGIPN
401  LQRTPENFS
```

References

[1] Kawashima, I. et al. (1991) FEBS Lett. 283, 199–202.
[2] Paul, S.R. et al. (1990) Proc. Natl Acad. Sci. USA 87, 7512–7516.
[3] Musashi, M. et al. (1991) Proc. Natl Acad. Sci. USA 88, 765–769.
[4] Quesniaux, V.F. et al. (1993) Int. Rev. Exp. Pathol. 34, 205–214.
[5] Musashi, M. et al. (1991) Blood 78, 1448–1451.
[6] Bruce, A.G. et al. (1992) Progr. Growth Factor Res. 4, 157–170.
[7] Gearing, A.J.H. et al. (1994) In The Cytokine Handbook, 2nd edn, Thomson, A.W. ed., Academic Press, London.
[8] Yang, Y.C. and Lin, T. (1992) Biofactors 4, 15–21.
[9] Morris, J.C. et al. (1996) Exp. Hematol. 24, 1369–1376.
[10] Hilton, D.J. et al. (1994) EMBO J. 13, 4765–4775.
[11] Nandurkar, H.H. et al. (1996) Oncogene 12, 585–593.

Other names
Natural killer cell stimulatory factor (NKSF), cytotoxic lymphocyte maturation factor (CLMF)[1].

THE MOLECULE

Interleukin 12 (IL-12) is a heterodimeric cytokine made up of two chains (p35 and p40) important in defence against intracellular pathogens[2]. It induces IFNγ production by T cells and NK cells, enhances NK and ADCC activity, and costimulates peripheral blood lymphocyte proliferation[1,3–7]. IL-12 also stimulates proliferation and induces the differentiation of the Th1 subset of T lymphocytes[8,9].

Crossreactivity
There is 70% homology between human and mouse p40, and 60% homology between the p35 chains of IL-12[10]. Human IL-12 is inactive on mouse cells, whereas mouse IL-12 is active on human cells[10].

Sources
B Lymphoblastoid cells, dendritic cells, B cells[11]. Dendritic cells are the most potent producers.

Bioassays
Proliferation of PHA-activated lymphoblasts in the presence of monoclonal antibodies to IL-2, or stimulation of IFNγ production by spleen cells[7].

Physicochemical properties of IL-12

	p35[a]		p40[a,b]	
Property	Human	Mouse	Human	Mouse
pI	6.5	8.2[c]	5.4	6[c]
Amino acids – precursor	253	215	328	335
– mature	196	193	306	313
M_r (K) – predicted	22.5	21.7	34.7	35.8
– expressed	30–33	?	35–44	?
Potential *N*-linked glycosylation sites	3	1	4	5
Disulfide bonds[d]	3	3	5	6

[a] Neither chain alone has biological activity.
[b] The p40 chain shares extensive sequence homology with the extracellular immunoglobulin and haematopoietin domains of the IL-6 receptor[12].
[c] Calculated pI.

3-D structure
Functional IL-12 is a heterodimer of the p35 and p40 subunits.

Gene structure

Human IL-12α (p35) is on chromosome 3q12–q13.2 and IL-12β (p40) is on chromosome 5q31.1–q33.2. Mouse IL-12α is on chromosome 3 and IL-12β on chromosome 11.

Amino acid sequence for human IL-12 p35 chain

Accession code: SwissProt P29459

-56 MWPPGSASQP PPSPAAATGL HPAARPVSLQ CRLSMCPARS LLLVATLVLL
-6 DHLSLA
1 RNLPVATPDP GMFPCLHHSQ NLLRAVSNML QKARQTLEFY PCTSEEIDHE
51 DITKDKTSTV EACLPLELTK NESCLNSRET SFITNGSCLA SRKTSFMMAL
101 CLSSIYEDLK MYQVEFKTMN AKLLMDPKRQ IFLDQNMLAV IDELMQALNF
151 NSETVPQKSS LEEPDFYKTK IKLCILLHAF RIRAVTIDRV TSYLNAS

Amino acid sequence for human IL-12 p40 chain

Accession code: SwissProt P29460

-22 MCHQQLVISW FSLVFLASPL VA
1 IWELKKDVYV VELDWYPDAP GEMVVLTCDT PEEDGITWTL DQSSEVLGSG
51 KTLTIQVKEF GDAGQYTCHK GGEVLSHSLL LLHKKEDGIW STDILKDQKE
101 PK**N**KTFLRCE AK**N**YSGRFTC WWLTTISTDL TFSVKSSRGS SDPQGVTCGA
151 ATLSAERVRG DNKEYEYSVE CQEDSACPAA EESLPIEVMV DAVHKLKYE**N**
201 YTSSFFIRDI IKPDPPKNLQ LKPLKNSRQV EVSWEYPDTW STPHSYFSLT
251 FCVQVQGKSK REKKDRVFTD KTSATVICRK **N**ASISVRAQD RYYSSSWSEW
301 ASVPCS

Amino acid sequence for mouse IL-12 p35 chain[10]

Accession code: SwissProt P43431

-22 MCQSRYLLFL ATLALLNHLS LA
1 RVIPVSGPAR CLSQSRNLLK TTDDMVKTAR EKLKHYSCTA EDIDHEDITR
51 DQTSTLKTCL PLELHK**N**ESC LATRETSSTT RGSCLPPQKT SLMMTLCLGS
101 IYEDLKMYQT EFQAINAALQ NHNHQQIILD KGMLVAIDEL MQSLNHNGET
151 LRQKPPVGEA DPYRVKMKLC ILLHAFSTRV VTINRVMGYL SSA

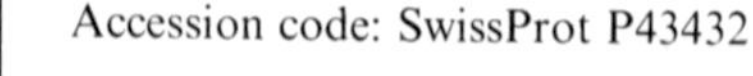

Amino acid sequence for mouse IL-12 p40 chain[10]

Accession code: SwissProt P43432

-22 MCPQKLTISW FAIVLLVSPL MA
1 MWELEKDVYV VEVDWTPDAP GETV**N**LTCDT PEEDDITWTS DQRHGVIGSG
51 KTLTITVKEF LDAGQYTCHK GGETLSHSHL LLHKKENGIW STEILKNFK**N**
101 KTFLKCEAP**N** YSGRFTCSWL VQRNMDLKFN IKSSSSPPDS RAVTCGMASL
151 SAEKVTLDQR DYEKYSVSCQ EDVTCPTAEE TLPIELALEA RQQNKYE**N**YS
201 TSFFIRDIIK PDPPKNLQMK PLKNSQVEVS WEYPDSWSTP HSYFSLKFFV
251 RIQRKKEKMK ETEEGCNQKG AFLVEKTSTE VQCKGGNVCV QAQDRYY**N**SS
301 CSKWACVPCR VRS

THE IL-12 RECEPTOR

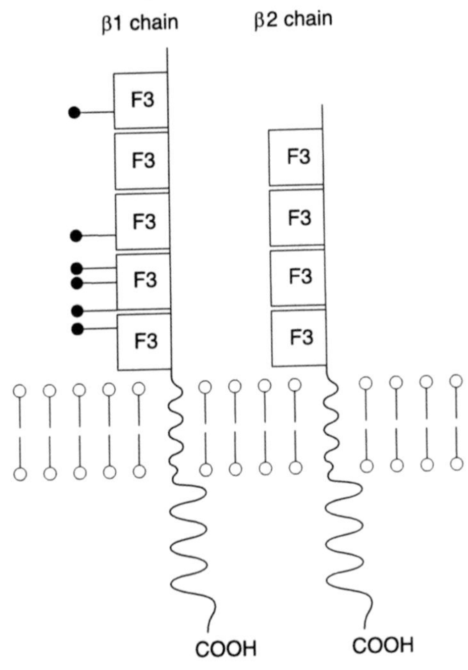

The functional high-affinity IL-12 receptor has two chains which both belong to the type I (haematopoietic) cytokine receptor superfamily. IL-12 p40 interacts with IL-12Rβ1[13] whilst IL-12 p35 interacts with IL-12Rβ2[14]. There may also be a third chain[15]. Neither is expressed on naïve T cells, but both are induced on developing Th1 cells.

Distribution

Activated T cells (Th1 cells express both IL-12Rβ1 and IL-12Rβ2, while Th2 cells express only IL-12Rβ1 and do not signal) and NK cells. In addition dendritic cells express a single class of high-affinity IL-12R.

Physicochemical properties of the IL-12 receptors

Property	IL-12Rβ1		IL-12Rβ2	
	Human	Mouse	Human	Mouse
Amino acids – precursor	662	738	862	874
– mature[a]	639	719	841	854
M_r (K) – predicted	73.108	81.661	97.13	98.196
– expressed	?	?	?	?
Potential *N*-linked glycosylation sites	6	13	8	14
Affinity K_d (M)	55×10^{-12}			

[a] After removal of predicted signal peptide.
[b] Coexpression of IL-12Rβ1 and IL-12Rβ2 ligand-binding affinities.

Signal transduction

Signal transduction involves Jak1[16] and STAT4[17], key regulators of Th1 cells.

Chromosomal location

The gene for IL-12Rβ1 is located on human chromosome 19at position p13.3, while human IL-12Rβ2 localizes to chromosome 1p31.2.

Amino acid sequence for human IL-12Rβ1

Accession code: SwissProt P42701

-23 MEPLVTWVVP LLFLFLLSRQ GAA
1 CRTSECCFQD PPYPDADSGS ASGPRDLRCY RISSDRYECS WQYEGPTAGV
51 SHFLRCCLSS GRCCYFAAGS ATRLQFSDQA GVSVLYTVTL WVESWAR**N**QT
101 EKSPEVTLQL YNSVKYEPPL GDIKVSKLAG QLRMEWETPD NQVGAEVQFR
151 HRTPSSPWKL GDCGPQDDDT ESCLCPLEMN VAQEFQLRRR QLGSQGSSWS
201 KWSSPVCVPP ENPPQPQVRF SVEQLGQDGR RRLTLKEQPT QLELPEGCQG
251 LAPGTEVTYR LQLHMLSCPC KAKATRTLHL GKMPYLSGAA YNVAVISSNQ
301 FGPGL**N**QTWH IPADTHTEPV AL**N**ISVGT**N**G TTMYWPARAQ SMTYCIEWQP
351 VGQDGGLATC SLTAPQDPDP AGMATYSWSR ESGAMGQEKC YYITIFASAH
401 PEKLTLWSTV LSTYHFGG**N**A SAAGTPHHVS VK**N**HSLDSVS VDWAPSLLST
451 CPGVLKEYVV RCRDEDSKQV SEHPVQPTET QVTLSGLRAG VAYTVQVRAD
501 TAWLRGVWSQ PQRFSIEVQV SDWLIFFASL GSFLSILLVG VLGYLGLNRA
551 ARHLCPPLPT PCASSAIEFP GGKETWQWIN PVDFQEEASL QEALVVEMSW
601 DKGERTEPLE KTELPEGAPE LALDTELSLE DGDRCKAKM

Disulfide bonds between cysteines 52 and 62. Variable splice sequence KAKM→DR at position 659–662 leads to the generation of an alternatively spliced shorter isoform.

Amino acid sequence for human IL-12Rβ2

Accession code: SwissProt Q99665

-21 MAHTFRGCSL AFMFIITWLL I
1 KAKIDACKRG DVTVKPSHVI LLGSTV**N**ITC SLKPRQGCFH YSRRNKLILY
51 KFDRRINFHH GHSLNSQVTG LPLGTTLFVC KLACINSDEI QICGAEIFVG
101 VAPEQPQ**N**LS CIQKGEQGTV ACTWERGRDT HLYTEYTLQL SGPK**N**LTWQK
151 QCKDIYCDYL DFGINLTPES PES**N**FTAKVT AVNSLGSSSS LPSTFTFLDI
201 VRPLPPWDIR IKFQKASVSR CTLYWRDEGL VLLNRLRYRP SNSRLWNMV**N**
251 VTKAKGRHDL LDLKPFTEYE FQISSKLHLY KGSWSDWSES LRAQTPEEEP
301 TGMLDVWYMK RHIDYSRQQI SLFWK**N**LSVS EARGKILHYQ VTLQELTGGK
351 AMTQ**N**ITGHT SWTTVIPRTG NWAVAVSAAN SKGSSLPTRI NIMNLCEAGL
401 LAPRQVSANS EGMDNILVTW QPPRKDPSAV QEYVVEWREL HPGGDTQVPL
451 NWLRSRPY**N**V SALISENIKS YICYEIRVYA LSGDQGGCSS ILGNSKHKAP
501 LSGPHINAIT EEKGSILISW NSIPVQEQMG CLLHYRIYWK ERDSNSQPQL
551 CEIPYRVSQN SHPINSLQPR VTYVLWMTAL TAAGESSHGN EREFCLQGKA
601 NWMAFVAPSI CIAIIMVGIF STHYFQQKVF VLLAALRPQW CSREIPDPAN
651 STCAKKYPIA EEKTQLPLDR LLIDWPTPED PEPLVISEVL HQVTPVFRHP
701 PCSNWPQREK GIQGHQASEK DMMHSASSPP PPRALQAESR QLVDLYKVLE
751 SRGSDPKPEN PACPWTVLPA GDLPTHDGYL PSNIDDLPSH EAPLADSLEE
801 LEPQHISLSV FPSSSLHPLT FSCGDKLTLD QLKMRCDSLM L

Amino acid sequence for mouse IL-12Rβ1

Accession code: SwissProt P42701

```
-19  MDMMGLPGTS KHITFLLLC
  1  QLGASGPGDG CCVEKTSFPE GASGSPLGPR NLSCRVSKTD YECSWQYDGP
 51  EDNVSHVLWC CFVPPNHTHT GQERCRYFSS GPDRTVQFWE QDGIPVLSKV
101  NFWVESRLGN RTMKSQKISQ YLYNWTKTTP PLGHIKVSQS HGQLRMDWNV
151  SEEAGAEVQF RRRMPTTNWT LGDCGPQVNS GSGVLGDICG SMSESCLCPS
201  ENMAQEIQIR RRRRLSSGAP GGPWSDWSMP VCVPPEVLPQ AKIKFLVEPL
251  NQGGRRRLTM QGQSPQLAVP EGCRGRPGAQ VKKHLVLVRM LSCRCQAQTS
301  KTVPLGKKLN LSGATYDLNV LAKTRFGRST IQKWHLPAQE LTETRALNVS
351  VGGNMTSMQW AAQAPGTTYC LEWQPWFQHR NHTHCTLIVP EEEDPAKMVT
401  HSWSSKPTLE QEECYRITVF ASKNPKNPML WATVLSSYYF GGNASRAGTP
451  RHVSVRNQTG DSVSVEWTAS QLSTCPGVLT QYVVRCEAED GAWESEWLVP
501  PTKTQVTLDG LRSRVMYKVQ VRADTARLPG AWSHPQRFSF EVQISRLSII
551  FASLGSFASV LLVGSLGYIG LNRAAWHLCP PLPTPCGSTA VEFPGSQGKQ
601  AWQWCNPEDF PEVLYPRDAL VVEMPGDRGD GTESPQAAPE CALDTRRPLE
651  TQRQRQVQAL SEARRLGLAR EDCPRGDLAH VTLPLLLGGV TQGASVLDDL
701  WRTHKTAEPG PPTLGQEA
```

Disulfide bonds between cysteines 53 and 63 of precursor sequence.

Amino acid sequence for mouse IL-12Rβ2

Accession code: SwissProt P97378

```
-20  MAQTVRECSL ALLFLFMWLL
  1  IKANIDVCKL GTVTVQPAPV IPLGSAANIS CSLNPKQGCS HYPSSNELIL
 51  LKFVNDVLVE NLHGKKVHDH TGHSSTFQVT NLSLGMTLFV CKLNCSNSQK
101  KPPVPVCGVE ISVGVAPEPP QNISCVQEGE NGTVACSWNS GKVTYLKTNY
151  TLQLSGPNNL TCQKQCFSDN RQNCNRLDLG INLSPDLAES RFIVRVTAIN
201  DLGNSSSLPH TFTFLDIVIP LPPWDIRINF LNASGSRGTL QWEDEGQVVL
251  NQLRYQPLNS TSWNMVNATN AKGKYDLRDL RPFTEYEFQI SSKLHLSGGS
301  WSNWSESLRT RTPEEEPVGI LDIWYMKQDI DYDRQQISLF WKSLNPSEAR
351  GKILHYQVTL QEVTKKTTLQ NTTRHTSWTR VIPRTGAWTA SVSAANSKGA
401  SAPTHINIVD LCGTGLLAPH QVSAKSENMD NILVTWQPPK KADSAVREYI
451  VEWRALQPGS ITKFPPHWLR IPPDNMSALI SENIKPYICY EIRVHALSES
501  QGGCSSIRGD SKHKAPVSGP HITAITEKKE RLFISWTHIP FPEQRGCILH
551  YRIYWKERDS TAQPELCEIQ YRRSQNSHPI SSLQPRVTYV LWMTAVTAAG
601  ESPQGNEREF CPQGKANWKA FVISSICIAI ITVGTFSIRY FRQKAFTLLS
651  TLKPQWYSRT IPDPANSTWV KKYPILEEKI QLPTDNLLMA WPTPEEPEPL
701  IIHEVLYHMI PVVRQPYYFK RGQGFQGYST SKQDAMYIAN PQATGTLTAE
751  TRQLVNLYKV LESRDPDSKL ANLTSPLTVT PVNYLPSHEG YLPSNIEDLS
801  PHEADPTDSF DLEHQHISLS IFASSSLRPL IFGGERLTLD RLKMGYDSLM
851  SNEA
```

References

[1] Gately, M.K. et al. (1991) J. Immunol. 147, 874–882.
[2] Locksley, R.M. (1993) Proc. Natl Acad. Sci. USA 90, 5879–5880.
[3] Wolf, S.F. et al. (1991) J. Immunol. 146, 3074–3081.
[4] Lieberman, M.D. et al. (1991) J. Surg. Res. 50, 410–415.
[5] Chan, S.H. et al. (1991) J. Exp. Med. 173, 869–879.
[6] Gubler, U. et al. (1991) Proc. Natl Acad. Sci. USA 88, 4143–4147.
[7] Stern, A. et al. (1990) Proc. Natl Acad. Sci. USA 87, 6808–6812.
[8] Hsieh, C.-S. et al. (1993) Science 260, 547–549.
[9] Manetti, R. et al. (1993) J. Exp. Med. 177, 1199–1204.
[10] Schoenhaut, D.S. et al. (1992) J. Immunol. 148, 3433.
[11] D'Andrea, A. et al. (1992) J. Exp. Med. 176, 1387–1398.
[12] Cella, M. et al. (1996) EMBO J. 184, 747–752.
[13] Chue, A.O. et al. (1994) J. Immunol. 153, 128–136.
[14] Presky, D.H. et al. (1996) Proc. Natl Acad. Sci. USA. 93, 14002–14007.
[15] Kawashima, T. et al. (1998) Cell. Immunol. 186, 39–44.
[16] Bacon, C.M. et al. (1995) J. Exp. Med. 181, 399–404.
[17] Trinchieri, G. (1998) Adv. Immunol. 70, 83–243.

Other names

Mouse IL-13 is also known as P600. Human IL-13 cDNA has also been referred to as NC30.

THE MOLECULE

Interleukin 13 (IL-13) is an IL-4-related cytokine which is produced mainly by activated Th2 cells, but also by mast cells, and NK cells. Recent studies using IL-13-deficient mice have established that IL-13 plays a central role in Th2 responses, mediating the expulsion of gastrointestinal parasites and in the development and pathology of asthma and allergy[1,2]. IL-13 also functions in the upregulation of IgE secretion by B cells and plays a role in the suppression of inflammatory responses through modulation of macrophage function. IL-13 inhibits the production of inflammatory cytokines (IL-1β, IL-6, TNFα, IL-8) by LPS-stimulated monocytes[3]. Human and mouse IL-13 induce CD23 expression on human B cells, promote B-cell proliferation in combination with anti-Ig or CD40 antibodies, and stimulate secretion of IgM, IgE and IgG4[4–6]. IL-13 has also been shown to prolong survival of human monocytes and increase surface expression of MHC class II and CD23[6].

Crossreactivity

There is 58% amino acid sequence identity between human and mouse IL-13 and significant cross-species reactivity.

Sources

Activated T cells, mast cells and NK cells.

Bioassays

Proliferation of human B cells costimulated with anti-IgM or CD40 antibody.

Physicochemical properties of IL-13

Property	Human	Mouse
pI	8.69	8.34
Amino acids – precursor	132	131
– mature[a]	112	113
M_r (K) – predicted	12.3	12.4
– expressed[b]	9/17	?
Potential *N*-linked glycosylation sites	4	3
Disulfide bonds	2	?

[a] After removal of signal peptide. Human IL-13 expressed in COS and Chinese hamster cells has N-terminal Gly. Mouse signal peptide is computer predicted.

[b] The 17 000 molecular weight form is glycosylated. Unglycosylated IL-13 has a predicted M_r of 12 000 (confirmed by electrospray mass spectrometry), but it migrates as a 9000 protein on SDS–PAGE.

3-D structure

IL-13 is a four antiparallel α-helical protein similar to IL-4.

Gene structure

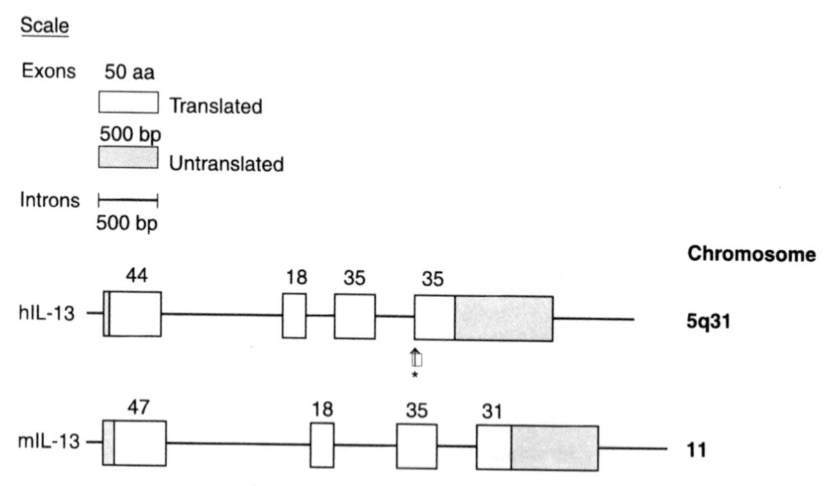

Human IL-13 has four exons of 44, 18, 35 and 35 amino acids and is on chromosome 5q31; mouse IL-13 is on chromosome 13 and has four exons of 47, 18, 35 and 31 amino acids. Alternative splicing gives rise to a second mRNA species in human T cells containing an additional codon for Gln at the 5′ end of exon 4[6].

Amino acid sequence for human IL-13[3,6]

Accession code: SwissProt P35225

```
-34  MHPLLNPLLL ALGL
-20  MALLLTTVIA LTCLGGFASP
  1  GPVPPSTALR ELIEELVNIT QNQKAPLCNG SMVWSINLTA GMYCAALESL
 51  INVSGCSAIE KTQRMLSGFC PHKVSAGQFS SLHVRDTKIE VAQFVKDLLL
101  HLKKLFREGR FN
```

There are two possible ATG initiation codons giving rise to Met −34 and Met −20. The second is the most likely initiation site. The protein secreted by transfected COS and Chinese hamster cells has Gly at the N-terminus[3]. Two mRNAs encoding one and two Gln residues at position 78 are expressed by activated T cells[6]. Disulfide bonds between Cys 28–56 and 44–70.

Amino acid sequence for mouse IL-13 (p600)[6,7]

Accession code: SwissProt P20109

```
-18  MALWVTAVLA LACLGGLA
  1  APGPVPRSVS LPLTLKELIE ELSNITQDQT PLCNGSMVWS VDLAAGGFCV
 51  ALDSLTNISN CNAIYRTQRI LHGLCNRKAP TTVSSLPDTK IEVAHFITKL
101  LSYTKQLFRH GPF
```

Computer-predicted signal peptide −1 to −18. Signal peptide in SwissProt database is given as −1 to −21 by similarity with human IL-13. Possible disulfide bonds between Cys33–61 and 49–75 by similarity with human IL-13.

THE IL-13 RECEPTOR

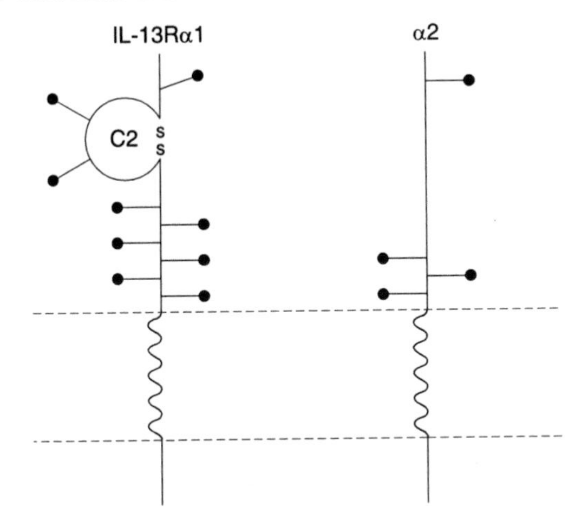

Human and murine IL-13 receptors have been cloned. Two human membrane-bound receptors have been identified, IL-13Rα1 and IL-13Rα2. IL-13Rα1 has a moderate affinity for IL-13 but requires IL-4Rα (CD124) to form a high-affinity receptor complex[8]. The IL-4Rα/IL-13Rα1 complex also functions as a functional IL-4 receptor (see IL-4 entry (page **58**)). Mouse and human IL-13Rα1 are members of the CKR-SF[9], with an Ig like domain distal to the membrane, a CKR domain with two four conserved cysteine residues and an FNIII like domain with the WSXWS motif. The murine receptor has a 60 amino acid cytoplasmic tail, whereas the human receptor has only 17 cytoplasmic amino acids. The human protein is 74% identical to the murine IL-13Rα1. IL-13Rα2 was discovered as a high-affinity IL-13-binding protein, but is incapable of signalling[10]. The relationship between IL-13Rα1 and IL-13Rα2 is unclear, but it has been suggested that IL-13Rα2 functions as an IL-13 antagonist. The γc receptor chain found in the IL-2, IL-4, IL-7, IL-9 and IL-15 receptor complexes does not participate in IL-13R signalling pathways[11].

Distribution

IL-13Rα1 is ubiquitously expressed, with highest levels in heart, liver, skeletal muscle and ovary. B cells, T cells and endothelial cells also express IL-13Rα1.

Physicochemical properties of the IL-13 receptors

	IL-13Rα1		IL-13Rα2	
Property	Human	Mouse	Human	Mouse
Amino acids – precursor	427	424	380	383
– mature[a]	406	399	354	
M_r (K) – predicted	48.759	48.402	44.176	
– expressed	?	56–68		
Potential *N*-linked glycosylation sites	10	4	4	4
Affinity K_d (M)	4×10^{-9}	$2–10 \times 10^{-9}$	45×10^{-11}	25×10^{-11}
+ IL-4Rα	30×10^{-12}	75×10^{-12}	45×10^{-11}	25×10^{-11}

[a] After removal of predicted signal peptide.

Signal transduction

Engagement of IL-13 with its receptor complex (IL-13Rα1/IL-4Rα) initiates Jak/STAT signalling pathways, leading to activation of STAT6[12] and STAT3[13] which may be downregulated by SOCS proteins.

Chromosomal location

The location of the human IL-13Rα1 gene is unknown. Human IL-13Rα2 is located on the X chromosome in the Xq24 region[14]. Mouse IL-13Rα1 and IL-13Rα2 both map to the X chromosome[15].

Amino acid sequence for human IL-13Rα1

Accession code: SwissProt P78552

-21 MEWPARLCGL WALLLCAGGG G
1 GGGGAAPTET QPPVT**N**LSVS VENLCTVIWT WNPPEGASS**N** CSLWYFSHFG
51 DKQDKKIAPE TRRSIEVPLN ERICLQVGSQ CST**N**ESEKPS ILVEKCISPP
101 EGDPESAVTE LQCIWH**N**LSY MKCSWLPGRN TSPDT**N**YTLY YWHRSLEKIH
151 QCENIFREGQ YFGCSFDLTK VKDSSFEQHS VQIMVKDNAG KIKPSFNIVP
201 LTSRVKPDPP HIK**N**LSFHND DLYVQWENPQ NFISRCLFYE VEVNNSQTET
251 HNVFYVQEAK CENPEFERNV E**N**TSCFMVPG VLPDTLNTVR IRVKTNKLCY
301 EDDKLWS**N**WS QEMSIGKKR**N** STLYITMLLI VPVIVAGAII VLLLYLKRLK
351 IIIFPPIPDP GKIFKEMFGD QNDDTLHWKK YDIYEKQTKE ETDSVVLIEN
401 LKKASQ

Disulfide bonds between cysteines 46–95, 134–144 and 173–185 by similarity. Conflict in sequence T→I at position 130 and G→D at position 358.

Amino acid sequence for human IL-13Rα2

Accession code: SwissProt Q14627

```
-26  MAFVCLAIGC LYTFLISTTF GCTSSS
  1  DTEIKVNPPQ DFEIVDPGYL GYLYLQWQPP LSLDHFKECT VEYELKYRNI
 51  GSETWKTIIT KNLHYKDGFD LNKGIEAKIH TLLPWQCTNG SEVQSSWAET
101  TYWISPQGIP ETKVQDMDCV YYNWQYLLCS WKPGIGVLLD TNYNLFYWYE
151  GLDHALQCVD YIKADGQNIG CRFPYLEASD YKDFYICVNG SSENKPIRSS
201  YFTFQLQNIV KPLPPVYLTF TRESSCEIKL KWSIPLGPIP ARCFDYEIEI
251  REDDTTLVTA TVENETYTLK TTNETRQLCF VVRSKVNIYC SDDGIWSEWS
301  DKQCWEGEDL SKKTLLRFWL PFGFILILVI FVTGLLLRKP NTYPKMIPEF
351  FCDT
```

Amino acid sequence for mouse IL-13Rα1

Accession code: SwissProt O09030

```
-25  MARPALLGEL LVLLLWTATV GQVAA
  1  ATEVQPPVTN LSVSVENLCT IIWTWSPPEG ASPNCTLRYF SHFDDQQDKK
 51  IAPETHRKEE LPLDEKICLQ VGSQCSANES EKPSPLVKKC ISPPEGDPES
101  AVTELKCIWH NLSYMKCSWL PGRNTSPDTH YTLYYWYSSL EKSRQCENIY
151  REGQHIACSF KLTKVEPSFE HQNVQIMVKD NAGKIRPSCK IVSLTSYVKP
201  DPPHIKHLLL KNGALLVQWK NPQNFRSRCL TYEVEVNNTQ TDRHNILEVE
251  EDKCQNSESD RNMEGTSCFQ LPGVLADAVY TVRVRVKTNK LCFDDNKLWS
301  DWSEAQSIGK EQNSTFYTTM LLTIPVFVAV AVIILLFYLK RLKIIIFPPI
351  PDPGKIFKEM FGDQNDDTLH WKKYDIYEKQ SKEETDSVVL IENLKKAAP
```

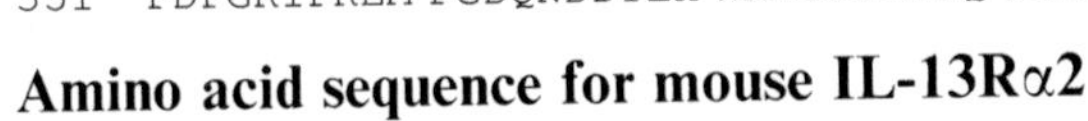

Amino acid sequence for mouse IL-13Rα2

Accession code: Genbank AAC33240

```
  1  MAFVHIRCLC FILLCTITGY SLEIKVNPPQ DFEILDPGLL GYLYLQWKPP
 51  VVIEKFKGCT LEYELKYRNV DSDSWKTIIT RNLIYKDGFD LNKGIEGKIR
101  THLSEHCTNG SEVQSPWIEA SYGISDEGSL ETKIQDMKCI YYNWQYLVCS
151  WKPGKTVYSD TNYTMFFWYE GLDHALQCAD YLQHDEKNVG CKLSNLDSSD
201  YKDFFICVNG SSKLEPIRSS YTVFQLQNIV KPLPPEFLHI SVENSIDIRM
251  KWSTPGGPIP PRCYTYEIVI REDDISWESA TDKNDMKLKR RANESEDLCF
301  FVRCKVNIYC ADDGIWSEWS EEECWEGYTG PDSKIIFIVP VCLFFIFLLL
351  LLCLIVEKEE PEPTLSLHVD LNKEVCAYED TLC
```

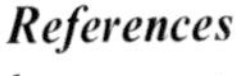

References

[1] McKenzie, G.J. et al. (1998) Curr. Biol. 8, 339–342.
[2] Wills-Karp, M. et al. (1998) Science 282, 2258–2261.
[3] Minty, A. et al. (1993) Nature 362, 248–250.
[4] Cocks, B.G. et al. (1993) Int. Immunol. 5, 657–663.
[5] Punnonen, J. et al. (1993) Proc. Natl Acad. Sci. USA 90, 3730–3734.
[6] McKenzie, A.N. et al. (1993) Proc. Natl Acad. Sci. USA 90, 3735–3739.
[7] Brown, K.D. et al. (1989) J. Immunol. 142, 679–687.
[8] Miloux, B. et al. (1997) FEBS Lett. 401, 163–166.

[9] Hilton, D.J. et al. (1996) Proc. Natl Acad. Sci. USA 93, 497–501.
[10] Caput, D. et al. (1996) J. Biol. Chem. 271, 16921–16926.
[11] Matthews, D.J. et al. (1995) Blood 85, 38–42.
[12] Lin, J.X. et al. (1995) Immunity 2, 331–339.
[13] Orchansky, P.L. et al. (1999) J. Biol. Chem. 274, 20818–20825.
[14] Guo, J. et al. (1997) Genomics 42, 141–145.
[15] Donaldson, D.D. et al. (1998) J. Immunol. 161, 2317–2324.

Other names
High-molecular-weight B-cell growth factor (HMW-BCGF)[1].

THE MOLECULE

Interleukin 14 (IL-14) is a cytokine which enhances the proliferation of activated B cells, and inhibits immunoglobulin synthesis[1–4]. IL-14 is unrelated to any other cytokines, but shares homology with complement factor Bb[2,3]. IL-14 has also been shown to selectively expand certain B-cell populations[1].

Crossreactivity
There are no reports of murine IL-14.

Sources
IL-14 is produced by T cells and some B-cell tumours[2].

Bioassays
Human IL-14 can be assayed by proliferation of *Staphylococcus aureus*-activated B cells[4].

Physicochemical properties of IL-14

Property	Human
pI	6.7–7.8
Amino acids – precursor	498
– mature	483
M_r (K) – predicted	53
– expressed	60
Potential *N*-linked glycosylation sites	3
Disulfide bonds[a]	?

[a] IL-14 has 34 Cys residues, some of which may form disulfide bonds.

3-D structure
No information.

Gene Structure

No information

Amino acid sequence for human IL-14[2]

Accession code: Genbank L15344

```
-15  MIRLLRSIIG LVLAL
  1  PKGRGSDAKG LLTSTSSHAK TFSRSLVSRQ EIPSSERPAR LCLGTPFLRS
 51  REVKVSLPEQ LGGPELDLSK PCSSRQMQFN DLGVLLVLSS EGQGGVIEPQ
101  SQTEKCSSHF CQLTVMAATD DSPSQDSSAH RQATDFICTH PHPQCKLHCT
151  LTGINAFARP CKMPYVRKTD PQKCQNCKCQ MPARSEHAWL QQWAAQHGSL
201  LWARLAGLPL PSQHDPTPGS LGPGGGGLLR PVWPDASVLG CSWVARRFCD
251  PGGAGCLMPQ CPSGLLSGPL SVREPWPPAL RSCTLLFRSL RSVCSARHSF
301  SSRWIFTCRP SSSLSRTVFS SAISSRALLL LSHRDRYMVV SFSSFLIFLV
351  IFSISCLNVV NTSLLLESVF WNSSNFSVYR ASCCFRWVSC CFISSHILWD
401  STASFRRKSF SRWCRSSASF SISWACWSLA STSCCCRSLC LKTLSICSSR
451  SSYCSISFLS LSASSMFSWR SLELRSLCCS ICS
```

THE IL-14 RECEPTOR

A single receptor type with an affinity of 203×10^{-9} M is detected on B cells and B-cell leukaemias[5,6]. IL-14 causes upregulation of its receptor on B cells. IL-14 receptors also bind complement fragment Bb[3].

Distribution

No information.

Signal transduction

IL-14 causes increases in intracellular cAMP, DAG and calcium[3].

Chromosomal location

No information.

Amino acid sequence

No information

References

1 Kehrl, J. et al. (1984) Immunol. Rev. 78, 75–96.
2 Ambrus, J.L. Jr et al. (1993) Proc. Natl Acad. Sci. USA 90, 6330–6334.
3 Ambrus, J.L. Jr et al. (1991) J. Biol. Chem. 266, 3702–3708.
4 Ambrus, J.L. Jr et al. (1990) J. Immunol. 145, 3949–3955.
5 Uckun, F. et al. (1989) J. Clin. Invest. 84, 1595–1608.
6 Ambrus, J.L. Jr et al. (1988) J. Immunol. 141, 861–869.

Other names

Known originally as IL-T, but now only IL-15 is used.

THE MOLECULE

Interleukin 15 (IL-15) is a 14–15-kDa member of the four α-helix bundle of cytokines which was originally isolated and cloned from the simian kidney epithelial cell line CV1/EBNA[1]. IL-15 stimulates the proliferation of T lymphocytes and shares many of the biological properties of IL-2, including stimulation of CTLL proliferation, and *in vitro* generation of alloantigen-specific cytotoxic T cells and nonantigen-specific lymphokine-activated killer (LAK) cells. One of the most critical roles of IL-15 is in the development, survival and activation of natural killer cells[1–4].

Crossreactivity

There is 97% sequence identity between human and simian IL-15. Human IL-15 is active on mouse CTLL cells.

Sources

IL-15 mRNA is found in a wide variety of human cell types including peripheral blood mononuclear cells (PBMCs), placenta, skeletal muscle, kidney, lung, liver, heart and the IMTLH bone marrow stromal cell line and is produced most abundantly by epithelial cells and monocytes. IL-15 is produced mainly by monocytes and dendritic cells[2].

Bioassays

Proliferation of activated T cells.

Physicochemical properties of IL-15

Property	Human	Mouse
pI	4.52	4.6
Amino acids – precursor	162	162
– mature	113	114
M_r (K) – predicted	12.77	13.25
– expressed	14–15	14–15
Potential *N*-linked glycosylation sites	1	1
Disulfide bonds[a]	2	2

[a] Disulfide bonds predicted by three-dimensional modelling.

3-D structure

Computer modelling of IL-15 indicates a four α-helical bundle structure similar to IL-2.

Gene structure

The gene for IL-15 is located on human chromosome 4q31 and the central region of mouse chromosome 8. The IL-15 gene consists of nine exons, unlike the number observed for IL-2, IL-4, IL-5 and IL-13[5].

Amino acid sequence for human IL-15[2]

Accession code: SwissProt P40933

```
-29  MRISKPHLRS ISIQCYLCLL LKSHFLTEA
  1  GIHVFILGCF SAGLPKTEAN WVNVISDLKK IEDLIQSMHI DATLYTESDV
 51  HPSCKVTAMK CFLLELQVIS HESGDTDIHD TVENLIILAN NILSSNGNIT
101  ESGCKECEEL EEKNIKEFLQ SFVHIVQMFI NTS
```

Disulfide bridges between Cys83–133 and 90–136, by similarity.

Amino acid sequence for mouse IL-15[5]

Accession code: SwissProt P48346

```
-29  MKILKPYMRN TSISCYLCFL LNSHFLTEA
  1  GIHVFILGCV SVGLPKTEAN WIDVRYDLEK IESLIQSIHI DTTLYTDSDF
 51  HPSCKVTAMN CFLLELQVIL HEYSNMTLNE TVRNVLYLAN STLSSNKNVA
101  ESGCKECEEL EEKTFTEFLQ SFIRIVQMFI NTS
```

Disulfide bridges between Cys83–133 and 90–136, by similarity.

THE IL-15 RECEPTOR

IL-15 uses two distinct receptor systems. On T cells and NK cells, the type I IL-15R system is used. This includes IL-2/15Rβ, which is shared with IL-2, IL-4, IL-7 and IL-9, the γc subunit and an IL-15-specific receptor subunit, IL-15Rα[3,6–8]. In addition, mast cells utilize a novel 60–65-kDa IL-15RX subunit which is unrelated to the IL-2R system. This is referred to as the type II IL-15R system[9]. IL-15Rα binds IL-15 with a very high affinity (K_d 10^{-11} M). In the absence of IL-15Rα , IL-2/15Rβ and γc acting together bind IL-15 with an intermediate affinity (K_d 10^{-9} M). IL-15RX also binds IL-15 with an intermediate affinity (K_d 10^{-11} M).

Chromosomal location

The gene for human IL-15Rα is located on chromosome 10p15–14, while that of mouse IL-15Rα is on chromosome 2. The chromosomal localization of IL-15RX has not yet been determined.

Distribution

IL-15Rα has a wide tissue distribution. It is expressed on T cells, B cells, macrophages, thymic stroma and bone marrow stroma cells.

Signal transduction

IL-15-mediated signalling pathways which involve the type I receptor system involve Jak1/Jak3 and STAT5/STAT3, while type II pathways involve Jak2/STAT5-dependent pathways.

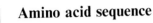

Amino acid sequence

No information

References

[1] Bamford, R.N. et al. (1994) Proc. Natl Acad. Sci. USA 91, 4940–4944.
[2] Grabstein, K.H. et al. (1994) Science 264, 965–968.
[3] Giri, J.G. et al. (1994) EMBO J. 13, 2822–2830.
[4] Tagaya, Y. et al. (1996) Immunity 4, 329–336.
[5] Anderson, D.M. et al. (1995) Genomics 25, 701–706.
[6] Carson, W.E. et al. (1994) J. Exp. Med. 180, 1395–1403.
[7] Giri, J.G. et al. (1995) EMBO J. 14, 3654–3663.
[8] Giri, J.G. et al. (1995) J. Leukoc. Biol. 57, 763–766.
[9] Tagaya, Y. et al. (1996) EMBO J. 15, 4928–4939.

Other names

Lymphocyte chemotactic factor (LCF).

THE MOLECULE

Interleukin 16 (IL-16) stimulates migration by CD4+ T cells, CD4+ monocytes and eosinophils[1,2]. It also induces IL-2 receptor and MHC class II expression on T lymphocytes. IL-16 binds to CD4 and has been shown to inhibit HIV replication. It has also been reported to inhibit CD3-dependent T-cell activation and proliferation[3]. IL-16 cDNA was originally reported to encode a 130-amino-acid protein (M_r 14 000) with no signal peptide. Anomalies in the cDNA sequence highlighted by Bazan and Schall, however, have suggested an ORF encoding a much larger (M_r 40 000) protein, but still lacking a hydrophobic signal peptide[4]. A larger protein of about 80 kDa has been identified in Western blots. More recently, a 2.6-kb mRNA encoding a pro-IL-16 peptide has been identified. The propeptide appears to be proteolytically cleaved to give rise to a 121-amino-acid active protein[5].

Crossreactivity

There is significant crossreactivity between human and mouse IL-16.

Sources

CD8+ T cells constitutively store bioactive protein. CD4+ T cells have IL-16 mRNA and release bioactive protein 18–24 hours after activation by antigen, mitogen or CD3 antibodies[2]. Eosinophils[6], mast cells and epithelial cells from asthmatics[2] also secrete IL-16. Low levels of IL-16 mRNA are present in brain, thymus, PBMCs and spleen.

Bioassays

T-cell chemotaxis.

Physicochemical properties of IL-16

Property	Human	Mouse
pI	9.1 (4.9)[a]	6.15
Amino acids – precursor	631/604[b]	624
– mature	121[c]	?
M_r (K) – predicted	63/67[d] (14)[e]	66.54
– expressed[c]	60/80[f] (17/50–60)[g]	?
Potential *N*-linked glycosylation sites	4/0[h]	?
Disulfide bonds	?	?

[a] 17-kDa protein. Calculated pI for the 121-amino-acid protein in parenthesis.
[b] Propeptide has two potential ATG initiation sites.
[c] Proteolytic cleavage of the propeptide is thought to give rise to a 121-amino-acid active IL-16 protein[5]. Previously active protein was said to be 130 amino acids.
[d] Predicted size of propeptide depending on ATG initiation site.
[e] Predicted size of 121-amino-acid protein in parenthesis.
[f] 60 000 product in PBMC, 80 000 product in COS cells transfected with 2.6 kb mRNA.
[g] The small monomeric form migrates in SDS–PAGE at 17 kDa. Apparent molecular weight of the bioactive tetramer is 50–60 kDa.
[h] Three in propeptide; none in 121-amino-acid product.

3-D structure

Amino acid sequence predicts a structure containing six β-sheets of which the β-2 sheet contains a GLGF sequence which might be important for the autoaggregation of the monomers[2]. Biologically active IL-16 is probably a homotetramer of the smaller 17-kDa peptides.

Gene structure

The gene for IL-16 is on human chromosome 15q26.1 and has a unique gene structure with no similarities to other human cytokine or chemokine genes.

Amino acid sequence for human IL-16 propeptide[5]

Accession code: SwissProt Q14005

1 MDYSFDTTAE DPWVRISDCI KNLFSPIMSE NHGHMPLQP**N** ASLNEEEGTQ
51 GHPDGTPPKL DTA**N**GTPKVY KSADSSTVKK GPPVAPKPAW FRQSLKGLRN
101 RASEPRGLPD PALSTQPAPA SREHLGSHIR ASSSSSSIRQ RISSFETFGS
151 SQLPDKGAQR LSLQPSSGEA AKPLGKHEEG RFSGLLGRGA APTLVPQQPE
201 QVLSSGSPAA SEARDPGVSE SPPPGRQPNQ KTFPPGPDPL LRLLSTQAEE
251 SQGPVLKMPS QRARSFPLTR SQSCETKLLD EKTSKLYSIS SQVSSAVMKS
301 LLCLPSSISC AQTPCIPKAG ASPTSSSNED SAA**N**GSAETS ALDTGFSLNL
351 SELREYTEGL TEAKEDDDGD HSSLQSGQSV ISLLSSEELK KLIEEVKVLD
401 EATLKQLDGI HVTILHKEEG AGLGFSLAGG ADLENKVITV HRVFPNGLAS
451 QEGTIQKGNE VLSINGKSLK GTTHHDALAI LRQAREPRQA VIVTRKLTPE
501 AMPDL**N**SSTD *SAASASAASD VSVESTAEAT VCTVTLEKMS AGLGFSLEGG*
551 *KGSLHGDKPL TINRIFKGAA SEQSETVQPG DEILQLGGTA MQGLTRFEAW*
601 NIIKALPDGP VTIVIRRKSL QSKETTAAGD S

The 121-amino-acid active peptide is shown in italics.

Amino acid sequence for mouse IL-16 propeptide

Accession code: SPTREMBLE O54824

1 MDYSFDITAE DPWVRISDCI KNLFSPIMSE NHSHRPLQPN TSLGEEDGTQ
51 GCPEGGLSKM DAANGAPRVY KSADGSTVKK GPPVAPKPAW FRQSLKGLRN
101 RAPDPRRPPE VASAIQPTPV SRDPPGPQPQ ASSSIRQRIS SFENFGSSQL
151 PDRGVQRLSL QPSSGETTKF PGKQDGGRFS GLLGQGATVT AKHRQTEVES
201 MSTTFPNSSE VRDPGLPESP PPGQRPSTKA LSPDPLLRLL TTQSEDTQGP
251 GLKMPSQRAR SFPLTRTQSC ETKLLDEKAS KLYSISSQLS SAVMKSLLCL
301 PSSVSCGQIT CIPKERVSPK SPCNNSSAAE GFGEAMASDT GFSLNLSELR
351 EYSEGLTEPG ETEDRNHCSS QAGQSVISLL SAEELEKLIE EVRVLDEATL
401 KQLDSIHVTI LHKEEGAGLG FSLAGGADLE NKVITVHRVF PNGLASQEGT
451 IQKGNEVLSI NGKSLKGATH NDALAILRQA RDPRQAVIVT RRTTVEATHD
501 LNSSTDSAAS ASAASDISVE SKEATVCTVT LEKTSAGLGF SLEGGKGSLH
551 GDKPLTINRI FKGDRTGEMV QPGDEILQLA GTAVQGLTRF EAWNVIKALP
601 DGPVTIVIRR TSLQCKQTTA SADS

THE IL-16 RECEPTOR

The receptor for IL-16 is probably CD4. The amino acid sequence, structure and distribution of CD4 is described in the *Leucocyte Antigen FactsBook*[7].

Signal transduction

IL-16 has been reported to stimulate transient inositol trisphosphate production and increases intracellular calcium, activation of PI-3 kinase and translocation of PKC as described for CD4[2,8].

References

[1] Cruikshank, W.W. et al. (1994) Proc. Natl Acad. Sci. USA 91, 5109–5113.
[2] Center, D.M. et al. (1996) Immunol. Today 17, 476–481.
[3] Cruikshank, W.W. et al. (1996) J. Immunol. 157, 5240–5248.
[4] Bazan, J.F. and Schall, T.J. (1996) Nature 381, 29–30.
[5] Baier, M. et al. (1997) Proc. Natl Acad. Sci. USA 94, 5273–5277.
[6] Lim, K.G. et al. (1996) J. Immunol. 156, 2566–2570.
[7] Barclay, A.N. et al. (1994) The Leucocyte Antigen FactsBook, Academic Press, London.
[8] Cruikshank, W.W. et al. (1991) J. Immunol. 146, 2928–2934.

Other names
CTLA-8 (cytotoxic T lymphocyte-associated antigen 8).

THE MOLECULE

Interleukin 17 (IL-17) is a T cell-derived cytokine that does not act directly on haematopoietic cells but stimulates stromal cell elements (epithelial cells, endothelial cells and fibroblasts) to secrete cytokines including IL-6, IL-8, G-CSF and prostaglandin E_2. It also enhances expression of ICAM-1[1-4]. Fibroblasts activated by IL-17 are able to support proliferation of CD34+ haematopoietic progenitors and their preferential differentiation into neutrophils[2]. Murine IL-17 has 57% identity with the predicted protein encoded by Herpesvirus saimiri ORF13[4,5]. The recombinant product of ORF13 (HVS13), also called vIL-17, binds to the IL-17 receptor and stimulates NFκB activity and IL-6 production by fibroblasts[5]. This is another example of a virus-captured gene with functional activity comparable to EBV BCRF1 and IL-10.

Crossreactivity
As far as is known, murine IL-17 does not act on human cells, whereas human IL-17 has activity on mouse cells comparable to murine IL-17. Viral IL-17 (vIL-17) has 57% homology with mouse IL-17 and 68% homology with human IL-17. vIL-17 is active on human and mouse cells.

Sources
IL-17 is produced mainly by CD4+ memory T cells.

Bioassays
Cytokine (IL-6 or IL-8) production by IL-17-activated epithelial cells or fibroblasts.

Physicochemical properties of IL-17

Property	Human	Mouse
PI (calculated)	8.6	8.8
Amino acids – precursor	155	158
– mature[a]	132	133
M_r (K) – predicted	15.1	15.0
– expressed	28, 31[b]	?
Potential *N*-linked glycosylation sites	1[c]	1[c]
Disulfide bonds	?[d]	?[d]

[a] After removal of signal peptide determined by N-terminal sequencing[2].
[b] Corresponding to glycosylated and non-glycosylated homodimers[2].
[c] Conserved in hIL-17, mIL-17 and vIL-17.
[d] There are six conserved Cys residues in human, mouse and viral IL-17.

3-D structure
Not known.

Gene structure[6]

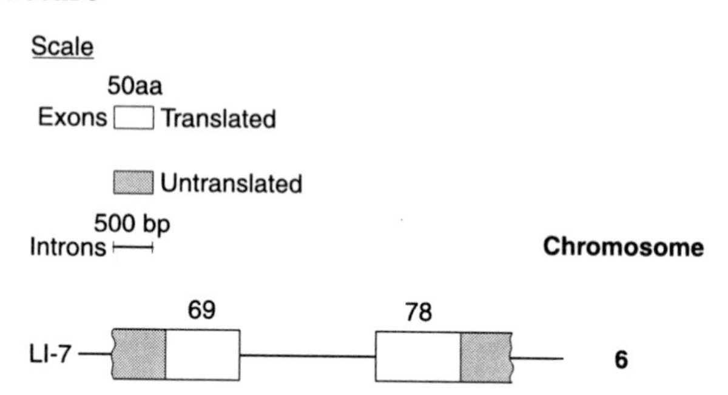

The gene for IL-17 is located on human chromosome 6 and contains two exons.

Amino acid sequence for human IL-17[2,3]

Accession code: SwissProt Q16552

```
-20  MTPGKTSLVS LLLLLSLEAI VKA
  1  GITIPRNPGC PNSEDKNFPR TVMVNLNIHN RNTNTNPKRS SDYYNRSTSP
 51  WNLHRNEDPE RYPSVIWEAK CRHLGCINAD GNVDYHMNSV PIQQEILVLR
101  REPPHCPNSF RLEKILVSVG CTCVTPIVHH VA
```

Amino acid sequence for mouse IL-17[6]

Accession code: SwissProt Q62386

```
-20  MSPGRASSVS LMLLLLLSLA ATVKA
  1  AAIIPQSSAC PNTEAKDFLQ NVKVNLKVFN SLGAKVSSRR PSDYLNRSTS
 51  PWTLHRNEDP DRYPSVIWEA QCRHQRCVNA EGKLDHHMNS VLIQQEILVL
101  KREPESCPFT FRVEKMLVGV GCTCVASIVR QAA
```

Note: Sequence reported originally for murine CTLA-8[4] was obtained from a mouse–rat hybridoma and is actually the sequence for rat CTLA-8. The mouse CTLA-8 sequence shown here is from Yao et al.[6].

THE IL-17 RECEPTOR

The IL-17 receptor has no significant homology with any other protein in either the Genbank or EMBL databases[5]. Murine IL-17 receptor has 12 cysteine residues in the extracellular domain but not in positions characteristic of the Ig-SF or the TNFR-SF. It does not have a WSXWS motif. It has a large cytoplasmic tail (521 amino acids) but no homology with any growth factor receptor known to be a tyrosine kinase. There is a segment (TPPPLRPRKVW) proximal to the transmembrane domain that is highly conserved in cytokine receptors.

Distribution

IL-17 receptor mRNA is widely distributed in murine tissues including spleen, kidney, lung and liver. It is also expressed in epithelial cell, fibroblast, B-cell, muscle, mast cell, and pre-B-cell lines and in triple negative thymic cells. Distribution of human IL-17 receptor is not yet known.

Physicochemical properties of the IL-17 receptor[5]

	Human	Mouse
Amino acids – precursor	866	864
– mature[a]	835	833
M_r (K) – predicted	112	97.8
– expressed	128	120[b]
Potential *N*-linked glycosylation sites[c]	7	8
Affinity K_d (M)[d]	10^7	?

[a] After removal of predicted signal peptide.
[b] Precipitated from transfected cells.
[c] Six sites conserved between human and mouse receptors.
[d] Low-affinity binding to cells transfected with cloned IL-17R. It is possible that high-affinity binding might depend on an as yet unidentified high-affinity converting subunit.

Signal transduction

IL-17 binding to IL-17 receptor on 3T3 cells activates NFκB[5]. The intracellular region has no homology with known receptor tyrosine kinases.

Chromosomal location

The IL-17R is on human chromosome 22 and mouse chromosome 6.

Amino acid sequence for human IL-17 receptor[7]

Accession code: SPTREMBLE O43844

```
-31  MGAARSPPSA VPGPLLGLLL LLLGVLAPGG A
  1  SLRLLDHRAL VCSQPGLNCT VKNSTCLDDS WIHPRNLTPS SPKDLQIQLH
 51  FAHTQQGDLF PVAHIEWTLQ TDASILYLEG AELSVLQLNT NERLCVRFEF
101  LSKLRHHHRR WRFTFSHFVV DPDQEYEVTV HHLPKPIPDG DPNHQSKNFL
151  VPDCEHARMK VTTPCMSSGS LWDPNITVET LEAHQLRVSF TLWNESTHYQ
201  ILLTSFPHME NHSCFEHMHH IPAPRPEEFH QRSNVTLTLR NLKGCCRHQV
251  QIQPFFSSCL NDCLRHSATV SCPEMPDTPE PIPDYMPLWV YWFITGISIL
301  LVGSVILLIV CMTWRLAGPG SEKYSDDTKY TDGLPAADLI PPPLKPRKVW
351  IIYSADHPLY VDVVLKFAQF LLTACGTEVA LDLLEEQAIS EAGVMTWVGR
401  QKQEMVESNS KIIVLCSRGT RAKWQALLGR GAPVRLRCDH GKPVGDLFTA
451  AMNMILPDFK RPACFGTYVV CYFSEVSCDG DVPDLFGAAP RYPLMDRFEE
501  VYFRIQDLEM FQPGRMHRVG ELSGDNYLRS PGGRQLRAAL DRFRDWQVRC
551  PDWFECENLY SADDQDAPSL DEEVFEEPLL PPGTGIVKRA PLVREPGSQA
601  CLAIDPLVGE EGGAAVAKLE PHLQPRGQPA PQPLHTLVLA AEEGALVAAV
651  EPGPLADGAA VRLALAGEGE ACPLLGSPGA GRNSVLFLPV DPEDSPLGSS
701  TPMASPDLLP EDVREHLEGL MLSLFEQSLS CQAQGGCSRP AMVLTDPHTP
751  YEEEQRQSVQ SDQGYISRSS PQPPEGLTEM EEEEEEEQDP GKPALPLSPE
801  DLESLRSLQR QLLFRQLQKN SGWDTMGSES EGPSA
```

Amino acid sequence for mouse IL-17 receptor[5]

Accession code: SPTREMBLE Q60943

```
-31  MAIRRCWPRV VPGPALGWLL LLLNVLAPGR A
  1  SPRLLDFPAP VCAQEGLSCR VKNSTCLDDW IHPKNLTPSS PKNIYINLSS
 51  VSSTQHGELV PVLHVEWTLQ TDASILYLEG AELSVLQLNT NERLCVKFQF
101  LSMLQHHRKR WRFSFSHFVV DPGQEYEVTV HHLPKPIPDG DPNHKSKIIF
151  VPDCEDSKMK MTTSCVSSGS LWDPNITVET LDTQHLRVDF TLWNESTPYQ
201  VLLESFSDSE NHSCFDVVKQ IFAPRQEEFH QRANVTFTLS KFHWCCHHHV
251  QVQPFFSSCL NDCLRHAVTV PCPVISNTTV PKPVADYIPL WVYGLITLIA
301  ILLVGSVIVL IICMTWRLSG ADQEKHGDDS KINGILPVAD LTPPPLRPRK
351  VWIVYSADHP LYVEVVLKFA QFLITACGTE VALDLLEEQV ISEVGVMTWV
401  SRQKQEMVES NSKIIILCSR GTQAKWKAIL GWAEPAVQLR CDHWKPAGDL
451  FTAAMNMILP DFKRPACFGT YVVCYFSGIC SERDVPDLFN ITSRYPLMDR
501  FEEVYFRIQD LEMFEPGRMH HVRELTGDNY LQSPSGRQLK EAVLRFQEWQ
551  TQCPDWFERE NLCLADGQDL PSLDEEVFED PLLPPGGGIV KQQPLVRELP
601  SDGCLVVDVC VSEEESRMAK LDPQLWPQRE LVAHTLQSMV LPAEQVPAAH
651  VVEPLHLPDG SGAAAQLPMT EDSEACPLLG VQRNSILCLP VDSDDLPLCS
701  TPMMSPDHLQ GDAREQLESL MLSVLQQSLS GQPLESWPRP EVVLEGCTPS
751  EEEQRQSVQS DQGYISRSSP QPPEWLTEEE ELELGEPVES LSPEELRSLR
801  KLQRQLFFWE LEKNPGWNSL EPRRPTPEEQ NPS
```

References

[1] Broxmeyer, H.E. (1996) J. Exp. Med. 183, 2411–2415.
[2] Fossiez, F. et al. (1996) J. Exp. Med. 183, 2593–2603.
[3] Yoo, Z. et al. (1995) J. Immunol. 155, 5483–5486.
[4] Rouvier, E. et al. (1993) J. Immunol. 150, 5445–5456.
[5] Yao, Z. et al. (1995) Immunity 3, 811–821.
[6] Yao, Z. et al. (1996) Gene 168, 223–225.
[7] Yao, Z. et al. (1997) Cytokine 9, 794–800.

Other names

Interferon gamma-inducing factor (IGIF), IL-1γ.

THE MOLECULE

IL-18 induces IFNγ production by mitogen (ConA) or CD3-activated T cells and enhances NK cell cytotoxicity by increasing Fas ligand expression. IL-18 also increases production of GM-CSF and several chemokines but has no effect or inhibits IL-4 production and promotes production of Th1 cells[1,2]. Its biological properties are rather like IL-12 and IL-18 and it acts in synergy with IL-12 by a mechanism involving induction of IL-18 receptors by IL-12[3]. IL-18 inhibits osteoclast formation by inducing GM-CSF[4]. It has no sequence homologies with any other cytokine but has a predicted tertiary structure similar to that of IL-1. Mature IL-18 is obtained by caspase-1 (ICE) cleavage of precursor molecule[5,6].

Crossreactivity

There is 65% amino acid sequence homology between mouse and human IL-18 and about 1% cross-species reactivity.

Sources

Mouse IL-18 is produced by Kuppfer cells after intraperitoneal injection of *Propionibacterium acnes*. Cellular source in humans has not yet been identified. IL-18 mRNA is expressed in human monocyte and macrophage cell lines. IL-18 is also produced by keratinocytes[7] and osteoblast stromal cells[5].

Bioassays

IFNγ production by nonadherent spleen cells suboptimally stimulated with ConA.

Physicochemical properties of IL-18

Property	Human	Mouse
pI	4.9	4.9
Amino acids – precursor	193	192
– mature[a]	157	157
M_r (K) – predicted	22.3	22.1
– expressed[a]	18.2	18.3[b]
Potential *N*-linked glycosylation sites	0	0
Disulfide bonds[c]	?	?

[a] IL-18 has no hydrophobic signal peptide-like sequences. Mature protein released after cleavage of precursor by caspase-1/IL-1-converting enzyme (ICE)[5,6].
[b] Purified from mouse liver extracts after *in vivo* stimulation with heat-killed *P. acnes*[1].
[c] Predicted mature protein has four (mouse) or three (human) Cys residues but biochemical evidence suggests that there are no disulfide bonds[1].

3-D structure

IL-18 is predicted to have a trefoil structure similar to that of IL-1 and FGF[8].

Gene structure

The gene for human IL-18 has five exons and six introns. Chromosomal location is not known. No information for mouse IL-18 is available.

Amino acid sequence for human IL-18[1]

Accession code: SwissProt Q14116

```
  1  MAAEPVEDNC INFVAMKFID NTLYFIAEDD ENLESDYFGK LESKLSVIRN
 51  LNDQVLFIDQ GNRPLFEDMT DSDCRDNAPR TIFIISMYKD SQPRGMAVTI
101  SVKCEKISTL SCENKIISFK EMNPPDNIKD TKSDIIFFQR SVPGHDNKMQ
151  FESSSYEGYF LACEKERDLF KLILKKEDEL GDRSIMFTVQ NED
```

Human IL-18 has no signal peptide. The propeptide (1–36) in italics is removed by caspase-1/ICE enzyme[4].

Amino acid sequence for mouse IL-18[9]

Accession code: SwissProt P70380

```
  1  MAAMSEDSCV NFKEMMFIDN TLYFIPEENG DLESDNFGRL HCTTAVIRNI
 51  NDQVLFVDKR QPVFEDMTDI DQSASEPQTR LIIYMYKDSE VRGLAVTLSV
101  KDSKMSTLSC KNKIISFEEM DPPENIDDIQ SDLIFFQKRV PGHNKMEFES
151  SLYEGHFLAC QKEDDAFKLI LKKKDENGDK SVMFTLTNLH QS
```

Mouse IL-18 has no signal peptide. The propeptide (1–35) shown in italics is removed by caspase-1/ICE enzyme[5].

THE IL-18 RECEPTOR

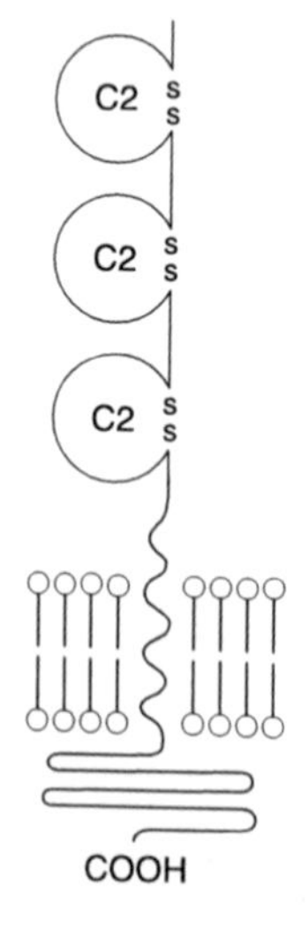

The IL-1R-related protein (IL-1Rrp), which was recently described as a member of IL-1R family but with no known ligand, has now been shown to be a receptor for IL-18[10,11]. It has three extracellular C2 Ig-like domains with some unusual features[11]. Other members of this family are the IL-1Rs, T1/ST2 and IL-1R accessory protein (IL-1R AcP) (see IL-1 entry (page **35**)). In addition to IL-18R, a second chain is required for signalling, termed AcPL (accessory protein-like), because of its homology with IL-1R AcP[12].

Distribution

IL-18R mRNA is present in spleen, thymus, leukocytes, liver, heart, lung, intestine, prostate and placenta but not in brain, skeletal muscle, kidney or pancreas. There is also widespread expression in haematopoietic cell lines and particularly high expression in Hodgkin's lymphoma cell lines.

Physicochemical properties of the IL-18 receptor

	Human	Mouse
Amino acids – precursor	541	537
– mature[a]	522	519
M_r (K) – predicted	60.2	59.6
– expressed	60–100	?
Potential *N*-linked glycosylation sites	8	10
Affinity K_d (M)	1.9×10^{-10}	?

[a] After removal of predicted signal peptide.

Signal transduction

IL-18 activates signalling pathways similar to IL-1[13]. Signals include the transcription factor NFκB, which regulates most of the IL-18-responsive genes, and activation of MAP kinases such as Jun N-terminal kinase[13,14]. IL-18 signalling requires both IL-18R and AcPL, which both contain Toll/IL-1 receptor (TIR) domains (placing them in the IL-1/Toll-like receptor family: see Chapter 4). The TIR domain-containing adapter MyD88 is recruited to this complex and leads to activation of the IL-1 receptor-associated kinase (IRAK) and the adapter Traf-6, in a manner analogous to IL-1 signalling. Mice deficient in MyD88 or IRAK do not respond to IL-18[15–17].

Chromosomal location

Human IL-18R (IL-1Rrp) is on chromosome 2q12–22.

Amino acid sequence for human IL-18 receptor[11]

Accession code: SPTREMBLE Q13478

```
-29  MNCRELPLTL WVLISVSTA
  1  ESCTSRPHIT VVEGEPFYLK HCSCSLAHEI ETTTKSWYKS SGSQEHVELN
 51  PRSSSRIALH DCVLEFWPVE LNDTGSYFFQ MKNYTQKWKL NVIRRNKHSC
101  FTERQVTSKI VEVKKFFQIT CENSYYQTLV NSTSLYKNCK KLLLENNKNP
151  TIKKNAEFED QGYYSCVHFL HHNGKLFNIT KTFNITIVED RSNIVPVLLG
201  PKLNHVAVEL GKNVRLNCSA LLNEEDVIYW MFGEENGSDP NIHEEKEMRI
251  MTPEGKWHAS KVLRIENIGE SNLNVLYNCT VASTGGTDTK SFILVRKADM
301  ADIPGHVFTR GMIIAVLILV AVVCLVTVCV IYRVDLVLFY RHLTRRDETL
351  TDGKTYDAFV SYLKECRPEN GEEHTFAVEI LPRVLEKHFG YKLCIFERDV
401  VPGGAVVDEI HSLIEKSRRL IIVLSKSYMS NEVRYELESG LHEALVERKI
451  KIILIEFTPV TDFTFLPQSL KLLKSHRVLK WKADKSLSYN SRFWKNLLYL
501  MPAKTVKPGR DEPEVLPVLS ES
```

Amino acid sequence for mouse IL-18 receptor[11]

Accession code: SPTREMBLE Q61098

-18 MHHEELILTL CILIVKSA
1 SKSCIHRSQI HVVEGEPFYL KPCGISAPVH R**N**ETATMRWF KGSASHEYRE
51 LN**N**RSSPRVT FHDHTLEFWP VEMEDEGTYI SQVGNDRR**N**W TL**N**VTKRNKH
101 SCFSDKLVTS RDVEV**N**KSLH ITCKNPNYEE LIQDTWLYKN CKEISKTPRI
151 LKDAEFGDEG YYSCVFSVHH **N**GTRY**N**ITKT V**N**ITVIEGRS KVTPAILGPK
201 CEKVGVELGK DVEL**N**CSASL NKDDLFYWSI RKEDSSDPNV QEDRKETTTW
251 ISEGKLHASK ILRFQKITEN YLNVLY**N**CTV ANEEAIDTKS FVLVRKEIPD
301 IPGHVFTGGV TVLVLASVAA VCIVILCVIY KVDLVLFYRR IAERDETLTD
351 GKTYDAFVSY LKECHPENKE EYTFAVETLP RVLEKQFGYK LCIFERDVVP
401 GGAVVEEIHS LIEKSRRLII VLSQSYLTNG ARRELESGLH EALVERKIKI
451 ILIEFTPASN ITFLPPSLKL LKSYRVLKWR ADSPSMNSRF WKNLVYLMPA
501 KAVKPWREES EARSVLSAP

References

[1] Ushio, S. et al. (1996) J. Immunol. 156, 4274–4279.
[2] Micallef, M.J. et al. (1996) Eur. J. Immunol. 26, 1647–1651.
[3] Okamura, H. et al. (1998) Adv. Immunol. 70, 281–312.
[4] Udagawa, N. et al. (1997) J. Exp. Med. 185, 1005–1012.
[5] Akita, K. et al. (1997) J. Biol. Chem. 272, 26595–26603.
[6] Gu, Y. et al. (1997) Science 275, 206–209.
[7] Stoll, S. et al. (1997) J. Immunol. 159, 298–302.
[8] Bazan, J.F. et al. (1996) Nature 379, 591.
[9] Okamura, H. et al. (1995) Nature 378, 88–91.
[10] Torigoe, K. et al. (1997) J. Biol. Chem. 272, 25737–25742.
[11] Parnet, P. et al. (1997) J. Biol. Chem. 271, 3967–3970.
[12] Born, T.L. et al. (1998) J. Biol. Chem. 273, 29445–29450.
[13] Thomassen, E. et al. (1998) Int. J. Cytokine Res. (1998) 18, 1077–1088.
[14] Matsumoto, S. et al. (1997) Biochem. Biophys. Res. Commun. 234, 454–457.
[15] Kojima, H. et al. (1998) Biochem. Biophys. Res. Commun. 244, 183–186.
[16] Adachi, O. et al. (1998) Immunity 9, 143–150.
[17] Kanakaraj, P. et al. (1999) J. Exp. Med. 189, 1129–1138.

Other Cytokines and Chemokines (in alphabetical order)

Other names
Erythroid differentiating factor (EDF), FSH-releasing protein (FRP), also called inhibin.

THE MOLECULE

Activins represent a family of TGFβ-like proteins which were initially discovered for their actions on the reproductive system but have subsequently been shown to play a role in a variety of developmental pathways. Activin is an endocrine hormone synthesized as a homo- or heterodimer of activin β-subunits[1] (also known as inhibin β-subunits as they dimerize with the inhibin α-subunit to form inhibin – which is antagonistic to activin). Five β-subunits have been identified thus far, β_A–β_E. Activin was first identified as a stimulator of pituitary FSH release[2]; inhibin, on the other hand, is an inhibitor in this regard. Subsequently, wide-ranging roles for activin in development and differentiation have been proposed, including embryonic axial development, photoreceptor development[3] and differentiation of human and mouse erythroleukaemic cell lines[4]. Overexpression of the activin β_A-subunit is also observed during the inflammatory process following cutaneous injury[5]. For the sake of brevity we will only discuss activin β_A-subunits here.

Crossreactivity
The mature C-terminal of the mouse activin subunit β_C shares 94% identity with its human homologue[6].

Sources
Expression of activin is principally found in the reproductive tissues, with low levels found in most areas of the brain, with the exception of the pituitary where they are high[7]. Expression is also observed in developing lung, kidney[8] and mesoderm[9].

Bioassays
Activin induces the accumulation of haemoglobin in human/mouse erythroleukaemic cell lines (K562 bioassay)[10]; FSH pituitary bioassay (FSH release stimulated)[11].

Physicochemical properties of activin β_A-chain

Property	Human	Mouse
pI[a]	7.07	7.07
Amino acids – precursor	426	424
– mature[b]	115	115
M_r (K) – predicted	12.9	12.9
– expressed[c]	15.5	?
Potential *N*-linked glycosylation sites	1	1
Disulfide bonds	5	5

[a] Predicted pI of mature processed protein.
[b] After removal of predicted signal sequence and propeptide.
[c] Activin is a homodimer of 25 kDa, the molecular weight of the monomer is 15.5 kDa.

3-D structure

The structure of activin is currently unknown.

Gene structure[12]

Inhibin/activin β_A-subunit is localized to human chromosome 7p15–p13.

Amino acid sequence for human activin β_A-subunit

Accession code: SwissProt P08476

-28 MPLLWLRGFL LASCWIIVRS *SPTPGSEG*
1 *HSAAPDCPSC ALAALPKDVP NSQPEMVEAV KKHILNMLHL KKRPDVTQPV*
51 *PKAALLNAIR KLHVGKVGEN GYVEIEDDIG RRAEMNELME QTSEIITFAE*
101 *SGTARKTLHF EISKEGSDLS VVERAEVWLF LKVPKA**N**RTR TKVTIRLFQQ*
151 *QKHPQGSLDT GEEAEEVGLK GERSELLLSE KVVDARKSTW HVFPVSSSIQ*
201 *RLLDQGKSSL DVRIACEQCQ ESGASLVLLG KKKKKEEEGE GKKKGGGEGG*
251 AGADEEKEQS HRPFLMLQAR QSEDHPHRRR RRGLECDGKV NICCKKQFFV
301 *SFKDIGWNDW IIAPSGYHAN YCEGECPS*HI AGTSGSSLSF HSTVINHYRM
351 RGHSPFANLK SCCVPTKLRP MSMLYYDDGQ NIIKKDIQNM IVEECGCS

Propeptide 29–310 is removed to form the mature protein (shown in italics). Conflicting sequence RMR→AC at position 377–379 of unprocessed protein. Disulfide bridges between Cys at position 314–322, 321–391, 350–423 and 354–425 by similarity.

Amino acid sequence for mouse activin β_A-subunit

Accession code: SwissProt Q04998

-20 MPLLWLRGFL LASCWIIVRS
1 *SPTPGSEGHG SAPDCPSCAL ATLPKDGPNS QPEMVEAVKK HILNMLHLKK*
51 *RPDVTQPVPK AALLNAIRKL HVGKVGENGY VEIEDDIGRR AEMNELMEQT*
101 *SEIITFAESG TARKTLHFEI SKEGSDLSVV ERAEVWLFLK VPKANRTRTK*
151 *VTIRLFQQQK HPQGSLDTGD EAEEMGLKGE RSELLLSEKV VDARKSTWHI*
201 *FPVSSSIQRL LDQGKSSLDV RIACEQCQES GASLVLLGKK KKKEVDGDGK*
251 *KKDGSDGGLE EEKEQSHRPF LMLQARQSED HPHRRRRRGL* ECDGKVNICC
301 KKQFFVSFKD IGWNDWIIAP SGYHANYCEG ECPSHIAGTS GSSLSFHSTV
351 INHYRMRGHS PFANLKSCCV PTKLRPMSML YYDDGQNIIK KDIQNMIVEE
401 CGCS

The propeptide sequence is removed to form the mature protein (shown in italics). Disulfide bridges between Cys at positions 312–320, 319–389, 348–421 and 388–388 by similarity.

THE ACTIVIN RECEPTOR

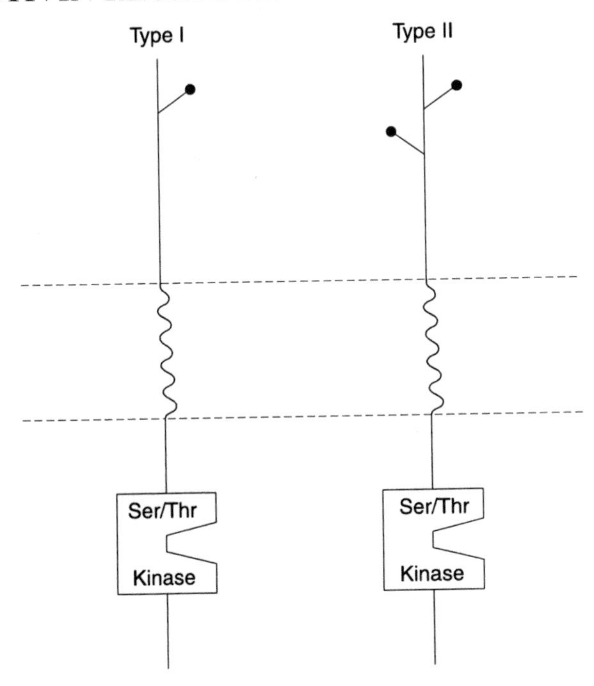

The receptor for activin represents a novel subclass of cell surface receptors with serine/threonine kinase activity. Activin binds both type I (50–55 kDa) and type II (70–75 kDa) receptors[13], both of which are members of the receptor serine/threonine kinase family[14], along with TGFβ and BMP. There are two type I receptors (ALK2 and ALK4)[14] and at least three type II receptors, type II, type IIB and type IIA-N[15]. Type I and II receptors form a stable complex after ligand binding, with the type I receptor being phosphorylated through the kinase activity of the type II receptor[14,16].

Distribution

Activin type I receptors are expressed in BMP-induced bone-forming tissues[17]. Type II receptors are expressed in pituitary adenomas and the breast cancer cell line MCF-7[18]. Activin type II and IIB receptor mRNAs have been reported in developing neural, muscular and exocrine glandular organs of human fetuses[19] and in the human placenta[20].

Physicochemical properties of activin receptors

	Type I		Type II		Type IIB	
	Human	Mouse	Human	Mouse	Human	Mouse
Amino acids – precursor	509	509	513	513	512	536
– mature[a]	489	489	497	497	492	518
M_r (K) – predicted	55.07	55.13	56.02	56.06	55.80	58.7
– expressed	?	?	?	?	?	?
Potential *N*-linked glycosylation sites	1	1	2	2	2	2
Affinity K_d (M)	?	?	3.6×10^{14}	?	?	?

[a] After removal of predicted signal peptide.

Signal transduction

Type I and II receptors form a stable complex after ligand binding, with the type I receptor being phosphorylated through the kinase activity of the type II[14,16]. There is some suggestion that the type II receptor acts as the primary activin receptor and the type I receptor as a downstream transducer of activin signals[16]. Downstream components of the pathway include Smad 2 and Smad 3, which associate with the activin receptor(s) and subsequently with Smad 4; Smad 7 inhibits at the level of Smad 2 and Smad 3[21].

Chromosomal location

The human activin receptor type II and IIB genes have been assigned to chromosome bands 2q22.2–q23.3 and 3p22[22].

Amino acid sequence for human activin receptor type I

Accession code: SwissProt Q04771

```
-20  MVDGVMILPV LIMIALPSPS
  1  MEDEKPKVNP KLYMCVCEGL SCGNEDHCEG QQCFSSLSIN DGFHVYQKGC
 51  FQVYEQGKMT CKTPPSPGQA VECCQGDWCN RNITAQLPTK GKSFPGTQNF
101  HLEVGLIILS VVFAVCLLAC LLGVALRKFK RRNQERLNPR DVEYGTIEGL
151  ITTNVGDSTL ADLLDHSCTS GSGSGLPFLV QRTVARQITL LECVGKGRYG
201  EVWRGSWQGE NVAVKIFSSR DEKSWFRETE LYNTVMLRHE NILGFIASDM
251  TSRHSSTQLW LITHYHEMGS LYDYLQLTTL DTVSCLRIVL SIASGLAHLH
301  IEIFGTQGKP AIAHRDLKSK NILVKKNGQC CIADLGLAVM HSQSTNQLDV
351  GNNPRVGTKR YMAPEVLDET IQVDCFDSYK RVDIWAFGLV LWEVARRMVS
401  NGIVEDYKPP FYDVVPNDPS FEDMRKVVCV DQQRPNIPNR WFSDPTLTSL
451  AKLMKECWYQ NPSARLTALR IKKTLTKIDN SLDKLKTDC
```

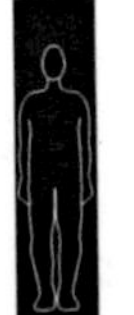

Amino acid sequence for human activin receptor type II

Accession code: SwissProt P27037

-19 MGAAAKLAFA VFLISCSSG
1 AILGRSETQE CLFFNANWEK DRT**N**QTGVEP CYGDKDKRRH CFATWK**N**ISG
51 SIEIVKQGCW LDDINCYDRT DCVEKKDSPE VYFCCCEGNM CNEKFSYFPE
101 MEVTQPTSNP VTPKPPYYNI LLYSLVPLML IAGIVICAFW VYRHHKMAYP
151 PVLVPTQDPG PPPPSPLLGL KPLQLLEVKA RGRFGCVWKA QLLNEYVAVK
201 IFPIQDKQSW QNEYEVYSLP GMKHENILQF IGAEKRGTSV DVDLWLITAF
251 HEKGSLSDFL KANVVSWNEL CHIAETMARG LAYLHEDIPG LKDGHKPAIS
301 HRDIKSKNVL LKNNLTACIA DFGLALKFEA GKSAGDTHGQ VGTRRYMAPE
351 VLEGAINFQR DAFLRIDMYA MGLVLWELAS RCTAADGPVD EYMLPFEEEI
401 GQHPSLEDMQ EVVVHKKKRP VLRDYWQKHA GMAMLCETIE ECWDHDAEAR
451 LSAGCVGERI TQMQRLTNII TTEDIVTVVT MVTNVDFPPK ESSL

Conflicting sequence L→V at position 13, GCV→PSL at position 204–206 and E→V at position 348.

Amino acid sequence for human activin receptor type IIB

Accession code: SwissProt Q13705

-18 MTAPWVALAL LWGSLWPG
1 SGRGEAETRE CIYYNANWEL ERT**N**QSGLER CEGEQDKRLH CYASWA**N**SSG
51 TIELVKKGCW LDDFNCYDRQ ECVATEENPQ VYFCCCEGNF CNERFTHLPE
101 AGGPEVTYEP PPTAPTLLTV LAYSLLPIGG LSLIVLLAFW MYRHRKPPYG
151 HVDIHEDPGP PPPSPLVGLK PLQLLEIKAR GRFGCVWKAQ LMNDFVAVKI
201 FPLQDKQSWQ SEREIFSTPG MKHENLLQFI AAEKRGSNLE VELWLITAFH
251 DKGSLTDYLK GNIITWNELC HVAETMSRGL SYLHEDVPWC RGEGHKPSIA
301 HRDFKSKNVL LKSDLTAVLA DFGLAVRFEP GKPPGDTHGQ VGTRRYMAPE
351 VLEGAINFQR DAFLRIDMYA MGLVLWELVS RCKAADGPVD EYMLPFEEEI
401 GQHPSLEELQ EVVVHKKMRP TIKDHWLKHP GLAQLCVTIE ECWDHDAEAR
451 LSAGCVEERV SLIRRSVNGT TSDCLVSLVT SVTNVDLPPK ESSI

Amino acid sequence for mouse activin receptor type I

Accession code: SwissProt P37172

-20 MVDGVMILPV LMMMAFPSPS
1 VEDEKPKVNQ KLYMCVCEGL SCGNEDHCEG QQCFSSLSIY DGFHVYQKGC
51 FQVYEQGKMT CKTPPSPGQA VECCQGDWCN R**N**ITAQLPTK GKSFPGTQNF
101 HLEVGLIILS VVFAVCLLAC ILGVALRKFK RRNQERLNPR DVEYGTIEGL
151 ITTNVGDSTL AELLDHSCTS GSGSGLPFLV QRTVARQITL LECVGKGRYG
201 EVWRGSWQGE NVAVKIFSSR DEKSWFRETE LYNTVMLRHE NILGFIASDM
251 TSRHSSTQLW LITHYHEMGS LYDYLQLTTL DTVSCLRIVL SIASGLAHLH
301 IEIFGTQGKS AIAHRDLKSK NILVKKNGQC CIADLGLAVM HSQSTNQLDV
351 GNNPRVGTKR YMAPEVLDET IQVDCFDSYK RVDIWAFGLV LWEVARRMVS
401 NGIVEDYKPP FYDVVPNDPS FEDMRKVVCV DQQRPNIPNR WFSDPTLTSL
451 AKLMKECWYQ NPSARLTALR IKKTLTKIDN SLDKLKTDC

Amino acid sequence for mouse activin receptor type II

Accession code: SwissProt P27038

-19 MGAAAKLAFA VFLISCSSG
1 AILGRSETQE CLFFNANWER DRT**N**QTGVEP CYGDKDKRRH CFATWK**N**ISG
51 SIEIVKQGCW LDDINCYDRT DCIEKKDSPE VYFCCCEGNM CNEKFSYFPE
101 MEVTQPTSNP VTPKPPYYNI LLYSLVPLML IAGIVICAFW VYRHHKMAYP
151 PVLVPTQDPG PPPPSPLLGL KPLQLLEVKA RGRFGCVWKA QLLNEYVAVK
201 IFPIQDKQSW QNEYEVYSLP GMKHENILQF IGAEKRGTSV DVDLWLITAF
251 HEKGSLSDFL KANVVSWNEL CHIAETMARG LAYLHEDIPG LKDGHKPAIS
301 HRDIKSKNVL LKNNLTACIA DFGLALKFEA GKSAGDTHGQ VGTRRYMAPE
351 VLEGAINFQR DAFLRIDMYA MGLVLWELAS RCTAADGPVD EYMLPFEEEI
401 GQHPSLEDMQ EVVVHKKKRP VLRDYWQKHA GMAMLCETIE ECWDHDAEAR
451 LSAGCVGERI TQMQRLTNII TTEDIVTVVT MVTNVDFPPK ESSL

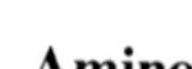

Amino acid sequence for mouse activin receptor type II

Accession code: SwissProt P27040

-18 MTAPWAALAL LWGSLCAG
1 SGRGEAETRE CIYYNANWEL ERT**N**QSGLER CEGEQDKRLH CYASWR**N**SSG
51 TIELVKKGCW LDDFNCYDRQ ECVATEENPQ VYFCCCEGNF CNERFTHLPE
101 PGGPEVTYEP PPTAPTLLTV LAYSLLPIGG LSLIVLLAFW MYRHRKPPYG
151 HVDIHEVRQC QRWAGRRDGC ADSFKPLPFQ DPGPPPPSPL VGLKPLQLLE
201 IKARGRFGCV WKAQLMNDFV AVKIFPLQDK QSWQSEREIF STPGMKHENL
251 LQFIAAEKRG SNLEVELWLI TAFHDKGSLT DYLKGNIITW NELCHVAETM
301 SRGLSYLHED VPWCRGEGHK PSIAHRDFKS KNVLLKSDLT AVLADFGLAV
351 RFEPGKPPGD THGQVGTRRY MAPEVLEGAI NFQRDAFLRI DMYAMGLVLW
401 ELVSRCKAAD GPVDEYMLPF EEEIGQHPSL EELQEVVVHK KMRPTIKDHW
451 LKHPGLAQLC VTIEECWDHD AEARLSAGCV EERVSLIRRS VNGTTSDCLV
501 SLVTSVTNVD LLPKESSI

Residues 124–131 from pro-sequence are missing in isoforms IIB3 and IIB4, while residues 175–198 are missing in isoforms IIB2 and IIB4.

References

[1] Schwall, R.H. et al. (1988) Mol. Endocrinol. 2, 1237–1242.
[2] Yu, J. et al. (1987) Nature 330, 765–767.
[3] Davis, A.A. et al. (2000) Mol. Cell. Neurosci. 15, 11–21.
[4] Murata, M. et al. (1988) Biochem. Biophys. Res. Commun. 151, 230–235.
[5] Hubner, G. et al. (1997) Lab. Invest. 77, 311–318.
[6] Schmitt, J. et al. (1996) Genomics 32, 358–366.
[7] Shintani, Y. et al. (1991) J. Immunol. Methods 137, 267–274.
[8] Ritvos, O. et al. (1995) Mech. Dev. 50, 229–245.
[9] Thomsen, G. et al. (1990) Cell 63, 485–493.
[10] Schwall, R.H. and Lai, C. (1991) Methods Enzymol. 198, 340–346.
[11] Balen, A.H. et al. (1995) In Vitro Cell Dev. Biol. Anim. 31, 316–322.
[12] Mason, A.J. et al. (1986) Biochem. Biophys. Res. Commun. 135, 957–964.
[13] Mathews, L.S. and Vale, W.W. (1993) J. Biol. Chem. 268, 19013–19018.

[14] Zimmerman, C.M. and Mathews, L.S. (1996) Biochem. Soc. Symp. 62, 25–38.
[15] Shoji, H. et al. (1998) Biochem. Biophys. Res. Commun. 246, 320–324.
[16] Attisano, L. et al. (1996) J. Mol Cell Biol. 16, 1066–1073.
[17] Takeda, K. (1994) Kokubyo Gakkai Zasshi 61, 512–526.
[18] Ying, S.Y. and Zhang, Z. (1996) Breast Cancer Res. Treat. 37, 151–160.
[19] Tuuri, T. et al. (1994) J. Clin. Endocrinol. Metab. 78, 1521–1524.
[20] Shinozaki, H. et al. (1995) Life Sci. 56, 1699–1706.
[21] Lebrun, J.J. et al. (1999) Mol. Endocrinol. 13, 15–23.
[22] Bondestam, J. et al. (1999) Cytogenet. Cell Genet. 87, 219–220.

Other names

AR, colorectum cell-derived growth factor (CRDGF), schwannoma-derived growth factor (SDGF).

THE MOLECULE

Amphiregulin is a 23-kDa bifunctional growth-modulating glycoprotein, which inhibits the growth of several human carcinoma cell lines in culture, while stimulating proliferation in some tumour cells[1]. Amphiregulin is related to EGF and mediates its biological action via the EGF receptor[2], and has been shown to increase EGFR tyrosine phosphorylation on binding to the receptor[3]. Amphiregulin also contains two putative nuclear localization sequences, and may play a role in nuclear growth regulation[4]. Amphiregulin contains cysteine residues in disulfide linkages which are essential for its biological activity. The molecule may also contain lipid and/or oligosaccharide moieties that are nonessential for activity.

Crossreactivity

Not known.

Sources

Expression of amphiregulin has been reported in ovarian carcinoma cell lines[4], psoriatic lesions, stomach and colon carcinomas[5], human pancreatic cell lines[6], the placenta[7] and in human mammary epithelial cell lines, being expressed to a greater degree in invasive breast carcinomas[2].

Bioassays

Stimulates DNA synthesis in MCF-10A human mammary epithelial cells[8].

Physicochemical properties of amphiregulin

Property	Human	Mouse
pI[a]	9.53[b]	9.65
Amino acids – precursor	252	248
– mature	83	148
M_r (K) – predicted	9.77	17.07
– expressed	23	?
Potential *N*-linked glycosylation sites	3	3
Disulfide bonds	3	2

[a] Predicted.
[b] Reported as approximately 7.8[9].

3-D structure

Not known.

Gene structure

The human amphiregulin gene is localized to chromosome 4q13–4q21 (a common breakpoint for acute lymphoblastic leukaemia)[10].

Amino acid sequence for human amphiregulin[10]

Accession code: SwissProt P15514

```
-19  MRAPLLPPAP VVLSLLILG
  1  SGHYAAGLDL NDTYSGKREP FSGDHSADGF EVTSRSEMSS GSEISPVSEM
 51  PSSSEPSSGA DYDYSEEYDN EPQIPGYIVD DSVRVEQVVK PPQNKTESEN
101  TSDKPKRKKK GGKNGKNRRN RKKKNPCNAE FQNFCIHGEC KYIEHLEAVT
151  CKCQQEYFGE RCGEKSMKTH SMIDSSLSKI ALAAIAAFMS AVILTAVAVI
201  TVQLRRQYVR KYEGEAEERK KLRQENGNVH AIA
```

The transmembrane region is underlined. The propeptide sequence (1–81) is shown in italics and is removed to generate the mature protein. Disulfide bonds between Cys at positions 146–159, 154–170 and 172–181 by similarity.

Amino acid sequence for mouse amphiregulin

Accession code: SwissProt P31955

```
  1  MRTPLLPLAR SVLLLLVLGS GHYAAALELN DPSSGKGESL SGDHSAGGLE
 51  LSVGREVSTI SEMPSGSELS TGDYDYSEEY DNEPQISGYI IDDSVRVEQV
101  IKPKKNKTEG EKSTEKPKRK KKGGKNGKGR RNKKKKNPCT AKFQNFCIHG
151  ECRYIENLEV VTCNCHQDYF GERCGEKSMK THSEDDKDLS KIAVVAVTIF
201  VSAIILAAIG IGIVITVHLW KRYFREYEGE TEERRRLRQE NGTVHAIA
```

The transmembrane region is underlined. The precise start of the propeptide sequence is unclear but it terminates at position 99 as shown in italics and is removed to generate the mature protein. A potential signal sequence is located upstream of the propeptide sequence. Disulfide bonds between Cys at positions 146–159, 154–170 and 172–181, 139–152, 147–163 and 165–174 by similarity.

THE AMPHIREGULIN RECEPTOR

Amphiregulin binds to the EGF receptor (see EGF entry (page **203**)).

Chromosomal location

No information.

References

[1] Modrell, B. et al. (1992) Growth Factors 7, 305–314.
[2] Salomon, D.S. et al. (1995) Breast Cancer Res. Treat. 33, 103–114.
[3] Johnson, G.R. et al. (1993) J. Biol. Chem. 268, 2924–2931.
[4] Johnson, G.R. et al. (1991) Biochem. Biophys. Res. Commun. 180, 481–488.
[5] Cook, P.W. et al. (1992) Cancer Res. 52, 3224–3227.
[6] Ebert, M. et al. (1994) Cancer Res. 54, 3959–3962.

[7] Lysiak, J.J. et al. (1995) Placenta 16, 359–366.
[8] Thompson, S.A. et al. (1996) J. Biol. Chem. 271, 17927–17931.
[9] Shoyab, M. et al. (1988) Proc. Natl Acad. Sci. USA 85, 6528–6532.
[10] Plowman, G.D. et al. (1990) Mol. Cell Biol. 10, 1969–1981.

Other names

None, but angiostatin is derived from plasminogen.

THE MOLECULE

Angiostatin is an angiogenesis inhibitor which potently blocks neovascularization and growth of tumour metastases *in vivo*[1–3]. The 38-kDa protein is generated by the cancer-mediated proteolytic cleavage of plasminogen[3,4]. Urokinase and free sulfhydryl donors (FSDs) have been implicated in this cleavage[4]. Angiostatin has been shown to induce metastatic dormancy (defined by a balance of apoptosis and proliferation), and systemic administration of human angiostatin inhibited the growth of human and murine primary carcinomas in mice[5]. Angiostatin contains four triple disulfide bond-containing kringle domains, which afford structural integrity. Kringles 1–3 exhibit inhibition of endothelial cell proliferation, whereas kringle 4 inhibits migration[6,7]. Angiostatin has been reported to cause a 4- to 5-fold increase in E-selectin expression[8].

Crossreactivity

Plasminogen (and thus angiostatin) has an inter-species homology of at least 75%.

Sources

The site of angiostatin biosynthesis is unknown; it has been suggested that cancer-mediated angiostatin production is effected through protease synthesis rather than direct plasminogen/angiostatin expression.

Bioassays

CAM assay[2]; cornea micropocket assay[2]; Lewis lung carcinoma spontaneous metatastic model[1,2]; various other primary tumour models[2,9].

Physicochemical properties of angiostatin

Property	Human	Mouse
pI[a]	7.74	5.46
Amino acids – precursor	810	812
– mature	?367	?338
M_r (K) – predicted	41.64	38.52
– expressed	38	?
Potential *N*-linked glycosylation sites	1	0
Disulfide bonds	13	10

[a] Predicted.

3-D structure

The crystal structure of angiostatin is not known. The crystal structures of individual kringle domains have, however, been elucidated. Human plasminogen kringle 4 has been solved to an R factor of 14.2% at 1.9 Å resolution; it is highly stabilized by an internal hydrophobic core and extensive hydrogen bonding[10].

Gene structure

The human plasminogen gene and hence angiostatin is localized to chromosome 6q26–6q27.

Amino acid sequence for human angiostatin

Accession code: SwissProt P00747

```
  1  MEHKEVVLLL LLFLKSGQGE PLDDYVNTQG ASLFSVTKKQ LGAGSIEECA
 51  AKCEEDEEFT CRAFQYHSKE QQCVIMAENR KSSIIIRMRD VVLFEKKVYL
101  SECKTGNGKN YRGTMSKTKN GITCQKWSST SPHRPRFSPA THPSEGLEEN
151  YCRNPDNDPQ GPWCYTTDPE KRYDYCDILE CEEECMHCSG ENYDGKISKT
201  MSGLECQAWD SQSPHAHGYI PSKFPNKNLK KNYCRNPDRE LRPWCFTTDP
251  NKRWELCDIP RCTTPPPSSG PTYQCLKGTG ENYRGNVAVT VSGHTCQHWS
301  AQTPHTHNRT PENFPCKNLD ENYCRNPDGK RAPWCHTTNS QVRWEYCKIP
351  SCDSSPVSTE QLAPTAPPEL TPVVQDCYHG DGQSYRGTSS TTTTGKKCQS
401  WSSMTPHRHQ KTPENYPNAG LTMNYCRNPD ADKGPWCFTT DPSVRWEYCN
451  LKKCSGTEAS VVAPPPVVLL PDVETPSEED CMFGNGKGYR GKRATTVTGT
501  PCQDWAAQEP HRHSIFTPET NPRAGLEKNY CRNPDGDVGG PWCYTTNPRK
551  LYDYCDVPQC AAPSFDCGKP QVEPKKCPGR VVGGCVAHPH SWPWQVSLRT
601  RFGMHFCGGT LISPEWVLTA AHCLEKSPRP SSYKVILGAH QEVNLEPHVQ
651  EIEVSRLFLE PTRKDIALLK LSSPAVITDK VIPACLPSPN YVVADRTECF
701  ITGWGETQGT FGAGLLKEAQ LPVIENKVCN RYEFLNGRVQ STELCAGHLA
751  GGTDSCQGDS GGPLVCFEKD KYILQGVTSW GLGCARPNKP GVYVRVSRFV
801  TWIEGVMRNN
```

Human angiostatin is derived from plasminogen. The protein sequence for angiostatin within that of plasminogen is shown in italics. (Italics run from residue 98, VVLFEKKVYL, through residue 465, …VVAPP.)

Amino acid sequence for mouse angiostatin

Accession code: SwissProt P20918

```
  1  MDHKEVILLF LLLLKPGQGD SLDGYISTQG ASLFSLTKKQ LAAGGVSDCL
 51  AKCEGETDFV CRSFQYHSKE QQCVIMAENS KTSSIIRMRD VILFEKRVYL
101  SECKTGIGNG YRGTMSRTKS GVACQKWGAT FPHVPNYSPS THPNEGLEEN
151  YCRNPDNDEQ GPWCYTTDPD KRYDYCNIPE CEEECMYCSG EKYEGKISKT
201  MSGLDCQAWD SQSPHAHGYI PAKFPSKNLK MNYCHNPDGE PRPWCFTTDP
251  TKRWEYCDIP RCTTPPPPPS PTYQCLKGRG ENYRGTVSVT VSGKTCQRWS
301  EQTPHRHNRT PENFPCKNLE ENYCRNPDGE TAPWCYTTDS QLRWEYCEIP
351  SCESSASPDQ SDSSVPPEEQ TPVVQECYQS DGQSYRGTSS TTITGKKCQS
401  WAAMFPHRHS KTPENFPDAG LEMNYCRNPD GDKGPWCYTT DPSVRWEYCN
451  LKRCSETGGS VVELPTVSQE PSGPSDSETD CMYGNGKDYR GKTAVTAAGT
501  PCQGWAAQEP HRHSIFTPQT NPRADLEKNY CRNPDGDVNG PWCYTTNPRK
551  LYDYCDIPLC ASASSFECGK PQVEPKKCPG RVVGGCVANP HSWPWQISLR
601  TRFTGQHFCG GTLIAPEWVL TAAHCLEKSS RPEFYKVILG AHEEYIRGLD
651  VQEISVAKLI LEPNNRDIAL LKLSRPATIT DKVIPACLPS PNYMVADRTI
701  CYITGWGETQ GTFGAGRLKE AQLPVIENKV CNRVEYLNNR VKSTELCAGQ
751  LAGGVDSCQG DSGGPLVCFE KDKYILQGVT SWGLGCARPN KPGVYVRVSR
801  FVDWIEREMR NN
```

Mouse angiostatin is derived from plasminogen. The protein sequence for angiostatin within that of plasminogen is shown in italics.

THE ANGIOSTATIN RECEPTOR

The angiostatin receptor remains to be identified.

References

[1] Sim, B.K. et al. (1997) Cancer Res. 57, 1329–1334.
[2] O Reilly, M.S. et al. (1994) Cell 79, 315–328.
[3] Gately, S. et al. (1996) Cancer Res. 56, 4887–4890.
[4] Gately, S. et al. (1997) Proc. Natl Acad. Sci. USA 94, 10868–10872.
[5] O Reilly, M.S. et al. (1996) Nature Med. 2, 689–692.
[6] Ji, W.R. et al. (1998) FASEB J. 12, 1731–1738.
[7] Cao, Y. et al. (1996) J. Biol. Chem. 271, 29461–29467.
[8] Luo, J. et al. (1998) Biochem. Biophys. Res. Commun. 245, 906–911.
[9] Kirsch, M. et al. (1998) Cancer Res. 58, 4654–4659.
[10] Mulichak, A.M. et al. (1991) Biochemistry 30, 10576–10588.

Other names

TRAIL (TNF-related apoptosis inducing ligand)[1], TL2[2] and TNFSF10.

THE MOLECULE

Apo2L/TRAIL is a newly discovered member of the TNF ligand family that is closely related to FasL. Apo2L was initially cloned from human heart and lymphocyte cDNA libraries[1,2]. Apo2L/TRAIL is a type II membrane protein, whose C-terminal extracellular domain shows clear homology to other TNF family members. Human Apo2L has been shown to be expressed as a cell surface protein, but a soluble form has also recently been described. The distinct biological function of Apo2L is still unknown, however both cell surface Apo2L and picomolar amounts of soluble Apo2L potently and rapidly induce caspase-dependent apoptosis in a wide variety of transformed cell lines of diverse origin, but not in most normal cell types[3] suggesting that Apo2L/TRAIL might be useful as an anti-cancer agent. Several lines of more recent evidence suggest that Apo2L may play a role during various immunological processes such as T-cell activation-induced cell death (in which FasL plays a prominent role[4]). Other postulated roles for Apo2L include target killing by cytotoxic lymphocytes[5] and in immune privilege of tumours. An additional more recently described aspect of Apo2L function relates to its role in the killing of dendritic cells by autologous CD4+ CTLs[6], suggesting a potentially unique role for Apo2L/TRAIL in modulation of the immune system[7].

Crossreactivity

Human Apo2L is 65% identical to mouse Apo2L at the amino acid level across the entire molecule and there is complete species crossreactivity[1].

Sources

Apo2L/TRAIL transcripts are widely expressed in a variety of tissues, most prominently in spleen, lung and prostate[1]. T cells upregulate Apo2L expression upon T-cell receptor stimulation[8].

Bioassays

Standard cytotoxicity or cell-viability assays. Standard methodologies for measuring apoptosis may also be employed.

Physicochemical properties of Apo2L ligand

Property	Human	Mouse
pI[a]	7.63	8.21
Amino acids – precursor	281	291
– mature[a]	281	291
M_r (K) – predicted	32.5	33.477
– expressed	32–33	
Potential *N*-linked glycosylation sites	1	?
Disulfide bonds	0	?

[a] Type II membrane glycoprotein with no signal peptide. A soluble form of Apo2L has also been identified which migrates with an apparent molecular weight of 24 kDa.

3-D structure

The crystal structure for human Apo2L/TRAIL is has recently been elucidated. The structure reveals that a unique frame insertion of 12–16 amino acids adopts a salient loop structure penetrating into the receptor-binding site, altering the common receptor-binding surface of the TNF family most likely for the specific recognition of cognate partners[9].

Gene structure

Apo2L/TRAIL is located on human chromosome 3 at position 3q26[1].

Amino acid sequence for human Apo2 ligand

Accession code: SwissProt P50591

```
  1  MAMMEVQGGP SLGQTCVLIV IFTVLLQSLC VAVTYVYFTN ELKQMQDKYS
 51  KSGIACFLKE DDSYWDPNDE ESMNSPCWQV KWQLRQLVRK MILRTSEETI
101  STVQEKQQNI SPLVRERGPQ RVAAHITGTR GRSNTLSSPN SKNEKALGRK
151  INSWESSRSG HSFLSNLHLR NGELVIHEKG FYYIYSQTYF RFQEEIKENT
201  KNDKQMVQYI YKYTSYPDPI LLMKSARNSC WSKDAEYGLY SIYQGGIFEL
251  KENDRIFVSV TNEHLIDMDH EASFFGAFLV G
```

Amino acid sequence for mouse Apo2 ligand

Accession code: SwissProt P50592

```
  1  MPSSGALKDL SFSQHFRMMV ICIVLLQVLL QAVSVAVTYM YFTNEMKQLQ
 51  DNYSKIGLAC FSKTDEDFWD STDGEILNRP CLQVKRQLYQ LIEEVTLRTF
101  QDTISTVPEK QLSTPPLPRG GRPQKVAAHI TGITRRSNSA LIPISKDGKT
151  LGQKIESWES SRKGHSFLNH VLFRNGELVI EQEGLYYIYS QTYFRFQEAE
201  ASKMVSKDK VRTKQLVQYI YKYTSYPDPI VLMKSARNSC WSRDAEYGLY
251  SIYQGGLFEL KKNDRIFVSV TNEHLMDLDQ EASFFGAFLI N
```

THE Apo2L/TRAIL RECEPTOR

Apo2L/TRAIL interacts with four human cell surface receptors that form a distinct group within the TNFRSF (TRAIL-R1-TRAIL-R4) (for a review see ref. 10). TRAIL-R1, also called DR4 (death receptor 4)[11] and TRAIL-R2 (also called DR5 (death receptor 5 and TRICK2 or KILLER)[12] have death domains (see FasL entry (page **225**)) that signal apoptosis. TRAIL-R3 (also called DcR1 (decoy receptor 1, TRID or LIT) is a GPI-linked cell surface protein that lacks a cytoplasmic domain and consequently a functional death domain. TRAIL-R3 can inhibit ligand-mediated apoptosis, thus functioning as a decoy receptor which competes with TRAIL-R1 and TRAIL-R2 for ligand binding[12]. TRAIL-R4 (also called DcR2 or TRUNDD[13]) has a substantially truncated death domain that does not signal apoptosis induction by Apo2L. DcR2 has a type I transmembrane topology, but contains a truncated, nonfunctional death domain in its cytoplasmic region. Furthermore, Apo2L/TRAIL has also been shown to bind a soluble member of the TNFRSF, osteoprotegerin (OPG)[14]. OPG is not related to the other Apo2L receptors, and it also binds another TNF ligand family member, OPGL (also called RANKL, TRANCE). All four TRAIL receptors contain closely related extracellular cysteine-rich domains (CRD domains), but only TRAIL-R1 and TRAIL-R2 contain cytoplasmic death domains. A murine receptor for Apo2L/TRAIL has recently been identified and shown to be an orthologue of TRAIL-R2/DR5[15].

Distribution

Expression of TRAIL-R1/DR4 mRNA has been detected in several human tissues, including spleen, peripheral blood leukocytes, small intestine, thymus and activated T cells in addition to a variety of transformed cell of diverse origin[11]. TRAIL-R2/DR5 mRNA has been detected in human fetal kidney, liver and lung, as well as adult peripheral blood lymphocytes, colon, small intestine, ovary, testis, prostate, thymus, spleen, pancreas, kidney, skeletal muscle, liver, lung and heart, in addition to a variety of transformed cell lines[12]. TRAIL-R3/DcR1 mRNA was detected in human heart, lung, liver, placenta, skeletal muscle, spleen and PBLs. TRAIL-R4/DcR2 mRNA was detected in fetal kidney, liver and lung as well as adult PBLs, colon, small intestine, pancreas, kidney, liver, lung and heart. It is interesting to note that TRAIL-R1 and TRAIL-R2 are widely expressed in most tumour cell lines so far tested, while expression of the decoy receptors is restricted to normal tissues and is more or less absent in transformed cell lines. This suggests a possible explanation for the observation that Apo2L can induce apoptosis through TRAIL-R1 and TRAIL-R2 in transformed tissue but not in normal tissue, where decoy receptors may inhibit this effect[11,12].

Physicochemical properties of Apo2L/TRAIL receptors

	TRAIL-R1		TRAIL-R2	
Property	Human	Mouse	Human	Mouse
Amino acids – precursor	468	?	411/440[b]	?
– mature[a]	445	?	358/387	?
M_r (K) – predicted		?	47.89	?
– expressed		?		
Potential *N*-linked glycosylation sites	0	?	0	?
Affinity K_d (M)[c]	$4–9 \times 10^{-11}$ (high-affinity sites) or 4×10^{-10} to 3.9×10^{-8} for low-affinity binding sites			

[a] After removal of predicted signal peptide. The cytosolic domain contains a death domain.
[b] Alternative splicing of 29 amino acid residues gives rise to two gene products of DR5.
[c] Two classes of binding site exist for all four members of the TRAIL receptor family.

	TRAIL-R3/DcR1		TRAIL-R4/DcR2	
Property	Human	Mouse	Human	Mouse
Amino acids – precursor	259	?	386	
– mature[a]	234	?	331	
M_r (K) – predicted	27.365		45.12	
– expressed		?	?	
Potential *N*-linked glycosylation sites	5	?	3	
Affinity K_d (M)	$4–9 \times 10^{-11}$ (high-affinity sites) or 4×10^{-10} to 3.9×10^{-8} for low-affinity binding sites			

[a] After removal of predicted signal peptide. The cytosolic domain contains a death domain.

Signal transduction

DR4-mediated signalling pathways leading to apoptosis are caspase-dependent[1]. However, the role of FADD, a death domain containing cytosolic protein which is used by Fas in caspase-mediated apoptosis induction in DR4 and DR5-mediated apoptosis induction is somewhat controversial, but recent evidence demonstrates the importance of FADD in this process. Upon overexpression, both DR4 and DR5 lead to activation of NFκB, however because Apo2L-mediated activation of this response is much weaker (you need 100- to 1000-fold higher concentrations of Apo2L than of TNF) the physiological significance of NFκB activation by DR4 and DR5 remains unclear[16–18].

Chromosomal location

The genes for all TRAIL receptors map to human chromosome 8p21–22.

Amino acid sequence for human TRAIL-R1/DR4[11]

Accession code: EMBL U90875

```
-23  MAPPPARVHL GAFLAVTPNP GSA
  1  ASGTEAAAAT PSKVWGSSAG RIEPRGGGRG ALPTSMGQHG PSARARAGRA
 51  PGPRPAREAS PRLRVHKTFK FVVVGVLLQV VPSSAATIKL HDQSIGTQQW
101  EHSPLGELCP PGSHRSERPG ACNRCTEGVG YTNASNNLFA CLPCTACKSD
151  EEERSPCTTT RNTACQCKPG TFRNDNSAEM CRKCSTGCPR GMVKVKDCTP
201  WSDIECVHKE SGNGHNIWVI LVVTLVVPLL LVAVLIVCCC IGSGCGGDPK
251  CMDRVCFWRL GLLRGPGAED NAHNEILSNA DSLSTFVSEQ QMESQEPADL
301  TGVTVQSPGE AQCLLGPAEA EGSQRRRLLV PANGADPTET LMLFFDKFAN
351  IVPFDSWDQL MRQLDLTKNE IDVVRAGTAG PGDALYAMLM KWVNKTGRNA
401  SIHTLLDALE RMEERHAKEK IQDLLVDSGK FIYLEDGTGS AVSLE
```

The precise location of the transmembrane region is not shown.

Amino acid sequence for human TRAIL-R2/DR5[19]

Accession code: SPTREMBLE O15531

```
-53  MEQRGQNAPA ASGARKRHGP GPREARGARP GLRVPKTLVL VVAAVLLLVS
-3   AES
  1  ALITQQDLAP QQRVAPQQKR SSPSEGLCPP GHHISEDGRD CISCKYGQDY
 51  STHWNDLLFC LRCTRCDSGE VELSPCTTTR NTVCQCEEGT FREEDSPEMC
101  RKCRTGCPRG MVKVGDCTPW SDIECVHKES GTKHSGEAPA VEETVTSSPG
151  TPASPCSLSG IIIGVTVAAV VLIVAVFVCK SLLWKKVLPY LKGICSGGGG
201  DPERVDRSSQ RPGAEDNVLN EIVSILQPTQ VPEQEMEVQE PAEPTGVNML
251  SPGESEHLLE PAEAERSQRR RLLVPANEGD PTETLRQCFD DFADLVPFDS
301  WEPLMRKLGL MDNEIKVAKA EAAGHRDTLY TMLIKWVNKT GRDASVHTLL
351  DALETLGERL AKQKIEDHLL SSGKFMYLEG NADSAMS
```

Two separate alternative splice products of DR5 are generated which contain 440 and 411 amino acids respectively. The longer isoform has 29 extra amino acids in the C-terminal portion of the extracellular domain. Following the transmembrane region, DR5 contains a death domain.

Amino acid sequence for human TRAIL-R3/DcR1[18]

Accession code: SPTREMBLE O14755

```
-25  MARIPKTLKF VVVIVAVLLP VLAYS
  1  ATTARQEEVP QQTVAPQQQR HSFKGEECPA GSHRSEHTGA CNPCTEGVDY
 51  TNASNNEPSC FPCTVCKSDQ KHKSSCTMTR DTVCQCKEGT FRNVNSPEMC
101  RKCSRCPSGE VQVSNCTSWD DIQCVEEFGA NATVETPAAE ETMNTSPGTP
151  APAAEETMNT SPGTPAPAAE ETMTTSPGTP APAAEETMTT SPGTPAPAAE
201  ETMTTSPGTP ASSHYLSCTI VGIIVLIVLL IVFV
```

A longer form of the protein with an extended N-terminus has also been reported. The predicted signal peptide is shown in italics. The hydrophobic C-terminus is reminiscent of the C-termini of proteins which are tethered to the cell surface by a GPI anchor.

Amino acid sequence for human TRAIL-R4/DcR2[20]

Accession code: SPTREMBLE Q9UBN6

```
-25  MGLWGQSVPT ASSARAGRYP GARTA
  1  SGTRPWLLDP KILKFVVFIV AVLLPVRVDS ATIPRQDEVP QQTVAPQQQR
 51  RSLKEEECPA GSHRSEYTGA CNPCTEGVDY TIASNNLPSC LLCTVCKSGQ
101  TNKSSCTTTR DTVCQCEKGS FQDKNSPEMC RTCRTGCPRG MVKVSNCTPR
151  SDIKCKNESA ASSTGKTPAA EETVTTILGM LASPYHYLII IVVLVIILAV
201  VVVGFSCRKK FISYLKGICS GGGGGPERVH RVLFRRRSCP SRVPGAEDNA
251  RNETLSNRYL QPTQVSEQEI QGQELAELTG VTVESPEEPQ RLLEQAEAEG
301  CQRRRLLVPV NDADSADIST LLDASATLEE GHAKETIQDQ LVGSEKLFYE
351  EDEAGSATSC L
```

Three allelic versions of DcR2 exist. These differ at amino acid residues 35 or 310 or both of the precursor peptide. Downstream of the transmembrane region, DcR2 contains a truncated death domain which is approximately one-third the size of a canonical death domain.

References

1. Wiley, S.R. et al. (1995) Immunity 3, 673–682.
2. Tan, K.B. et al. (1997) Gene 204, 35–46.
3. Marsters, S.A. et al. (1999) Recent Prog. Horm. Res. 54, 225–234.
4. Nagata, S. (1997) Cell 88, 355–365.
5. Thomas, W.D. and Hersey, P. (1998) J. Immunol. 161, 2195–2200.
6. Wang, J. et al. (1999) Cell 98, 47–58.
7. Fanger, N.A. et al. (1999) J. Exp. Med. 190, 1155–1164.
8. Jeremias, I. et al. (1998) Eur. J. Immunol. 28, 143–152.
9. Cha, S.S. et al. (1999) Immunity 11, 253–261.
10. Ashkenazi, A. and Dixit, V.M. (1999) Curr. Opin. Cell Biol. 11, 255–260.
11. Pan, G. et al. (1997) Science 276, 111–113.
12. Pan, G. et al. (1997) Science 277, 815–818.
13. Degli-Esposti, M.A. et al. (1997) Immunity 7, 813–820.
14. Simonet, W.S. et al. (1997) Cell 89, 309–319.

[15] Wu, G.S. et al. (1999) Cancer Res. 59, 2770–2775.
[16] Sprick, M.R. et al. (2000) Immunity 12, 599–609.
[17] Schneider, P. et al. (1997) Immunity 7, 831–836.
[18] Schneider, P. et al. FEBS Lett. 416, 329–334.
[19] Screaton, G.R. et al. (1997) Curr. Biol. 7, 693–696.
[20] Marsters, S.A. et al. (1997) Curr. Biol. 7, 1003–1006.

Other names

TRDL-1α (tumour necrosis factor-related death ligand 1α), TNFSF13, (TNF superfamily member 13).

THE MOLECULE

Human APRIL (a proliferation-inducing ligand) was cloned following screening of public databases using an improved profile search based on an optimal alignment of all currently known TNF ligand family members[1]. It shows highest homology to FasL (CD95). APRIL is unique among the TNF superfamily, as it is abundantly expressed in tumour cells. Overexpression of a truncated soluble APRIL (lacking 110 amino acids from its N-terminus) has been shown to accelerate the growth of transformed cells *in vitro* and *in vivo*, while another study demonstrates the ability of full-length APRIL to induce apoptosis[2].

Crossreactivity

No information.

Sources

APRIL mRNA is expressed at low abundance in normal human tissues, but is expressed at high levels in human cancers of colon, thyroid and lymphoid tissues *in vivo*[1].

Bioassays

APRIL stimulates the growth of various tumour cell lines, both of haematopoietic and nonhaematopoietic origin. It also enhances the *in vivo* growth of NIH 3T3 cells in nude mice. It has also been shown to induce cell death in Jurkat T cells with similar kinetics to that of FasL[2].

Physicochemical properties of APRIL

Property	Human	Mouse
pI	9.67[a]	?
Amino acids – precursor	250	?
– mature[b]	250	?
M_r (K) – predicted	27.43	?
– expressed	40	?
Potential *N*-linked glycosylation sites	1	?
Disulfide bonds	?	?

[a] This represents the predicted pI of APRIL/TRDL-1α.

3-D structure

The crystal structure for APRIL is unknown, but is predicted to contain eight β-strands in its extracellular domain, which are folded into an antiparallel β-sandwich structure with a 'jelly-roll' topology (three monomers associating about a 3-fold axis of symmetry forming a bell-shaped trimer), based on its homology with lymphotoxin α, TNFα and CD40L[3–5]. The absence of a signal peptide confirms that APRIL is a type II membrane protein that is typical of the TNF ligand family.

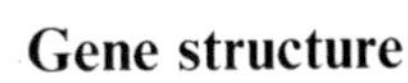

Gene structure

APRIL is localized to human chromosome 17p13.3[2] adjacent to the gene for TWEAK, another TNF superfamily member.

Amino acid sequence for human APRIL[2]

Accession code: Genbank AF046888

```
  1 MPASSPFLLA PKGPPGNMGG PVREPALSVA LWLSWGAALG AVACAMALLT
 51 QQTELQSLRR EVSRLQGTGG PSQNGEGYPW QSLPEQSSDA LEAWENGERS
101 RKRRAVLTQK QKKQHSVLHL VPINATSKDD SDVTEVMWQP ALRRGRGLQA
151 QGYGVRIQDA GVYLLYSQVL FQDVTFTMGQ VVSREGQGRQ ETLFRCIRSM
201 PSHPDRAYNS CYSAGVFHLH QGDILSVIIP RARAKLNLSP HGTFLGFVKL
```

Human APRIL contains a single *N*-linked glycosylation site (Asn124) within the receptor-binding motif. Traditional cDNA library screening has also identified two additional splice variants, designated TRDL-1β and TRDL-1γ, which differ from TRDL-1α (APRIL) by the deletion of two small regions within the protein coding region[2]. TRDL-1β is identical to TRDL-1α, with the exception of a 48-bp deletion that removed 16 amino acids corresponsing to residues 113–128 of TRDL-1α. TRDL-1γ contains a 3′ deletion of 181 bp that results in substitution of four C-terminal residues of TRDL-1α with a single leucine residue[2].

THE APRIL RECEPTOR

The signalling receptor for APRIL is unknown, but recent *in vitro* binding experiments demonstrated that APRIL coprecipitated Fas and HVEM, suggesting that it may function as an alternate ligand for these receptors[2]. Further work is needed to define the cognate ligand for APRIL and to determine under what conditions APRIL binds Fas and/or HVEM.

References

[1] Hahne, M. et al. (1998) J. Exp. Med. 188, 1185–1190.
[2] Kelly, K. et al. (2000) Cancer Res. 60, 1021–1027.
[3] Eck, M.J. and Sprang, S.R. (1989) J. Biol. Chem. 264, 17595–17605.
[4] Eck, M.J. et al. (1992) J. Biol. Chem. 267, 2119–2122.
[5] Banner, D.W. et al. (1993) Cell 73, 431–445.

Other names

TALL-1 (TNF and apoptosis ligand-related leukocyte-expressed ligand-1[1]) or BLyS (B-lymphocyte stimulator), THANK or zTNF4.

THE MOLECULE

Human B cell-activating factor (BAFF) belongs to the TNF family and was cloned following screening of public databases using an improved profile search[2] based on an optimal alignment of all currently known TNF ligand family members[3]. BAFF, like other members of this family, lacks a signal peptide suggesting that it is a type II membrane protein. Membrane-bound BAFF is processed and secreted through the action of the furin family of proprotein convertases. Both membrane-bound and soluble BAFF induce the proliferation of anti-immunoglobulin M-stimulated peripheral blood B lymphocytes. In addition, transgenic mice overexpressing BAFF have severe B-cell hyperplasia and lupus-like autoimmune disease. BAFF therefore plays an important role in the control of B-cell function[3]. More recently, BAFF has been shown to mediate the survival of peripheral immature B lymphocytes[4,5].

Crossreactivity

Human and murine BAFF share 68% identity and 98% homology in their β-strand-rich ectodomain, suggesting that the BAFF gene has been highly conserved during evolution[3].

Sources

Human BAFF mRNA is expressed on T cells, monocytes, macrophages and dendritic cells[3].

Bioassays

Membrane-bound and soluble BAFF induce the proliferation of anti-immunoglobulin M-stimulated peripheral blood B cells[3].

Physicochemical properties of BAFF

Property	Human	Mouse
pI[a]	5.92	6.1
Amino acids – precursor	285	309
– mature[b]	285	309
M_r (K) – predicted	27.43	34.192
– expressed	18	?
Potential *N*-linked glycosylation sites	2	?
Disulfide bonds	?	0

[a] This represents the predicted pI of BAFF.
[b] Human and mouse BAFF lack signal peptides.

3-D structure

The crystal structure for BAFF is unknown, but is predicted to resemble that of APRIL, lymphotoxin α, TNFα and CD40L. The absence of a signal peptide confirms that BAFF is a type II membrane protein that is typical of the TNF ligand family.

Gene structure

BAFF is localized to human chromosome 13q32–34[3].

Amino acid sequence for human BAFF[3]

Accession code: SPTREMBLE Q9Y275

```
  1  MDDSTEREQS RLTSCLKKRE EMKLKECVSI LPRKESPSVR SSKDGKLLAA
 51  TLLLALLSCC LTVVSFYQVA ALQGDLASLR AELQGHHAEK LPAGAGAPKA
101  GLEEAPAVTA GLKIFEPPAP GEGNSSQNSR NKRAVQGPEE TVTQDCLQLI
151  ADSETPTIQK GSYTFVPWLL SFKRGSALEE KENKILVKET GYFFIYGQVL
201  YTDKTYAMGH LIQRKKVHVF GDELSLVTLF RCIQNMPETL PNNSCYSAGI
251  AKLEEGDELQ LAIPRENAQI SLDGDVTFFG ALKLL
```

The extracellular domain of BAFF is predicted to contain two potential *N*-glycosylation sites, but only N124 was shown to be glycosylated[3].

Amino acid sequence for mouse BAFF[6]

Accession code: SPTREMBLE Q9WU72

```
  1  MDESAKTLPP PCLCFCSEKG EDMKVGYDPI TPQKEEGAWF GICRDGRLLA
 51  ATLLLALLSS SFTAMSLYQL AALQADLMNL RMELQSYRGS ATPAAAGAPE
101  LTAGVKLLTP AAPRPHNSSR GHRNRRAFQG PEETEQDVDL SAPPAPCLPG
151  CRHSQHDDNG MNLRNIIQDC LQLIADSDTP TIRKGTYTFV PWLLSFKRGN
201  ALEEKENKIV VRQTGYFFIY SQVLYTDPIF AMGHVIQRKK VHVFGDELSL
251  VTLFRCIQNM PKTLPNNSCY SAGIARLEEG DEIQLAIPRE NAQISRNGDD
301  TFFGALKLL
```

Murine BAFF encodes a slightly longer protein than human BAFF, due to insertion between the transmembrane region and the first of several β-strands which constitute the receptor-binding domain in all TNF family members[6].

THE BAFF RECEPTOR

B-cell maturation antigen (BCMA) was first identified as part of a translocation event in a malignant T-cell lymphoma patient[7]. Subsequent identification of the mouse BCMA gene and further motif analysis led to the prediction that BCMA is a member of the TNF receptor superfamily[8]. Recently it has been shown that BCMA is the receptor for BAFF[9]. Characterization of BCMA showed that it is primarily expressed in immune organs and mature B-cell lines[7,10]. TNF receptor family members are generally type I membrane proteins with a leader sequence and a cysteine-rich extracellular domain. The extracellular domains are organized as a series of alternative A and B modules, which are stabilized by internal disulfide bridges. A single C module, not involved in ligand binding, is found in the fourth cysteine repeat of TNFR type I. BCMA appears to be distantly related to the TNF receptor family, because it is devoid of a signal sequence and contains a single A and one C module instead of multiple A and B modules[8].

Distribution

The BAFF receptor (BCMA) is expressed primarily on mature B cells[9] and B-cell lines[7,10].

Physiochemical properties of the BAFF receptor

Property	Human	Mouse
Amino acids – precursor	184	185
– mature[a]	?	?
M_r (K) – predicted	201.38	204.42
– expressed	?	?
Potential *N*-linked glycosylation sites	?	?
Affinity K_d (M)	?	?

[a] After removal of predicted signal peptide.

Signal transduction

Overexpression of BCMA activates NFκB, and this activation is potentiated by BAFF. Moreover, BCMA-mediated NFκB activation is inhibited by dominant negative mutants of TNF receptor-associated factor 5 (TRAF5), TRAF6, NFκB-inducing kinase (NIK) and IκB kinase (IKK). These data indicate that BCMA is a receptor for BAFF and BCMA activates NFκB through a TRAF5-, TRAF6-, NIK- and IKK-dependent pathway[11].

Chromosomal location

The gene for BCMA is located on mouse and human chromosome 16. The gene for BCMA on human chromosome 16 is fused to the IL-2 gene by a t(4;16)(q26;p13) translocation in a form of T-cell acute lymphoblastic leukaemia.

Amino acid sequence for human BCMA[7]

Accession code: SwissProt Q02223

```
  1  MLQMAGQCSQ NEYFDSLLHA CIPCQLRCSS NTPPLTCQRY CNASVTNSVK
 51  GTNAILWTCL GLSLIISLAV FVLMFLLRKI SSEPLKDEFK NTGSGLLGMA
101  NIDLEKSRTG DEIILPRGLE YTVEECTCED CIKSKPKVDS DHCFPLPAME
151  EGATILVTTK TNDYCKSLPA ALSATEIEKS ISAR
```

The putative transmembrane region is underlined.

Amino acid sequence for mouse BCMA[8]

Accession code: SwissProt O88472

```
  1  MAQQCFHSEY FDSLLHACKP CHLRCSNPPA TCQPYCDPSV TSSVKGTYTV
 51  LWIFLGLTLV LSLALFTISF LLRKMNPEAL KDEPQSPGQL DGSAQLDKAD
101  TELTRIRAGD DRIFPRSLEY TVEECTCEDC VKSKPKGDSD HFFPLPAMEE
151  GATILVTTKT GDYGKSSVPT ALQSVMGMEK PTHTR
```

References

[1] Shu, H.B. et al. (1999) J. Leukoc. Biol. 65, 680–683.
[2] Bucher, P. et al. (1996) Comput. Chem. 20, 3–23.
[3] Schneider, P. et al. (1999) J. Exp. Med. 189, 1747–1756.
[4] Batten, M. et al. (2000) J. Exp. Med. 192, 1453–1466.
[5] Thompson, J.S. et al. (2000) J. Exp. Med. 192, 129–135.
[6] Banner, D.W. et al. (1993) Cell 73, 431–445.
[7] Laabi, Y. et al. (1992) EMBO J. 11, 3897–3904.
[8] Madry, C. et al. (1998) Int. Immunol. 10, 1693–1702.
[9] Thompson, J.S. et al. (2000) J. Exp. Med. 192, 129–136.
[10] Laabi, Y. et al. (1994) Nucleic Acids Res. 22, 1147–1154.
[11] Shu, H.B. and Johnson, H. (2000) Proc. Natl Acad. Sci. USA 97, 9156–9161.

Other names
CD137L and TNFSF9.

THE MOLECULE

The 4-1BB ligand (4-1BBL) was first identified on activated macrophages and mature B cells and was later found to be expressed by mitogenically activated T cells[1–3]. 4-1BBL induces the proliferation of activated peripheral blood T cells and may have a role in activation-induced cell death (AICD). Furthermore, 4-1BBL may participate in cognate interactions between T cells and B cells/macrophages. Mouse 4-1BBL is a 50-kDa, 309-amino-acid residue transmembrane glycoprotein that is the largest of the TNF superfamily members[4].

Crossreactivity
Human and mouse 4-1BBL share 37% identity at the amino acid level[5]. This level of cross-species conservation is much lower than that found in other members of the TNF superfamily[6].

Sources
Cells known to express 4-1BBL include B cells, dendritic cells and macrophages[7,8].

Bioassays
Proliferation of activated peripheral blood T cells.

Physicochemical properties of 4-1BBL

Property	Human	Mouse
pI[a]	6.52	5.48
Amino acids – precursor	254	309
– mature[b]	254	309
M_r (K) – predicted	26.624	33.853
– expressed	50	
Potential *N*-linked glycosylation sites	0	3
Disulfide bonds	0	0

[a] Human and mouse 4-1BBL both lack a signal peptide.

3-D structure
The crystal structure for 4-1BBL is unknown, but is predicted to resemble that of TNFα.

Gene structure
Human 4-1BBL maps to chromosome 19p13.3.

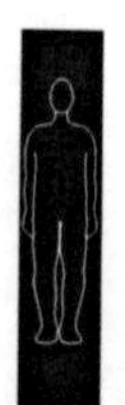

Amino acid sequence for human 4-1BBL[6]

Accession code: SwissProt P41273

```
  1  MEYASDASLD PEAPWPPAPR ARACRVLPWA LVAGLLLLLL LAAACAVFLA
 51  CPWAVSGARA SPGSAASPRL REGPELSPDD PAGLLDLRQG MFAQLVAQNV
101  LLIDGPLSWY SDPGLAGVSL TGGLSYKEDT KELVVAKAGV YYVFFQLELR
151  RVVAGEGSGS VSLALHLQPL RSAAGAAALA LTVDLPPASS EARNSAFGFQ
201  GRLLHLSAGQ RLGVHLHTEA RARHAWQLTQ GATVLGLFRV TPEIPAGLPS
251  PRSE
```

Human 4-1BBL lacks a signal peptide. The putative signal-anchor (transmembrane sequence) is underlined.

Amino acid sequence for mouse 4-1BBL[4]

Accession code: SwissProt P41274

```
  1  MDQHTLDVED TADARHPAGT SCPSDAALLR DTGLLADAAL LSDTVRPTNA
 51  ALPTDAAYPA VNVRDREAAW PPALNFCSRH PKLYGLVALV LLLLIAACVP
101  IFTRTEPRPA LTITTSPNLG TRENNADQVT PVSHIGCPNT TQQGSPVFAK
151  LLAKNQASLC NTTLNWHSQD GAGSSYLSQG LRYEEDKKEL VVDSPGLYYV
201  FLELKLSPTF TNTGHKVQGW VSLVLQAKPQ VDDFDNLALT VELFPCSMEN
251  KLVDRSWSQL LLLKAGHRLS VGLRAYLHGA QDAYRDWELS YPNTTSFGLF
301  LVKPDNPWE
```

Murine 4-1BBL lacks a signal peptide. The putative signal-anchor (transmembrane sequence) is underlined.

THE 4-1BBL RECEPTOR

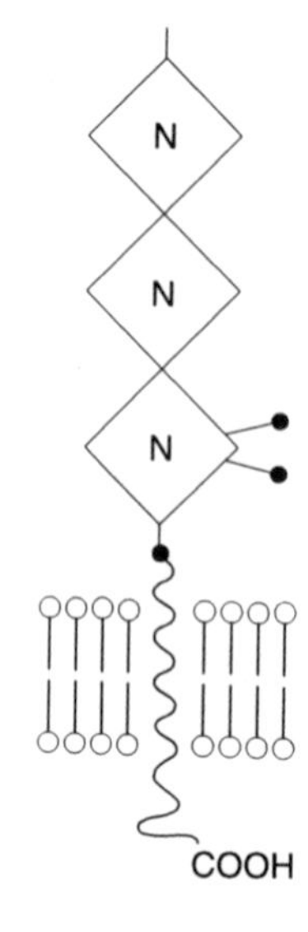

The receptor for 4-1BBL was originally named ILA for induced by lymphocyte activation in humans and 4-1BB in the mouse. 4-1BB is a member of the tumour necrosis factor receptor superfamily. Both human and mouse are also referred to as CD137 and occur as either a monomer or dimer on the surface of cells. 4-1BB was cloned by differential screening in the course of a search for cDNAs expressed preferentially in murine cytolytic and helper T-cell clones. The human homologue was cloned from a cDNA library derived from activated human T-cell leukemia virus type 1-transformed human T lymphocytes. The receptor functions mainly as a costimulatory molecule in T lymphocytes. In addition, several lines of evidence have shown that interactions between 4-1BB and its ligand are involved in the antigen-presentation process and the generation of cytotoxic T cells. Recent studies, however, have demonstrated that 4-1BB plays more diverse roles: Signals through 4-1BB are important for long-term survival of CD8+ T cells and the induction of helper T cell anergy. Clinically, there is great interest in 4-1BB, because T-cell activation induced by anti-4-1BB monoclonal antibodies is highly efficient in the eradication of established tumour cells in mice. Now, since mice deficient in 4-1BB or the 4-1BB ligand are available, subtle roles played by 4-1BB may be revealed in the near future (for a review see ref. 9). Interactions of 4-1BBL with 4-1BB may also lead to T-cell apoptosis as a result of reverse signalling through 4-1BBL. Thus 4-1BB/4-1BBL interactions appear to function in a complex manner in the regulation of T-cell survival and death. There is also recent evidence in favour of a role for 4-1BB during Th1 and Th2, Tc1 and Tc2 development[10]. During T-cell activation, when acute activation occurs via the TCR, 4-1BB has been shown to replace CD28 in the stimulation of naïve T cells[11,12].

Distribution

4-1BB is preferentially expressed on the surface of a variety of T-cell populations, including activated CD4+, CD8+ and NK cells[2]. 4-1BB mRNA has also been detected in several other tissues, including B cells, monocytes, fibroblasts, epithelial cells and various transformed cell lines.

Physiochemical properties of the 4-1BBL receptor

Property	Human	Mouse
Amino acids – precursor	255	256
– mature[a]	237	232
M_r (K) – predicted	27.89	27.59
– expressed	35	30–35
Potential *N*-linked glycosylation sites	2	2
Affinity K_d (M)	1×10^{-9}	1×10^{-9}

[a] After removal of predicted signal peptide.

Signal transduction

Interactions between 4-1BB and its cognate ligand 4-1BBL lead to activation of NFκB. The cytosolic domain of 4-1BB, like that of other members of the TNF receptor superfamily, interacts with TNF receptor-associated factors. TRAF1, 2 and 3 have been shown to associate with the runs of acidic residues within the cytoplasmic domain of 4-1BB and to transduce signals from 4-1BB to NFκB activation[13].

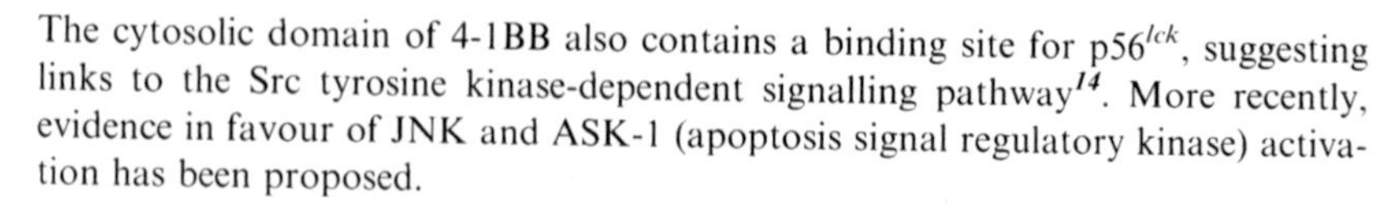

The cytosolic domain of 4-1BB also contains a binding site for p56^{lck}, suggesting links to the Src tyrosine kinase-dependent signalling pathway[14]. More recently, evidence in favour of JNK and ASK-1 (apoptosis signal regulatory kinase) activation has been proposed.

Chromosomal location

The gene for 4-1BB is located on mouse chromosome 4, spanning 13 kb.

Amino acid sequence for human 4-1BB[6]

Accession code: SwissProt Q07011

-17 MGNSCYNIVA TLLLVLN
1 FERTRSLQDP CSNCPAGTFC DNNRNQICSP CPPNSFSSAG GQRTCDICRQ
51 CKGVFRTRKE CSSTSNAECD CTPGFHCLGA GCSMCEQDCK QGQELTKKGC
101 KDCCFGTFND QKRGICRPWT **N**CSLDGKSVL V**N**GTKERDVV CGPSPADLSP
151 GASSVTPPAP AREPGHSPQI ISFFLALTST ALLFLLFFLT LRFSVVKRGR
201 KKLLYIFKQP FMRPVQTTQE EDGCSCRFPE EEEGGCEL

The transmembrane region is underlined. There is a conflict in the sequence at position 107 K→R.

Amino acid sequence for mouse 4-1BB

Accession code: SwissProt P20334

-24 *MGNNCYNVVV IVLLLVGCEK VGAV*
1 QNSCDNCQPG TFCRKYNPVC KSCPPSTFSS IGGQPNCNIC RVCAGYFRFK
51 KFCSSTHNAE CECIEGFHCL GPQCTRCEKD CRPGQELTKQ GCKTCSLGTF
101 NDQ**N**GTGVCR PWT**N**CSLDGR SVLKTGTTEK DVVCGPPVVS FSPSTTISVT
151 PEGGPGGHSL QVLTLFLALT SALLLALIFI TLLFSVLKWI RKKFPHIFKQ
201 PFKKTTGAAQ EEDACSCRCP QEEEGGGGGY EL

The transmembrane region is underlined.

References

1. Hurtado, J.C. et al. (1995) J. Immunol. 155, 3360–3367.
2. Vinay, D.S. and Kwon, B.S. (1998) Semin. Immunol. 10, 481–489.
3. Gramaglia, I. et al. (2000) Eur. J. Immunol. 30, 392–402.
4. Goodwin, R.G. et al. (1993) Eur. J. Immunol. 23, 2631–2641.
5. Zhou, Z. et al. (1995) Immunol. Lett. 45, 67–73.
6. Alderson, M.R. et al. (1994) Eur. J. Immunol. 24, 2219–2227.
7. DeBenedette, M.A. et al. (1997) J. Immunol. 158, 551–559.
8. Pollok, K.E. et al. (1994) Eur. J. Immunol. 24, 367–374.
9. Kwon, B. et al. (2000) Mol. Cells 10, 119–126.
10. Vinay, D.S. and Kwon, B.S. (1999) Cell. Immunol. 192, 63–71.
11. Vinay, D.S. and Kwon, B.S. (1999) Immunobiology 200, 246–263.
12. Watts, T.H. and DeBenedette, M.A. (1999) Curr. Opin. Immunol. 11, 286–293.
13. Arch, R.H. and Thompson, C.B. (1998) Mol. Cell Biol. 18, 558–565.
14. Kim, Y.J. et al. (1993) J. Immunol. 151, 1255–1262.

Other names
BCA-1 is also known as B lymphocyte chemoattractant (BLC).

THE MOLECULE

B cell-attracting chemokine 1 (BCA-1) is a potent chemoattractant for B lymphocytes, and is involved in the formation of B-cell areas of lymphoid organs[1–3]. BCA-1 is inactive on T cells, neutrophils and monocytes[2,3].

Crossreactivity
Only the human form has been described.

Sources
BCA-1 mRNA is constitutively expressed in liver, spleen, lymph nodes, appendix and stomach[2,3].

Bioassays
Chemotaxis of peripheral blood B cells.

Physicochemical properties of BCA-1

Property	Human	Mouse
Amino acids – precursor	109	?
– mature	87	?
M_r (K) – predicted	12.7	?
– expressed	?	?
Potential *N*-linked glycosylation sites	1	?
Disulfide bonds	2	?

Chemically synthesized BCA-1 was active, therefore glycosylation is not essential.

3-D structure
No information.

Amino acid sequence for human BCA-1

Accession code: SwissProt O43927

```
-22  MKFISTSLLL MLLVSSLSPV QG
  1  VLEVYYTSLR CRCVQESSVF IPRRFIDRIQ ILPRGNGCPR KEIIVWKKNK
 51  SIVCVDPQAE WIQRMMEVLR KRSSSTLPVP VFKRKIP
```

THE BCA-1 RECEPTOR, CXCR5

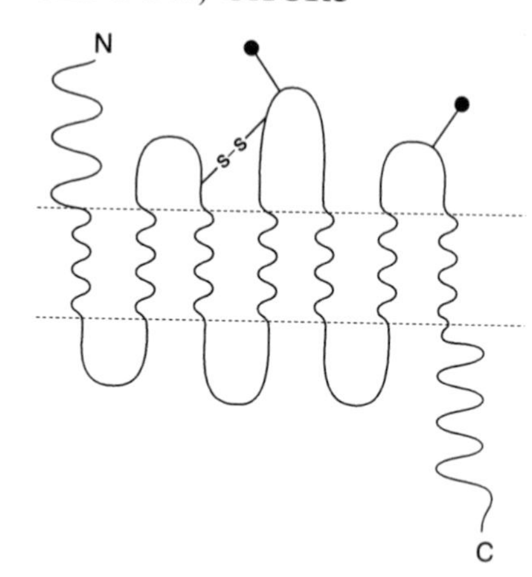

The BCA-1 receptor was identified as the BRL1 or MDR-15 gene, now known as CXCR5. BRL1 and MDR-15 are splice variants of the same gene[4–6]. Knockout mice lacking the BRL1 gene lack inguinal lymph nodes, have few Peyer's patches with an abnormal architecture and no germinal centres in the spleen, consistent with a deficit in B-cell migration[1].

Distribution

CXCR5 is expressed on B lymphocytes and Burkitt's lymphoma cells, but not plasma cells, some T helper memory cells and cerebellar neurons[4–6].

Physicochemical properties of the receptor

Property	Human	Mouse
Amino acids – precursor	372[a]	374
– mature	?	?
M_r (K) – predicted	42.0	42.1
– expressed	?	?
Potential *N*-linked glycosylation sites	2	2
Affinity K_d (M)[b]	1×10^{-8} to 10^{-6}	?

[a] The MDR-15 splice variant lacks the N-terminal 45 amino acids.
[b] Values are IC_{50} for stimulation of B-cell migration.

Signal transduction

BCA-1 stimulates migration of normal human B cells in the absence of a Ca^{2+} flux, but does mediate a Ca^{2+} flux in murine B cells transfected with the human receptor[2,3]. Its mode of signal transduction is therefore unclear.

Chromosomal location

No information.

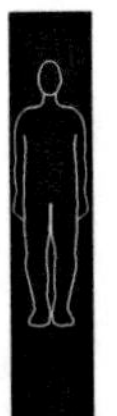

Amino acid sequence for human receptor

Accession code: SwissProt P32302

```
  1  MNYPLTLEMD LENLEDLFWE LDRLDNYNDT SLVENHLCPA TEGPLMASFK
 51  AVFVPVAYSL IFLLGVIGNV LVLVILERHR QTRSSTETFL FHLAVADLLL
101  VFILPFAVAE GSVGWVLGTF LCKTVIALHK VNFYCSSLLL ACIAVDRYLA
151  IVHAVHAYRH RRLLSIHITC GTIWLVGFLL ALPEILFAKV SQGHHNNSLP
201  RCTFSQENQA ETHAWFTSRF LYHVAGFLLP MLVMGWCYVG VVHRLRQAQR
251  RPQRQKAVRV AILVTSIFFL CWSPYHIVIF LDTLARLKAV DNTCKLNGSL
301  PVAITMCEFL GLAHCCLNPM LYTFAGVKFR SDLSRLLTKL GCTGPASLCQ
351  LFPSWRRSSL SESENATSLT TF
```

Residue 344 can be G to S. Residues in italics are absent from the MDR-15 splice variant. Potential disulfide bond Cys122–202.

Amino acid sequence for mouse BCA-1 receptor

Accession code: SwissProt Q04683

```
  1  MNYPLTLDMG SITYNMDDLY KELAFYSNST EIPLQDSNFC STVEGPLLTS
 51  FKAVFMPVAY SLIFLLGMMG NILVLVILER HRHTRSSTET FLFHLAVADL
101  LLVFILPFAV AEGSVGWVLG TFLCKTVIAL HKINFYCSSL LVACIAVDRY
151  LAIVHAVHAY RRRRLLSIHI TCTAIWLAGF LFALPELLFA KVGQPHNNDS
201  LPQCTFSQEN EAETRAWFTS RFLYHIGGFL LPMLVMGWCY VGVVHRLLQA
251  QRRPQRQKAV RVAILVTSIF FLCWSPYHIV IFLDTLERLK AVNSSCELSG
301  YLSVAITLCE FLGLAHCCLN PMLYTFAGVK FRSDLSRLLT KLGCAGPASL
351  CQLFPNWRKS SLSESENATS LTTF
```

Potential disulfide bond Cys124–204.

References

[1] Forster, R. et al. (1996) Cell 87, 1037–1047.
[2] Legler, D.F. et al. (1998) J. Exp. Med. 187, 655–660.
[3] Gunn, M.D. et al. (1998) Nature 391, 799–803.
[4] Forster, R. et al. (1994) Blood 84, 830–840.
[5] Dobner, T. et al. (1992) Eur. J. Immunol. 22, 2795–2799.
[6] Kaiser, E. et al. (1993) Eur. J. Immunol. 23, 2532–2539.

Other names
None.

THE MOLECULE

Brain-derived neurotrophic factor (BDNF) is important in the development and maintenance of the vertebrate nervous system. It promotes the survival of neuronal populations located either in the central nervous system or directly connected to it, and helps to maintain neurons and their differentiated phenotype in the adult. It is similar to NGF and NT-3 but with its own neuronal specificities[1–3].

Crossreactivity
The amino acid sequences of mature BDNF from mouse, human and pig are identical, with complete cross-species reactivity.

Sources
High levels of BDNF mRNA are found in pyramidal and granule cells of the hippocampus, and specific regions of the cortex, and cerebellum of the CNS. Low levels are detectable in the spinal cord[4]. Also in heart, lung and skeletal muscle in adult.

Bioassays
Survival and outgrowth of neurites from embryo chicken dorsal root ganglia. Survival and differentiation of trkB-transfected cell lines.

Physicochemical properties of BDNF

Property	Human	Mouse
pI	9.99	9.99
Amino acids – precursor	247	249
– mature[a]	119	119
M_r (K) – predicted	13.5	13.5
– expressed[b]	27	27
Potential *N*-linked glycosylation sites[c]	0	0
Disulfide bonds[d]	3	3

[a] After removal of propeptide.
[b] Non-covalently linked dimer.
[c] One *N*-linked glycosylation site in the propeptide.
[d] Conserved between NGF, BDNF, NT-3 and NT-4.

3-D structure
The molecule has 70% β-sheet and is expressed as a tightly associated homodimer[5]. Crystal structure is not yet known, but is likely to be similar to NGF.

Gene structure[6]

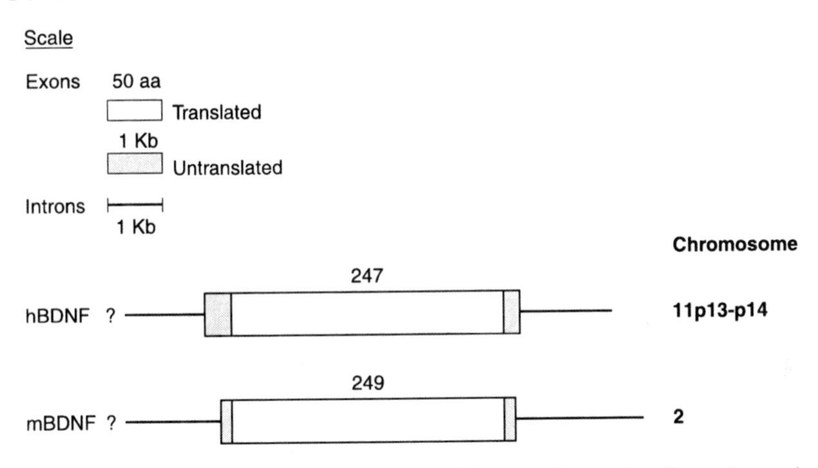

There is some evidence for more than one mRNA species arising from alternative splicing and other ATG initiation sites on additonal upstream exons (similar to NGF).

Amino acid sequence for human BDNF[6,7]

Accession code: SwissProt P23560

-18 MTILFLTMVI SYFGCMKA
1 *APMKEANIRG QGGLAYPGVR THGTLESVNG PKAGSRGLTS LADTFEHVIE*
51 *ELLDEDQKVR PNEENNKDAD LYTSRVMLSS QVPLEPPLLF LLEEYKNYLD*
101 *AA***N***MSMRVRR* HSDPARRGEL SVCDSISEWV TAADKKTAVD MSGGTVTVLE
151 KVPVSKGQLK QYFYETKCNP MGYTKEGCRG IDKRHWNSQC RTTQSYVRAL
201 TMDSKKRIGW RFIRIDTSCV CTLTIKRGR

Variant sequence V→M at position 48. Mature BDNF is formed by removal of a predicted signal peptide and propeptide (in italics, 1–110). Other precursor forms with extended N-terminal sequences similar to NGF may also exist. Disulfide bonds between Cys123–190, 168–219 and 178–221 by similarity.

Amino acid sequence for mouse BDNF[4]

Accession code: SwissProt P21237

-18 MTILFLTMVI SYFGCMKA
1 *APMKEVNVHG QGNLAYPGVR THGTLESVNG PRAGSRGLTT TSLADTFEHV*
51 *IEELLDEDQK VRPNEENHKD ADLYTSRVML SSQVPLEPPL LFLLEEYKNY*
101 *LDAA***N***MSMRV RR*HSDPARRG ELSVCDSISE WVTAADKKTA VDMSGGTVTV
151 LEKVPVSKGQ LKQYFYETKC NPMGYTKEGC RGIDKRHWNS QCRTTQSYVR
201 ALTMDSKKRI GWRFIRIDTS CVCTLTIKRG R

Mature BDNF is formed by removal of a predicted signal peptide and propeptide (in italics, 1–112). Other precursor forms with extended N-terminal sequences similar to NGF may also exist. Disulfide bonds between Cys125–192, 170–221 and 180–223 by similarity.

THE BDNF RECEPTORS

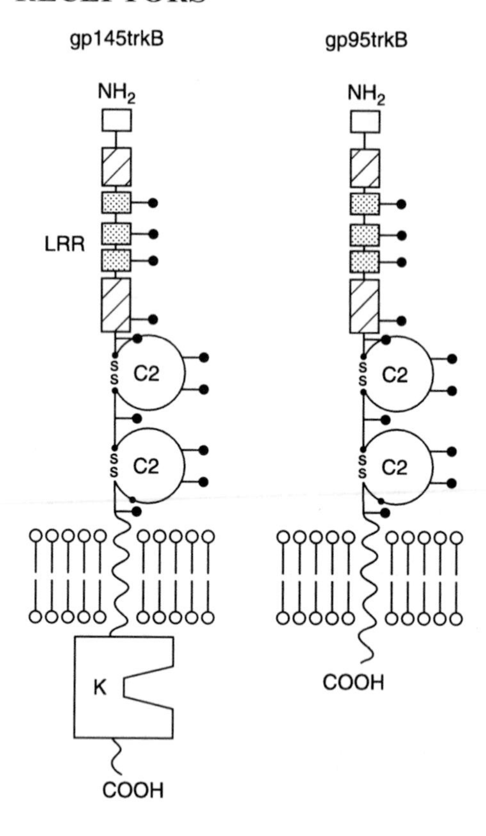

BDNF binds to low-affinity and high-affinity cell surface receptors. The low-affinity receptor is LNGFR, which also binds the other members of the NGF neurotrophin family, NGF, NT-3 and NT-4[8-10] (see NGF entry (page **397**)). Cells expressing LNGFR do not respond to BDNF. Rather, BDNF mediates its neurotrophic properties through gp145^{trkB} (trkB)[3,11-13]. The trkB receptor is a member of the trk family of tyrosine kinases described in the entry for NGF receptors. It exists in dynamic equilibrium between monomeric (low-affinity) and dimeric (high-affinity) forms. Two classes of trkB receptors have been identified. These are gp145^{trkB} and a truncated isoform gp95^{trkB}, which lacks the cytoplasmic kinase domain[14,15]. Additional trkB receptors may also exist[15,16]. Both gp145^{trkB} and gp95^{trkB} have identical extracellular and transmembrane domains, but gp95^{trkB} has a short 23-amino-acid cytoplasmic domain including 11 C-terminal residues, which are different to those on the gp145^{trkB} receptor[15,16]. gp145^{trkB} is also a functional receptor for NT-4 and NT-3 but not NGF[17,18].

Distribution

TrkB mRNA has been shown by *in situ* hybridization studies to be expressed widely throughout the central and peripheral nervous system including non-neuronal cells such as glia and Schwann cells[19]. It has also been shown to be present by Northern blotting in brain, lung, muscle and ovaries[14].

Physicochemical properties of the BDNF receptor (trkB)

	Mouse		Human	
Property	gp145^{trkB}	gp95^{trkB}	gp145^{trkB}	gp95^{trkB}
Amino acids – precursor	821	476	822	477
– mature[a]	790	445	791	446
M_r (K) – predicted	88.6	52	88.3	49
– expressed	145	95	145	95
Potential *N*-linked glycosylation sites	11	11	11	11
Affinity K_d (M)[b,c]	10^{-9}	ND	10^{-9}	?
Kinase activity	yes	no	yes	no

[a] After removal of predicted signal peptide.
[b] TrkB binds NT-3 with similar affinity[12]. This low-affinity binding was measured for trkB expressed in NIH 3T3 pJM8 cells, which are unresponsive to BDNF and NT-3. It is possible that high-affinity binding would be observed in responsive cells similar to NGF binding to trkA.
[c] High-affinity receptor is homodimer.

Signal transduction by gp145^{trkB}

gp145^{trkB} has intrinsic protein tyrosine kinase activity. Both BDNF and NT-3 stimulate receptor autophosphorylation[3,12], and phosphorylation of PLC-γ1 resulting in PIP_2 hydrolysis with release of IP_3 and DAG[20]. LNGFR is not required for signal transduction[17,21].

Chromosomal location

LNGFR is on chromosome 17q21–q22. *trkB* is on human chromosome 9q22.1.

Amino acid sequence for human BDNF receptor (trkB)[22–24]

Accession code: SwissProt Q16620

-31 MSSWIRWHGP AMARLWGFCW LVVGFWRAAF A

1 CPTSCKCSAS RIWCSDPSPG IVAFPRLEPN SVDPE**N**ITEI FIANQKRLEI

51 INEDDVEAYV GLR**N**LTIVDS GLKFVAHKAF LKNSNLQHI**N** FTRNKLTSLS

101 RKHFRHLDLS ELILVGNPFT CSCDIMWIKT LQEAKSSPDT QDLYCL**N**ESS

151 KNIPLANLQI PNCGLPSANL AAP**N**LTVEEG KSITLSCSVA GDPVPNMYWD

201 VGNLVSKHM**N** ETSHTQGSLR IT**N**ISSDDSG KQISCVAENL VGEDQDSV**N**L

251 TVHFAPTITF LESPTSDHHW CIPFTVKGNP KPALQWFYNG AIL**N**ESKYIC

301 TKIHVT**N**HTE YHGCLQLDNP THMNNGDYTL IAKNEYGKDE KQISAHFMGW

351 PGIDDGANPN YPDVIYEDYG TAANDIGDTT **N**RSNEIPSTD VTDKTGREHL

401 SVYAVVVIAS VVGFCLLVML FLLKLARHSK FGMKGPASVI SNDDDSASPL

451 HHISNGSNTP SSSEGGPDAV IIGMTKIPVI ENPQYFGITN SQLKPDTFVQ

501 HIKRHNIVLK RELGEGAFGK VFLAECYNLC PEQDKILVAV KTLKDASDNA

551 RKDFHREAEL LTNLQHEHIV KFYGVCVEGD PLIMVFEYMK HGDLNKFLRA

601 HGPDAVLMAE GNPPTELTQS QMLHIAQQIA AGMVYLASQH FVHRDLATRN

651 CLVGENLLVK IGDFGMSRDV YSTDYYRVGG HTMLPIRWMP PESIMYRKFT

701 TESDVWSLGV VLWEIFTYGK QPWYQLSNNE VIECITQGRV LQRPRTCPQE

751 VYELMLGCWQ REPHMRKNIK GIHTLLQNLA KASPVYLDIL G

Variable splicing at position 436 (PASVISNDDDS→FVLFHKIPLDG) gives rise to a gp95^{trkB} variant missing the cytoplasmic kinase domain. Tyr675 is site of autophosphorylation.

Amino acid sequence for mouse BDNF receptor (trkB)[14]

Accession code: SwissProt P15209

-31 MSPWLKWHGP AMARLWGLCL LVLGFWRASL A

1 CPTSCKCSSA RIWCTEPSPG IVAFPRLEPN SVDPE**N**ITEI LIANQKRLEI

51 INEDDVEAYV GLR**N**LTIVDS GLKFVAYKAF LKNSNLRHI**N** FTRNKLTSLS

101 RRHFRHLDLS DLILTGNPFT CSCDIMWLKT LQETKSSPDT QDLYCL**N**ESS

151 KNMPLANLQI PNCGLPSARL AAP**N**LTVEEG KSVTLSCSVG GDPLPTLYWD

201 VGNLVSKHM**N** ETSHTQGSLR IT**N**ISSDDSG KQISCVAENL VGEDQDSV**N**L

251 TVHFAPTITF LESPTSDHHW CIPFTVRGNP KPALQWFYNG AIL**N**ESKYIC

301 TKIHVT**N**HTE YHGCLQLDNP THMNNGDYTL MAKNEYGKDE RQISAHFMGR

351 PGVDYETNPN YPEVLYEDWT TPTDIGDTT**N** KSNEIPSTDV ADQSNREHLS

401 VYAVVVIASV VGFCLLVMLL LLKLARHSKF GMKGPASVIS NDDDSASPLH

451 HISNGSNTPS SSEGGPDAVI IGMTKIPVIE NPQYFGITNS QLKPDTFVQH

501 IKRHNIVLKR ELGEGAFGKV FLAECYNLCP EQDKILVAVK TLKDASDNAR

551 KDFHREAELL TNLQHEHIVK FYGVCVEGDP LIMVFEYMKH GDLNKFLRAH

601 GPDAVLMAEG NPPTELTQSQ MLHIAQQIAA GMVYLASQHF VHRDLATRNC

651 LVGENLLVKI GDFGMSRDVY STDYYRVGGH TMLPIRWMPP ESIMYRKFTT

701 ESDVWSLGVV LWEIFTYGKQ PWYQLSNNEV IECITQGRVL QRPRTCPQEV

751 YELMLGCWQR EPHTRKNIKS IHTLLQNLAK ASPVYLDILG

Variable splicing at position 435 (PASVISNDDDS→FVLFHKIPLDG) gives rise to a gp95^{trkB} variant missing the cytoplasmic kinase domain[15]. Tyr674 is site of autophosphorylation.

The sequences for human and rat LNGFR are given in the NGF entry (page **397**).

References

1 Yancopoulos, G.D. et al. (1990) Cold Spring Harbor Symp. Quant. Biol. LV, 371–379.
2 Ebendal, T. (1992) J. Neurosci. Res. 32, 461–470.
3 Friedman, W.J. and Greene, L.A. (1999) Exp. Cell Res. 253, 131–142.
4 Hofer, M. et al. (1990) EMBO J. 9, 2459–2464.
5 Radziejewski, C. et al. (1992) Biochemistry 31, 4431–4436.
6 Maisonpierre, P.C. et al. (1991) Genomics 10, 558–568.
7 Jones, K.R. and Reichardt, L.F. (1990) Proc. Natl Acad. Sci. USA 87, 8060–8064.
8 Rodriguez-Tebar, A. et al. (1992) EMBO J. 11, 917–922.
9 Chao, M.V. (1991) Microbiol. Immunol. 165, 39–53.
10 Meakin, S.O. and Shooter, E.M. (1992) Trends Neurosci. 15, 323–331.
11 Squinto, S.P. et al. (1991) Cell 65, 885–893.
12 Soppet, D. et al. (1991) Cell 65, 895–903.
13 Klein, R. et al. (1991) Cell 66, 395–403.
14 Klein, R. (1989) EMBO J. 8, 3701–3709.
15 Klein, R. et al. (1990) Cell 61, 647–656.
16 Middlemas, D.S. et al. (1991) Mol. Cell Biol. 11, 143–153.
17 Glass, D.J. et al. (1991) Cell 66, 405–413.
18 Klein, R. et al. (1992) Neuron 8, 947–956.
19 Klein, R. et al. (1990) Development 109, 845–850.
20 Widmer HR et al. (1992) J. Neurochem. 59, 2113–2124.
21 Barker, P.A. and Murphy, R.A. (1992) Mol. Cell Biochem. 110: 1–15.
22 Allen, S.J. et al. (1994) Neuroscience 60, 825–834.
23 Nakagawara, A. et al. (1995) Genomics 20,25, 538–546.
24 Shelton, D.L. et al. (1995) J. Neurosci. 15, 477–491.

Other names

None.

THE MOLECULE

Betacellulin is a member of the EGF family of growth factors which functions as a potent mitogen for retinal pigment epithelial and vascular smooth muscle cells[1]. Human and mouse betacellulin are 32-kDa glycoproteins which are processed from larger transmembrane precursors by proteolytic cleavage[1,2]. Betacellulin is a ligand for the EGF receptor and acts as a potent mitogen for retinal pigment epithelial cells and vascular smooth muscle cells. There is emerging evidence that betacellulin from beta cells could play a role in vascular complications associated with diabetes[3].

Crossreactivity

The amino acid sequence of the human betacellulin precursor protein exhibits 78% similarity with that of the mouse precursor protein[2]. The C-terminal domain of murine betacellulin has 50% sequence similarity with that of rat transforming growth factor α[1]. Bovine and rat betacellulin have also more recently been described.

Sources

Betacelluin was found to be expressed mainly in human pancreas and small intestine, while murine betacellulin is expressed in kidney, liver and beta tumour cell lines, with no expression in human or mouse brain. The restricted pattern of expression suggests that betacellulin possesses some specific function distinct from those of other members of the EGF family.

Bioassays

Proliferation of retinal pigment epithelial and vascular smooth muscle cells[1].

Physicochemical properties of betacellulin

Property	Human	Mouse
pI[a]	8.18	8.69
Amino acids – precursor	178	177
– mature[b]	80	80
M_r (K) – predicted	8.98	9.052
– expressed	32	32
Potential *N*-linked glycosylation sites	1	3
Disulfide bonds	3	3

[a] This represents the predicted pI of human and mouse betacellulin.
[b] Both human and mouse betacellulin contain putative signal peptides (residues 1–31) in addition to propeptides (residues 112–178 (human) and 112–177 (mouse)) which are removed in the mature molecules.

3-D structure

Not known.

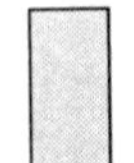

Gene structure

Betacellulin is localized to human chromosome 4q13–q21[2] and mouse chromosome 5[1].

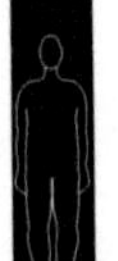

Amino acid sequence for human betacellulin

Accession code: SwissProt P35070

-31 MDRAARCSGA SSLPLLLALA LGLVILHCVV A
1 DGNSTRSPET NGLLCGDPEE NCAATTTQSK RKGHFSRCPK QYKHYCIKGR
51 CRFVVAEQTP SCVCDEGYIG ARCERVDLFY *LRGDRGQILV ICLIAVMVVF*
101 *IILVIGVCTC CHPLRKRRKR KKKEEEMETL GKDITPINED IEETNIA*

Disulfide bridges between Cys69–82, 77–93 and 95–104 or unprocessed precursor.

Amino acid sequence for mouse betacellulin

Accession code: SwissProt Q05928

-31 MDPTAPGSSV SSLPLLLVLA LGLAILHCVV A
1 DGNTTRTPET NGSLCGAPGE NCTGTTPRQK VKTHFSRCPK QYKHYCIHGR
51 CRFVVDEQTP SCICEKGYFG ARCERVDLFY *LQQDRGQILV VCLIVVMVVF*
101 *IILVIGVCTC CHPLRKHRKK KKEEKMETLD KDKTPISEDI QETNIA*

THE BETACELLULIN RECEPTOR

Betacellulin binds to the EGF receptor (also known as c-*erbB*) with equivalent avidity to that of EGF (K_d 0.5 nM)[4]. See EGF entry (page **203**) for details.

References

[1] Shing, Y. et al. (1993) Science 259, 1604–1607.
[2] Sasada, R. et al. (1993) Biochem. Biophys. Res. Commun. 190, 1173–1179.
[3] Sasada, R. and Igarashi, K. (1993) Nippon Rinsho 51, 3308–3317.
[4] Watanabe, T. et al. (1994) J. Biol. Chem. 269, 9966–9973.

Other names

Members of the BMP family are also known by other names, such as osteogenin (BMP-3)[1], osteogenic protein-1, OP-1 (BMP-7)[2], growth/differentiation factor GDF-11 (BMP-11) and cartilage-derived morphogenetic protein 2 (CDMP-2) (BMP-13).

THE MOLECULE

Bone morphogenetic proteins (BMPs) represent a family of proteins that initiate, promote and maintain cartilage and bone morphogenesis, differentiation and regeneration in both the developing embryo and adult[3–5]. There are more than 30 known BMPs, 15 of which are found in mammals. BMPs belong to the transforming growth factor β (TGFβ) superfamily which includes TGFβs, activins/inhibins, Mullerian-inhibiting substance (MIS) and glial cell line-derived neurotrophic factor[6]. Alignment of mammalian BMPs demonstrates that BMPs 2–15 share a conserved motif which is distinct from the structure of BMP-1. BMP-1, is not a BMP family member, rather it is a procollagen C proteinase related to *Drosophila* tolloid, which has been postulated to regulate BMP activity via proteolysis of BMP antagonists/binding proteins[5]. BMP-2 is involved in cardiac morphogenesis, BMP-4 in mesoderm development in developing embryos while BMP 7 is critical for kidney morphogenesis. For the sake of brevity, only BMP-2 will be further discussed in detail below.

Crossreactivity

Unknown.

Sources

BMPs were initially identified, purified and cloned from bone. BMP-2 is particularly abundant in lung, spleen, colon and in low but significant amounts in heart, brain, liver, skeletal muscle, kidney, pancreas, prostate, ovaries and small intestine[4,5].

Bioassays

In vivo bone induction bioassay[5].

Physicochemical properties of BMP-2

Property	Human	Mouse
pI (calculated)	8.21	8.2
Amino acids – precursor	396	394
– mature[a]	114	146
M_r (K) – predicted	12.90	12.9
– expressed[b]	30–38	30–38
Potential *N*-linked glycosylation sites	4	4
Disulfide bonds	4	4

[a] After removal of signal peptide and propeptide.
[b] The BMPs are dimeric as a result of disulfide bonds which are critical for activity.

3-D structure

The 3-D structure of BMP-2 has been determined at 2.7 Å resolution and shares many common features with TGFβ (i.e. the cycteine-knot motif with two finger-like double-stranded sheets). In contrast to TGFβ, however, the structure of BMP-2 shows differences in the flexibility of the N-terminus and the orientation of the central α-helix as well as two external loops at the fingertips with respect to the scaffold[7].

Gene structure

BMP-2 is located on human chromosome 20 and mouse chromosome 2[8,9].

Amino acid sequence for human BMP-2

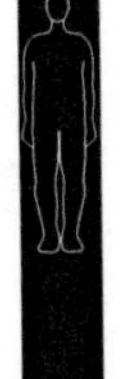

Accession code: SwissProt P12643

```
  1  MVAGTRCLLA LLLPQVLLGG AAGLVPELGR RKFAAASSGR PSSQPSDEVL
 51  SEFELRLLSM FGLKQRPTPS RDAVVPPYML DLYRRHSGQP GSPAPDHRLE
101  RAASRANTVR SFHHEESLEE LPETSGKTTR RFFFNLSSIP TEEFITSAEL
151  QVFREQMQDA LGNNSSFHHR INIYEIIKPA TANSKFPVTR LLDTRLVNQN
201  ASRWESFDVT PAVMRWTAQG HANHGFVVEV AHLEEKQGVS KRHVRISRSL
251  HQDEHSWSQI RPLLVTFGHD GKGHPLHKRE KRQAKHKQRK RLKSSCKRHP
301  LYVDFSDVGW NDWIVAPPGY HAFYCHGECP FPLADHLNST NHAIVQTLVN
351  SVNSKIPKAC CVPTELSAIS MLYLDENEKV VLKNYQDMVV EGCGCR
```

BMP-2 has a potential signal sequence upstream of a propeptide sequence (?-282) shown in italics. The precise locations of these are unclear but they span residues 1–282. Disulfide bonds between Cys296–361, 325–393, 329–395 in human BMP-2 and an interchain disulfide bond between dimers at residue 360, by similarity.

Amino acid sequence for mouse BMP-2[8]

Accession code: SwissProt P21274

```
  1  MVAGTRCLLV LLLPQVLLGG AAGLIPELGR KKFAAASSRP LSRPSEDVLS
 51  EFELRLLSMF GLKQRPTPSK DVVVPPYMLD LYRRHSGQPG APAPDHRLER
101  AASRANTVRT FHQLEAVEEL PEMSGKTARR FFFNLSSVPS DEFLTSAELQ
151  IFREQIQEAL GNSSFQHRIN IYEIIKPAAA NLKFPVTRLL DTRLVNQNTS
201  QWESFDVTPA VMRWTTQGHT NHGFVVEVAH LEENPGVSKR HVRISRSLHQ
251  DEHSWSQIRP LLVTFGHDGK GHPLHKREKR QAKHKQRKRL KSSCKRHPLY
301  VDFSDVGWND WIVAPPGYHA FYCHGECPFP LADHLNSTNH AIVQTLVNSV
351  NSKIPKACCV PTELSAISML YLDENEKVVL KNYQDMVVEG CGCR
```

Mouse BMP-2 has a potential signal sequence upstream of a propeptide sequence. The precise locations of these are unclear but they span residues 1–280 (shown in italics). Disulfide bonds between Cys294–359, 323–391, 327–393 and an interchain dilsulfide bond between dimers at residue 358 by similarity. Conflicting sequence T→S at 110, QL→HE at 113 and G→R at 271[8].

THE BMP RECEPTOR

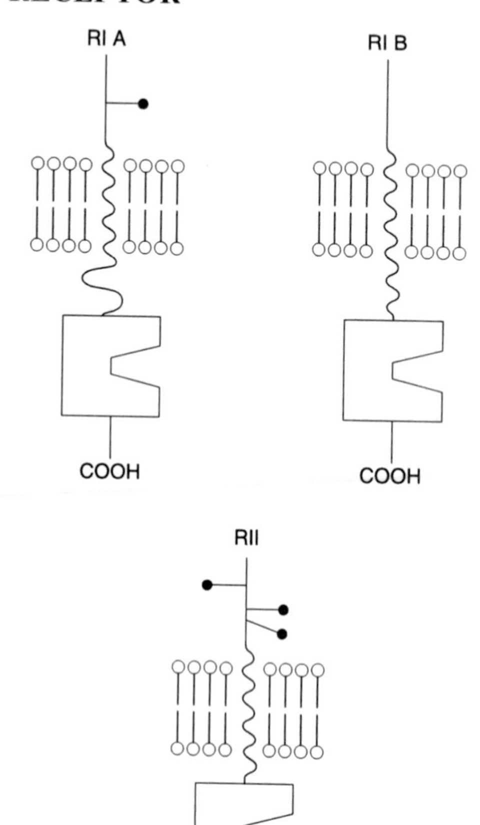

The biological actions of BMPs are mediated by binding to BMP receptor type I and type II (M_r 115.3 kDa), which are members of the TGFβ receptor superfamily. There are two BMP type I receptors: BMPR-1A (M_r 60.2 kDa) (also known as activin receptor-like kinase-3, ALK-3) and BMPR-1B (M_r 56.93 kDa) in mammals. A mammalian BMP type II has also been identified (M_r 115.3 kDa)[10]. Type II receptors bind ligands in the absence of type I receptors, but they require their respective type I receptors for signalling, whereas the type I receptors require their respective type II receptors for ligand binding. Both type I and type II BMP receptors are serine/threonine kinases.

Distribution

BMPR-IB is widely expressed in mammalian tissues, while BMPR-IA is highly expressed in skeletal muscle. BMPR type II is widely distributed in different tissues including heart, brain, placenta, lung, liver, skeletal muscle, kidney, pancreas, prostate, testis, ovary and small intestine[10].

Physicochemical properties of BMP receptors

Property	Type IA		Type IB		Type II	
	Human	Mouse	Human	Mouse	Human	Mouse
Amino acids – precursor	532	532	502	502	1038	?
– mature[a]	508	508	489	489	1012	?
M_r (K) – predicted	60.201	60.06	56.93	56.94	115.3	?
– expressed	?	?	?	?	?	?
Potential *N*-linked glycosylation sites	1	1	0	0	0	0
Affinity K_d (M)	?	?	?	?	?	?

[a] After removal of predicted signal peptide.

Signal transduction

The interaction of the BMP family ligands with the type I and type II receptor serine/threonine kinases results in activation of the type I receptor kinase. The type I and II receptors interact with ligand concurrently. Activation of BMPR-I induces phosphorylation of a set of BMP receptor-regulated acceptor proteins called Smads (Smad 1, 5 and 8). Phosphorylation of Smads facilitates their interaction with co-Smads, which then translocate to the nucleus and bind to target sequences in BMP-regulated genes[11,12].

Chromosomal location

Not known.

Amino acid sequence for human BMPR-IA

Accession code: SwissProt P36894

```
-23  MTQLYIYIRL LGAYLFIISR VQG
  1  QNLDSMLHGT GMKSDSDQKK SENGVTLAPE DTLPFLKCYC SGHCPDDAIN
 51  NTCITNGHCF AIIEEDDQGE TTLASGCMKY EGSDFQCKDS PKAQLRRTIE
101  CCRTNLCNQY LQPTLPPVVI GPFFDGSIRW LVLLISMAVC IIAMIIFSSC
151  FCYKHYCKSI SSRRRYNRDL EQDEAFIPVG ESLKDLIDQS QSSGSGSGLP
201  LLVQRTIAKQ IQMVRQVGKG RYGEVWMGKW RGEKVAVKVF FTTEEASWFR
251  ETEIYQTVLM RHENILGFIA ADIKGTGSWT QLYLITDYHE NGSLYDFLKC
301  ATLDTRALLK LAYSAACGLC HLHTEIYGTQ GKPAIAHRDL KSKNILIKKN
351  GSCCIADLGL AVKFNSDTNE VDVPLNTRVG TKRYMAPEVL DESLNKNHFQ
401  PYIMADIYSF GLIIWEMARR CITGGIVEEY QLPYYNMVPS DPSYEDMREV
451  VCVKRLRPIV SNRWNSDECL RAVLKLMSEC WAHNPASRLT ALRIKKTLAK
201  MVESQDVKI
```

Amino acid sequence for human BMPR-IB

Accession code: SwissProt O00238

-13 MLLRSAGKLN VGT
1 KKEDGESTAP TPRPKVLRCK CHHHCPEDSV NNICSTDGYC FTMIEEDDSG
51 LPVVTSGCLG LEGSDFQCRD TPIPHQRRSI ECCTERNECN KDLHPTLPPL
101 KNRDFVDGPI HHRALLISVT VCSLLLVLII LFCYFRYKRQ ETRPRYSIGL
151 EQDETYIPPG ESLRDLIEQS QSSGSGSGLP LLVQRTIAKQ IQMVKQIGKG
201 RYGEVWMGKW RGEKVAVKVF FTTEEASWFR ETEIYQTVLM RHENILGFIA
251 ADIKGTGSWT QLYLITDYHE NGSLYDYLKS TTLDAKSMLK LAYSSVSGLC
301 HLHTEIFSTQ GKPAIAHRDL KSKNILVKKN GTCCIADLGL AVKFISDTNE
351 VDIPPNTRVG TKRYMPPEVL DESLNRNHFQ SYIMADMYSF GLILWEVARR
401 CVSGGIVEEY QLPYHDLVPS DPSYEDMREI VCIKKLRPSF PNRWSSDECL
451 RQMGKLMTEC WAHNPASRLT ALRVKKTLAK MSESQDIKL

Amino acid sequence for human BMPR-II

Accession code: SPTREMBLE Q13873

-26 MTSSLQRPWR VPWLPWTILL VSTAAA
1 SQNQERLCAF KDPYQQDLGI GESRISHENG TILCSKGSTC YGLWEKSKGD
51 INLVKQGCWS HIGDPQECHY EECVVTTTPP SIQNGTYRFC CCSTDLCNVN
101 FTENFPPPDT TPLSPPHSFN RDETIIIALA SVSVLAVLIV ALCFGYRMLT
151 GDRKQGLHSM NMMEAAASEP SLDLDNLKLL ELIGRGRYGA VYKGSLDERP
201 VAVKVFSFAN RQNFINEKNI YRVPLMEHDN IARFIVGDER VTADGRMEYL
251 LVMEYYPNGS LCKYLSLHTS DWVSSCRLAH SVTRGLAYLH TELPRGDHYK
301 PAISHRDLNS RNVLVKNDGT CVISDFGLSM RLTGNRLVRP GEEDNAAISE
351 VGTIRYMAPE VLEGAVNLRD CESALKQVDM YALGLIYWEI FMRCTDLFPG
401 ESVPEYQMAF QTEVGNHPTF EDMQVLVSRE KQRPKFPEAW KENSLAVRSL
451 KETIEDCWDQ DAEARLTAQC AEERMAELMM IWERNKSVSP TVNPMSTAMQ
501 NERNLSHNRR VPKIGPYPDY SSSSYIEDSI HHTDSIVKNI SSEHSMSSTP
551 LTIGEKNRNS INYERQQAQA RIPSPETSVT SLSTNTTTTN TTGLTPSTGM
601 TTISEMPYPD ETNLHTTNVA QSIGPTPVCL QLTEEDLETN KLDPKEVDKN
651 LKESSDENLM EHSLKQFSGP DPLSSTSSSL LYPLIKLAVE ATGQQDFTQT
701 ANGQACLIPD VLPTQIYPLP KQQNLPKRPT SLPLNTKNST KEPRLKFGSK
751 HKSNLKQVET GVAKMNTINA AEPHVVTVTM NGVAGRNHSV NSHAATTQYA
801 NRTVLSGQTT NIVTHRAQEM LQNQFIGEDT RLNINSSPDE HEPLLRREQQ
851 AGHDEGVLDR LVDRRERPLE GGRTNSNNNN SNPCSEQDVL AQGVPSTAAD
901 PGPSKPRRAQ RPNSLDLSAT NVLDGSSIQI GESTQDGKSG SGEKIKKRVK
951 TPYSLKRWRP STWVISTESL DCEVNNNGSN RAVHSKSSTA VYLAEGGTAT
1001 TMVSKDIGMN CL

Conflict in sequence R→G at position 828 of precursor sequence.

Amino acid sequence for mouse BMPR-IA

Accession code: SwissProt P36895

```
-23  MTQLYTYIRL LGACLFIISH VQG
  1  QNLDSMLHGT GMKSDLDQKK PENGVTLAPE DTLPFLKCYC SGHCPDDAIN
 51  NTCITNGHCF AIIEEDDQGE TTLTSGCMKY EGSDFQCKDS PKAQLRRTIE
101  CCRTNLCNQY LQPTLPPVVI GPFFDGSIRW LVVLISMAVC IVAMIIFSSC
151  FCYKHYCKSI SSRGRYNRDL EQDEAFIPVG ESLKDLIDQS QSSGSGSGLP
201  LLVQRTIAKQ IQMVRQVGKG RYGEVWMGKW RGEKVAVKVF FTTEEASWFR
251  ETEIYQTVLM RHENILGFIA ADIKGTGSWT QLYLITDYHE NGSLYDFLKC
301  ATLDTRALLK LAYSAACGLC HLHTEIYGTQ GKPAIAHRDL KSKNILIKKN
351  GSCCIADLGL AVKFNSDTNE VDIPLNTRVG TKRYMAPEVL DESLNKNHFQ
401  PYIMADIYSF GLIIWEMARR CITGGIVEEY QLPYYNMVPS DPSYEDMREV
451  VCVKRLRPIV SNRWNSDECL RAVLKLMSEC WAHNPASRLT ALRIKKTLAK
501  MVESQDVKI
```

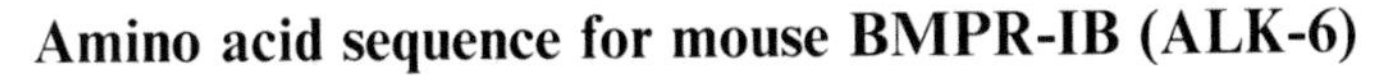

Amino acid sequence for mouse BMPR-IB (ALK-6)

Accession code: SwissProt P36898

```
-13  MLLRSSGKLN VGT
  1  KKEDGESTAP TPRPKILRCK CHHHCPEDSV NNICSTDGYC FTMIEEDDSG
 51  MPVVTSGCLG LEGSDFQCRD TPIPHQRRSI ECCTERNECN KDLHPTLPPL
101  KDRDFVDGPI HHKALLISVT VCSLLLVLII LFCYFRYKRQ EARPRYSIGL
151  EQDETYIPPG ESLRDLIEQS QSSGSGSGLP LLVQRTIAKQ IQMVKQIGKG
201  RYGEVWMGKW RGEKVAVKVF FTTEEASWFR ETEIYQTVLM RHENILGFIA
251  ADIKGTGSWT QLYLITDYHE NGSLYDYLKS TTLDAKSMLK LAYSSVSGLC
301  HLHTEIFSTQ GKPAIAHRDL KSKNILVKKN GTCCIADLGL AVKFISDTNE
351  VDIPPNTRVG TKRYMPPEVL DESLNRNHFQ SYIMADMYSF GLILWEIARR
401  CVSGGIVEEY QLPYHDLVPS DPSYEDMREI VCMKKLRPSF PNRWSSDECL
451  RQMGKLMTEC WAQNPASRLT ALRVKKTLAK MSESQDIKL
```

References

[1] Luyten, F.P. et al. (1989) J. Biol. Chem. 264, 13377–13380.
[2] Ozkaynak, E. et al. (1991) Biochem. Biophys. Res. Commun. 179, 116–123.
[3] Harland, R.M.(1994) Proc. Natl Acad. Sci. USA 91, 10243–10246.
[4] Maiti, S.K. and Singh, G.R. (1998) Ind. J. Exp. Biol. 36, 237–244.
[5] Wozney, J.M. et al. (1988) Science 242, 1528–1534.
[6] Kingsley, D.M. (1994) Trends Genet. 10, 16–21.
[7] Scheufler, C. et al. (1999) J. Mol. Biol. 287, 103–115.
[8] Feng, J.Q. et al. 1994) Biochim. Biophys. Acta 1218, 221–224.
[9] Ghosh-Choudhury, N. et al. (1994) C. R. Eukaryot. Gene Expr. 4, 345–355.
[10] Rosenzweig, B.L. et al. (1995) Proc. Natl Acad. Sci. USA 92, 7632–7636.
[11] Jonk, L.J. et al. (1998) J. Biol. Chem. 273, 21145–21152.
[12] Kawabata, M. et al. (1998) Cytokine Growth Factor Rev. 9, 49–61.

Other names

CD70, Ki-24 antigen.

THE MOLECULE

CD27 ligand (CD27L) is a lymphocyte-specific member of the TNFR family that is highly induced upon activation of T and B cells. CD27L induces the proliferation of costimulated T cells and enhances the generation of cytolytic T cells. CD27L was origionally identified by the use of the Ki-24 monoclonal antibody and human CD27L cDNA was subsequently isolated using a probe formed from the extra-cellular domain of CD27L fused to the constant domain of the human immunoglobulin molecule[1]. More recently, mouse CD27L has been described[2]. CD27L functions in the early immune response which has been implicated in plasma cell differentiation, and may regulate the size and function of antigen-primed lymphocyte populations[3].

Crossreactivity

Murine CD27L shares 62% homology with its human counterpart[2].

Sources

Expression of CD27L is predominantly confined to activated lymphocytes including NK cells[4], B cells[5], CD45RO+, CD4+ and CD8+ T cells[6], γδ T cells[7] and certain types of leukaemic B cells[8].

Bioassays

Induces the proliferation of PHA-costimulated T cells[9].

Physicochemical properties of CD27L

Property	Human	Mouse
pI[a]	8.93	8.14
Amino acids – precursor	193	195
– mature[b]	193	195
M_r (K) – predicted	21.1	22.0
– expressed	50	?
Potential *N*-linked glycosylation sites	2	3
Disulfide bonds	0	0

[a] Predicted pI of complete sequence.
[b] CD27L has no signal peptide.

3-D structure

A crystal structure for CD27L is unknown, but is predicted to be similar to that of TNFα.

Gene structure

CD27L is localized to human chromosome 19p13[1]. Murine CD27L consists of three exons spanning approximately 4 kb of DNA and is localized on chromosome 17[2].

Amino acid sequence for human CD27L[1]

Accession code: SwissProt P32970

1 MPEEGSGCSV RRRPYGCVLR AALVPLVAGL VICLVVCIQR FAQAQQQLPL
51 ESLGWDVAEL QL**N**HTGPQQD PRLYWQGGPA LGRSFLHGPE LDKGQLRIHR
101 DGIYMVHIQV TLAICSSTTA SRHHPTTLAV GICSPASRSI SLLRLSFHQG
151 CTIVSQRLTP LARGDTLCT**N** LTGTLLPSRN TDETFFGVQW VRP

Conflict in sequence V→A at position 154.

Amino acid sequence for mouse CD27L[2]

Accession code: SPTREMBLE O55237

1 MPEEGRPCPW VRWSGTAFQR QWPWLLLVVF ITVFCCWFHC SGLLSKQQQR
51 LLEHPEPHTA ELQLNLTVPR KDPTLRWGAG PALGRSFTHG PELEEGHLRI
101 HQDGLYRLHI QVTLANCSSP GSTLQHRATL AVGICSPAAH GISLLRGRFG
151 QDCTVALQRL TYLVHGDVLC TNLTLPLLPS RNADETFFGV QWICP

THE CD27L RECEPTOR

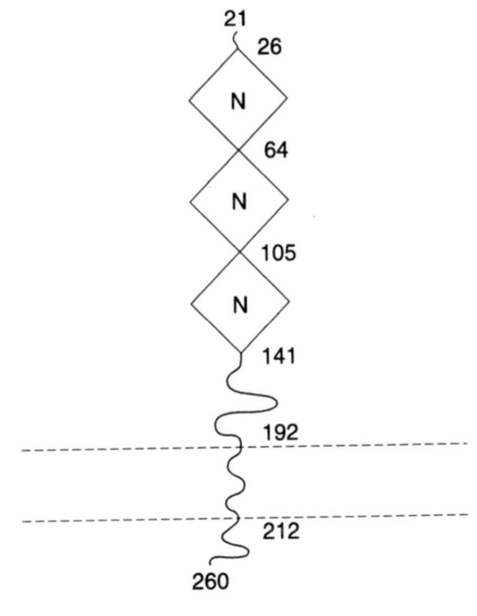

CD27L binds CD27, a dimeric transmembrane receptor present on the majority of human T lymphocytes[10]. Human CD27 has been found only on lymphocytes. In T cells, its expression strongly increases in a transient fashion upon antigenic stimulation in agreement with its role in T-cell activation. Murine CD27 shows an identity of 65% compared with human CD27, and 29% identity to 4-1BB, another lymphocyte-specific member of the receptor family defined only at the cDNA level. Human CD27 is a 50–55 kDa type I transmembrane glycoprotein[11] with a predicted molecular weight of 27 kDa. Expression studies identified murine CD27 mRNA in thymus and spleen, but not in nonlymphoid tissues[12]. Murine CD27 and 4-1BB mRNA are expressed with different kinetics during T-cell activation, suggesting that these molecules play different roles in T-cell activation. Peptide antisera identified murine CD27 as a 45-kDa protein on thymocytes and activated T cells[12]. A soluble form of CD27 has also been detected in both blood and urine, most likely the result of proteolytic processing.

Distribution

Cells known to express CD27 include NK cells, B cells, CD4+, CD8+ T cells and thymocytes[10,12].

Physiochemical Properties of the CD27 Receptor

Property	Human	Mouse
Amino acids – precursor	260	250
– mature[a]	240	230
M_r (K) – predicted	29.156	28.164
– expressed	50–55	45
Potential *N*-linked glycosylation sites	1	2
Affinity K_d (M)[b]	1.83×10^{-8}	1.58×10^{-9}

[a] After removal of predicted signal peptide..
[b] Scatchard analysis of soluble CD27-Fc binding to EBV transformed MP-1 cells demonstrates low- and high-affinity binding sites for CD27L, while additional studies using recombinant soluble CD27 binding to CD27L transfected 3T3 cells demonstrates K_d values of 1.14 × 10–8 and 1.25×10^{-9} [1,6].

Signal transduction

Signalling pathways initiated following CD27L–CD27 interactions result in activation of NFκB. This NFκB activation signal is inhibited by dominant negative Traf-2 or intact Traf-3, indicating that Traf-2 and Traf-3 works as a mediator and an inhibitor, respectively.

Chromosomal location

The gene for CD27 maps to human chromosome 19p13[1].

Amino acid sequence for human CD27

Accession code: SwissProt P26842

```
-20  MARPHPWWLC VLGTLVGLSA
  1  TPAPKSCPER HYWAQGKLCC QMCEPGTFLV KDCDQHRKAA QCDPCIPGVS
 51  FSPDHHTRPH CESCRHCNSG LLVRNCTITA NAECACRNGW QCRDKECTEC
101  DPLPNPSLTA RSSQALSPHP QPTHLPYVSE MLEARTAGHM QTLADFRQLP
151  ARTLSTHWPP QRSLCSSDFI RILVIFSGMF LVFTLAGALF LHQRRKYRSN
201  KGESPVEPAE PCRYSCPREE EGSTIPIQED YRKPEPACSP
```

Mature CD27 is formed after removal of the signal peptide. Conflict in sequence A→T at position 59 of precursor sequence.

Amino acid sequence for mouse CD27

Accession code: SwissProt P41272

```
-20  MAWPPPYWLC MLGTLVGLSA
  1  TLAPNSCPDK HYWTGGGLCC RMCEPGTFFV KDCEQDRTAA QCDPCIPGTS
 51  FSPDYHTRPH CESCRHCNSG FLIRNCTVTA NAECSCSKNW QCRDQECTEC
101  DPPLNPALTR QPSETPSPQP PPTHLPHGTE KPSWPLHRQL PNSTVYSQRS
151  SHRPLCSSDC IRIFVTFSSM FLIFVLGAIL FFHQRRNHGP NEDRQAVPEE
201  PCPYSCPREE EGSAIPIQED YRKPEPAFYP
```

Mature CD27 is formed after removal of the signal peptide.

References

[1] Goodwin, R.G. et al. (1993) Cell 73, 447–456.
[2] Tesselaar, K. et al. (1997) J. Immunol. 159, 4959–4965.
[3] Lens, S.M. et al. (1998) Semin. Immunol. 10, 491–499.
[4] Yang, F.C. et al. (1996) Immunology 88, 289–293.
[5] Lens, S.M. et al. (1996) Eur. J. Immunol. 26, 2964–2971.
[6] Agematsu, K. et al. (1995) Eur. J. Immunol. 25, 2825–2829.
[7] Orengo, A.M. et al. (1997) Clin. Exp. Immunol. 107, 608–613.
[8] Ranheim, E.A. et al. (1995) Blood 85, 3556–3565.
[9] Bowman, M.R. et al. (1994) J. Immunol. 152, 1756–1761.
[10] Camerini, D. et al. (1991) J. Immunol. 147, 3165–3169.
[11] Loenen, W.A. et al. (1992) J. Immunol. 149, 3937–3943.
[12] Gravestein, L.A. et al. (1993) Eur. J. Immunol. 23, 943–950.

Other names
TNFSF8.

THE MOLECULE

CD30 ligand (CD30L) is a 40-kDa (26 kDa in the human) type II transmembrane glycoprotein that belongs to the TNF ligand family which binds CD30. Using a chimeric probe consisting of the extracellular domain of CD30 fused to truncated immunoglobulin heavy chains, CD30L was expression cloned from the murine T-cell clone 7B9[1]. The encoded protein is a 239-amino-acid type II membrane protein whose C-terminal domain shows significant homology to TNFα, TNFβ and CD40L. Unlike other members of the TNF ligand family, CD30L is extensively glycosylated. CD30L enhances the proliferation of CD3-activated T cells yet induces differential responses, including cell death, in several CD30+ lymphoma-derived clones. Thus, engagement of CD30L with its cognate receptor CD30 may result in either cell proliferation or cell death depending on the cell type involved and the microenvironmental conditions in which the signal is delivered[1]. CD30L is postulated to play a role in thymic-negative selection based on CD30L knockout studies, where knockout mice have impaired negative selection[2].

Crossreactivity
Human and mouse CD30 share 72% identity at the amino acid level[1].

Sources
Cellular sources of CD30L include monocytes and macrophages[1], B cells plus activated CD4+ and CD8+ T cells[3], neutrophils, megakaryocytes, resting CD2+ T cells, erythroid precursors[4] and eosinophils[5].

Bioassays
Bioassays for CD30L include induction of proliferation of CD4+ T-cell clones and ATL cell lines[6,7].

Physicochemical properties of CD30 ligand

Property	Human	Mouse
pI	7.62	8.43
Amino acids – precursor	234	239
– mature[a]	234	238
M_r (K) – predicted	26	26.5
– expressed	26	40
Potential *N*-linked glycosylation sites	5	6
Disulfide bonds	0	1

[a] CD30L lacks a signal peptide.

3-D structure
The crystal structure for CD30L is unknown, but is predicted to be similar to that of TNFα.

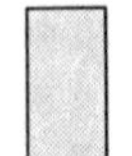

Gene structure

The human and murine CD30L genes map to 9q33 and the proximal region of chromosome 4, respectively[1].

Amino acid sequence for human CD30 ligand

Accession code: SwissProt P32971

```
  1  MDPGLQQALN GMAPPGDTAM HVPAGSVASH LGTTSRSYFY LTTATLALCL
 51  VFTVATIMVL VVQRTDSIPN SPDNVPLKGG NCSEDLLCIL KRAPFKKSWA
101  YLQVAKHLNK TKLSWNKDGI LHGVRYQDGN LVIQFPGLYF IICQLQFLVQ
151  CPNNSVDLKL ELLINKHIKK QALVTVCESG MQTKHVYQNL SQFLLDYLQV
201  NTTISVNVDT FQYIDTSTFP LENVLSIFLY SNSD
```

Amino acid sequence for mouse CD30 ligand

Accession code: SwissProt P32972

```
  1  MEPGLQQAGS CGAPSPDPAM QVQPGSVASP WRSTRPWRST SRSYFYLSTT
 51  ALVCLVVAVA IILVLVVQKK DSTPNTTEKA PLKGGNCSED LFCTLKSTPS
101  KKSWAYLQVS KHLNNTKLSW NEDGTIHGLI YQDGNLIVQF PGLYFIVCQL
151  QFLVQCSNHS VDLTLQLLIN SKIKKQTLVT VCESGVQSKN IYQNLSQFLL
201  HYLQVNSTIS VRVDNFQYVD TNTFPLDNVL SVFLYSSSD
```

THE CD30L RECEPTOR CD30

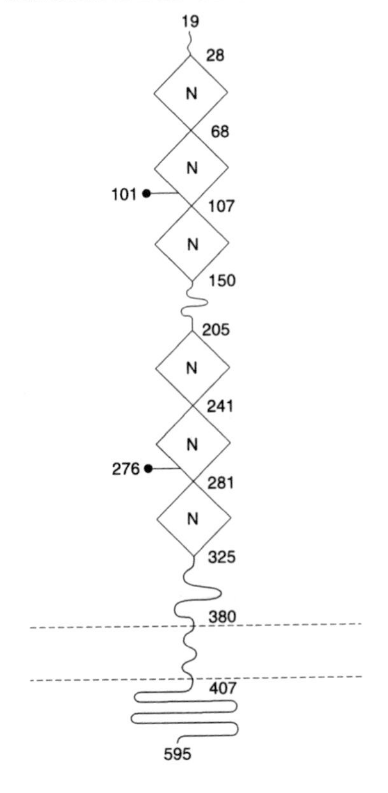

Human CD30 (also known as Ki-1) is a 105–120-kDa transmembrane glycoprotein often associated with Reed-Sternberg cells of Hodgkin's disease[8–10]. CD30 was initially discovered using a monoclonal antibody Ki-1, which was raised against a Hodgkin's disease-derived cell line[11]. Unlike other members of the TNFR superfamily, there are considerable differences between human and mouse CD30, in terms of overall length. Mature human CD30 is 577 amino acid residues long, with a 365-amino-acid extracellular ligand-binding domain and a 188-amino-acid intracellular cytoplasmic domain[9]. Murine CD30, however, has a 90-amino-acid deletion in the extracellular domain, making this region considerably smaller than its human homologue[12]. Overall, there is about 60% amino acid identity between human and mouse CD30. An 85-kDa soluble form of CD30 has been detected in the serum of CD30+ lymphomas, which is proteolytically cleaved from cell surface CD30. In activated T cells, CD30 is preferentially associated with Th2 effector populations, since its expression is partly dependent on the Th2-specific cytokine IL-4.

Distribution

Cells known to express CD30 include Reed-Sternberg cells[13], CD4+[14] and CD8+ T cells[12].

Physiochemical properties of the CD30L

Property	Human	Mouse
Amino acids – precursor	595	480
– mature[a]	573	462
M_r (K) – predicted	63.74	53.21
– expressed	105–120	?
Potential *N*-linked glycosylation sites	2	?
Affinity K_d (M)	?	2.5×10^{-9}

[a] After removal of predicted signal peptide.

Signal transduction

CD30L-CD30 interactions lead to activation of NFκB at levels similar to that observed for TNFα[15]. The cytosolic domain of CD30, like that of TNFR II and CD40, interacts with TNF receptor-associated factors . Traf-2 plays a key role in transducing signals from CD30 to NFκB activation[15]. The region of CD30 which interacts with Traf-2 is also responsible for Traf-1 interactions, thus suggesting that Traf-1 and/or Traf-2 play essential roles in cell death as well as cell proliferation[16]. Binding of Traf-5 to CD30 has also recently been demonstrated[17].

Chromosomal location

The gene for human CD30 is located on the short arm of chromosome 1 at position 1p36[9].

Amino acid sequence for human CD30

Accession code: SwissProt P28908

-18 MRVLLAALGL LFLGALRA
1 FPQDRPFEDT CHGNPSHYYD KAVRRCCYRC PMGLFPTQQC PQRPTDCRKQ
51 CEPDYYLDEA DRCTACVTCS RDDLVEKTPC AWNSSRVCEC RPGMFCSTSA
101 VNSCARCFFH SVCPAGMIVK FPGTAQKNTV CEPASPGVSP ACASPENCKE
151 PSSGTIPQAK PTPVSPATSS ASTMPVRGGT RLAQEAASKL TRAPDSPSSV
201 GRPSSDPGLS PTQPCPEGSG DCRKQCEPDY YLDEAGRCTA CVSCSRDDLV
251 EKTPCAWNSS RTCECRPGMI CATSATNSCA RCVPYPICAA ETVTKPQDMA
301 EKDTTFEAPP LGTQPDCNPT PENGEAPAST SPTQSLLVDS QASKTLPIPT
351 SAPVALSSTG KPVLDAGPVL FWVILVLVVV VGSSAFLLCH RRACRKRIRQ
401 KLHLCYPVQT SQPKLELVDS RPRRSSTQLR SGASVTEPVA EERGL***M***SQPL
451 METCHSVGAA YLESLPLQDA SPAGGPSSPR DLPEPRVSTE HTNNKIEKIY
501 IMKADTVIVG TVKAELPEGR GLAGPAEPEL EEELEADHTP HYPEQETEPP
551 LGSCSDVMLS VEEEGKEDPL PTAASGK

The shorter cytoplasmic form of CD30 (CD30v), which is only expressed in alveolar macrophages, is produced as a result of an alternative initiation codon (the second initiation codon is indicated in bold italics and is in the same reading frame as CD30).

Amino acid sequence for mouse CD30

Accession code: SPTREMBLE Q60846

-18 MSALLTAAGL LFLGMLQA
1 FPTDRPLKTT CAGDLSHYPG EAARNCCYQC PSGLSPTQP CPRGPAHCRK
51 QCAPDYYVNE DGKCTACVTC LPGLVEKAPC SGNSPRICE CQPGMHCCTP
101 AVNSCARCKL HCSGEEVVKS PGTAKKDTIC ELPSSGSGP NCSNPGDRKT
151 LTSHATPQAM PTLESPANDS ARSLLPMRVT NLVQEDATE LVKVPESSSS
201 KAREPSPDPG NAEKNMTLEL PSPGTLPDIS TSENSKEPA STASTLSLVV
251 DAWTSSRMQP TSPLSTGTPF LDPGPVLFWV AMVVLLVGS GSFLLCYWKA
301 CRRRFQQKFH LDYLVQTFQP KMEQTDSCPT EKLTQPQRS GSVTDPSTGH
351 KLSPVSPPPA VETCASVGAT YLENLPLLDD SPAGNPFSP REPPEPRVST
401 EHTNNRIEKI YIMKADTVIV GSVKTEVPEG RAPAGSTES ELEAELEVDH
451 APHYPEQETE PPLGSCTEVM FSVEEGGKED HGPTTVSEK

References

[1] Smith, C.A. et al. (1993) Cell 73, 1349–1360.
[2] Amakawa, R. et al. (1996) Cell 84, 551–562.
[3] Younes, A. et al. (1996) Br. J. Haematol. 93, 569–571.
[4] Gattei, V. et al. (1997) Blood 89, 2048–2059.
[5] Pinto, A. et al. (1996) Blood 88, 3299–3305.
[6] Gruss, H.J. et al. (1994) Blood 83, 2045–2056.
[7] Del Prete, G. et al. (1995) Faseb J. 9, 81–86.
[8] Falini, B. et al. (1992) Br. J. Haematol. 82, 38–45.
[9] Durkop, H. et al. (1997) Cell 68, 421–427.
[10] Gruss, H.J. et al. (1995) Eur. J. Immunol. 25, 2083–2089.
[11] Andreesen, R. et al. (1989) Am. J. Pathol. 134, 187–192.

[12] Bowen, M.A. et al. (1996) J. Immunol. 156, 442–449.
[13] Schwab, U. et al. (1982) Nature 299, 65–67.
[14] Alzona, M. et al. (1994) J. Immunol. 153, 2861–2867.
[15] Lee, S.Y. et al. (1996) Proc. Natl Acad. Sci. USA 93, 9699–9703.
[16] Lee, S.Y. et al. (1996) J. Exp. Med. 183, 669–674.
[17] Gedrich, R.W. et al. (1996) J. Biol. Chem. 271, 12852–12858.

Other names

CD154, gp39, T-BAM or TRAP (TNF-related activating protein).

THE MOLECULE

CD40 ligand (CD40L) is a 39-kDa type II transmembrane glycoprotein that was origionally identified on the surface of CD4+ T cells after T-cell antigen receptor activation[1]. CD40L is a member of the TNF ligand family. Although usually considered to be a membrane-bound protein, natural, proteolytically cleaved 15–18-kDa soluble forms of CD40L with full biological activity have also been described[2–4]. CD40L binds to its receptor CD40 on B cells and dendritic cells and thereby provides a critical helper T cell signal needed for germinal centre formation, isotype class switching and production of immunoglobulin antibodies[5–8]. Defects in CD40L are the cause of an X-linked immunodeficiency with hyperIgM (HIGM1), an immunoglobulin switch defect characterized by elevated concentrations of IgM and decreased amounts of all other isotypes[9,10]. CD40L knockout mice fail to mount secondary antibody responses to T cell-dependent antigens and undergo isotype class switching[11]. CD40L may also transmit a signal back to T cells, resulting in short-term T-cell proliferation. A role for CD40L in proliferation of epithelial, fibroblast and smooth muscle cells has also been demonstrated (see refs 12–14 for reviews on CD40L function).

Crossreactivity

Human and mouse CD40L are 73% identical at the amino acid sequence level and mouse CD40L can act on human cells[15].

Sources

Cellular sources of CD40L include B cells, CD4+ and CD8+ T cells[16], mast cells and basophils[17], eosinophils[18], dendritic cells, and monocytes, NK cells, and γδ T cells.

Bioassays

Induction of immunoglobulin class switching in B cells[1].

Physicochemical properties of CD40 ligand

Property	Human	Mouse
pI	8.53	8.26
Amino acids – precursor	261	260
– mature[a]	239	238
M_r (K) – predicted	29.27	29.39
– expressed	39	39
Potential *N*-linked glycosylation sites	1	1
Disulfide bonds	1	1

[a] After removal of the signal peptide.

3-D structure

A crystal structure of the extracellular domain of CD40L has been described[19] and a 3-D structure has been produced showing how CD40L interacts with the CD40 receptor[20].

Gene structure

CD40L is expressed on human chromosome Xq24[21,22] and mouse X-proximal chromosome.

Amino acid sequence for human CD40 ligand[1]

Accession code: SwissProt P29965

```
  1  MIETYNQTSP RSAATGLPIS MKIFMYLLTV FLITQMIGSA LFAVYLHRRL
 51  DKIEDERNLH EDFVFMKTIQ RCNTGERSLS LLNCEEIKSQ FEGFVKDIML
101  NKEETKKENS FEMQKGDQNP QIAAHVISEA SSKTTSVLQW AEKGYYTMSN
151  NLVTLENGKQ LTVKRQGLYY IYAQVTFCSN REASSQAPFI ASLCLKSPGR
201  FERILLRAAN THSSAKPCGQ QSIHLGGVFE LQPGASVFVN VTDPSQVSHG
251  TGFTSFGLLKL
```

Putative disulfide bond between amino acid residues 178 and 218 of the precursor protein. The putative signal anchor (transmembrane region) is underlined.

Amino acid sequence for mouse CD40 ligand[15]

Accession code: SwissProt P27548

```
  1  MIETYSQPSP RSVATGLPAS MKIFMYLLTV FLITQMIGSV LFAVYLHRRL
 51  DKVEEEVNLH EDFVFIKKLK RCNKGEGSLS LLNCEEMRRQ FEDLVKDITL
101  NKEEKKENSF EMQRGDEDPQ IAAHVVSEAN SNAASVLQWA KKGYYTMKSN
151  LVMLENGKQL TVKREGLYYV YTQVTFCSNR EPSSQRPFIV GLWLKPSIGS
201  ERILLKAANT HSSSQLCEQQ SVHLGGVFEL QAGASVFVNV TEASQVIHRV
251  GFSSFGLLKL
```

Putative disulfide bond between amino acid residues 177 and 217 of the precursor protein. Mature CD40 ligand formed after removal of the signal peptide. The putative signal anchor (transmembrane region) is underlined.

THE CD40L RECEPTOR CD40

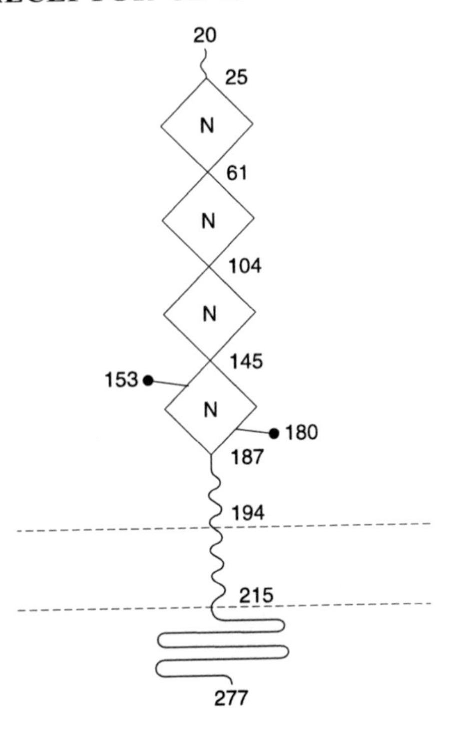

CD40 is a 50-kDa, 277-amino-acid transmembrane glycoprotein, most often associated with B-cell proliferation and differentiation[5]. It was cloned using mAb G28-5[23]. Human CD40 cDNA encodes a 20-amino-acid signal sequence, a 713-amino-acid extracellular region, a 22-amino-acid transmembrane region and a 62-amino-acid cytoplasmic domain[23]. Mouse CD40 is 62% identical to human CD40 at the amino acid level, however it contains an extra 28 amino acid residues in the cytoplasmic region of the molecule[24]. CD40 ligation is associated with the induction of apoptosis, which occurs as a consequence of FasL upregulation (rather than occurring via death domain-mediated pathways) which primes cells for subsequent FasL-mediated cell death[25].

Distribution

Cells known to express CD40 include B cells[24], monocytes and basophils (but not mast cells)[26], eosinophils[40], endothelial cells[27], interdigitating dendritic cells[28], Langerhans cells[29], blood dendritic cells[41], fibroblasts[27], keratinocytes[30], Reed-Sternberg cells of Hodgkin's disease, and Kaposi's sarcoma cells[31].

Physiochemical properties of the CD40

Property	Human	Mouse
Amino acids – precursor	277	289
– mature[a]	258	270
M_r (K) – predicted	30.61	32.11
– expressed	50	50–60
Potential *N*-linked glycosylation sites	2	1
Affinity K_d (M)	?	?

[a] After removal of predicted signal peptide.

Signal transduction

Currently it is believed that the normal signalling pathway following CD40 ligation involves activation of NFκB in a diversity of cell types[32,33], resulting in CD40-induced antibody secretion and ICAM-1/CD54 upregulation. In B cells, CD40-mediated NFκB activation helps to subvert the BCR-mediated cell death[34]. Ligation of CD40 has also been shown to activate members of the MAP kinase family including JNK[35], p38 and p44/p42 MAP kinase (ERK1/2)[36] in a cell-specific manner. Dimerization at least of CD40 is a prerequisite to signalling. Signal transduction via CD40 has also been shown to involve activation of lyn kinase, PI-3 kinase and phosphorylation of phospholipase Cγ2[37]. In addition, association of Traf family members 1, 2, 3 and 6 to CD40 has also been shown[38].

Chromosomal location

Southern blot analysis demonstrates that murine CD40 is a single-copy gene that maps in the distal region of mouse chromosome 2[39].

Amino acid sequence for human CD40[23]

Accession code: SwissProt P25942

```
-19  MVRLPLQCVL WGCLLTAVH
  1  PEPPTACREK QYLINSQCCS LCQPGQKLVS DCTEFTETEC LPCGESEFLD
 51  TWNRETHCHQ HKYCDPNLGL RVQQKGTSET DTICTCEEGW HCTSEACESC
101  VLHRSCSPGF GVKQIATGVS DTICEPCPVG FFSNVSSAFE KCHPWTSCET
151  DLVVQQAGTN KTDVVCGPQD RLRALVVIPI IFGILFAILL VLVFIKKVAK
201  KPTNKAPHPK QEPQEINFPD DLPGSNTAAP VQETLHGCQP
251  VTQEDGKESR ISVQERQ
```

Putative disulfide bond between amino acid residues 26–37, 38–51, 41–59, 62–77, 83–103, 105–119, 111–116, 125–143 of the precursor protein. Mature CD40 formed after removal of the signal peptide.

Amino acid sequence for mouse CD40[24]

Accession code: SwissProt P27512

```
-19  MVSLPRLCAL WGCLLTAVH
  1  LGQCVTCSDK QYLHDGQCCD LCQPGSRLTS HCTALEKTQC HPCDSGEFSA
 51  QWNREIRCHQ HRHCEPNQGL RVKKEGTAES DTVCTCKEGQ HCTSKDCEAC
```

```
101  AQHTPCIPGF GVMEMATETT DTVCHPCPVG FFSNQSSLFE KCYPWTSCED
151  KNLEVLQKGT SQTNVICGLK SRMRALLVIP VVMGILITIF GVFLYIKKVV
201  KKPKDNEMLP PAARRQDPQE MEDYPGHNTA APVQETLHGC QPVTQEDGKE
251  SRISVQERQV TDSIALRPLV
```

Mature CD40 is formed after removal of the signal peptide.

References

[1] Hollenbaugh, D. et al. (1992) EMBO J. 11, 4313–4321.
[2] Pietravalle, F. et al. (1996) Eur. J. Immunol. 26, 725–728.
[3] Pietravalle, F. et al. (1996) J. Biol. Chem. 271, 5965–5967.
[4] Graf, D. et al. (1995) Eur. J. Immunol. 25, 1749–1754.
[5] Van Kooten, C. and Banchereau, J. (1996) Adv. Immunol. 61, 1–77.
[6] Roy, M. et al. (1993) J. Immunol. 151, 2497–2510.
[7] Nonoyama, S. et al. (1993) J. Exp. Med. 178, 1097–1102.
[8] Marshall, L.S. et al. (1993) J. Clin. Immunol. 13, 165–174.
[9] Aruffo, A. et al. (1993) Cell 72, 291–300.
[10] Callard, R.E. et al. (1994) J. Immunol. 153, 3295–3306.
[11] Yu, P. et al. (1999) Eur. J. Immunol. 29, 615–625.
[12] Clark, L.B. et al. (1996) Adv. Immunol. 63, 43–78.
[13] Foy, T.M. et al. (1996) Annu. Rev. Immunol. 14, 591–617.
[14] Foy, T.M. et al. (1994) Semin. Immunol. 6, 259–266.
[15] Armitage, R.J. et al. (1992) Nature 357, 80–82.
[16] Desai-Mehta, A. et al. (1996) J. Clin. Invest. 97, 2063–2073.
[17] Gedrich, R.W. et al. (1996) J. Biol. Chem. 271, 12852–12858.
[18] Gauchat, J.F. et al. (1995) Eur. J. Immunol. 25, 863–865.
[19] Karpusas, M. et al. (1995) Structure 3, 1426.
[20] Singh, J. et al. (1998) Protein Sci. 7, 1124–1135.
[21] Callard, R.E. et al. (1993) Immunol. Today 14, 559–564.
[22] DiSanto, J.P. et al. (1993) Nature 361, 541–543.
[23] Stamenkovic, I. et al. (1989) EMBO J. 8, 1403–1410.
[24] Torres, R.M. and Clark, E.A. (1992) J. Immunol. 148, 620–626.
[25] Schattner, E.J. et al. (1995) J. Exp. Med. 182, 1557–1565.
[26] Agis, H. et al. (1996) Immunology 87, 535–543.
[27] Yellin, M.J. et al. (1995) J. Exp. Med. 182, 1857–1864.
[28] Van Den Berg, T.K. et al. (1996) Immunology 88, 294–300.
[29] Peguet-Navarro, J. et al. (1995) J. Immunol. 155, 4241–4247.
[30] Gaspari, A.A. et al. (1996) Eur. J. Immunol. 26, 1371–1377.
[31] Carbone, A. et al. (1995) Am. J. Pathol. 146, 780–781.
[32] Berberich, I. et al. (1996) EMBO J. 15, 92–101.
[33] Hess, S. et al. (1995) J. Immunol. 155, 4588–4595.
[34] Schauer, S.L. et al. (1996) J. Immunol. 157, 81–86.
[35] Karmann, K. et al. (1996) J. Exp. Med. 184, 173–182.
[36] Sutherland, C.L. et al. (1996) J. Immunol. 157, 3381–3390.
[37] Ren, C.L. et al. (1994) J. Exp. Med. 179, 673–680.
[38] Pullen, S.S. et al. (1998) Biochemistry 37, 11836–11845.
[39] Grimaldi, J.C. et al. (1992) J. Immunol. 149, 3921–3926.
[40] Ohkawara, Y. et al. (1996) J. Clin. Invest. 97, 1761–1766.
[41] McLellan, A.D. (1996) Eur. J. Immunol. 26, 1204–1210.

Other names

Secondary lymphoid-tissue chemokine (SLC), TCA4, exodus-2.

THE MOLECULE

6Ckine is a chemoattractant for T and B lymphocytes, and promotes adhesion to endothelium via B2 and α4β7 integrins[1–7]. It forms a distinct subfamily of the CC chemokines, as it has a C-terminal extension of about 30 amino acids which contains an extra two Cys residues, in addition to the normal four cysteines in other CC chemokines. It is also unusual in binding to both a CC receptor, CCR7, and the CXCR3 receptor, which also binds the CXC chemokines γIP-10, MIG and I-TAC. 6Ckine has been shown to share at least some of the properites of the other CXCR3 ligands, including angiostasis[8].

Crossreactivity

Mouse and human 6Ckine share 86% amino acid identity; human 6Ckine is active on mouse lymphocytes.

Sources

Human 6Ckine is highly expressed in lymph nodes by high endothelial venules, possibly by interdigitating dendritic cells, and is also found in appendix and spleen[1,2,4,6]. Murine 6Ckine is also found in testis, lung, liver, kidney and heart[1,3,4].

Bioassays

Chemotaxis of IL-2-activated lymphocytes.

Physicochemical properties of 6Ckine

Property	Human	Mouse
Amino acids – precursor	134	133
– mature	111	110
M_r (K) – predicted	14.6	14.6
– expressed	15.0	?
Potential *N*-linked glycosylation sites	0	0
Disulfide bonds[a]	2	2

[a] It is not clear if Cys80 and Cys98 form a disulfide bond in the C-terminal extension.

3-D structure

No information.

Gene structure

The gene for human 6Ckine is on chromosome 9p13[2].

Amino acid sequence for human 6Ckine

Accession code: SwissProt O00585

```
-23  MAQSLALSLL ILVLAFGIPR TQG
  1  SDGGAQDCCL KYSQRKIPAK VVRSYRKQEP SLGCSIPAIL FLPRKRSQAE
 51  LCADPKELWV QQLMQHLDKT PSPQKPAQGC RKDRGASKTG KKGKGSKGCK
101  RTERSQTPKG P
```

Amino acid sequence for mouse 6Ckine

Accession code: SwissProt O09006

```
-23  MAQMMTLSLL SLDLALCIPW TQG
  1  SDGGGQDCCL KYSQKKIPYS IVRGYRKQEP SLGCPIPAIL FLPRKHSKPE
 51  LCANPEEGWV QNLMRRLDQP PAPGKQSPGC RKNRGTSKSG KKGKGSKGCK
101  RTEQTQPSRG
```

Potential dibasic cleavage site is in bold italics.

THE RECEPTORS, CCR7 and CXCR3

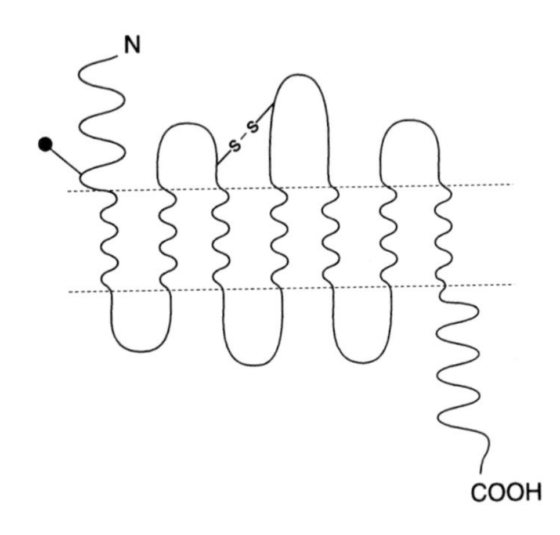

6Ckine binds to both CCR7 and CXCR3 with high affinity, causing a Ca flux, chemotaxis and enhanced adhesion[7–11]. For description of CXCR3, see γIP-10 entry (page **328**). Both receptors are selectively expressed on lymphocytes, CCR7 on B and T cells, and CXCR3 on Th1 cells. CCR7 is also known as Epstein–Barr virus-induced gene 1 (EBI1) or Burkitt's lymphoma receptor 2 (BLR2). CCR7 expression is stimulated by EBV infection of B cells and herpesvirus 6 and infection of T cells[12,13]. It is not clear which receptors 6Ckine predominantly signals through in each of its target cell types.

Distribution

CCR7 is expressed on T and B lymphocytes.

Physicochemical properties of the CCR7 receptor

Property	Human	Mouse
Amino acids – precursor	378	378
– mature	?	?
M_r (K) – predicted	42.9	42.9
– expressed	?	?
Potential *N*-linked glycosylation sites	1	1
Affinity K_d (M)	?	?

Signal transduction

6Ckine binds to CCR7 causing a pertussis toxin-sensitive Ca^{2+} flux with an EC_{50} of between 1 and 10 nM, causing chemotaxis in T lymphocytes with an EC_{50} of less than 1 nM[9–11]. 6Ckine binds to CXCR3 causing a Ca^{2+} flux in transfected HEK293 cells at concentrations below 100 nM, and was angiostatic in a rat corneal assay at 10 nM[8].

Chromosomal location

The human gene for CCR7 is located on chromosome 17q12–q21.2[14].

Amino acid sequence for human CCR7

Accession code: SwissProt P32248

1 MDLGKPMKSV LVVALLVIFQ VCLCQDEVTD DYIGD**N**TTVD YTLFESLCSK
51 KDVRNFKAWF LPIMYSIICF VGLLGNGLVV LTYIYFKRLK TMTDTYLLNL
101 AVADILFLLT LPFWAYSAAK SWVFGVHFCK LIFAIYKMSF FSGMLLLLCI
151 SIDRYVAIVQ AVSAHRHRAR VLLISKLSCV GIWILATVLS IPELLYSDLQ
201 RSSSEQAMRC SLITEHVEAF ITIQVAQMVI GFLVPLLAMS FCYLVIIRTL
251 LQARNFERNK AIKVIIAVVV VFIVFQLPYN GVVLAQTVAN FNITSSTCEL
301 SKQLNIAYDV TYSLACVRCC VNPFLYAFIG VKFRNDLFKL FKDLGCLSQE
351 QLRQWSSCRH IRRSSMSVEA ETTTTFSP

Disulfide bond links Cys129–210.

Amino acid sequence for mouse CCR7

Accession code: SwissProt P47774

1 MDPGKPRKNV LVVALLVIFQ VCFCQDEVTD DYIGE**N**TTVD YTLYESVCFK
51 KDVRNFKAWF LPLMYSVICF VGLLGNGLVI LTYIYFKRLK TMTDTYLLNL
101 AVADILFLLI LPFWAYSEAK SWIFGVYLCK GIFGIYKLSF FSGMLLLLCI
151 SIDRYVAIVQ AVSRHRHRAR VLLISKLSCV GIWMLALFLS IPELLYSGLQ
201 KNSGEDTLRC SLVSAQVEAL ITIQVAQMVF GFLVPMLAMS FCYLIIIRTL
251 LQARNFERNK AIKVIIAVVV VFIVFQLPYN GVVLAQTVAN FNITNSSCET
301 SKQLNIAYDV TYSLASVRCC VNPFLYAFIG VKFRSDLFKL FKDLGCLSQE
351 RLRHWSSCRH VRNASVSMEA ETTTTFSP

Disulfide bond links Cys129–210.

References

[1] Hedrick, J.A. and Zlotnik, A. (1997) J. Immunol. 159, 1589–1593.
[2] Nagira, M. et al. (1997) J. Biol. Chem. 272,19518–19524.
[3] Hromas, R. et al. (1997) J. Immunol. 159, 2554–2558.
[4] Willimann, K. et al. (1998) Eur. J. Immunol. 28, 2025–2034.
[5] Nagira, M. et al. (1998) Eur. J. Immunol. 28, 1516–1523.
[6] Gunn, M.D. et al. (1998) Proc. Natl Acad. Sci. USA 95, 258–263.
[7] Pachynski, R.K. et al. (1998) J. Immunol. 161, 952–956.
[8] Soto, H. et al. (1998) Proc. Natl Acad. Sci. USA 95, 8205–8210.
[9] Yoshida, R. et al. (1998) J. Biol. Chem. 273, 7118–7122.
[10] Campbell, J.J. et al. (1998) J. Cell Biol. 141, 1053–1059.
[11] Yoshie, O. et al. (1997) J. Leuk. Biol. 62, 634–644.
[12] Burgstahler, R. et al. (1995) Biochem. Biophys. Res. Commun. 215, 737–743.
[13] Hasegawa, H. et al. (1994) J. Virol. 68, 5326–5329.
[14] Schweickart, V.L. et al. (1994) Genomics 23, 643–650.

Other names
None.

THE MOLECULE

Ciliary neurotrophic factor (CNTF) promotes the survival and/or differentiation of sympathetic neurons, primary sensory neurons, motor neurons, hippocampal neurons, basal forebrain neurons and type 2 astrocytes. CNTF has also been shown to have effects other than neuronal – it is an endogenous pyrogen and induces acute phase protein expression in hepatocytes. It has no homology with the other neurotrophic growth factors, NGF, BDNF and NT-3. The absence of a signal peptide and *N*-linked glycosylation sites is consistent with CNTF being a cytosolic protein, but how CNTF is secreted is unclear.

Crossreactivity
There is 84% homology between human and rat CNTF, and both human and rat CNTF are active on neurons from chicken dorsal root ganglia.

Sources
CNTF has been purified from embryonic chick eye, and rat and rabbit sciatic nerve.

Bioassays
Activity of CNTF can be determined by its ability to maintain neurons from embryonic chickens.

Physicochemical properties of CNTF

Property	Human	Mouse
pI	6.0	6.0
Amino acids – precursor	200	200
– mature[a]	200	200
M_r (K) – predicted	22.9	22.9
– expressed	22	22.5
Potential *N*-linked glycosylation sites	0	0
Disulfide bonds	0	0

[a] CNTF has no signal peptide.

3-D structure
The 3-D structure of CNTF is not known, but it has significant homologies with IL-6, LIF, oncostatin M and G-CSF. It is thought that these molecules share a four-helix bundle structure[1]. The X-ray structure reveals that CNTF is a dimer.

Gene structure[2]

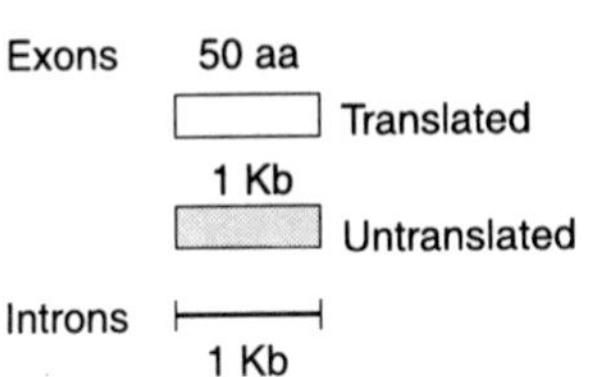

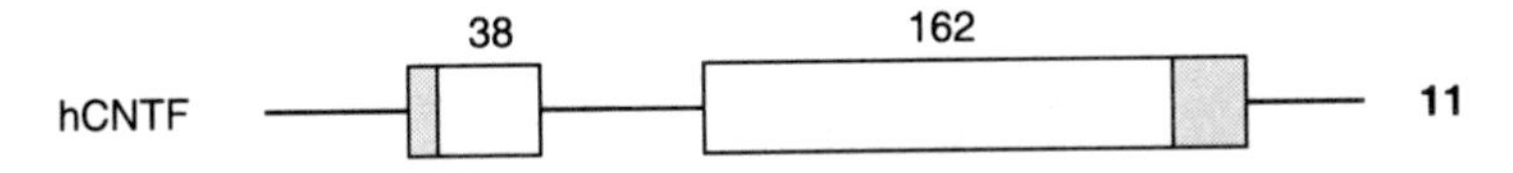

Human CNTF has two exons of 38 and 162 amino acids and is found on chromosome 11q12.

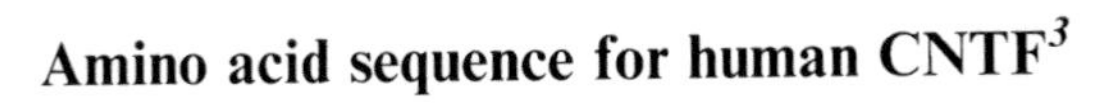

Amino acid sequence for human CNTF[3]

Accession code: SwissProt P26441

```
  1  MAFTEHSPLT PHRRDLCSRS IWLARKIRSD LTALTESYVK HQGLNKNINL
 51  DSADGMPVAS TDQWSELTEA ERLQENLQAY RTFHVLLARL LEDQQVHFTP
101  TEGDFHQAIH TLLLQVAAFA YQIEELMILL EYKIPRNEAD GMPINVGDGG
151  LFEKKLWGLK VLQELSQWTV RSIHDLRFIS SHQTGIPARG SHYIANNKKM
```

Amino acid sequence for mouse CNTF[4]

Accession code: SwissProt P51642

```
  1  MAFAEQSPLT LHRRDLCSRS IWLARKIRSD LTALMESYVK HQGLNKNISL
 51  DSVDGVPVAS TDRWSEMTEA ERLQENLQAY RTFQGMLTKL LEDQRVHFTP
101  TEGDFHQAIH TLTLQVSAFA YQLEELMALL EQKVPEKEAD GMPVTIGDGG
151  LFEKKLWGLK VLQELSQWTV RSIHDLRVIS SHHMGISAHE SHYGAKQM
```

THE CNTF RECEPTOR

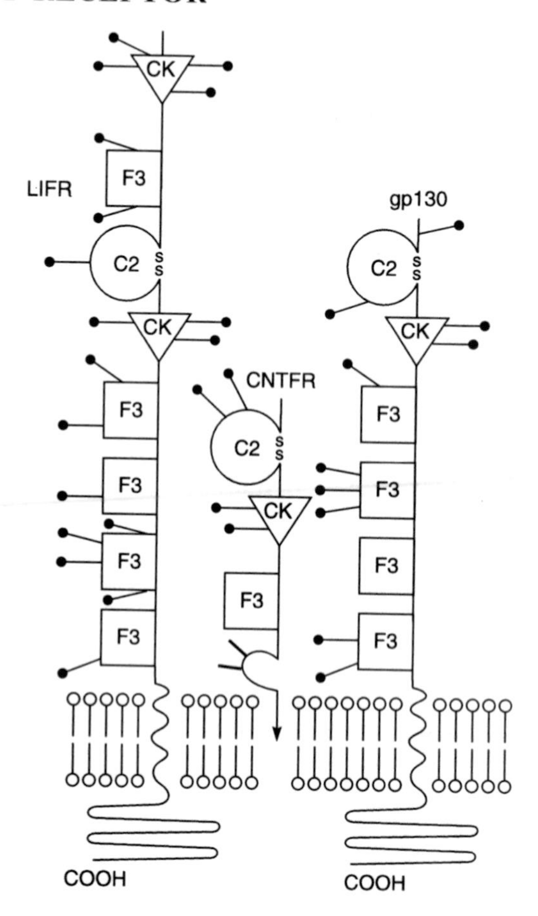

The CNTF receptor has homology to the IL-6R and is unrelated to receptors for NGF, BDNF and NT-3[5]. Unlike all other cytokine receptors, the CNTF receptor is anchored to cell membranes by a glycosylphosphatidylinositol (GPI) linkage[6]. Associated chains are the LIFR and gp130 (see IL-6 entry (page **69**)) which is shared with receptors for IL-6, LIF and oncostatin M[7–11]. Complexes of CNTF and soluble CNTF receptor can act through gp130/LIFR to activate cells which do not express CNTF receptor[12].

Distribution

The CNTF receptor is expressed mostly within the nervous system and skeletal muscle. It is expressed at lower levels in sciatic nerve, adrenal gland, skin, liver, kidney and testes.

Physicochemical properties of the CNTF receptor

Property	Human	Mouse
Amino acids – precursor	372	372
– mature[a]	352	352
M_r (K) – predicted	38.9	40.83
– expressed	72	?
Potential *N*-linked glycosylation sites	4	?
Affinity K_d (M)	?	?

[a] After removal of predicted signal peptide. Actual cleavage site is not known.

Signal transduction

The signal transduction components of the CNTFR (gp130 and LIFR) are tyrosine phosphorylated in response to CNTF binding[9,10]. Signalling probably requires heterodimerization of gp130 and LIFR[9]. May also involve activation of PKC[13].

Chromosomal location

Not known.

Amino acid sequence for human CNTF receptor[6]

Accession code: SwissProt P26992

```
-20  MAAPVPWACC AVLAAAAAVV
  1  YAQRHSPQEA PHVQYERLGS DVTLPCGTAN WDAAVTWRVN GTDLAPDLLN
 51  GSQLVLHGLE LGHSGLYACF HRDSWHLRHQ VLLHVGLPPR EPVLSCRSNT
101  YPKGFYCSWH LPTPTYIPNT FNVTVLHGSK IMVCEKDPAL KNRCHIRYMH
151  LFSTIKYKVS ISVSNALGHN ATAITFDEFT IVKPDPPENV VARPVPSNPR
201  RLEVTWQTPS TWPDPESFPL KFFLRYRPLI LDQWQHVELS DGTAHTITDA
251  YAGKEYIIQV AAKDNEIGTW SDWSVAAHAT PWTEEPRHLT TEAQAAETTT
301  STTSSLAPPP TTKICDPGEL GSGGGPSAPF LVSVPITLAL AAAAATASSL
351  LI
```

The exact cleavage site for the signal peptide has not been established. The C-terminal hydrophobic sequence (in italics, 327–352) is typical of GPI-linked proteins and may be removed although the exact cleavage site is not known.

Amino acid sequence for mouse CNTF receptor

Accession code: SPTREMBLE O88507

```
-20  MTASVPWACC AVLAAAAAV
  1  YTQKHSPQEA PHVQYERLGA DVTLPCGTAS WDAAVTWRVN GTDLAPDLLN
 51  GSQLILRSLE LGHSGLYACF HRDSWHLRHQ VLLHVGLPPR EPVLSCRSNT
101  YPKGFYCSWH LPTPTYIPNT FNVTVLHGSK IMVCEKDPAL KNRCHIRYMH
151  LFSTIKYKVS ISVSNALGHN TTAITFDEFT IVKPDPPENV VARPVPSNPR
201  RLEVTWQTPS TWPDPESFPL KFFLRYRPLI LDQWQHVELS DGTAHTITDA
251  YAGKEYIIQV AAKDNEIGTW SDWSVAAHAT PWTEEPRHLT TEAQAPETTT
301  STTSSLAPPP TTKICDPGEL GSGGGPSILF LTSVPVTLVL AAAAATANNL
351  LI
```

References

[1] Bazan, J.F. (1991) Neuron 7, 197–208.
[2] Lam, A. et al. (1991) Gene 102, 271–276.
[3] Negro, A. et al. (1991) Eur. J. Biochem. 201, 289–294.
[4] Saotome, Y et al. (1995) Gene 152, 233–238..
[5] Davis, S. and Yancopoulos, G.D. (1993) Curr. Opin. Cell. Biol. 5, 281–285.
[6] Davis, S. et al. (1991) Science 253, 59–63.
[7] Kishimoto, T. et al. (1992) Science 258, 593–597.
[8] Ip, N.Y. et al. (1992) Cell 69, 1121–1132.
[9] Davis, S. et al. (1993) Science 260, 1805–1808.
[10] Stahl, N. et al. (1993) J. Biol. Chem. 268, 7628–7631.
[11] Baumann, H. et al. (1993) J. Biol. Chem. 268, 8414–8417.
[12] Davis, S. et al. (1993) Science 259, 1736–1739.
[13] Kalberg, C. et al. (1993) J. Neurochem. 60, 145–152.

Other names

None.

THE MOLECULE

Cardiotrophin-1 (CT-1) is the most recently described member of the IL-6 family of cytokines, which utilize gp130 in their receptor complexes. CT-1 was initially identified because of its ability to induce cardiac myocyte hypertrophy[1] and subsequently shown to stimulate haematopoietic, neuronal and developmental activities. A protective role for CT-1 in some TNF-mediated diseases has been suggested[2], along with potential anti-inflammatory properties[3]. CT-1 signalling is mediated via the gp130 receptor subunit and the LIF receptor[1,4], leading to activation of both the p42/p44 MAP kinase and Jak/STAT pathways[5].

Crossreactivity

Human CT-1 is 80% identical to mouse CT-1[6].

Sources

CT-1 mRNA is expressed in adult human heart, skeletal muscle, ovary, colon, prostate and testis and also in fetal kidney and lung[6].

Bioassays

Induction of cardiac myocyte hypertrophy[1].

Physicochemical properties of CT-1

Property	Human	Mouse
pI[a]	9.18	6.52
Amino acid – precursor	201	203
– mature[b]	201	203
M_r (K) – predicted	21.2	21.5
– expressed	?	?
Potential *N*-linked glycosylation sites	0	1
Disulfide bonds	0	0

[a] Predicted.
[b] Cardiotrophin-1 does not have a signal sequence.

3-D structure

The 3-D structure of CT-1 is unknown but is predicted to form a classical four α-helix bundle.

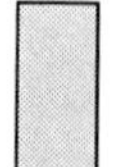

Gene structure[6]

The coding region of CT-1 is contained in three exons separated by two introns located on human chromosome 16p11.1-11.2.

Amino acid sequence for human CT-1

Accession code: SwissProt Q16619

```
  1  MSRREGSLED PQTDSSVSLL PHLEAKIRQT HSLAHLLTKY AEQLLQEYVQ
 51  LQGDPFGLPS FSPPRLPVAG LSAPAPSHAG LPVHERLRLD AAALAALPPL
101  LDAVCRRQAE LNPRAPRLLR RLEDAARQAR ALGAAVEALL AALGAANRGP
151  RAEPPAATAS AASATGVFPA KVLGLRVCGL YREWLSRTEG DLGQLLPGGS
201  A
```

Althouh there is no N-terminal signal sequence, CT-1 is efficiently secreted in the culture medium by an as yet unknown mechanism. Human CT-1 has two cysteine residues and no *N*-glycosylation sites.

Amino acid sequence for mouse CT-1

Accession code: SwissProt Q60753

```
  1  MSQREGSLED HQTDSSISFL PHLEAKIRQT HNLARLLTKY AEQLLEEYVQ
 51  QQGEPFGLPG FSPPRLPLAG LSGPAPSHAG LPVSERLRQD AAALSVLPAL
101  LDAVRRRQAE LNPRAPRLLR SLEDAARQVR ALGAAVETVL AALGAAARGP
151  GPEPVTVATL FTANSTAGIF SAKVLGFHVC GLYGEWVSRT EGDLGQLVPG
201  GVA
```

Mouse CT-1 has one cysteine residue and one potential *N*-glycosylation site.

THE CT-1 RECEPTOR

The receptor for CT-1 has been partially characterized. Several studies suggest that CT-1 binds to and activates the LIF receptor. More recent studies have confirmed this observation and demonstrate a K_d of 2 nM for CT-1 binding to gp190, which is identical to that observed for LIF binding.

gp130 acts as a high-affinity converter for CT-1 binding to gp190. For details of the LIF receptor see the LIF entry (page **346**) and for the gp130 subunit see the IL-6 entry (page **69**). The existence of a third receptor component for CT-1 (in the form of a GPI-linked protein) has also been postulated more recently.[7]

References

[1] Pennica, D. et al. (1995) J. Biol. Chem. 270, 10915–10922.
[2] Benigni, F. et al. (1996) Am J Pathol 149, 1847–18450.
[3] Pulido, E.J. et al. (1999) J. Surg. Res. 84, 240–246.
[4] Sheng, Z. et al. (1996) Development 122, 419–428.
[5] Latchman, D.S. (2000) Pharmacol. Ther. 85, 29–37.
[6] Pennica, D. et al. (1996) Cytokine 8, 183–189.
[7] Robledo, O. et al. (1997) J. Biol. Chem. 272, 4855–4863.

Other names

CCL27, Eskine, interleukin 11 receptor RT alpha-locus chemokine (ILC), ALP[1–4].

THE MOLECULE

Cutaneous T cell-attracting chemokine (CTACK) is a keratinocyte-derived CC chemokine that selectively attracts cutaneous lymphocyte antigen (CLA) positive memory T cells into the skin[5]. The murine form is also highly expressed in placenta, although a splice variant lacking the signal sequence and carrying a nuclear localization sequence is differentially expressed in brain and testes[4].

Crossreactivity

Human CTACK is 78% identical to the murine form and both act across species.

Sources

CTACK is constitutively produced by keratinocytes, but can be upregulated by IL-1/TNF stimulation[5].

Bioassays

Chemotaxis of CLA-positive T cells.

Physicochemical properties of CTACK

Property	Human	Mouse
Amino acids – precursor	112	120
– mature	88	95
M_r (K) – predicted	12.6	13.4
– expressed	?	?
Potential *N*-linked glycosylation sites	0	1
Disulfide bonds	2	2

3-D structure

No information.

Gene structure

The gene for human CTACK is on chromosome 9p13, and for mouse on chromosome 4[5].

Amino acid sequence for human CTACK

Accession code: SwissProt Q9Y4X3

```
-24  MKGPPTFCSL LLLSLLLSPD PTAA
  1  FLLPPSTACC TQLYRKPLSD KLLRKVIQVE LQEADGDCHL QAFVLHLAQR
 51  SICIHPQNPS LSQWFEHQER KLHGTLPKLN FGMLRKMG
```

Amino acid sequence for mouse CTACK

Accession code: SwissProt Q9Z1X0

```
-25   MMEGLSPASS LPLLLLLLSP APEAA
  1   LPLPSSTSCC TQLYRQPLPS RLLRRIVHME LQEADGDCHL QAVVLHLARR
 51   SVCVHPQNRS LARWLERQGK RLQGTVPSLN LVLQKKMYSN PQQQN
```

A splice variant of murine CTACK (PESKY) has been described that replaces the signal sequence with a nuclear targeting sequence[4].

```
-32   MSPTSQRLSL EAPSLPLRSW HPWNKTKQKQ EA
```

THE RECEPTOR, CCR10

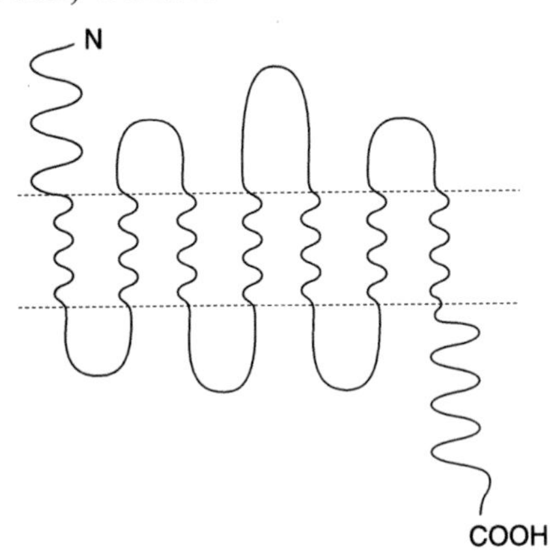

The CTACK receptor, CCR10, is also known as GPR2[6]. CCR10 also binds to mucosae-associated epithelial chemokine (MEC), which is 40% homologous to CTACK[7].

Distribution

CCR10 is predominantly expressed in small intestine, colon, skin, lymph nodes, Peyer s patches, thymus and spleen[1,8].

Physicochemical properties of the CCR10 CTACK receptor

Property	Human	Mouse
Amino acids – precursor	362	362
– mature		
M_r (K) – predicted	38.4	38.8
– expressed	?	?
Potential *N*-linked glycosylation sites	0	0
Affinity K_d (M)	?	?

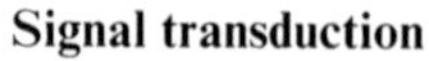

Signal transduction

CTACK binds specifically to CCR10 and mediates a pertussis toxin-sensitive Ca^{2+} flux with an EC_{50} of between 10 and 25 nM, and causes chemotaxis of CCR10-transfected cells with an ED_{50} of around 50–100 nM[1,8]. CTACK binds to CCR10, mediating polarization of CD4+ T cells with a maximal response at 1 ng/ml.

Chromosomal location

The human CCR10 gene is on chromosome 17q21.1–17q21.3[6].

Amino acid sequence for human CCR10 receptor

Accession code: SwissProt P46092

```
  1 MGTEATEQVS WGHYSGDEED AYSAEPLPEL CYKADVQAFS RAFQPSVSLT
 51 VAALGLAGNG LVLATHLAAR RAARSPTSAH LLQLALADLL LALTLPFAAA
101 GALQGWSLGS ATCRTISGLY SASFHAGFLF LACISADRYV AIARALPAGP
151 RPSTPGRAHL VSVIVWLLSL LLALPALLFS QDGQREGQRR CRLIFPEGLT
201 QTVKGASAVA QVALGFALPL GVMVACYALL GRTLLAARGP ERRRALRVVV
251 ALVAAFVVLQ LPYSLALLLD TADLLAARER SCPASKRKDV ALLVTSGLAL
301 ARCGLNPVLY AFLGLRFRQD LRRLLRGGSS PSGPQPRRGC PRRPRLSSCS
351 APTETHSLSW DN
```

Amino acid sequence for mouse CCR10 receptor

Accession code: SwissProt Q9JIP1

```
  1 MGTKPTEQVS WGLYSGYDEE AYSVGPLPEL CYKADVQAFS RAFQPSVSLM
 51 VAVLGLAGNG LVLATHLAAR RTTRSPTSVH LLQLALADLL LALTLPFAAA
101 GALQGWNLGS TTCRAISGLY SASFHAGFLF LACISADRYV AIARALPAGQ
151 RPSTPSRAHL VSVFVWLLSL FLALPALLFS RDGPREGQRR CRLIFPESLT
201 QTVKGASAVA QVVLGFALPL GVMAACYALL GRTLLAARGP ERRRALRVVV
251 ALVVAFVVLQ LPYSLALLLD TADLLAARER SCSSSKRKDL ALLVTGGLTL
301 VRCSLNPVLY AFLGLRFRRD LRRLLQGGGC SPKPNPRGRC PRRLRLSSCS
351 APTETHSLSW DN
```

References

[1] Homey, B. et al. (2000) J. Immunol. 164, 3465–3470.

[2] Ishikawa-Mochizuki, I. (1999) FEBS Lett. 460, 544–548.

[3] Hromas, R. et al. (1999) Biochem. Biophys. Res. Commun. 258, 737–740.

[4] Baird, J.W. et al. (1999) J. Biol. Chem. 274, 33496–33503

[5] Morales, J. et al. (1999) Proc. Natl Acad. Sci. USA 96, 14470–14475.

[6] Marchese, A. et al. (1994) Genomics 23, 609–618.

[7] Pan, J. et al. (2000) J. Immunol. 165, 2943–2949.

[8] Jarmin, D.I. et al. (2000) J. Immunol. 164, 3460–3464.

Other names
β-Urogastrone[1].

THE MOLECULE

Epidermal growth factor (EGF) is a 53-amino-acid cytokine which is proteolytically cleaved from a large integral membrane protein precursor[2]. EGF acts to stimulate growth of epithelial cells, accelerates tooth eruption and eyelid opening in mice, inhibits gastric acid secretion and is involved in wound healing[3].

Crossreactivity
There is about 70% homology between mature 53-amino-acid human and mouse EGF[4–6]. Neither factor acts across species. EGF is closely structurally related to TGFα and to vaccinia growth factor, both of which bind to the EGF receptor[7,8,9].

Sources
Ectodermal cells, monocytes, kidney and duodenal glands.

Bioassays
Proliferation of the A431 carcinoma line.

Physicochemical properties of EGF

Property	Human	Mouse
pI[a]	4.6	4.89
Amino acids – precursor[b]	1207	1217
– mature	53	53
M_r (K) – predicted	6	6.045
– expressed	6	6
Potential *N*-linked glycosylation sites[c]	0	0
Disulfide bonds[d]	3	3

[a] Represents the pI of mature processed EGF.
[b] The full-length protein is composed of eight extracellular EGF-like domains, only one of which has EGF activity.
[c] Processed mature molecule.
[d] In active EGF, disulfide bonds link residues 6–20, 14–31 and 33–42 in human and mouse EGF (numbering for processed active molecule only beginning at amino acid Asn949 of precursor for human and Asn977 for mouse).

3-D structure
Structure determined by NMR shows a major antiparallel β-sheet from residues 19–23 and 28–32, and a minor antiparallel β-sheet from residues 37–38 and 44–45. Residues 41 and 47 are involved in receptor binding (numbering for processed active molecule only beginning at amino acid Asn949 for human and Asn977 for mouse precursor).

Gene structure

The gene for human EGF is on chromosome 4q25.

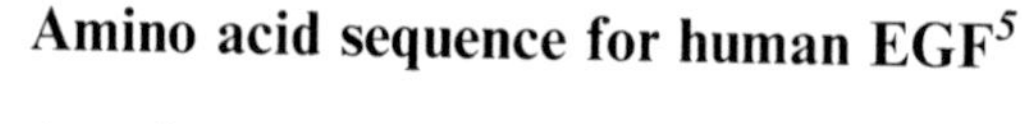

Amino acid sequence for human EGF[5]

Accession code: SwissProt P01133

```
 -22 MLLTLIILLP VVSKFSFVSL SA
   1 PQHWSCPEGT LAGNGNSTCV GPAPFLIFSH GNSIFRIDTE GTNYEQLVVD
  51 AGVSVIMDFH YNEKRIYWVD LERQLLQRVF LNGSRQERVC NIEKNVSGMA
 101 INWINEEVIW SNQQEGIITV TDMKGNNSHI LLSALKYPAN VAVDPVERFI
 151 FWSSEVAGSL YRADLDGVGV KALLETSEKI TAVSLDVLDK RLFWIQYNRE
 201 GSNSLICSCD YDGGSVHISK HPTQHNLFAM SLFGDRIFYS TWKMKTIWIA
 251 NKHTGKDMVR INLHSSFVPL GELKVVHPLA QPKAEDDTWE PEQKLCKLRK
 301 GNCSSTVCGQ DLQSHLCMCA EGYALSRDRK YCEDVNECAF WNHGCTLGCK
 351 NTPGSYYCTC PVGFVLLPDG KRCHQLVSCP RNVSECSHDC VLTSEGPLCF
 401 CPEGSVLERD GKTCSGCSSP DNGGCSQLCV PLSPVSWECD CFPGYDLQLD
 451 EKSCAASGPQ PFLLFANSQD IRHMHFDGTD YGTLLSQQMG MVYALDHDPV
 501 ENKIYFAHTA LKWIERANMD GSQRERLIEE GVDVPEGLAV DWIGRRFYWT
 551 DRGKSLIGRS DLNGKRSKII TKENISQPRG IAVHPMAKRL FWTDTGINPR
 601 IESSSLQGLG RLVIASSDLI WPSGITIDFL TDKLYWCDAK QSVIEMANLD
 651 GSKRRRLTQN DVGHPFAVAV FEDYVWFSDW AMPSVIRVNK RTGKDRVRLQ
 701 GSMLKPSSLV VVHPLAKPGA DPCLYQNGGC EHICKKRLGT AWCSCREGFM
 751 KASDGKTCLA LDGHQLLAGG EVDLKNQVTP LDILSKTRVS EDNITESQHM
 801 LVAEIMVSDQ DDCAPVGCSM YARCISEGED ATCQCLKGFA GDGKLCSDID
 851 ECEMGVPVCP PASSKCINTE GGYVCRCSEG YQGDGIHCLD IDECQLGVHS
 901 CGENASCTNT EGGYTCMCAG RLSEPGLICP DSTPPPHLRE DDHHYSVRNS
 951 DSECPLSHDG YCLHDGVCMY IEALDKYACN CVVGYIGERC QYRDLKWWEL
1001 RHAGHGQQQK VIVVAVCVVV LVMLLLLSLW GAHYYRTQKL LSKNPKNPYE
1051 ESSRDVRSRR PADTEDGMSS CPQPWFVVIK EHQDLKNGGQ PVAGEDGQAA
1101 DGSMQPTSWR QEPQLCGMGT EQGCWIPVSS DKGSCPQVME RSFHMPSYGT
1151 QTLEGGVEKP HSLLSANPLW QQRALDPPHQ MELTQ
```

Soluble mature EGF amino acids Asn949–Arg1001.

Amino acid sequence for mouse EGF[6]

Accession code: SwissProt P01132

```
   1 MPWGRRPTWL LLAFLLVFLK ISILSVTAWQ TGNCQPGPLE RSERSGTCAG
  51 PAPFLVFSQG KSISRIDPDG TNHQQLVVDA GISADMDIHY KKERLYWVDV
 101 ERQVLLRVFL NGTGLEKVCN VERKVSGLAI DWIDDEVLWV DQQNGVITVT
 151 DMTGKNSRVL LSSLKHPSNI AVDPIERLMF WSSEVTGSLH RAHLKGVDVK
 201 TLLETGGISV LTLDVLDKRL FWVQDSGEGS HAYIHSCDYE GGSVRLIRHQ
 251 ARHSLSSMAF FGDRIFYSVL KSKAIWIANK HTGKDTVRIN LHPSFVTPGK
 301 LMVVHPRAQP RTEDAAKDPD PELLKQRGRP CRFGLCERDP KSHSSACAEG
 351 YTLSRDRKYC EDVNECATQN HGCTLGCENT PGSYHCTCPT GFVLLPDGKQ
 401 CHELVSCPGN VSKCSHGCVL TSDGPRCICP AGSVLGRDGK TCTGCSSPDN
 451 GGCSQICLPL RPGSWECDCF PGYDLQSDRK SCAASGPQPL LLFANSQDIR
 501 HMHFDGTDYK VLLSRQMGMV FALDYDPVES KIYFAQTALK WIERANMDGS
 551 QRERLITEGV DTLEGLALDW IGRRIYWTDS GKSVVGGSDL SGKHHRIIIQ
 601 ERISRPRGIA VHPRARRLFW TDVGMSPRIE SASLQGSDRV LIASSNLLEP
```

```
651  SGITIDYLTD TLYWCDTKRS VIEMANLDGS KRRRLIQNDV GHPFSLAVFE
701  DHLWVSDWAI PSVIRVNKRT GQNRVRLQGS MLKPSSLVVV HPLAKPGADP
751  CLYRNGGCEH ICQESLGTAR CLCREGFVKA WDGKMCLPQD YPILSGENAD
801  LSKEVTSLSN STQAEVPDDD GTESSTLVAE IMVSGMNYED DCGPGGCGSH
851  ARCVSDGETA ECQCLKGFAR DGNLCSDIDE CVLARSDCPS TSSRCINTEG
901  GYVCRCSEGY EGDGISCFDI DECQRGAHNC AENAACTNTE GGYNCTCAGR
951  PSSPGRSCPD STAPSLLGED GHHLDRNSYP GCPSSYDGYC LNGGVCMHIE
1001 SLDSYTCNCV IGYSGDRCQT RDLRWWELRH AGYGQKHDIM VVAVCMVALV
1051 LLLLLGMWGT YYYRTRKQLS NPPKNPCDEP SGSVSSSGPD SSSGAAVASC
1101 PQPWFVVLEK HQDPKNGSLP ADGTNGAVVD AGLSPSLQLG SVHLTSWRQK
1151 PHIDGMGTGQ SCWIPPSSDR GPQEIEGNSH LPSYRPVGPE KLHSLQSANG
1201 SCHERAPDLP RQTEPVK
```

Signal sequence is not known. Soluble mature EGF amino acids Asn977–Arg1029.

THE EGF RECEPTOR

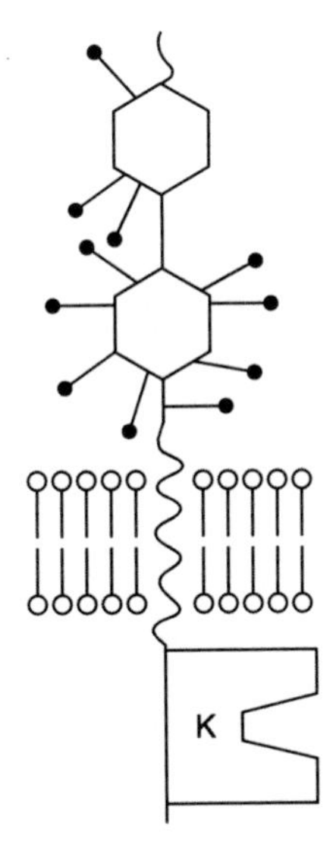

The EGF receptor (also known as c-erbB) is a class I receptor tyrosine kinase[10,11]. The receptor is also shared with TGFα, and with vaccinia virus growth factor. A viral oncogene v-*erbB* encodes a truncated EGF receptor lacking most of the extracellular domains. The EGF receptor is involved in bacterial and viral invasion of cells[12,13].

Distribution

Most cells.

Physicochemical properties of the EGF receptor

Property	Human	Mouse
Amino acids – precursor	1210	1210
– mature	1186	1186
M_r (K) – predicted	132	134.85
– expressed	170	170
Potential *N*-linked glycosylation sites	11	10
Affinity K_d (M)	$3\text{–}10 \times 10^{-9}$	?

Signal transduction

Binding of EGF to its receptor triggers oligomerization and an increase in receptor affinity. The tyrosine kinase domain then autophosphorylates the receptor[11]. Subsequent events include Tyr phosphorylation of other proteins, including the Neu p185 receptor, breakdown of inositol lipids, increase in calcium concentration and activation of Ser/Thr protein kinases. Ser/Thr phosphorylation of the receptor, which can occur by transmodulation from the PDGF receptor, downregulates its affinity.

Chromosomal location

The human receptor gene is found on chromosome 7.

Amino acid sequence for human EGF receptor[10]

Accession code: SwissProt P00533

-24 MRPSGTAGAA LLALLAALCP ASRA
1 LEEKKVCQGT SNKLTQLGTF EDHFLSLQRM FNNCEVVLGN LEITYVQRNY
51 DLSFLKTIQE VAGYVLIALN TVERIPLENL QIIRGNMYYE NSYALAVLSN
101 YDA**N**KTGLKE LPMRNLQEIL HGAVRFSNNP ALCNVESIQW RDIVSSDFLS
151 **N**MSMDFQNHL GSCQKCDPSC P**N**GSCWGAGE ENCQKLTKII CAQQCSGRCR
201 GKSPSDCCHN QCAAGCTGPR ESDCLVCRKF RDEATCKDTC PPLMLYNPTT
251 YQMDVNPEGK YSFGATCVKK CPRNYVVTDH GSCVRACGAD SYEMEEDGVR
301 KCKKCEGPCR KVCNGIGIGE FKDSLSI**N**AT NIKHFK**N**CTS ISGDLHILPV
351 AFRGDSFTHT PPLDPQELDI LKTVKEITGF LLIQAWPE**N**R TDLHAFENLE
401 IIRGRTKQHG QFSLAVVSL**N** ITSLGLRSLK EISDGDVIIS GNKNLCYANT
451 INWKKLFGTS GQKTKIISNR GENSCKATGQ VCHALCSPEG CWGPEPRDCV
501 SCR**N**VSRGRE CVDKCKLLEG EPREFVENSE CIQCHPECLP QAM**N**ITCTGR
551 GPDNCIQCAH YIDGPHCVKT CPAGVMGE**N**N TLVWKYADAG HVCHLCHP**N**C
601 TYGCTGPGLE GCPTNGPKIP SIATGMVGAL LLLLVVALGI GLFMRRRHIV
651 RKRTLRRLLQ ERELVEPLTP SGEAPNQALL RILKETEFKK IKVLGSGAFG
701 TVYKGLWIPE GEKVKIPVAI KELREATSPK ANKEILDEAY VMASVDNPHV
751 CRLLGICLTS TVQLITQLMP FGCLLDYVRE HKDNIGSQYL LNWCVQIAKG
801 MNYLEDRRLV HRDLAARNVL VKTPQHVKIT DFGLAKLLGA EEKEYHAEGG
851 KVPIKWMALE SILHRIYTHQ SDVWSYGVTV WELMTFGSKP YDGIPASEIS
901 SILEKGERLP QPPICTIDVY MIMVKCWMID ADSRPKFREL IIEFSKMARD
951 PQRYLVIQGD ERMHLPSPTD SNFYRALMDE EDMDDVVDAD EYLIPQQGFF

1001 SSPSTSRTPL LSSLSATSNN STVACIDRNG LQSCPIKEDS FLQRYSSDPT
1051 GALTEDSIDD TFLPVPEYIN QSVPKRPAGS VQNPVYHNQP LNPAPSRDPH
1101 YQDPHSTAVG NPEYLNTVQP TCVNSTFDSP AHWAQKGSHQ ISLDNPDYQQ
1151 DFFPKEAKPN GIFKGSTAEN AEYLRVAPQS SEFIGA

Tyr1068, Tyr1148, Tyr1173, Tyr1086 and Thr654 are phosphorylated.

Amino acid sequence for mouse EGF receptor

Accession code: SwissProt Q01279

-24 MRPSGTARTT LLVLLTALCA AGGA
1 LEEKKVCQGT SNRLTQLGTF EDHFLSLQRM YNNCEVVLGN LEITYVQRNY
51 DLSFLKTIQE VAGYVLIALN TVERIPLENL QIIRGNALYE NTYALAILSN
101 YGT**N**RTGLRE LPMRNLQEIL IGAVRFSNNP ILCNMDTIQW RDIVQNVFMS
151 **N**MSMDLQSHP SSCPKCDPSC P**N**GSCWGGGE ENCQKLTKII CAQQCSHRCR
201 GRSPSDCCHN QCAAGCTGPR ESDCLVCQKF QDEATCKDTC PPLMLYNPTT
251 YQMDVNPEGK YSFGATCVKK CPRNYVVTDH GSCVRACGPD YYEVEEDGIR
301 KCKKCDGPCR KVCNGIGIGE FKDTLSI**N**AT NIKHFKYCTA ISGDLHILPV
351 AFKGDSFTRT PPLDPRELEI LKTVKEITGF LLIQAWPD**N**W TDLHAFENLE
401 IIRGRTKQHG QFSLAVVGL**N** ITSLGLRSLK EISDGDVIIS GNRNLCYANT
451 INWKKLFGTP NQKTKIMNNR AEKDCKAVNH VCNPLCSSEG CWGPEPRDCV
501 SCQ**N**VSRGRE CVEKCNILEG EPREFVENSE CIQCHPECLP QAM**N**ITCTGR
551 GPDNCIQCAH YIDGPHCVKT CPAGIMGE**N**N TLVWKYADAN NVCHLCHA**N**C
601 TYGCAGPGLQ GCEVWPSGPK IPSIATGIVG GLLFIVVVAL GIGLFMRRRH
651 IVRKRTLRRL LQERELVEPL TPSGEAPNQA HLRILKETEF KKIKVLGSGA
701 FGTVYKGLWI PEGEKVKIPV AIKELREATS PKANKEILDE AYVMASVDNP
751 HVCRLLGICL TSTVQLITQL MPYGCLLDYV REHKDNIGSQ YLLNWCVQIA
801 KGMNYLEDRR LVHRDLAARN VLVKTPQHVK ITDFGLAKLL GAEEKEYHAE
851 GGKVPIKWMA LESILHRIYT HQSDVWSYGV TVWELMTFGS KPYDGIPASD
901 ISSILEKGER LPQPPICTID VYMIMVKCWM IDADSRPKFR ELILEFSKMA
951 RDPQRYLVIQ GDERMHLPSP TDSNFYRALM DEEDMEDVVD ADEYLIPQQG
1001 FFNSPSTSRT PLLSSLSATS NNSTVACINR NGSCRVKEDA FLQRYSSDPT
1051 GAVTEDNIDD AFLPVPEYVN QSVPKRPAGS VQNPVYHNQP LHPAPGRDLH
2001 YQNPHSNAVG NPEYLNTAQP TCLSSGFNSP ALWIQKGSHQ MSLDNPDYQQ
2051 DFFPKETKPN GIFKGPTAEN AEYLRVAPPS SEFIGA

Conflict in sequence C→S at position 19, C→W at position 539 and HP→DR at position 1116. Serine-rich domain at position 1028–1071 of unprocessed precursor and kinase domain at position 714–981.

References

[1] Gregory, H. (1975) Nature 257, 325–327.
[2] Carpenter, G. and Cohen, S. (1990) J. Biol. Chem. 265, 7709–7712.
[3] Burgess, A.W. (1989) In Br. Med. Bull. 45, Growth Factors, Waterfield, M.D. ed., Churchill Livingstone, Edinburgh, pp. 401–424.
[4] Bell, G.I. et al. (1986) Nucleic Acids Res. 14, 8427–8446.
[5] Gray, A.et al. (1983) Nature 303, 722–725.
[6] Scott, J. et al. (1983) Science 221, 236–240.
[7] Montelione, G.T. et al. (1986) Proc. Natl Acad. Sci. USA 83, 8594–8598.

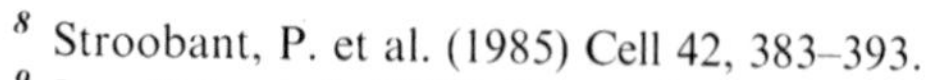

[8] Stroobant, P. et al. (1985) Cell 42, 383–393.
[9] Lee, C.L. et al. (1985) Nature 313, 489–491.
[10] Ullrich, A. et al. (1984) Nature 309, 418–425.
[11] Ullrich, A. and Schlessinger, J. (1990) Cell 61, 203–212.
[12] Galan, J.E. et al. (1992) Nature 357, 588–589.
[13] Eppstein, D.A. et al. (1985) Nature 318, 663–665.
[14] Petch, L.A. et al. (1990) Mol. Cell. Biol. 10, 2973–2982.

Other names
MIP-3β, CKB-11.

THE MOLECULE

EBI-1 ligand chemokine (ELC) is a chemoattractant for T and B lymphocytes and myeloid progenitor cells[1–6]. It is constitutively expressed in thymus and lymph nodes and may play a role in lymphocyte migration to these areas.

Crossreactivity
Mouse and human share 78% amino acid sequence homology; human ELC is active on murine cells.

Sources
ELC is expressed by dendritic cells in thymus and lymph nodes, and in activated bone marrow stromal cells[5,6].

Bioassays
Chemotaxis of lymphocytes.

Physicochemical properties of ELC

Property	Human	Mouse
pI	10.1	?
Amino acids – precursor	98	108
– mature	77	83
M_r (K) – predicted	11.0	12.2
– expressed	8.8	9.0
Potential *N*-linked glycosylation sites	0	1
Disulfide bonds	2	2

3-D structure
No information.

Gene structure

The gene for human ELC is on chromosome 9p13.

Amino acid sequence for human ELC

Accession code: SwissProt Q99731

```
-21  MALLLALSLL VLWTSPAPTL S
  1  GTNDAEDCCL SVTQKPIPGY IVRNFHYLLI KDGCRVPAVV FTTLRGRQLC
 51  APPDQPWVER IIQRLQRTSA KMKRRSS
```

Amino acid sequence for mouse ELC

Accession code: SwissProt O70460

```
-25  MAPRVTPLLA FSLLVLWTFP APTLG
  1  GANDAEDCCL SVTQRPIPGN IVKAFRYLLN EDGCRVPAVV FTTLRGYQLC
 51  APPDQPWVDR IIRRLKKSSA KNKGNSTRRS PVS
```

THE ELC RECEPTOR, CCR7

The ELC receptor, CCR7, is shared with the related 6Ckine chemokine. For a full description of the receptor, see the 6Ckine entry (page **188**). ELC binds to CCR7 with high affinity, causing a pertussis toxin-sensitive Ca^{2+} flux with an ED_{50} of 0.9–3.0 nm, and stimulating chemotaxis with an ED_{50} of about 100 nM[1,6–9].

References

1. Yoshida, R. et al. (1997) J. Biol. Chem. 272, 13803–13809.
2. Rossi, D.L. et al. (1997) J. Immunol. 158, 1033–1036.
3. Yoshida, R. et al. (1998) J. Exp. Med. 188, 181–191.
4. Kim, C.H. et al. (1998) J. Immunol. 160, 2418–2424.
5. Kim, C.H. et al. (1998) J. Immunol. 161, 2580–2585.
6. Ngo, V.N. et al. (1998) J. Exp. Med. 188, 181–191.
7. Yoshida, R. et al. (1998) Int. Immunol. 10, 901–910.
8. Yanagihara, S. et al. (1998) J. Immunol. 161, 3096–3102.
9. Campbell, J.J. et al. (1998) Science 279, 381–384.

Other names

None.

THE MOLECULE

Epithelial neutrophil chemoattractant 78 (ENA-78) is a neutrophil chemoattractant and activating factor and angiogenic factor[1–5]. ENA-78 is widely expressed in inflammatory disease tissues such as rheumatoid arthritis, pancreatitis, lung inflammation and inflammatory bowel disease[6–8].

Crossreactivity

There is no murine homologue of ENA-78.

Sources

Platelets, epithelial cells, endothelial cells, mast cells and monocytes[1,6,7,9].

Bioassays

Neutrophil chemoattraction.

Property	Human
Amino acids – precursor	114
– mature	78[a]
M_r (K) – predicted	12
– expressed	8
Potential *N*-linked glycosylation sites	0
Disulfide bonds	2

[a] Proteolytic cleavage with cathepsin G or chymotrypsin yields products 5–78 and 9–78 with enhanced potency[10].

Physicochemical properties of ENA-78

3-D structure

No information; could be modelled on IL-8.

Gene structure

The gene for human ENA-78 is on chromosome 4q13–q21[11].

Amino acid sequence for human ENA-78[1–4]

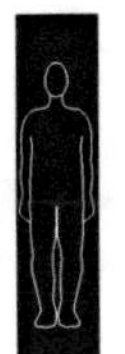

Accession code: SwissProt P42830

```
-36  MSLLSSRAAR VPGPSSSLCA LLVLLLLLTQ PGPIAS
  1  AGPAAAVLRE LRCVCLQTTQ GVHPKMISNL QVFAIGPQCS KVEVVASLKN
 51  GKEICLDPEA PFLKKVIQKI LDGGNKEN
```

THE ENA-78 RECEPTOR, CXCR2

CXCR2 is a common receptor for ENA-78 (EC_{50} 11 nm)[12], IL-8, Groα, β and γ as well as NAP-2. See IL-8 entry (page **80**).

References

1. Power, C.A et al. (1994) Gene 151, 333–334.
2. Chang, M.S. et al. (1994) J. Biol. Chem. 269, 25277–25282.
3. Corbett, M.S. et al. (1994) Biochem. Biophys. Res. Commun. 205, 612–617.
4. Walz, A. et al. (1991) J. Exp. Med. 174, 1355–1362.
5. Arenberg, D.A. et al. (1998) J. Clin. Invest. 102, 465–472.
6. Lukacs, N.W. et al. (1998) J. Leukoc. Biol. 63, 746–751.
7. Walz, A. et al. (1997) J. Leukoc. Biol. 62, 604–611.
8. Koch, A. et al. (1994) J. Clin. Invest. 94, 1012–1018.
9. Imaizumi, T. et al. (1997) Am. J. Resp. Cell Mol. Biol. 17, 181–192.
10. Nufer, O. et al. (1997) Biochemistry 38, 636–642.
11. O'Donovan, N. et al. (1999) Cytogen. Cell Genetics 84, 39–42.
12. Ahuja, S.K. and Murphy, P.M. (1996) J. Biol. Chem. 271, 20545–20550.

Other names
None.

THE MOLECULE

Eotaxin 1 is a potent selective chemoattractant for eosinophils, basophils and Th2 lymphocytes which is released by activated endothelial cells, epithelial cells, eosinophils and macrophages[1–9]. It is readily isolated from allergic tissues, and appears to be a major factor in recruiting eosinophils into tissues[6,7]. Eotaxin also promotes the generation of mast cell and myeloid progenitor cells[10].

Crossreactivity
Mouse and human eotaxin share 59% amino acid sequence homology. Eotaxin is 66% homologous to MCP-1 but shares none of its biological activities.

Sources
Eotaxin is produced by IFNγ-stimulated endothelial cells, complement-activated eosinophils, TNF-activated monocytes and dermal fibroblasts and in response to IL-4 stimulation *in vivo*. Eotaxin mRNA is constitutively expressed in small intestine, colon, lung and heart[1–4,8,9].

Bioassays
Chemotaxis of eosinophils.

Physicochemical properties of eotaxin 1

Property	Human	Mouse
Amino acids – precursor	97	97
– mature	74	74
M_r (K) – predicted	10.7	10.9
– expressed	9.0, 9.3	?
Potential *N*-linked glycosylation sites	0	0
Disulfide bonds	2	2

Eotaxin 1 is variably *O*-glycosylated on Thr71[12].

3-D structure
A solution structure of human eotaxin 1 has been obtained by NMR spectroscopy[11]. Under physiological conditions eotaxin 1 is likely to exist as a monomer, with a typical chemokine fold of a 3-stranded β-sheet with overlying α-helix. Eotaxin 1 differs from MCP-1 and RANTES predominantly at the N-terminus which is unstructured in eotaxin 1.

Gene structure

The gene for human eotaxin 1 (*SCYA11*) is on chromosome 17q21.1–q21.2[13].

Amino acid sequence for human eotaxin 1

Accession code: SwissProt P51671

```
-23  MKVSAALLWL LLIAAAFSPQ GLA
  1  GPASVPTTCC FNLANRKIPL QRLESYRRIT SGKCPQKAVI FKTKLAKDIC
 51  ADPKKKWVQD SMKYLDQKSP TPKP
```

Amino acid sequence for mouse eotaxin 1

Accession code: SwissProt P48298

```
-23  MQSSTALLFL LLTVTSFTSQ VLA
  1  HPGSIPTSCC FIMTSKKIPN TLLKSYKRIT NNRCTLKAIV FKTRLGKEIC
 51  ADPKKKWVQD ATKHLDQKLQ TPKP
```

THE RECEPTOR, CCR3

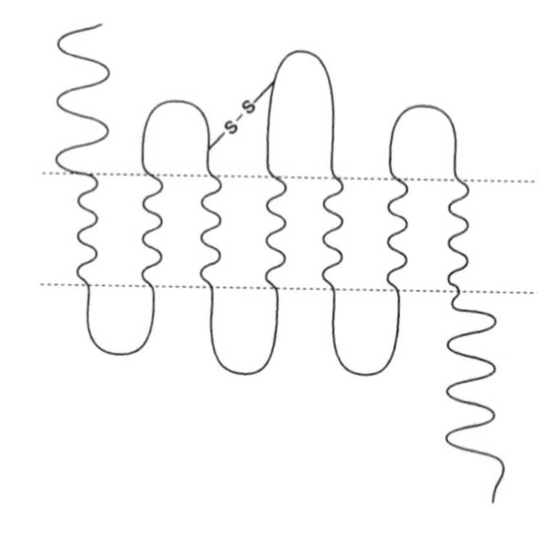

The eotaxin 1 receptor, CCR3 (also known as CC CKR3, CKR-3 and MIP-1α receptor-like 2), is selectively expressed on eosinophils, basophils and Th2 lymphocytes[1,14–16,18,19]. CCR3 also binds eotaxin 2, RANTES and MCP-2, -3 and -4, but with lower affinity and signal transduction. CCR3 also acts as an HIV co-receptor[20]. It is not yet clear if the effects of eotaxin 1 on myeloid progenitors are mediated by CCR3 or a novel receptor[10].

Distribution

CCR3 is expressed by eosinophils, basophils and Th2 lymphocytes.

Physicochemical properties of the eotaxin 1 receptor, CCR3

Property	Human	Mouse
Amino acids – precursor	355	359
M_r (K) – predicted	41.0	41.8
– expressed	45–55	?
Potential *N*-linked glycosylation sites	0	0
Affinity K_d (M)	1×10^{-10} M	?

Western blotting of the human CCR3 reveals several bands between 45 and 55 kDa, possibly *O*-glycosylated forms.

Signal transduction

Binding of eotaxin 1 to cells expressing CCR3 causes a pertussis toxin inhibitable calcium flux, with an ED_{50} of 0.3 nM [14].

Chromosomal location

The human gene is located on chromosome 3[17].

Amino acid sequence for human CCR3

Accession code: SwissProt P51677

```
  1  MTTSLDTVET FGTTSYYDDV GLLCEKADTR ALMAQFVPPL YSLVFTVGLL
 51  GNVVVVMILI KYRRLRIMTN IYLLNLAISD LLFLVTLPFW IHYVRGHNWV
101  FGHGMCKLLS GFYHTGLYSE IFFIILLTID RYLAIVHAVF ALRARTVTFG
151  VITSIVTWGL AVLAALPEFI FYETEELFEE TLCSALYPED TVYSWRHFHT
201  LRMTIFCLVL PLLVMAICYT GIIKTLLRCP SKKKYKAIRL IFVIMAVFFI
251  FWTPYNVAIL LSSYQSILFG NDCERSKHLD LVMLVTEVIA YSHCCMNPVI
301  YAFVGERFRK YLRHFFHRHL LMHLGRYIPF LPSEKLERTS SVSPSTAEPE
351  LSIVF
```

Amino acid sequence for mouse CCR3

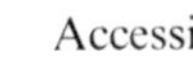

Accession code: SwissProt P51678

```
  1  MAFNTDEIKT VVESFETTPY EYEWAPPCEK VRIKELGSWL LPPLYSLVFI
 51  IGLLGNMMVV LILIKYRKLQ IMTNIYLFNL AISDLLFLFT VPFWIHYVLW
101  NEWGFGHYMC KMLSGFYYLA LYSEIFFIIL LTIDRYLAIV HAVFALRART
151  VTFATITSII TWGLAGLAAL PEFIFHESQD SFGEFSCSPR YPEGEEDSWK
201  RFHALRMNIF GLALPLLVMV ICYSGIIKTL LRCPNKKKHK AIRLIFVVMI
251  VFFIFWTPYN LVLLFSAFHR TFLETSCEQS KHLDLAMQVT EVIAYTHCCV
301  NPVIYAFVGE RFRKHLRLFF HRNVAVYLGK YIPFLPGEKM ERTSSVSPST
351  GEQEISVVF
```

References

[1] Kitaura, M. et al. (1996) J. Biol. Chem. 271, 7725–7730.
[2] Rothenberg, M.E. et al. (1995) Proc. Natl Acad. Sci. USA 92, 8960–8964.
[3] Ganzalo, J.A. et al. (1996) Immunity 4, 1–14.
[4] Garcia-Zepeda, E.A. et al. (1996) Nature Med. 2, 449–456.
[5] Yamada, H. et al. (1997) Biochem. Biophys. Res. Commun. 231, 365–368.
[6] Rothenberg, M.E. et al. (1997) J. Exp. Med. 185, 785–790.
[7] Jose, P.J. et al. (1994) J. Exp. Med. 179, 881–887.
[8] Nakamura, H. et al. (1998) Am. J. Physiol. 275, 601–610.
[9] Nakajima, T. et al. (1998) FEBS Lett. 434, 226–230.
[10] Quackenbush, E.J. et al. (1998) Blood 92, 1887–1897.
[11] Crump, P. et al. (1998) J. Biol. Chem. 273, 22471–22479.
[12] Noso, N. et al. (1998) Eur. J. Biochem. 253, 114–122.
[13] Garcia-Zepeda, E.A. et al. (1997) Genomics 41, 471–476.
[14] Daugherty, B.L. et al. (1996) J. Exp. Med. 183, 2349–2354.
[15] Ponath, P.D. et al. (1996) J. Exp. Med. 183, 2347–2448.
[16] Post, T.W. et al. (1995) J. Immunol. 155, 5299–5305.
[17] Sampson, M. et al. (1996) Genomics 15, 522–526.
[18] Sallusto, F. et al. (1997) Science 277, 2005–2007.
[19] Gao, J.L. et al. (1996) Biochem. Biophys. Res. Commun. 223, 679–684.
[20] Choe, H. et al. (1996) Cell 85, 1135–1148.

Eotaxin 2

Other names
CkB-6, myeloid progenitor inhibitory factor 2 (MPIF-2).

THE MOLECULE

Eotaxin 2 is a potent selective chemoattractant for eosinophils, basophils and T cells, and inhibits the proliferation of multipotential haematopoietic progenitor cells[1–3]. Eotaxin 2 shares the CCR3 receptor with eotaxin 1.

Crossreactivity
No information on mouse eotaxin 2. Eotaxin 2 shares only 39% amino acid homology with eotaxin 1.

Sources
Eotaxin 2 is produced by activated monocytes and T lymphocytes[2].

Bioassays
Chemotaxis of eosinophils

Physicochemical properties of eotaxin 2

Property	Human	Mouse
Amino acids – precursor	119	?
– mature	93	?
M_r (K) – predicted	11.2	?
– expressed	8.8, 11.2	?
Potential *N*-linked glycosylation sites	1	?
Disulfide bonds	2	?

Eotaxin 2 has been isolated as a variety of C-terminally truncated variants ending at residues A78, A76, R75 and R73.

3-D structure
No information.

Gene structure

The gene for human eotaxin 2 (*SCYA24*) is on chromosome 7q11.23[4].

Amino acid sequence for human eotaxin 2

Accession code: SwissProt O00175

```
-26  MAGLMTIVTS LLFLGVCAHH IIPTGS
  1  VVIPSPCCMF FVSKRIPENR VVSYQLSSRS TCLKGGVIFT TKKGQQSCGD
 51  PKQEWVQRYM KNLDAKQKKA SPRARAVAVK GPVQRYPGNQ TTC
```

Amino acid sequence for mouse eotaxin 2

No information.

THE RECEPTOR, CCR3

The eotaxin 2 receptor, CCR3 (also known as CC CKR3, CKR-3 and MIP-1α receptor-like 2), is selectively expressed on eosinophils, basophils and Th2 lymphocytes. For more detailed description see eotaxin 1 entry (page **213**). Eotaxin 2 binds with high affinity to CCR3 on eosinophils, mediating a pertussis toxin-sensitive Ca^{2+} flux with an ED_{50} between 10 and 100 ng/ml. Eotaxin 2 causes an increase in reactive oxygen species with an ED_{50} of 50 ng/ml, histamine and leukotriene C4 release with an ED_{50} of between 10 and 100 nM, and mediates chemotaxis in eosinophils and resting T lymphocytes with an ED_{50} of 50 nM and 10 ng/ml respectively[1–3,5]. In general, eotaxin 2 is about 10-fold less potent than eotaxin 1 in its CCR3-mediated effects. Eotaxin 2 inhibits the proliferation of high-proliferative potential colony-forming cells with an ED_{50} of 10 ng/ml. It is unclear but possible that this response is mediated by a novel receptor[2].

References

[1] White, J.R. et al. (1997) J. Leukoc. Biol. 62, 667–675.
[2] Patel, V.P. et al. (1997) J. Exp. Med. 185, 1163–1172.
[3] Forssmann, U. et al. (1997) J. Exp. Med. 185, 2171–2176.
[4] Nomiyama, H. et al. (1998) Genomics 49, 339–340.
[5] Elsner, J. et al. (1998) Eur. J. Immunol. 28, 2152–2158.

Other names
CCL26.

THE MOLECULE

Eotaxin 3 is a chemoattractant for eosinophils, basophils, T cells and monocytes[1–3]. It shares the CCR3 receptor with eotaxin 1 and 2.

Crossreactivity
No information.

Sources
Eotaxin 3 is produced by endothelial cells stimulated with IL-4 or IL-13. This is distinct from eotaxin 1 which is stimulated by TNF or IFNγ[2]. mRNA is expressed in normal heart and ovary[3].

Bioassays
Chemotaxis of eosinophils.

Physicochemical properties of eotaxin 3

Property	Human
Amino acids – precursor	94
– mature	71
M_r (K) – predicted	10.6
– expressed	12
Potential *N*-linked glycosylation sites	0
Disulfide bonds	2

3-D structure
No information.

Gene structure

The gene for human eotaxin 3 (*SCYA26*) is on chromosome 7q11.23[1,3].

Amino acid sequence for human eotaxin 3

Accession code: SwissProt Q9Y258

```
-23  MMGLSLASAV LLASLLSLHL GTA
  1  TRGSDISKTC CFQYSHKPLP WTWVRSYEFT SNSCSQRAVI FTTKRGKKVC
 51  THPRKKWVQK YISLLKTPKQ L
```

Amino acid sequence for mouse eotaxin 3

No information.

THE RECEPTOR, CCR3

The eotaxin 3 receptor, CCR3 (also known as CC CKR3, CKR-3 and MIP-1α receptor-like 2), is selectively expressed on eosinophils, basophils and Th2 lymphocytes. For more detailed description see eotaxin 1 entry (page **213**). Eotaxin 3 binds to CCR3, mediating a pertussis toxin-sensitive Ca flux with an EC_{50} of 3 nM, about 10-fold less active than eotaxin 1[3]. Eotaxin 3 caused chemotaxis of eosinophils and basophils at greater than 300–1000 nM, which is about 10-fold less potent than eotaxin 1.

References

[1] Guo, R.F. et al. (1999) Genomics 58, 313–317.
[2] Shinkai, A. et al. (1999) J. Immunol. 163, 1602–1610.
[3] Kitaura, M. et al. (1999) J. Biol. Chem. 274, 27975–27980.

Other names

None.

THE MOLECULE

Erythropoietin (Epo) is an unusual cytokine in that it is not produced by haematopoietic cells, only by kidney or liver. It acts as a true hormone, stimulating erythroid precursors to generate red blood cells. It also stimulates platelet generation[1–4].

Crossreactivity

There is about 80% homology between human and mouse Epo[4]. Human Epo is active on murine cells and vice versa.

Sources

Kidney or liver in response to hypoxia or anaemia[5].

Bioassays

Bone marrow erythroid colony formation in methylcellulose, or stimulation of proliferation of TF-1 cell line[6,7].

Physicochemical properties of Epo

Property	Human	Mouse
pI	4–5	4–5
Amino acids – precursor	193	192
– mature[a]	166	165
M_r (K) – predicted	18.4	18.6
– expressed	36	34
Potential *N*-linked glycosylation sites[b]	3	3
Disulfide bonds[c]	2	2

[a] Mature human Epo has C-terminal arginine removed post-translationally.
[b] Human Epo also has one *O*-linked site (Ser126).
[c] Disulfide bonds link Cys7–161 and 29–33.

3-D structure

The crystal structure of Epo is unknown. Epo is a globular protein with a structure comprising four antiparallel α-helical bundle with one short and two long intervening loops and several stretches of β-sheets.

Gene structure[8]

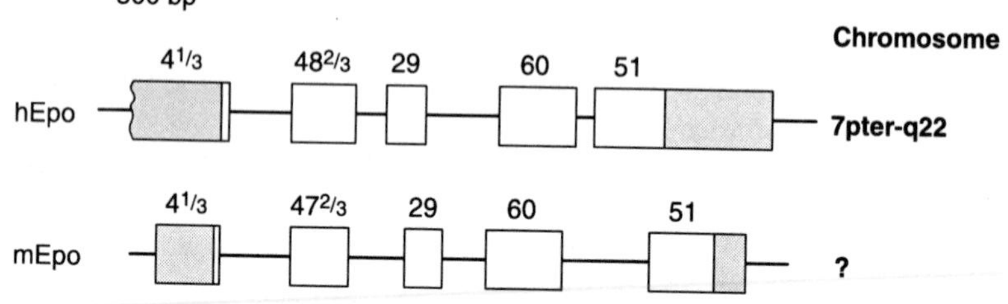

Human Epo is found on chromosome 7pter–q22 and mouse Epo is on chromosme 5. Human Epo has five exons of $4^{1/3}$, $49^{2/3}$, 29, 60 and 51 amino acids. Mouse Epo also contains five exons of $4^{1/3}$, $47^{2/3}$, 29, 60 and 51 amino acids.

Amino acid sequence for human Epo[9]

Accession code: SwissProt P01588

```
-27  MGVHECPAWL WLLLSLLSLP LGLPVLG
  1  APPRLICDSR VLERYLLEAK EAENITTGCA EHCSLNENIT VPDTKVNFYA
 51  WKRMEVGQQA VEVWQGLALL SEAVLRGQAL LVNSSQPWEP LQLHVDKAVS
101  GLRSLTTLLR ALGAQKEAIS PPDAASAAPL RTITADTFRK LFRVYSNFLR
151  GKLKLYTGEA CRTGDR
```

Amino acid sequence for mouse Epo[4]

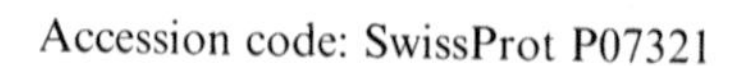

Accession code: SwissProt P07321

```
-26  MGVPERPTLL LLLSLLLIPL GLPVLC
  1  APPRLICDSR VLERYILEAK EAENVTMGCA EGPRLSENIT VPDTKVNFYA
 51  WKRMEVEEQA IEVWQGLSLL SEAILQAQAL LANSSQPPET LQLHIDKAIS
101  GLRSLTSLLR VLGAQKELMS PPDTTPPAPL RTLTVDTFCK LFRVYANFLR
151  GKLKLYTGEV CRRGDR
```

THE EPO RECEPTOR

A single class of receptor for Epo, which is a member of the haematopoietin receptor family, has been identified on mouse and human cells[10–12]. The recombinant receptor alone forms high- and low-affinity structures on COS cells, probably by self-association.

Distribution

The Epo receptor is expressed not only by erythroid cells, but also by embryonic stem cells, endothelial cells and neural cells.

Physicochemical properties of the Epo receptor

Property	Human	Mouse
Amino acids – precursor	508	507
– mature	484	483
M_r (K) – predicted	52.5	52.5
– expressed	85–100	85–100
Potential *N*-linked glycosylation sites	1	1
Affinity K_d (M)	$0.1–1.1 \times 10^{-9}$	$0.3–2 \times 10^{-10}$

Signal transduction

The Epo receptor couples directly to the Jak2 kinase to cause tyrosine phosphorylation[13].

Chromosomal location

The gene for the human Epo receptor is on chromosome 19p[14].

Amino acid sequence for human Epo receptor[10]

Accession code: SwissProt P19235

```
-24  MDHLGASLWP QVGSLCLLLA GAAW
  1  APPPNLPDPK FESKAALLAA RGPEELLCFT ERLEDLVCFW EEAASAGVGP
 51  GNYSFSYQLE DEPWKLCRLH QAPTARGAVR FWCSLPTADT SSFVPLELRV
101  TAASGAPRYH RVIHINEVVL LDAPVGLVAR LADESGHVVL RWLPPPETPM
151  TSHIRYEVDV SAGNGAGSVQ RVEILEGRTE CVLSNLRGRT RYTFAVRARM
201  AEPSFGGFWS AWSEPVSLLT PSDLDPLILT LSLILVVILV LLTVLALLSH
251  RRALKQKIWP GIPSPESEFE GLFTTHKGNF QLWLYQNDGC LWWSPCTPFT
301  EDPPASLEVL SERCWGTMQA VEPGTDDEGP LLEPVGSEHA QDTYLVLDKW
351  LLPRNPPSED LPGPGGSVDI VAMDEGSEAS SCSSALASKP SPEGASAASF
401  EYTILDPSSQ LLRPWTLCPE LPPTPPHLKY LYLVVSDSGI STDYSSGDSQ
```

Residues 112–128 are also predicted to be a transmembrane spanning sequence. Disulfide bonds between Cys28–38 and 67–83.

Amino acid sequence for mouse Epo receptor[11]

Accession code: SwissProt P14753

```
-24  MDKLRVPLWP RVGPLCLLLA GAAW
  1  APSPSLPDPK FESKAALLAS RGSEELLCFT QRLEDLVCFW EEAASSGMDF
 51  NYSFSYQLEG ESRKSCSLHQ APTVRGSVRF WCSLPTADTS SFVPLELQVT
101  EASGSPRYHR IIHINEVVLL DAPAGLLARR AEEGSHVVLR WLPPPGAPMT
151  THIRYEVDVS AGNRAGGTQR VEVLEGRTEC VLSNLRGGTR YTFAVRARMA
201  EPSFSGFWSA WSEPASLLTA SDLDPLILTL SLILVLISLL LTVLALLSHR
251  RTLQQKIWPG IPSPESEFEG LFTTHKGNFQ LWLLQRDGCL WWSPGSSFPE
301  DPPAHLEVLS EPRWAVTQAG DPGADDEGPL LEPVGSEHAQ DTYLVLDKWL
351  LPRTPCSENL SGPGGSVDPV TMDEASETSS CPSDLASKPR PEGTSPSSFE
401  YTILDPSSQL LCPRALPPEL PPTPPHLKYL YLVVSDSGIS TDYSSGGSQG
451  VHGDSSDGPY SHPYENSLVP DSEPLHPGYV ACS
```

Residues 111–127 are also predicted to be a transmembrane spanning sequence. Disulfide bonds between Cys28–38 and 66–82.

References

1. Krantz, S.B. (1991) Blood 77, 419–434.
2. Spivak, J.L. (1989) Nephron 52, 289–294.
3. Jacobs, K. et al. (1985) Nature 313, 806–810.
4. Shoemaker, C.B. and Mitsock, L.D. (1986) Mol. Cell. Biol. 6, 849–858.
5. Schuster, S.J. et al. (1987) Blood 70, 316–318.
6. Krumwieh, D. et al. (1988) In Cytokines: Laboratory and Clinical Evaluation., Vol. 69, Developments in Biological Standardisation, Gearing, A.J.H. and Hennessen, W. eds, Karger, Basle, pp. 15–22.
7. Gearing, A.J.H. et al. (1994) In The Cytokine Handbook, Thomson, A.W. ed., Academic press, London.
8. Law, M.L. et al. (1986) Proc. Natl Acad. Sci. USA 83, 6920–6924.
9. Lin, F.-K. et al. (1985) Proc. Natl Acad. Sci. USA 82, 7580–7584.
10. Jones, S.S. et al. (1990) Blood 76, 31–35.
11. D Andrea, A.D. et al. (1989) Cell 57, 277–285.
12. Kuramochi, S. et al. (1990) J. Mol. Biol. 216, 567–575.
13. Witthuhn, B.A. et al. (1993) Cell 74, 227–236.
14. Winkelmann, J.C. et al. (1990) Blood 76, 24–30.

Other names
Apo1 ligand, CD95L.

THE MOLECULE

Fas ligand (FasL) is a highly conserved type II membrane glycoprotein belonging to the TNF family of cytokines, which induces apoptosis when it binds to cells expressing Fas[1,2]. It is expressed on activated T cells and NK cells and can be released as a soluble protein by metalloprotease-dependent cleavage from the cell surface. Soluble FasL functions as a homotrimer. The *gld* (generalized lymphoproliferative disease) mutation in FasL results in lymphoproliferation and lymphadenopathy with high levels of immunoglobulin in homozygous *gld*/*gld* mice[3,4]. Mutations in Fas give rise to a similar phenotype[5]. Mutations in human Fas also give rise to lymphoproliferative disease and autoimmunity[6].

Crossreactivity
There is 76.9% amino acid sequence homology between human and mouse FasL and complete cross-species reactivity.

Sources
Activated T cells and NK cells.

Bioassays
Apoptosis of Fas-expressing lymphocytes.

Physicochemical properties of FasL

Property	Human	Mouse
pI	9.41	8.97
Amino acids – precursor	281	279
– mature[a]	281	279
M_r (K) – predicted	31.8	31.5
– expressed	40/26[b]	?
Potential *N*-linked glycosylation sites	3	4
Disulfide bonds[c]	1	1

[a] Type II membrane glycoprotein and no signal peptide.
[b] FasL is a 40-kDa transmembrane protein, when proteolytically cleaved FasL is a 70-kDa homotrimer of 26-kDa monomers with full biological activity.
[c] Disulfide bonds between Cys58 and 105 (human) and 72 and 120 (mouse).

3-D structure
The crystal structure of FasL is unknown. However, it is predicted to resemble that of TNF.

Gene structure[3,7]

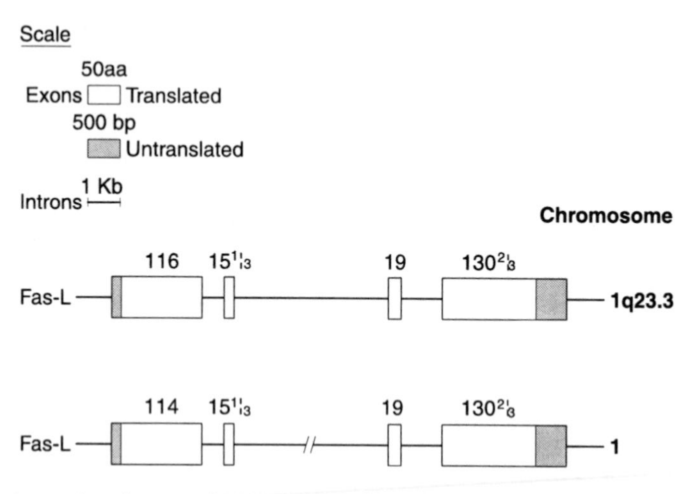

FasL is located on human chromosome 1q23 and mouse chromosome 1.

Amino acid sequence for human FasL[3,7]

Accession code: SwissProt P48023

```
  1  MQQPFNYPYP QIYWVDSSAS SPWAPPGTVL PCPTSVPRRP GQRRPPPPPP
 51  PPPLPPPPPP PPLPPLPLPP LKKRGNHSTG LCLLVMFFMV LVALVGLGLG
101  MFQLFHLQKE LAELRESTSQ MHTASSLEKQ IGHPSPPPEK KELRKVAHLT
151  GKSNSRSMPL EWEDTYGIVL LSGVKYKKGG LVINETGLYF VYSKVYFRGQ
201  SCNNLPLSHK VYMRNSKYPQ DLVMMEGKMM SYCTTGQMWA RSSYLGAVFN
251  LTSADHLYVN VSELSLVNFE ESQTFFGLYK L
```

Disulfide bond between Cys 202 and 233 by similarity.

Amino acid sequence for mouse FasL[4]

Accession code: SwissProt P41047

```
  1  MQQPMNYPCP QIFWVDSSAT SSWAPPGSVF PCPSCGPRGP DQRRPPPPPP
 51  PVSPLPPPSQ PLPLPPLTPL KKKDHNTNLW LPVVFFMVLV ALVGMGLGMY
101  QLFHLQKELA ELREFTNQSL KVSSFEKQIA NPSTPSEKKE PRSVAHLTGN
151  PHSRSIPLEW EDTYGTALIS GVKYKKGGLV INETGLYFVY SKVYFRGQSC
201  NNQPLNHKVY MRNSKYPEDL VLMEEKRLNY CTTGQIWAHS SYLGAVFNLT
251  SADHLYVNIS QLSLINFEES KTFFGLYKL
```

Variant amino acid 273 F→L in *gld* mutation prevents binding to Fas. Disulfide bond between Cys200–231 by similarity.

THE FASL RECEPTOR, FAS/APO1

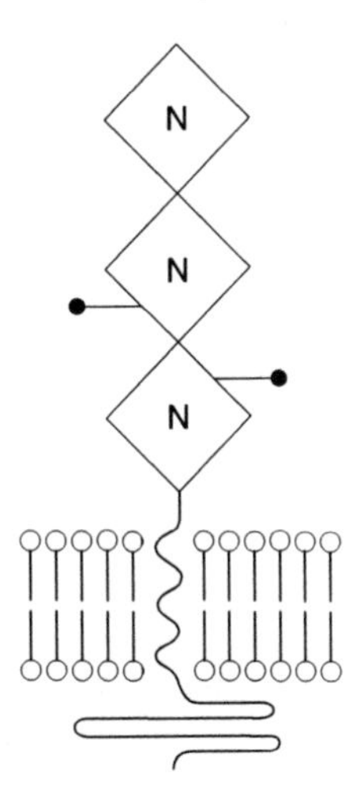

The receptor for Fas ligand is Fas/Apo1 (CD95) (apoptosis 1), a member of the TNFR superfamily. There is 50% amino acid identity between mouse and human Fas. A 68-amino-acid portion of the signal-transducing domain significantly conserved in the Fas antigen as well as in the type I tumour necrosis factor receptor is required for apoptosis and is called the death domain (DD). Fas has additional effects besides its apoptosis-inducing ability. On fibroblasts, Fas ligation can lead to proliferation or apoptosis depending on the level of Fas expression[8]. A soluble form of Fas is also generated as a result of alternative mRNA splicing. In addition to Fas, another protein has been shown to bind FasL with an affinity that is comparable to that of Fas, namely decoy receptor 3 (DcR3), a novel secreted member of the TNFR superfamily which is closely related to OPG[9,10]. DcR3 lacks an intracellular cytoplasmic domain and thus is thought to mop up FasL, thus inhibiting FasL-mediated apoptosis by competing with Fas for FasL binding as has been shown for LIGHT, another TNF-SF member, which also binds DcR3. DcR3 therefore inhibits LIGHT and FasL-mediated apoptosis (see LIGHT entry (page **351**)) by competing with the endogenous transmembrane receptors for these ligands[11].

Distribution

Fas expression is widespread. Unlike mouse thymocytes, Fas expression is weak on human thymocytes. Fas is highly expressed on activated lymphocytes.

Physicochemical properties of FasL receptor

Property	Fas/Apo1 Human	Fas/Apo1 Mouse	DcR3 Human	DcR3 Mouse
Amino acids – precursor	335	327	300	?
– mature[a]	319	306	277	?
M_r (K) – predicted	36	35	32.69	?
– expressed	43	45	37	?
Potential *N*-linked glycosylation sites	2	2	1	?
Affinity K_d (M)	1×10^{-9}	1×10^{-9}	1×10^{-9}	?

[a] After removal of predicted signal peptide.

Signal transduction

Ligation of the extracellular domain of the cell surface receptor Fas/Apo1 with FasL elicits a characteristic apoptotic response in susceptible cells. Fas associates via its DD with intracellular adapter molecules, such as MORT/FADD[12] and FLICE[13,14], via DD–DD interactions. FADD then recruits caspase-8 to the activated receptor, resulting in activation of a caspase cascade culminating in apoptosis (for a review see ref. 17). Fas ligation also results in partial hydrolysis of sphingomyelin and increase in ceramide and downstream activation of protein kinases.

Chromosomal location

The gene for human Fas/Apo1 is on chromosome 10q23 and mouse Fas/Apo1 gene is on chromosome 19.

Amino acid sequence for human Fas/Apo1[15]

Accession code: SwissProt P25445

```
-16  MLGIWTLLPL VLTSVA
  1  RLSSKSVNAQ VTDINSKGLE LRKTVTTVET QNLEGLHHDG QFCHKPCPPG
 51  ERKARDCTVN GDEPDCVPCQ EGKEYTDKAH FSSKCRRCRL CDEGHGLEVE
101  INCTRTQNTK CRCKPNFFCN STVCEHCDPC TKCEHGIIKE CTLTSNTKCK
151  EEGSRSNLGW LCLLLLPIPL IVWVKRKEVQ KTCRKHRKEN QGSHESPTLN
201  PETVAINLSD VDLSKYITTI AGVMTLSQVK GFVRKNGVNE AKIDEIKNDN
251  VQDTAEQKVQ LLRNWHQLHG KKEAYDTLIK DLKKANLCTL AEKIQTIILK
301  DITSDSENSN FRNEIQSLV
```

Death domain amino acids 214–298

Amino acid sequence for mouse Fas/Apo1[16]

Accession code: SwissProt P25446

```
-21  MLWIWAVLPL VLAGSQLRVH T
  1  QGTNSISESL KLRRRVHETD KNCSEGLYQG GPFCCQPCQP GKKKVEDCKM
 51  NGGTPTCAPC TEGKEYMDKN HYADKCRRCT LCDEEHGLEV ETNCTLTQNT
101  KCKCKPDFYC DSPGCEHCVR CASCEHGTLE PCTATSNTNC RKQSPRNRLW
151  LLTILVLLIP LVFIYRKYRK RKCWKRRQDD PESRTSSRET IPMNASNLSL
201  SKYIPRIAED MTIQEAKKFA RENNIKEGKI DEIMHDSIQD TAEQKVQLLL
251  CWYQSHGKSD AYQDLIKGLK KAECRRTLDK FQDMVQKDLG KSTPDTGNEN
301  EGQCLE
```

Variant amino acid 225 I→N in *lpr* mutation and death domain amino acids 201–285

Amino acid sequence for human DcR3

Accession code: SPTREMBLE O95407

```
-29  MRALEGPGLS LLCLVLALPA LLPVPAVRG
  1  VAETPTYPWR DAETGERLVC AQCPPGTFVQ RPCRRDSPTT CGPCPPRHYT
 51  QFWNYLERCR YCNVLCGERE EEARACHATH NRACRCRTGF FAHAGFCLEH
101  ASCPPGAGVI APGTPSQNTQ CQPCPPGTFS ASSSSSEQCQ PHRNCTALGL
151  ALNVPGSSSH DTLCTSCTGF PLSTRVPGAE ECERAVIDFV AFQDISIKRL
201  QRLLQALEAP EGWGPTPRAG RAALQLKLRR RLTELLGAQD GALLVRLLQA
251  LRVARMPGLE RSVRERFLPV H
```

References

1. Suda, T. et al. (1993) Cell 75, 1169–1178.
2. Lynch, D.H. et al. (1995) Immunol. Today 16, 569–574.
3. Lynch, D.H. et al. (1994) Immunity 1, 131–136.
4. Takahashi, T. et al. (1994) Cell 76, 969–976.
5. Watanabe-Fukunaga, R. et al. (1992) Nature 356, 314–317.
6. Rieux-Laucat, F. et al. (1995) Science 268, 1347–1349.
7. Takahashi, T. et al. (1994) Int. Immunol. 6, 1567–1574.
8. Freiberg, R.A. et al. (1997) J. Invest. Dermatol. 108, 215–219.
9. Pitti, R. M. et al. (1998) Nature 396, 699–703.
10. Ashkenazi, A. and Dixit, V.M. (1999) Curr. Opin. Cell Biol. 11, 255–260.
11. Yu, K.Y. et al. (1999) J. Biol. Chem. 274, 13733–13736.
12. Chinnaiyan, A.M. et al. (1995) Cell 81, 505–512.
13. Vincenz, C. and Dixit, V.M. (1997) J. Biol. Chem. 272, 6578–6583.
14. Muzio, M. et al. (1996) Cell 85, 817–827.
15. Itoh, N. et al. (1991) Cell 66, 233–243.
16. Watanabe-Fukunaga, R. et al. (1992) J. Immunol. 148, 1274–1279.
17. Green, D.R. and Ware, C.F. (1997) Proc. Natl Acad. Sci. USA 94, 5986–5990.

Other names
aFGF: Heparin-binding growth factor (HBGF-1), endothelial cell growth factor (ECGF), embryonic kidney-derived angiogenesis factor I, astroglial growth factor I, retina-derived growth factor, eye derived growth factor II, prostatropin.

bFGF: Leukaemia growth factor, macrophage growth factor, embryonic kidney-derived angiogenesis factor 2, prostatic growth factor, astroglial growth factor 2, endothelial growth factor, tumour angiogenesis factor, hepatoma growth factor, chondrosarcoma growth factor, cartilage-derived growth factor, eye-derived growth factor I, heparin-binding growth factors class II, myogenic growth factor, human placenta purified factor, uterine-derived growth factor, embryonic carcinoma-derived growth factor, human pituitary growth factor, pituitary-derived chondrocyte growth factor, adipocyte growth factor, prostatic osteoblastic factor, mammary tumour-derived growth factor.

THE MOLECULES

Basic fibroblast growth factor (bFGF) and acidic FGF (aFGF) are modulators of cell proliferation, motility, differentiation and angiogenesis[1]. Both factors have a high affinity for heparin and are found associated with the extracellular matrix[1–3]. Several cellular proto-oncogenes with sequence homology (40–50%) to aFGF and bFGF have been identified in humans[4,5] and mice[6] (see FGF family below). Some of these, such as int-2, are expressed primarily during embryogenesis and are under tight temporal control, suggesting that they play an important role in early development. Only aFGF and bFGF are described in detail below.

Crossreactivity
There is 95% homology between human and mouse aFGF and bFGF, and complete cross-species reactivity.

Sources
aFGF: Detected in large amounts only in the brain; lesser amounts in retina, bone matrix, osteoblasts, astrocytes, kidney, endothelial cells and fetal vascular smooth muscle cells.

bFGF: Brain, retina, pituitary, kidney, placenta, testis, corpus luteum, adrenal glands, monocytes, prostate, bone, liver, cartilage, endothelial cells, epithelial cells.

Bioassays
Proliferation of vascular endothelial cells.

THE FGF FAMILY

Name	Amino acids (human)[a]	Secreted form	Major tissue localization[b]	Human chromosome	Known receptors	SwissProt accession codes	
						Human	Mouse
aFGF (FGF-1)	155	Not secreted	Adult tissue	5q31.3–q33.2	FGFR-1,2,3,4	P05230 P07502	P10935
bFGF (FGF-2)	155	Not secreted	Adult and embryonic tissue	4q25–q27	FGFR-1,2	P09038	P15655
INT-2 (FGF-3)	239	ND	Embryonic tissue	11q13		P11487	P05524
K-FGF/HST (FGF-4)	206	176	Embryonic tissue	11q13.3	FGFR-1,2	P08620	P11403
FGF-5	267	ND	Neonatal brain	4q21		P12034	P15656
FGF-6	208	ND	Embryonic	12p13	FGFR-2 (variant)	P10767	P21658
KGF (FGF-7)	194	163	Dermis, kidney, gastrointestinal tract	Unassigned		P21781	Unassigned

[a] Primary translation product.
[b] References 4, 5.

Physicochemical properties of acidic and basic FGF

Property	aFGF		bFGF	
	Human	Mouse	Human	Mouse
pI	5.4	5.4	9.6	9.6
Amino acids – precursor	155	155	155	154
			196[a]	
			201[a]	
			210[a]	
– mature[b]	155	155	155	154
	140	140	146[c]	145
M_r (K) – predicted[d]	17.5	17.4	17.3	17.2
– expressed	16	16–18	16.5[e]	16.5[e]
			18	
			22.5[a]	
			23.1[a]	
			24.2[a]	
Potential *N*-linked glycosylation sites	1	1	0	0
Disulfide bonds	0	0	0	0

[a] Long forms arising by translation from upstream CUG initiation sites (see sequence).
[b] The predominant form of aFGF is 155 amino acids in length and the predominant form of bFGF is 155 (human), 154 (mouse) amino acids. There is no typical signal peptide. Removal of N-terminal propeptide (15 amino acids in aFGF and 9 amino acids in bFGF) gives rise to truncated forms which are still active. Another truncated form coded for by exon 1 with an additional four C-terminal amino acids has been described and shown to be an antagonist to the full length form[7] (see sequence).
[c] Smaller truncated forms which retain biological activity have been isolated from different tissues, but it is not known whether these are due to *in vivo* processing or artefacts of extraction and purification.
[d] Primary translation product.
[e] For 146 and 145 amino acid product.

3-D structure

X-Ray crystallography to a resolution of 1.8 Å has shown that bFGF is composed entirely of a β-sheet structure with a 3-fold repeat of a four stranded antiparallel β-meander which forms a barrel-like structure with three loops. The topology of bFGF is very similar to that of IL-1. The receptor and heparin-binding sites are adjacent but separate determinants on the β-barrel[8,9].

Gene structure[10–13]

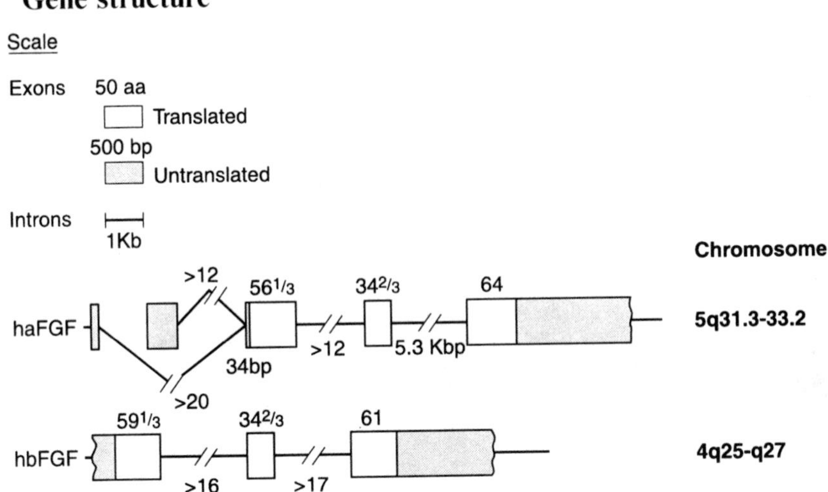

Human FGFa has five exons and is on chromosome 5q31.3–33.2 and FGFb has three exons and is on chromosome 4q25-q27. Mouse FGFa is on chromosome 18 and FGFb is on chromosome 3. Different sized aFGF mRNA species containing alternative 5' untranslated sequences coded for by additional upstream exons have been described and shown to be expressed in a tissue-specific manner[14].

Note: A new amino acid numbering system for the FGFs has been proposed in which the amino acid immediately following the initiation Met is designated as amino acid number 1. For consistency with the rest of this book we have not followed this proposal but designated the initiation Met itself as number 1.

Amino acid sequence for human aFGF (FGF-1)[10–13]

Accession code: SwissProt P05230, P07502

```
  1  MAEGEITTFT ALTEKFNLPP GNYKKPKLLY CSNGGHFLRI LPDGTVDGTR
 51  DRSDQHIQLQ LSAESVGEVY IKSTETGQYL AMDTDGLLYG SQTPNEECLF
101  LERLEENHYN TYISKKHAEK NWFVGLKKNG SCKRGPRTHY GQKAILFLPL
151  PVSSD
```

aFGF has no signal sequence. Truncated forms of 140 and 134 amino acids in length arise by cleavage between amino acids Lys15/Phe16 and Gly21/Asn22. An FGF antagonist has been identified consisting of the first 56 amino acids coded for by exon 1 and an additional four amino acids (TDTK) forming the C-terminus[7]. Cys residues at positions 31 and 98 are conserved in all members of the FGF family but do not form disulfide bonds.

Amino acid sequence for human bFGF (FGF-2)[15,16]

Accession code: SwissProt P09038

```
-55  LGDRGRGRAL PGGRLGGRGR GRAPQRVGGR GRGRGTAAPR AAPAARGSRP
 -5  GPAGT
  1  MAAGSITTLP ALPEDGGSGA FPPGHFKDPK RLYCKNGGFF LRIHPDGRVD
 51  GVREKSDPHI KLQLQAEERG VVSIKGVCAN RYLAMKEDGR LLASKCVTDE
101  CFFFERLESN NYNTYRSRKY TSWYVALKRT GQYKLGSKTG PGQKAILFLP
151  MSAKS
```

bFGF has no signal sequence. High molecular weight forms (M_r 24 200, 23 100, 22 500) arise by initiation of translation at CUG triplets coding for Leu residues (bold) at positions −55, −46, and −41[17,18]. Truncated form arises by cleavage between amino acids Lys9/Pro10. It is not clear whether these are formed *in vivo* or are artefacts of purification. Cys residues at positions 34 and 101 are conserved in all members of the FGF family, but do not form disulfide bonds.

Amino acid sequence for mouse aFGF (FGF-1)[6,19]

Accession code: SwissProt P10935

```
  1  MAEGEITTFA ALTERFNLPL GNYKKPKLLY CSNGGHFLRI LPDGTVDGTR
 51  DRSDQHIQLQ LSAESAGEVY IKGTETGQYL AMDTEGLLYG SQTPNEECLF
101  LERLEENHYN TYTSKKHAEK NWFVGLKKNG SCKRGPRTHY GQKAILFLPL
151  PVSSD
```

aFGF has no signal sequence. Truncated form arises by cleavage between amino acids Arg15/Phe16. Cys residues at positions 31 and 98 are conserved in all members of the FGF family, but do not form disulfide bonds.

Amino acid sequence for mouse bFGF (FGF-2)[6]

Accession code: SwissProt P15655

```
  1  MAASGITSLP ALPEDGGAAF PPGHFKDPKR LYCKNGGFFL RIHPDGRVDG
 51  VREKSDPHVK LQLQAEERGV VSIKGVCANR YLAMKEDGRL LASKCVTEEC
101  FFFERLESNN YNTYRSRKYS SWYVALKRTG QYKLGSKTGP GQKAILFLPM
151  SAKS
```

bFGF has no signal sequence. Truncated form arises by cleavage between amino acids Lys9/Pro10. Cys33 and 100 are conserved in all members of the FGF family, but do not form disulfide bonds.

THE FGF RECEPTORS

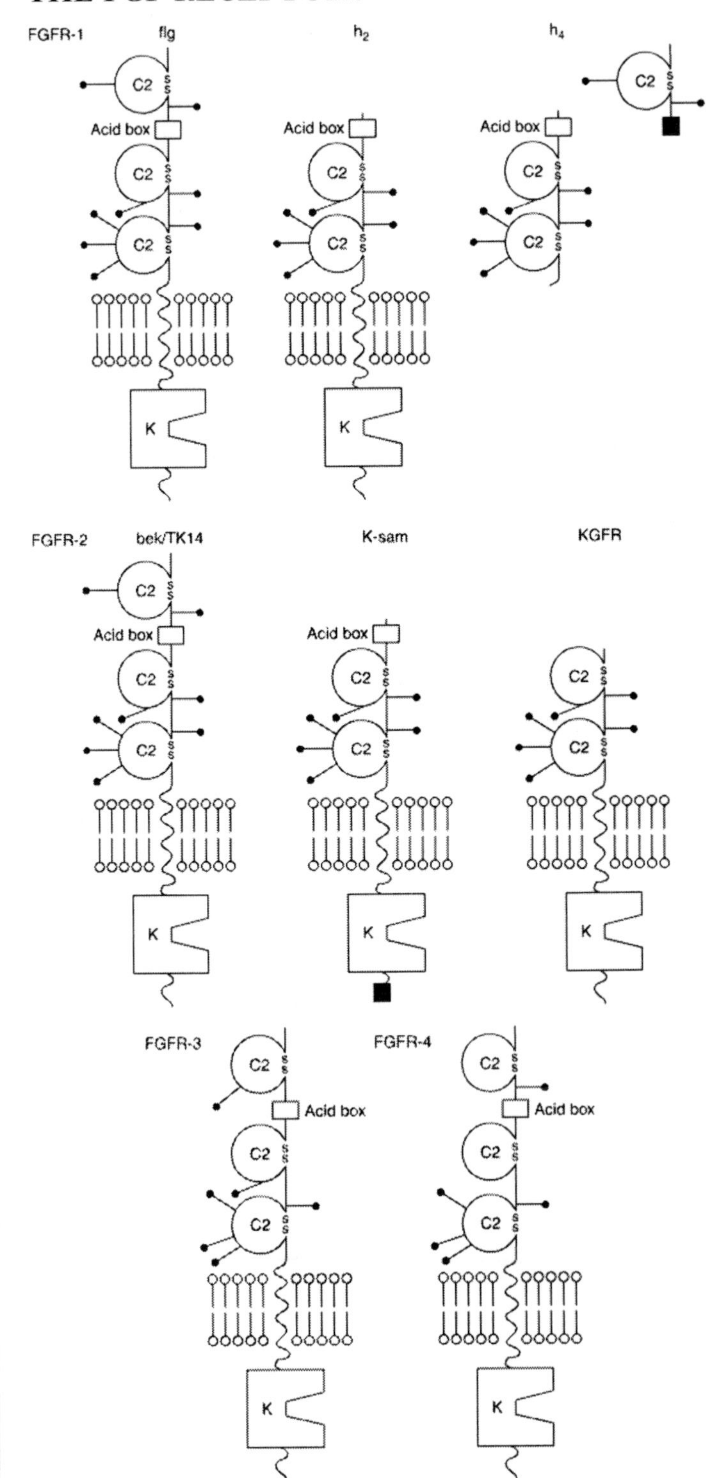

FGF binds to both low-affinity and high-affinity receptors[20]. Low-affinity receptors are heparan sulfate proteoglycans. Binding of FGF to heparin and other glycosaminoglycans (GAGs) protects FGF from degradation and can hold FGF in the extracellular matrix as a reservoir. FGF also binds to cell surface heparan sulfate proteoglycans, but it is not clear whether these can signal. FGF binds simultaneously to low-affinity heparan sulfate proteoglycan and to the high-affinity receptor[21]. High-affinity FGF receptors are members of a complex family of receptors characterized by between 1 and 3 extracellular Ig-SF domains determined by alternative mRNA splicing, and an intracellular tyrosine kinase split by a short inserted sequence of 14 amino acids[20]. Between the first and second Ig-SF domains there is a unique motif that has not been seen in other growth factor receptors consisting of eight consecutive acidic residues called the acid box. There are at least four distinct receptors, FGFR-1 (*flg*), FGFR-2 (*bek*), FGFR-3 and FGFR-4, coded for by separate genes. For FGFR-1 and FGFR-2, at least, alternative splicing gives rise to multiple forms, including soluble receptors[22]. The different names given to these receptors and their splice variants are shown in the table below and in the receptor diagram. The receptors have distinct patterns of expression on different tissues. On binding aFGF or bFGF the receptors form dimers. Dimerization is required for signal transduction involving auto- or interchain phosphorylation[23]. FGF receptors have been shown to be receptors for herpes simplex virus type 1[24].

Physicochemical properties of the FGF receptors

	FGFR-1 (*flg*)		FGFR-2 (*bek*)		FGFR-3		FGFR-4	
Properties	Human	Mouse	Human	Mouse	Human	Mouse	Human	Mouse
Amino acids								
–precursor[a]	822	822	821	820	806	801	802	808
– mature	801	802	800	800	784	781	778	790
M_r (K)								
– predicted[b]	89.4	89.7	89.7	89.8	85.8	85.8	85.5	87.8
– expressed[b]	160	145	135	150 125	135 125 97	?	140	?
Potential *N*-linked glycosylation sites	8	8	8	9	6	6	5	5
Affinity K_d (M)	Usually between 23×10^{-9} and 53×10^{-10}							

[a] There are multiple forms of FGFR-1 and FGFR-2 which arise from alternative mRNA splicing (see sequence).

[b] M_r values given are for full-length forms. Variants of human FGFR-3 are due to glycosylation.

Names given to the different FGF Receptors

FGFR-1	FGFR-2	FGFR-3	FGFR-4
flg	bek	cek2	FGFR-4
bFGFR	Cek3	FGFR-3	
Cek1	K-SAM		
N-bFGFR	TK14		
h2, h3	TK25		
h4, h5	KGFR		
FGFR-1	FGFR-2		

There is a very high degree of homology between human and mouse FGF receptors: 98% for FGFR-1 with only three amino acid differences in the intracellular domain.

Distribution[25,26]

FGFR-1 (*flg*) mRNA is highly expressed in human fetal brain, skin, growth plates of developing bones and calvarial bones with lesser amounts in other tissues. FGFR-2 (*bek*) mRNA is expressed in brain, choroid plexus, liver, lung, intestine, kidney, skin, growth plates of developing bone and calvarial bone. FGFR-3 mRNA is expressed in brain, lung, intestine, kidney, skin, growth plates of developing bones and calvarial bone. FGFR-4 mRNA is expressed in human fetal adrenals, lung, kidney, liver, pancreas, intestine, striated muscle and spleen.

Signal transduction

The FGF receptors belong to the split tyrosine kinase receptor family which includes PDGFR and c-*kit*. Ligand binding induces dimerization and interchain and/or autophosphorylation[27]. The receptor binds to and activates PLCγ through Tyr766 (FGFR-1) and stimulates the PI second messenger pathway. Receptors are internalized, and degradation is slow and partial.

Chromosomal location

	Human	Mouse
FGFR-1	8p12	8
FGFR-2	10q25.3–q26	7
FGFR-3	4p16.3	5
FGFR-4	5q33–qter	13

Amino acid sequence for human FGFR-1 (flg)[28–31]

Accession code: SwissProt P11362, P17049

```
-21  MWSWKCLLFW AVLVTATLCT A
  1  RPSPTLPEQA QPWGAPVEVE SFLVHPGDLL QLRCRLRDDV QSINWLRDGV
 51  QLAESNRTRI TGEEVEVQDS VPADSGLYAC VTSSPSGSDT TYFSVNVSDA
101  LPSSEDDDDD DDSSSEEKET DNTKPNRMPV APYWTSPEKM EKKLHAVPAA
```

```
151  KTVKFKCPSS GTPNPTLRWL KNGKEFKPDH RIGGYKVRYA TWSIIMDSVV
201  PSDKGNYTCI VENEYGSINH TYQLDVVERS PHRPILQAGL PANKTVALGS
251  NVEFMCKVYS DPQPHIQWLK HIEVNGSKIG PDNLPYVQIL KTAGVNTTDK
301  EMEVLHLRNV SFEDAGEYTC LAGNSIGLSH HSAWLTVLEA LEERPAVMTS
351  PLYLEIIIYC TGAFLISCMV GSVIVYKMKS GTKKSDFHSQ MAVHKLAKSI
401  PLRRQVTVSA DSSASMNSGV LLVRPSRLSS SGTPMLAGVS EYELPEDPRW
451  ELPRDRLVLG KPLGEGCFGQ VVLAEAIGLD KDKPNRVTKV AVKMLKSDAT
501  EKDLSDLISE MEMMKMIGKH KNIINLLGAC TQDGPLYVIV EYASKGNLRE
551  YLQARRPPGL EYCYNPSHNP EEQLSSKDLV SCAYQVARGM EYLASKKCIH
601  RDLAARNVLV TEDNVMKIAD FGLARDIHHI DYYKKTTNGR LPVKWMAPEA
651  LFDRIYTHQS DVWSFGVLLW EIFTLGGSPY PGVPVEELFK LLKEGHRMDK
701  PSNCTNELYM MMRDCWHAVP SQRPTFKQLV EDLDRIVALT SNQEYLDLSM
751  PLDQYSPSFP DTRSSTCSSG EDSVFSHEPL PEEPCLPRHP AQLANGGLKR
801  R
```

There are several alternative forms of the FGFR-1 receptor derived by alternative splicing[20] (see receptor diagram). Two full-length forms vary only by the presence or absence of two amino acids (Arg and Met) at positions 127–128. Short forms are missing amino acids 31–119 comprising the first Ig domain. These also vary by the presence or absence of the Arg–Met dipeptide at positions 127–128. Soluble forms have been described in which the C-terminal 510 amino acids have been replaced by the unique 79-amino-acid C-terminal sequence:

```
VIMAPVFVGQ STGKETTVSG AQVPVGRLSC PRMGSFLTLQ AHTLHLSRDL
ATSPRTSNRG HKVEVSWEQR AAGMGGAGL
```

Another soluble form has a complete Ig domain 1 followed by 32 unique amino acids and a stop codon. Conflicting amino acid sequence ALFDRI→IYLTGS at positions 650–655 and G→R at position 796. Phosphorylated tyrosines at positions 632, 745. Tyr745 is required for PLCγ1 association and activation.

Amino acid sequence for human FGFR-2 (*bek*)[29]

Accession code: SwissProt P21802

```
-21  MVSWGRFICL VVVTMATLSL A
  1  RPSFSLVEDT TLEPEEPPTK YQISQPEVYV AAPGESLEVR CLLKDAAVIS
 51  WTKDGVHLGP NNRTVLIGEY LQIKGATPRD SGLYACTASR TVDSETWYFM
101  VNVTDAISSG DDEDDTDGAE DFVSENSNNK RAPYWTNTEK MEKRLHAVPA
151  ANTVKFRCPA GGNPMPTMRW LKNGKEFKQE HRIGGYKVRN QHWSLIMESV
201  VPSDKGNYTC VVENEYGSIN HTYHLDVVER SPHRPILQAG LPANASTVVG
251  GDVEFVCKVY SDAQPHIQWI KHVEKNGSKY GPDGLPYLKV LKAAGVNTTD
301  KEIEVLYIRN VTFEDAGEYT CLAGNSIGIS FHSAWLTVLP APGREKEITA
351  SPDYLEIAIY CIGVFLIACM VVTVILCRMK NTTKKPDFSS QPAVHKLTKR
401  IPLRRQVTVS AESSSSMNSN TPLVRITTRL SSTADTPMLA GVSEYELPED
451  PKWEFPRDKL TLGKPLGEGC FGQVVMAEAV GIDKDKPKEA VTVAVKMLKD
501  DATEKDLSDL VSEMEMMKMI GKHKNIINLL GACTQDGPLY VIVEYASKGN
551  LREYLRARRP PGMEYSYDIN RVPEEQMTFK DLVSCTYQLA RGMEYLASQK
601  CIHRDLAARN VLVTENNVMK IADFGLARDI NNIDYYKKTT NGRLPVKWMA
```

```
651  PEALFDRVYT HQSDVWSFGV LMWEIFTLGG SPYPGIPVEE LFKLLKEGHR
701  MDKPANCTNE LYMMMRDCWH AVPSQRPTFK QLVEDLDRIL TLTTNEEYLD
751  LSQPLEQYSP SYPDTRSSCS SGDDSVFSPD PMPYEPCLPQ YPHINGSVKT
```

There are several alternative forms of the FGFR-2 receptor derived by alternative splicing using different exons[32] (see receptor diagram). One of the alternative forms is a receptor for KGF and aFGF (FGFR-2 does not bind KGF) and is expressed only on epithelial cells. It is identical to FGFR-2 except for two missing amino acids (Ala–Ala) at positions 293–294 followed immediately by a divergent 38 amino acid sequence (HSGINSSNAE VLALFNVTEA DAGEYICKVS NYIGQANQ) from position 295–332 corresponding to the second half of the third Ig domain, an insert of three amino acids (Lys–Gln–Gln) between residues 340 and 341, and a G→R substitution at position 592. The second variant (K-SAM)[33] has a deletion of 87 amino acids from positions 18–104 which includes the first Ig domain, and a deletion of two amino acids (Val–Thr) at positions 407–408. As with the KGF receptor, it has two missing amino acids (Ala–Ala) at positions 293–294, followed by the divergent 38 amino acid sequence from positions 295–332, and the Lys–Gln–Gln insert between residues 340 and 341. It also has a truncated C-terminus ending with a divergent 12 amino acid sequence (PPNPSLMSIF RK) from position 740. Tyr635 is autophosphorylated.

Amino acid sequence for human FGFR-3[34,35]

Accession code: SwissProt P22607

```
-22  MGAPACALAL CVAVAIVAGA SS
  1  ESLGTEQRVV GRAAEVPGPE PGQQEQLVFG SGDAVELSCP PPGGGPMGPT
 51  VWVKDGTGLV PSERVLVGPQ RLQVLNASHE DSGAYSCRQR LTQRVLCHFS
101  VRVTDAPSSG DDEDGEDEAE DTGVDTGAPY WTRPERMDKK LLAVPAANTV
151  RFRCPAAGNP TPSISWLKNG REFRGEHRIG GIKLRHQQWS LVMESVVPSD
201  RGNYTCVVEN KFGSIRQTYT LDVLERSPHR PILQAGLPAN QTAVLGSDVE
251  FHCKVYSDAQ PHIQWLKHVE VNGSKVGPDG TPYVTVLKTA GANTTDKELE
301  VLSLHNVTFE DAGEYTCLAG NSIGFSHHSA WLVVLPAEEE LVEADEAGSV
351  YAGILSYGVG FFLFILVVAA VTLCRLRSPP KKGLGSPTVH KISRFPLKRQ
401  VSLESNASMS SNTPLVRIAR LSSGEGPTLA NVSELELPAD PKWELSRARL
451  TLGKPLGEGC FGQVVMAEAI GIDKDRAAKP VTVAVKMLKD DATDKDLSDL
501  VSEMEMMKMI GKHKNIINLL GACTQGGPLY VLVEYAAKGN LREFLRARRP
551  PGLDYSFDTC KPPEEQLTFK DLVSCAYQVA RGMEYLASQK CIHRDLAARN
601  VLVTEDNVMK IADFGLARDV HNLDYYKKTT NGRLPVKWMA PEALFDRVYT
651  HQSDVWSFGV LLWEIFTLGG SPYPGIPVEE LFKLLKEGHR MDKPANCTHD
701  LYMIMRECWH AAPSQRPTFK QLVEDLDRVL TVTSTDEYLD LSAPFEQYSP
751  GGQDTPSSSS SGDDSVFAHD LLPPAPPSSG GSRT
```

Tyr625 is site of autophosphorylation by similarity.

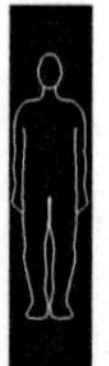

Amino acid sequence for human FGFR-4[25,35]

Accession code: SwissProt P22455

```
-24  MRLLLALLGV LLSVPGPPVL SLEA
  1  SEEVELEPCL APSLEQQEQE LTVALGQPVR LCCGRAERGG HWYKEGSRLA
 51  PAGRVRGWRG RLEIASFLPE DAGRYLCLAR GSMIVLQNLT LITGDSLTSS
101  NDDEDPKSHR DPSNRHSYPQ QAPYWTHPQR MEKKLHAVPA GNTVKFRCPA
151  AGNPTPTIRW LKDGQAFHGE NRIGGIRLRH QHWSLVMESV VPSDRGTYTC
201  LVENAVGSIR YNYLLDVLER SPHRPILQAG LPANTTAVVG SDVELLCKVY
251  SDAQPHIQWL KHIVINGSSF GAVGFPYVQV LKTADINSSE VEVLYLRNVS
301  AEDAGEYTCL AGNSIGLSYQ SAWLTVLPEE DPTWTAAAPE ARYTDIILYA
351  SGSLALAVLL LLAGLYRGQA LHGRHPRPPA TVQKLSRFPL ARQFSLESGS
401  SGKSSSSLVR GVRLSSSGPA LLAGLVSLDL PLDPLWEFPR DRLVLGKPLG
451  EGCFGQVVRA EAFGMDPARP DQASTVAVKM LKDNASDKDL ADLVSEMEVM
501  KLIGRHKNII NLLGVCTQEG PLYVIVECAA KGNLREFLRA RRPPGPDLSP
551  DGPRSSEGPL SFPVLVSCAY QVARGMQYLE SRKCIHRDLA ARNVLVTEDN
601  VMKIADFGLA RGVHHIDYYK KTSNGRLPVK WMAPEALFDR VYTHQSDVWS
651  FGILLWEIFT LGGSPYPGIP VEELFSLLRE GHRMDRPPHC PPELYGLMRE
701  CWHAAPSQRP TFKQLVEALD KVLLAVSEEY LDLRLTFGPY SPSGGDASST
751  CSSSDSVFSH DPLPLGSSSF PFGSGVQT
```

Tyr618 is potential site of autophosphorylation by similarity.

Amino acid sequence for mouse FGFR-1 (*flg*)[36–38]

Accession code: SwissProt P16092

```
-20  MWGWKCLLFW AVLVTATLCT
  1  ARPAPTLPEQ AQPWGVPVEV ESLLVHPGDL LQLRCRLRDD VQSINWLRDG
 51  VQLVESNRTR ITGEEVEVRD SIPADSGLYA CVTSSPSGSD TTYFSVNVSD
101  ALPSSEDDDD DDDSSSEEKE TDNTKPNRRP VAPYWTSPEK MEKKLHAVPA
151  AKTVKFKCPS SGTPNPTLRW LKNGKEFKPD HRIGGYKVRY ATWSIIMDSV
201  VPSDKGNYTC IVENEYGSIN HTYQLDVVER SPHRPILQAG LPANKTVALG
251  SNVEFMCKVY SDPQPHIQWL KHIEVNGSKI GPDNLPYVQI LKTAGVNTTD
301  KEMEVLHLRN VSFEDAGEYT CLAGNSIGLS HHSAWLTVLE ALEERPAVMT
351  SPLYLEIIIY CTGAFLISCM LGSVIIYKMK SGTKKSDFHS QMAVHKLAKS
401  IPLRRQVTVS ADSSASMNSG VLLVRPSRLS SSGTPMLAGV SEYELPEDPR
451  WELPRDRLVL GKPLGEGCFG QVVLAEAIGL DKDKPNRVTK VAVKMLKSDA
501  TEKDLSDLIS EMEMMKMIGK HKNIINLLGA CTQDGPLYVI VEYASKGNLR
551  EYLQARRPPG LEYCYNPSHN PEEQLSSKDL VSCAYQVARG MEYLASKKCI
601  HRDLAARNVL VTEDNVMKIA DFGLARDIHH IDYYKKTTNG RLPVKWMAPE
651  ALFDRIYTHQ SDVWSFGVLL WEIFTLGGSP YPGVPVEELF KLLKEGHRMD
701  KPSNCTNELY MMMRDCWHAV PSQRPTFKQL VEDLDRIVAL TSNQEYLDLS
751  IPLDQYSPSF PDTRSSTCSS GEDSVFSHEP LPEEPCLPRH PTQLANSGLK
801  RR
```

Splice variants using three alternative exons for the third Ig domain give rise to different receptor forms (a) with a divergent sequence (SGINSSDAEV LTLFNVTEAQ SGEYVCKVSN YIGEANQSAW LTV) from position 296–338 and (b) with a divergent sequence (VLLTSFLG) followed by a stop codon, which have different tissue expression and binding affinities for bFGF[39]. The second of these (b) codes for a secreted receptor that lacks transmembrane and cytoplasmic sequences including part of the third Ig domain. In another splice variant, 89 amino acids (11–99), including the first Ig domain, are missing[38]. Tyr633 is a potential site of autophosphorylation by similarity.

Conflicting amino acid sequences T→S at position 209, ILQ→HPS at positions 236–238, G→A at position 250, G→A at position 420, I→M at position 524, VL→LV at positions 609–610, R→H at position 736, and E→D at position 745. In addition, the dipeptide Arg–Arg at positions 128 and 129 is missing in one sequence[37]. This may be a splice variant similar to that described for human FGFR-1.

Amino acid sequence for mouse FGFR-2 (*bek*)[40]

Accession code: GenEMBL X5544 and SwissProt P21803

-21 MVSWGRFICL VLVTMATLSL A
1 RPSFSLVEDT TLEPEEPPTK YQISQPEAYV VVPRGSLELQ CMLKDAAVIS
51 WTKDGVHLGP N**N**RTVLIGRY LQIKGATPRD SGLYACTAAR TVDSETWYFM
101 V**N**VTDAISSG DDEDDTDSSE RVSE**N**RSNQR APYWTNTEKM EKRLHAVPAA
151 NTVKFRCPAG GNPTPTMRWL KNGKEFKQEH RIGGYKVRNQ HRSLIMESVV
201 PSDKG**N**ITCL VENEYGSI**N**H TYHLDVVERS PHRPILQAGL PA**N**ASTVVGG
251 DVRFVCKVYS DAQPHIQWIK HVEK**N**GSKYG PDGLPYLKVL KAAGV**N**TTDK
301 EIEVLYIR**N**V TFEDAGEYTC LAGNSIGISF HSAWLTVLPA PVREKEITAS
351 PDYLEIAIYC IGVFLIACMV VTVIFCRMKT TTKKPDFSSQ PAVHKLTKRI

401 PLRRQVTVSA ESSSSMNSNT PLVRITTRLS STADTPMLAG VSEYELPEDP
451 KWEFPRDKLT LGKPLGEGCF GQVVMAEAVG IDKDKPKEAV TVAVKMLKDD
501 ATEKDLSDLV SEMEMMKMIG KHKNIINLLG ACTQDGPLYV IVEYASKGNL
551 REYLRARRPP GMEYSYDINR VPEEQMTFKD LVSCTYQLAR GMEYLASQKC
601 IHRDLAARNV LVTENNVMKI ADFGLARDIN NIDYYKKTTN GRLPVKWMAP
651 EALFDRVYTH QSDVWSFGVL MWEIFTLGGS PYPGIPVEEL FKLLKEGHRM
701 DKPTNCTNEL YMMMRDCWHA VPSQRPTFKQ LVEDLDRILT LTTNEEYLDL
751 TQPLEQYSPS YPDTSSSCSS GDDSVFSPDP MPYEPCLPQY PHINGSVKT

There is a variant similar to human K-SAM (see human FGFR-2 above)[32] which is missing 114 amino acids including the first Ig domain at positions 19–132, and has an I→Y substitution at position 207, an R→E substitution at position 253, two missing amino acids (Ala–Ala) at positions 292–293 followed by a divergent 38 amino acid sequence (HSGINSSNAE VLALFNVTEM DAGEYICKVS NYIGQANQ) from positions 294–331, and a Lys–Gln–Gln insert between amino acids 339 and 340. Tyr634 is a potential site of autophosphorylation by similarity.

Amino acid sequence for mouse FGFR-3[41]

Accession code: GenEMBL M81342

-20 MVVPACVLVF CVAVVAGATS
1 EPPGPEQRVV RRAAEVPGPE PSQQEQVAFG SGDTVELSCH PPGGAPTGPT
51 VWAKDGTGLV ASHRILVGPQ RLQVL**N**ASHE DAGVYSCQHR LTRRVLCHFS
101 VRVTDAPSSG DDEDGEDVAE DTGAPYWTRP ERMDKKLLAV PAANTVRFRC
151 PAAGNPTPSI SWLKNGKEFR GEHRIGGIKL RHQQWSLVME SVVPSDRG**N**Y
201 TCVVENKFGS IRQTYTLDVL ERSPHRPILQ AGLPA**N**QTAI LGSDVEFHCK
251 VYSDAQPHIQ WLKHVEV**N**GS KVGPDGTPYV TVLKTAGA**N**T TDKELEVLSL
301 H**N**VTFEDAGE YTCLAGNSIG FSHHSAWLVV LPAEEELMET DEAGSVYAGV
351 LSYGVVFFLF ILVVAAVILC RLRSPPKKGL GSPTVHKVSR FPLKRQVSLE
401 SNSSMNSNTP LVRIARLSSG EGPVLANVSE LELPADPKWE LSRTRLTLGK
451 PLGEGCFGQV VMAEAIGIDK DRTAKPVTVA VKMLKDDATD KDLSDLVSEM
501 EMMKMIGKHK NIINLLGACT QGGPLYVLVE YAAKGNLREF LRARRPPGMD
551 YSFDACRLPE EQLTCKDLVS CAYQVARGME YLASQKCIHR DLAARNVLVT
601 EDNVMKIADF GLARDVHNLD YYKKTTNGRL PVKWMAPEAL FDRVYTHQSD
651 VWSFGVLLWE IFTPGGPSPY PGIPVEELFK LLKEGHRMDK PASCTHDLYM
701 IMRECWHAVP SQRPTFKQLV EDLDRILTVT STDEYLDLSV PFEQYSPGGQ
751 DTPSSSSSGD DSVFTHDLLP PGPPSNGGPR T

Tyr621 is a potential site of autophosphorylation by similarity.

Amino acid sequence for mouse FGFR-4[42]

Accession code: GenEMBL X59927

-18 MWLLLALLSI FQGTPALS
1 LEASEEMEQE PCLAPILEQQ EQVLTVALGQ PVRLCCGRTE RGRHWYKEGS
51 RLASAGRVRG WRGRLEIASF LPEDAGRYLC LARGSMTVVH **N**LTLLMDDSL
101 TSISNDEDPK TLSSSSSGHV YPQQAPYWTH PQRMEKKLHA VPAGNTVKFR
151 CPACRNPMPT IHWLKDGQAF HGENRIGGIR LRHQHWSLVM ESVVPSDRGT
201 YTCLVENSLG SIRYSYLLDV LERSPHRPIL QAGLPA**N**TTA VVGSDVELLC
251 KVYSDAQPHI QWLKHVVI**N**G SSFGADGFPY VQVLKTTDI**N** ISEVQVLYLR
301 **N**VSAEDAGEY TCLAGNSIGL SYQSAWLTVL PEEDLTWTTA TPEARYTDII
351 LYVSGSLVLL VLLLLAGVYH RQVIRGHYSR QPVTIQKLSR FPLARQFSLE
401 SRSSGKSSLS LVRGVRLSSS GPPLLTGLVN LDLPLDPLWE FPRDRLVLGK
451 PLGEGCFGQV VRAEAFGQVV RAEAFGMDPS RPDQTSTVAV KMLKDNASDK
501 DLADLVSEME VMKLIGRHKN IINLLGVCTQ EGPLYVIVEC AAKGNLREFL
551 RARRPPGPDL SPDGPRSSEG PLSFPALVSC AYQVARGMQY LESRKCIHRD
601 LAARNVLVTE DDVMKIADFG LARGVHHIDY YKKTSNGRLP VKWMAPEALF
651 DRVYTHQSDV WSFEILLWEI FTLGGSPYPG IPVEELFSLL REGHRMERPP
701 NCPSELYGLM RECWHAAPSQ RPTFKQLVEA LDKVLLAVSE EYLDLRLTFG
751 PFSPSNGDAS STCSSSDSVF SHDPLPLEPS PFPFSDSQTT

Tyr630 is a potential site of autophosphorylation by similarity.

References

[1] Basilico, C. and Moscatelli, D. (1992) Adv. Cancer Res. 59, 115–165.
[2] Baird, A. and Bohlen, P. (1990) In Peptide Growth Factors and their Receptors I, Sporn, M.B. and Roberts, A.B. eds, Springer-Verlag, New York, pp. 369–418.
[3] Gospodarowicz, D. (1990) Clin. Orthop. 257, 231–248.
[4] Klagsbrun, M. (1989) Prog. Growth Factor Res. 1, 207–235.
[5] Benharroch, D. and Birnbaum, D. (1990) Isr. J. Med. Sci. 26, 212–219.
[6] Hebert, J.M. et al. (1990) Dev. Biol. 138, 454–463.
[7] Yu, Y.-L. et al. (1992) J. Exp. Med. 175, 1073–1080.
[8] Eriksson, A.E. et al. (1991) Proc. Natl Acad. Sci. USA 88, 3441–3445.
[9] Zhang, J. et al. (1991) Proc. Natl Acad. Sci. USA 88, 3446–3450.
[10] Mergia, A. et al. (1989) Biochem. Biophys. Res. Commun. 164, 1121–1129.
[11] Jaye, M. et al. (1986) Science 233, 541–545.
[12] Wang, W.-P. et al. (1989) Mol. Cell. Biol. 9, 2387–2395.
[13] Crumley, G. et al. (1990) Biochem. Biophys. Res. Commun. 171, 7–13.
[14] Myers, R.L. et al. (1993) Oncogene 8, 341- 349.
[15] Abraham, J.A. et al. (1986) Cold Spring Harbor Symp. Quant. Biol. LI, 657–668.
[16] Abraham, J.A. et al. (1986) EMBO J. 5, 2523- 2528.
[17] Prats, H. et al. (1989) Proc. Natl Acad. Sci. USA 86, 1836–1840.
[18] Florkiewicz, R.Z. and Sommer, A. (1989) Proc. Natl Acad. Sci. USA 86, 3978–3981.
[19] Goodrich, S.P. et al. (1989) Nucleic Acids Res. 17, 2867.
[20] Johnson, D.E. and Williams, L.T. (1993) Adv. Cancer. Res. 60, 1–41.
[21] Klagsbrun, M. and Baird, A. (1991) Cell 67, 229–231.
[22] Dionne, C.A. et al. (1991) Ann. N.Y. Acad. Sci. 638, 161–166.
[23] Robinson, C.J. (1991) Trends Pharmacol. Sci. 12, 123–124.
[24] Kaner, R.J. et al. (1990) Science 248, 1410–1413.
[25] Partanen, J. et al. (1991) EMBO J. 10, 1347- 1354.
[26] Korhonen, J. et al. (1991) Ann. N.Y. Acad. Sci. 638, 403–405.
[27] Bellot, F. et al. (1991) EMBO J. 10, 2849–2854.
[28] Isacchi, A. et al. (1990) Nucleic Acids Res. 18, 1906.
[29] Dionne, C.A. et al. (1990) EMBO J. 9, 2685- 2692.
[30] Itoh, N. et al. (1990) Biochem. Biophys. Res. Commun. 169, 680–685.
[31] Wennstrom, S. et al. (1991) Growth Factors 4, 197–208.
[32] Miki, T. et al. (1992) Proc. Natl Acad. Sci. USA 89, 246–250.
[33] Hattori, Y. et al. (1990) Proc. Natl Acad. Sci. USA 87, 5983–5987.
[34] Keegan, K. et al. (1991) Proc. Natl Acad. Sci. USA 88, 1095–1099.
[35] Partanen, J. et al. (1990) Proc. Natl Acad. Sci. USA 87, 8913–8917.
[36] Reid, H.H. et al. (1990) Proc. Natl Acad. Sci. USA 87, 1596–1600.
[37] Safran, A. et al. (1990) Oncogene 5, 635–643 .
[38] Mansukhani, A. et al. (1990) Proc. Natl Acad. Sci. USA 87, 4378–4382.
[39] Werner, S. et al. (1992) Mol. Cell. Biol. 12, 82–88.
[40] Raz, V. et al. (1991) Oncogene 6, 753–760.
[41] Ornitz, D.M. and Leder, P. (1992) J. Biol. Chem. 267, 16305–16311.
[42] Stark, K.L. et al. (1991) Development 113, 641–651.

Other names
Ligand for fetal liver kinase 2 receptor.

THE MOLECULE

Flt3 ligand (Flt3L) belongs to a small family of growth factors including stem cell factor (steel factor or *kit* ligand) and CSF-1 that regulate proliferation of haematopoietic precursors[1]. Alternative splicing of exon 6 gives rise to soluble and type I membrane-bound forms[2,3]. Flt3L binds to tyrosine kinase receptor Flt3 expressed mainly on cells of the haematopoietic lineage, including very primitive progenitor cells. Like SCF, Flt3L does not stimulate proliferation by itself but synergizes with many other growth factors and interleukins to enhance proliferation. Unlike SCF, it has no activity on mast cells. Flt3L-deficient mice develop normally but have reduced numbers of pro-B cells and are less able to reconstitute T cells and myeloid cells in irradiated mice[4].

Crossreactivity
There is 72% identity between human and mouse Flt3 ligand and complete cross-species activity.

Sources
Human and mouse Flt3L mRNA is expressed widely in adult and fetal tissue particularly spleen, lung and peripheral blood mononuclear cells. Soluble form is produced by mouse thymic stromal cells[2].

Bioassays
Costimulation of haematopoietic precursors.

Physicochemical properties of Flt3 ligand

Property	Human	Mouse
pI	8.17	
Amino acids – precursor	235	232
– mature[a]	209	206
M_r (K) – predicted[b]	23.7	23.2
– expressed[b]	30	30
Potential *N*-linked glycosylation sites	2	2
Disulfide bonds[c]	?	?

[a] Mature transmembrane form after removal of signal peptide.
[b] Transmembrane form.
[c] Shares conserved Cys residues with SCF and CSF-1 which are believed to form intramolecular disulfide bonds.

3-D structure
Four helix bundle similar to SCF and CSF-1. Native ligand reported to form a homodimer.

Gene structure[5]

Scale

50aa
Exons ☐ Translated
500 bp
Untranslated

500 bp
Introns

Chromosome

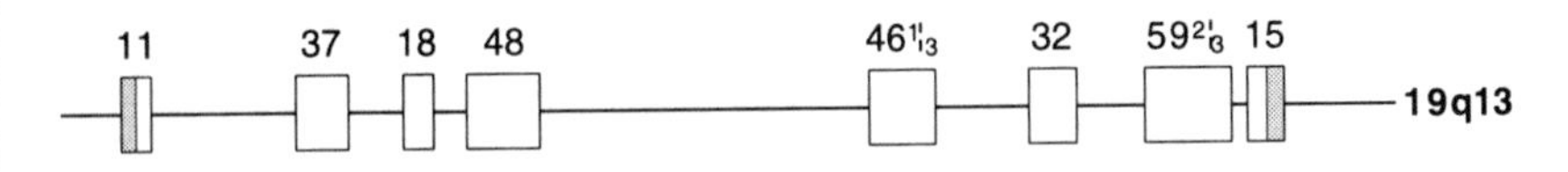

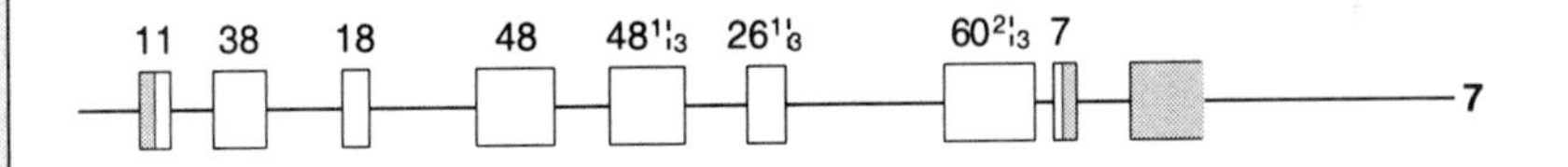

Alternative splicing to include exon 6 gives rise to soluble forms. Similar alternative splicing of exon 6 has been reported for SCF and CSF-1 (reviewed in ref. 3).

Amino acid sequence for human Flt3 ligand[2,6]

Accession code: SwissProt P49771

```
-26  MTVLAPAWSP TTYLLLLLLL SSGLSG
  1  TQDCSFQHSP ISSDFAVKIR ELSDYLLQDY PVTVASNLQD EELCGGLWRL
 51  VLAQRWMERL KTVAGSKMQG LLERVNTEIH FVTKCAFQPP PSCLRFVQTN
101  ISRLLQETSE QLVALKPWIT RQNFSRCLEL QCQPDSSTLP PPWSPRPLEA
151  TAPTAPQPPL LLLLLLPVGL LLLAAAWCLH WQRTRRRTPR PGEQVPPVPS
201  PQDLLLVEH
```

Alternative splicing (soluble) variant diverging sequence from D135[2,3]. Conflicting sequence G→A at position 46.

Amino acid sequence for mouse Flt3 ligand[2,7]

Accession code: SwissProt P49772

```
-26  MTVLAPAWSP NSSLLLLLLL LSPCLR
  1  GTPDCYFSHS PISSNFKVKF RELTDHLLKD YPVTVAVNLQ DEKHCKALWS
 51  LFLAQRWIEQ LKTVAGSKMQ TLLEDVNTEI HFVTSCTFQP LPECLRFVQT
101  NISHLLKDTC TQLLALKPCI GKACQNFSRC LEVQCQPDSS TLLPPRSPIA
151  LEATELPEPR PRQLLLLLLL LLPLTLVLLA AAWGLRWQRA RRRGELHPGV
201  PLPSHP
```

Alternative splicing (soluble) variants have diverging sequence from D138[2,3]. Conflicting sequence A→G at position 115 and L at position 172 missing.

THE FLT3 LIGAND RECEPTOR FLT3/FLK2 (CD135)

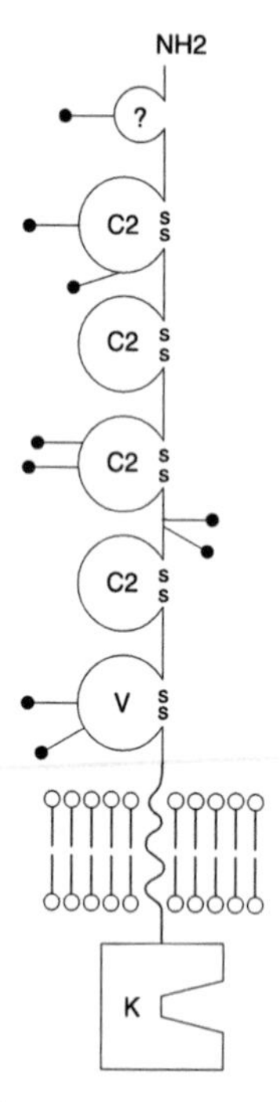

Flt3, also known as Flk2 or STK1, is a class III receptor tyrosine kinase consisting of five extracellular Ig domains a single transmembrane domain and an intracellular tyrosine kinase domain split by an inserted sequence of about 100 amino acids. It is structurally related to the PDGF receptors, the M-CSF receptor (c-*fms*) and the SCF receptor (c-*kit*)[8–10].

Distribution

Mainly cells of the haematopoietic lineage, also nervous system, gonads and placenta[8,11].

Physicochemical properties of the Flt3L receptor (Flt3/Flk2)

Property	Human	Mouse
Amino acids – precursor	993	1000
– mature[a]	967	975
M_r (K) – predicted	110	110.6
– expressed	155[b]	132/155[c]
Potential *N*-linked glycosylation sites	9	9
Affinity K_d (M)	?	3×10^{-8}

[a] After removal of predicted signal peptide.
[b] Reduced.
[c] Two molecular weight isoforms.

Signal transduction

Ligand binding to Flt3 results in receptor dimerization and association and/or phosphorylation of cytoplasmic proteins including Ras GTPase activating protein, PLCγ, VAV, SHC, PI-3 kinase and GRB2/Ras (reviewed in ref. 8).

Chromosomal location

Flt3 is found on human chromosome 13q12 and mouse chromosome 5 and is closely linked to Flt1, the receptor for VEGF.

Amino acid sequence for human Flt3[9,12]

Accession code: SwissProt P36888

-26 MPALARDAGT VPLLVVFSAM IFGTIT

1 NQDLPVIKCV LINHKN**N**DSS VGKSSSYPMV SESPEDLGCA LRPQSSGTVY

51 EAAAVEVDVS ASITLQVLVD APG**N**ISCLWV FKHSSLNCQP HFDLQNRGVV

101 SMVILKMTET QAGEYLLFIQ SEATNYTILF TVSIRNTLLY TLRRPYFRKM

151 ENQDALVCIS ESVPEPIVEW VLCDSQGESC KEESPAVVKK EEKVLHELFG

201 TDIRCCARNE LGRECTRLFT IDLNQTPQTT LPQLFLKVGE PLWIRCKAVH

251 VNHGFGLTWE LENKALEEGN YFEMSTYST**N** RTMIRILFAF VSSVAR**N**DTG

301 YYTCSSSKHP SQSALVTIVG KGFI**N**AT**N**SS EDYEIDQYEE FCFSVRFKAY

351 PQIRCTWTFS RKSFPCEQKG LDNGYSISKF CNHKHQPGEY IFHAENDDAQ

401 FTKMFTLNIR RKPQVLAEAS ASQASCFSDG YPLPSWTWKK CSDKSP**N**CTE

451 EITEGVWNRK ANRKVFGQWV SSSTL**N**MSEA IKGFLVKCCA YNSLGTSCET

501 ILLNSPGPFP FIQDNISFYA TIGVCLLFIV VLTLLICHKY KKQFRYESQL

551 QMVQVTGSSD NEYFYVDFRE YEYDLKWEFP RENLEFGKVL GSGAFGKVMN

601 ATAYGISKTG VSIQVAVKML KEKADSSERE ALMSELKMMT QLGSHENIVN

651 LLGACTLSGP IYLIFEYCCY GDLLNYLRSK R*EKFHRTWTE* *IFKEHNFSFY*

701 *PTFQSHPNSS* *MPGSREVQIH* *PDSDQISGLH* *GNSFHSEDEI* *EYENQKRLEE*

751 *EEDL*NVLTFE DLLCFAYQVA KGMEFLEFKS CVHRDLAARN VLVTHGKVVK

801 ICDFGLARDI MSDSNYVVRG NARLPVKWMA PESLFEGIYT IKSDVWSYGI

851 LLWEIFSLGV NPYPGIPVDA NFYKLIQNGF KMDQPFYATE EIYIIMQSCW

901 AFDSRKRPSF PNLTSFLGCQ LADAEEAMYQ NVDGRVSECP HTYQNRRPFS

951 REMDLGLLSP QAQVEDS

Tyr943 is required for association with GRB2 and Tyr929 with PI-3 kinase **[Q1]**. Protein kinase domain amino acids 584–917, kinase insert region 682–755 in italics. Conflicting sequence A→G at position −19, TV→QL at position −17 and −18, A→R at position 52, T→M at position 201 and G→E at position 320.

Amino acid sequence for mouse Flt3[10]

Accession code: SwissProt Q00342

-27 MRALAQRSDR RLLLLVVLSV MILET
1 VTNQDLPVIK CVLISHEN**N**G SSAGKPSSYR MVRGSPEDLQ CTPRRQSEGT
51 VYEAATVEVA ESGSITLQVQ LATPGDLSCL WVFKHSSLGC QPHFDLQNRG
101 IVSMAIL**N**VT ETQAGEYLLH IQSEAA**N**YTV LFTVNVRDTQ LYVLRRPYFR
151 KMENQDALLC ISEGVPEPTV EWVLCSSHRE SCKEEGPAVV RKEEKVLHEL
201 FGTDIRCCAR NALGRESTKL FTIDLNQAPQ STLPQLFLKV GEPLWIRCKA
251 IHVNHGFGLT WELEDKALEE GSYFEMSTYS T**N**RTMIRILL AFVSSVGR**N**D
301 TGYYTCSSSK HPSQSALVTI LEKGFI**N**ATS SQEEYEIDPY EKFCFSVRFK
351 AYPRIRCTWI FSQASFPCEQ RGLEDGYSIS KFCDHKNKPG EYIFYAENDD
401 AQFTKMFTLN IRKKPQVLA**N** ASASQASCSS DGYPLPSWTW KKCSDKSP**N**C
451 TEEIPEGVWN KKANRKVFGQ WVSSSTL**N**MS EAGKGLLVKC CAYNSMGTSC
501 ETIFLNSPGP FPFIQDNISF YATIGLCLPF IVVLIVLICH KYKKQFRYES
551 QLQMIQVTGP LDNEYFYVDF RDYEYDLKWE FPRENLEFGK VLGSGAFGRV
601 MNATAYGISK TGVSIQVAVK MLKEKADSCE KEALMSELKM MTHLGHHDNI
651 VNLLGACTLS GPVYLIFEYC CYGDLLNYLR SKREKFHRTW TEIFKEHNFS
701 FYPTFQAHSN SSMPGSREVQ LHPPLDQLSG FNGNSIHSED EIEYENQKRL
751 AEEEEEDLNV LTFEDLLCFA YQVAKGMEFL EFKSCVHRDL AARNVLVTHG
801 KVVKICDFGL ARDILSDSSY VVRGNARLPV KWMAPESLFE GIYTIKSDVW
851 SYGILLWEIF SLGVNPYPGI PVDANFYKLI QSGFKMEQPF YATEGIYFVM
901 QSCWAFDSRK RPSFPNLTSF LGCQLAEAEE AMYQNMGGNV PEHPSIYQNR
951 RPLSREAGSE PPSPQAQVKI HRERS

Tyr947 is required for association with GRB2 and Tyr933 with PI-3 kinase. Protein kinase domain amino acids 584–919.

Note: The sequence given above is the confirmed sequence for murine Flt3 originally reported by Rosnet et al.[10]. The SwissProt sequence is that of Flk2 reported by Matthews et al.[13]. This sequence differs from the Flt3 sequence at several positions and by a number of C-terminal amino acids but it has never been confirmed and the validity of the Flk2 sequence is unclear.

References

[1] Lyman, S.D. (1995) Int. J. Haematol. 62, 63–73.
[2] Hannum, C. et al. (1994) Nature 368, 643–648.
[3] Lyman, S.D. et al. (1995) Oncogene 10, 149–157.
[4] Mackarehtschian, K. et al. (1995) Immunity 3, 147–161.
[5] Lyman, S.D. et al. (1995) Oncogene 11, 1165–1172.
[6] Lyman, S.D. et al. (1994) Blood 83, 2795–2801.
[7] Lyman, S.D. et al. (1993) Cell 75, 1157–1167.
[8] Rosnet, O. et al. (1996) Acta Haematol. 95, 218–223.
[9] Rosnet, O. et al. (1993) Blood 82, 1110–1119.
[10] Rosnet, O. et al. (1991) Oncogene 6, 1641–1650.
[11] Maroc, N. et al. (1993) Oncogene 8, 909–918.
[12] Small, D. et al. (1994) Proc. Natl Acad. Sci. USA 91, 459–463.
[13] Matthews, W. et al. (1991) Cell 65, 1143–1152.

Other names

Neurotactin.

THE MOLECULE

Fractalkine is the first member of a novel family of chemokines (the CX3C family) in which the first pair of cysteines is separated by three amino acids to give the CXXXC motif[1–3]. In addition, the chemokine domain of fractalkine is part of a larger structure with a mucin-like domain, transmembrane and intracytoplasmic domain. The chemokine domain is functional in the cell surface-linked molecule and can also be shed by proteolysis to give a functional chemokine. Both forms are chemoattractants for monocytes, microglia and T cells. The cell surface form of fractalkine is capable of directly mediating cell adhesion to cells bearing its receptor[4,5].

Crossreactivity

Mouse and human fractalkine share 67% amino acid identity. Murine fractalkine acts on human cells[1].

Sources

Fractalkine is produced by microglia, activated endothelial cells and neurons, and mRNA is expressed in brain, kidney, lung and heart[1–3].

Bioassays

Chemotaxis or adhesion of lymphocytes.

Physicochemical properties of fractalkine

Property	Human	Mouse
Amino acids – precursor	397	395
– mature[a]	373	373
M_r (K) – predicted	42.2	42.0
– expressed	95.0, 50.0	?
Potential *N*-linked glycosylation sites[b]	1	1
Disulfide bonds	2	2

[a] A soluble form of fractalkine can be generated by cleavage at a dibasic RR site proximal to the transmembrane domain.

[b] Fractalkine has 17 mucin-like repeats, associated with several potential *O*-glycosylated Ser and Thr residues.

3-D structure

No information.

Gene structure

The gene for human fractalkine is on chromosome 16q13, and murine fractalkine on the long arm of chromosome 8[1,3,6].

Amino acid sequence for human fractalkine

Accession code: SwissProt P78423

-24 MAPISLSWLL RLATFCHLTV LLAG
1 *QHHGVTKC***N***I TCSKMTSKIP VALLIHYQQN QASCGKRAII LETRQHRLFC*
51 *ADPKEQWVKD AMQHLDRQAA ALTRNG*GTFE KQIGEVKPRT TPAAGGMDES
101 VVLEPEATGE SSSLEPTPSS QEAQRALGTS PELPTGVTGS SGTRLPPTPK
151 AQDGGPVGTE LFRVPPVSTA ATWQSSAPHQ PGPSLWAEAK TSEAPSTQDP
201 STQASTASSP APEENAPSEG QRVWGQGQSP RPENSLEREE MGPVPAHTDA
251 FQDWGPGSMA HVSVVPVSSE GTPSREPVAS GSWTPKAEEP IHATMDPQRL
301 GVLITPVPDA QAAT**RR**QAVG LLAFLGLLFC LGVAMFTYQS LQGCPRKMAG
351 EMAEGLRYIP RSCGSNSYVL VPV

Chemokine domain is in italics. Dibasic RR cleavage site is in bold.

Amino acid sequence for mouse fractalkine

Accession code: SwissProt O35933

-24 MAPSPLAWLL RLAAFFHLCT LLPG
1 *QHLGMTKCEI MCDKMTSRIP VALLIRYQLN QESCGKRAIV LETTQHRRFC*
51 *ADPKEKWVQD AMKHLDHQAA ALTKNG*GKFE KRVD**N**VTPGI TLATRGLSPS
101 ALTKPESATL EDLALELTTI SQEARGTMGT SQEPPAAVTG **SS**LSTSEAQD
151 AGLTAKPQSI GSFEAADIST TVWPSPAVYQ SGSSSWAEEK ATESPSTTAP
201 SPQVSTTSPS TPEENVGSEG QPPWVQGQDL SPEKSLGSEE INPVHTDNFQ
251 ERGPGNTVHP SVAPISSEET PSPELVASGS QAPKIEEPIH ATADPQKLSV
301 LITPVPDTQA AT**RR**QAVGLL AFLGLLFCLG VAMFAYQSLQ GCPRKMAGEM
351 VEGLRYVPRS CGSNSYVLVP V

Chemokine domain is in italics. Dibasic RR cleavage site (which generates soluble fractalkine) is shown in bold.

THE FRACTALKINE RECEPTOR, CX3R1

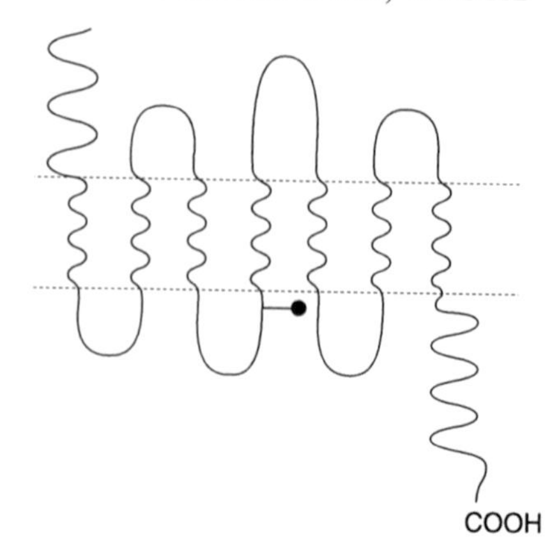

The specific fractalkine receptor CX3R1, also known as V28 or CMKBRL1, is a typical 7-TM spanning chemokine receptor[5,7–9]. CX3R1 binds to fractalkine with high affinity, mediating a calcium flux which is pertussis toxin sensitive, and chemotaxis. Binding of the membrane form of fractalkine to cells bearing the receptor mediates strong adhesion, which is not dependent on G protein signalling[4,5]. CX3R1 has also been shown to act as a co-receptor with CD4 for HIV-1 and HIV-2[7,10].

Distribution

CX3R1 is found on NK cells, T cells, monocytes, neurons and microglia[5,7–9,11,12].

Physicochemical properties of the fractalkine receptor

Property	Human	Mouse
Amino acids – precursor	355	366
– mature	?	?
M_r (K) – predicted	40.4	42.0
– expressed	?	?
Potential *N*-linked glycosylation sites	1	1
Affinity K_d (M)[a]	$30–50 \times 10^{-12}$	?

[a] K_d for human lymphocytes and monocytes.

Signal transduction

Fractalkine causes a pertussis toxin-sensitive calcium flux with an EC_{50} of 2 nM in transfected K562 cells and 20 nM in transfected 3.2 cells. Fractalkine stimulation of hippocampal neurons resulted in a modulation of Ca^{2+} currents, and activation of the MAP kinase pathway[11,12].

Chromosomal location

The human gene is located on chromosome 3p21–3pter.

Amino acid sequence for human receptor

Accession code: SwissProt P49238

```
  1  MDQFPESVTE NFEYDDLAEA CYIGDIVVFG TVFLSIFYSV IFAIGLVGNL
 51  LVVFALTNSK KPKSVTDIYL LNLALSDLLF VATLPFWTHY LINEKGLHNA
101  MCKFTTAFFF IGFFGSIFFI TVISIDRYLA IVLAANSMNN RTVQHGVTIS
151  LGVWAAAILV AAPQFMFTKQ KENECLGDYP EVLQEIWPVL RNVETNFLGF
201  LLPLLIMSYC YFRIIQTLFS CKNHKKAKAI KLILLVVIVF FLFWTPYNVM
251  IFLETLKLYD FFPSCDMRKD LRLALSVTET VAFSHCCLNP LIYAFAGEKF
301  RRYLYHLYGK CLAVLCGRSV HVDFSSSESQ RSRHGSVLSS NFTYHTSDGD
351  ALLLL
```

Disulfide bond links Cys102–175.

Amino acid sequence for mouse receptor

Accession code: SPTREMBLE Q60943

```
  1  MSTSFPELDL ENFEYDDSAE ACYLGDIVAF GTIFLSVFYA LVFTFFTGLV
 51  GNLLVVLALT NSRKPKSITD IYLLNLALSD LLFVATLPFW THYLISHEGL
101  HNAMCKFTLT TAFFFIGFFG GIFFITVISI DRYLAIVLAA NSMNNRTVQH
151  GVTISLGVWA AAILVASFTP QFMFTKRKDN ECLGDYPEVL QEMWPVLRNS
201  EVNILGFALP LLIMSFCYFR IIQTLFSCFT KNRKKARAVR LILLVVFAFF
251  LFWTPYNIMI FLETLKFYNF FPSCDMKRDL RLALSVTETF TVAFSHCCLN
301  PFIYAFAGEK FRRYLGHLYR KCLAVLCGHP VHTGFSPESQ RSRQDSILSS
351  FTFTHYTSEG DGSLLL
```

Disulfide bonds link Cys105–182.

References

[1] Pan, Y. et al. (1997) Nature 387, 611–617.
[2] Bazan, J.F. et al. (1997) Nature 385, 640–644.
[3] Rossi, D.L. et al. (1998) Genomics 47, 163–170.
[4] Fong, Am et al. (1998) J. Exp. Med. 188, 1413–1419.
[5] Imai, T. et al. (1997) Cell 91, 521–530.
[6] Nomiyama, H. et al. (1998) Cytogen. Cell Genet. 81, 10–11.
[7] Combadiere, C. et al. (1998) J. Biol. Chem. 273, 23799–23804.
[8] Combadiere, C. et al. (1998) Biochem. Biophys. Res. Commun. 253, 728–732.
[9] Combadiere, C. et al. (1995) DNA Cell Biol. 14, 673–680.
[10] Reeves, J.D. et al. (91997) Virology 231, 130–134.
[11] Harrison, J.K. et al. (1998) Proc. Natl Acad. Sci. USA 95, 10896–10901.
[12] Meucci, O. et al. (1998) Proc. Natl Acad. Sci. USA 95, 14500–14505.

Other names

SCYB6.

THE MOLECULE

Granulocyte chemotactic protein 2 (GCP-2) is a potent neutrophil chemoattractant and anti-angiogenic chemokine produced by connective tissue cells[1–5].

Crossreactivity

Human and murine GCP-2 share 61% amino acid sequence identity and act across species[5].

Sources

Osteosarcoma cells stimulated with IL-1, TNF and LPS, fibroblasts stimulated with LPS and double-stranded RNA, epithelial cells stimulated with PMA and endometrial cells stimulated with IFN or IFNα make GCP-2. IFNγ inhibits osteosarcoma cell production of GCP-2.

Bioassays

Chemotaxis of neutrophils.

Physicochemical properties of GCP-2

Property	Human	Mouse
Amino acids – precursor	114	?
– mature	77–67	78–67
M_r (K) – predicted	11.9	?
– expressed	5–6.5	7–8
Potential *N*-linked glycosylation sites	0	0
Disulfide bonds	2	2

Human GCP-2 can be N-terminally processed by CD26 to remove the first two GP residues[6]. Natural purified GCP-2 exists in at least four forms 1–77, 3–77, 6–77 and 9–77. All of these forms appear to be effective neutrophil chemoattractants. Purified natural GCP-2 also lacks the C-terminal two KN residues. Murine GCP-2 exists as multiple N-terminally processed forms lacking each of the first 10 residues; the shorter forms seem to be more potent chemoattractants.

3-D structure

No information.

Gene structure

No information.

Amino acid sequence for human GCP-2

Accession code: SwissProt P80162

```
-37  MSLPSSRAAR VPGPSGSLCA LLALLLLLTP PGPLASA
  1  GPVSAVLTEL RCTCLRVTLR VNPKTIGKLQ VFPAGPQCSK VEVVASLKNG
 51  KQVCLDPEAP FLKKVIQKIL DSGNKKN
```

Amino acid sequence for mouse GCP-2

Accession code: none registered

```
  1  APSSVIAATE LRCVCLTVTP KINPKLIANL EVIPAGPQCP TVEVIAKLKN
 51  QKEVCLDPEA PVIKKIIQKI LGSDKKKA
```

THE GCP-2 RECEPTORS, CXCR1 and CXCR2

GCP-2 binds and signals through the CXCR1 and CXCR2 receptors[7–9]. These receptors are also used by IL-8. See IL-8 entry (page **80**). GCP-2 binds with high affinity to both receptors causing a pertussis toxin-sensitive calcium flux and subsequent chemotaxis, ED_{50} ~20 nM for CXCR1 and 1 nM for CXCR2. Interestingly, although these receptors are expressed on lymphocytes and monocytes as well as neutrophils, GCP-2 does not cause chemotaxis of mononuclear cells.

References

1. Rovai, L.E. et al. (1997) J. Immunol. 158, 5257–5266.
2. Froyen, G. et al. (1997) Eur. J. Biochem. 243, 762–769.
3. Proost, P. et al. (1993) Biochemistry 32, 10170–10177.
4. Wuyts, A. et al. (1996) J. Immunol. 157, 1736–1743.
5. Van Damme, J. et al. (1997) J. Leukoc. Biol. 62, 563–569.
6. Proost, P. et al. (1998) J. Biol. Chem. 273, 7222–7227.
7. Wuyts, A. et al. (1997) Biochemistry 36, 2716–2723.
8. Wolf, M. et al. (1998) Eur. J. Immunol. 28, 164–170.
9. Wuyts, A. et al. (1998) Eur. J. Biochem. 255, 67–73.

Other names

CSFβ, pluripoietin.

THE MOLECULE

Granulocyte colony-stimulating factor is a growth, differentiation and activating factor for neutrophils and their precursors. It also synergizes with IL-3 to stimulate growth of haematopoietic progenitors, and causes proliferation and migration of endothelial cells[1–3].

Crossreactivity

There is about 73% homology between human and mouse G-CSF, which both act across species[5–8].

Sources

Macrophages, fibroblasts, endothelial cells, bone marrow stroma[1].

Bioassays

Bone marrow colony formation in soft agar, proliferation of NFS60 cell line[9].

Physicochemical properties of G-CSF

Property	Human	Mouse
Amino acids – precursor	207	208
– mature	177 (174)[a]	178
M_r (K) – predicted	19	19
– expressed	21	21
Potential *N*-linked glycosylation sites[b]	0	0
Disulfide bonds[c]	2	2

[a] The 177 and 174 amino acid forms are alternatively spliced variants, the 174 amino acid form is predominant. The 177 amino acid form is 20-fold less biologically active.
[b] Human G-CSF has one *O*-linked site.
[c] Disulfide bonds between Cys39–45 and 67–77 (hG-CSF).

3-D structure

G-CSF forms a four α-helical bundle structure similar to growth hormone, GM-CSF, IFNβ, IL-4 and IL-2[10,11].

Gene structure[8,12]

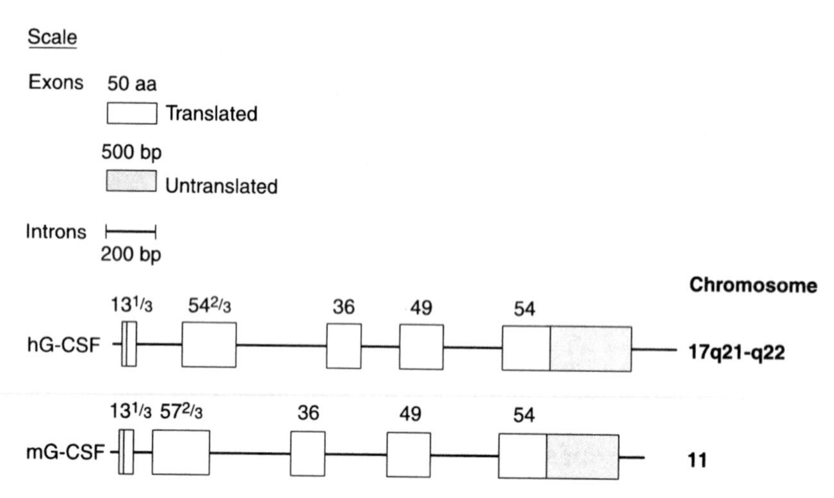

The gene for human G-CSF is located on chromosome 17q21–22 and consists of five exons of $13^{1/3}$, $54^{2/3}$, 36, 49 and 54 amino acids. This position corresponds to a translocation found in acute promyelocytic leukaemias[12]. The gene for mouse G-SCF is located on chromosome 11, containing five exons of $13^{1/3}$, $57^{2/3}$, 36, 49 and 54 amino acids.

Amino acid sequence for human G-CSF[4–6]

Accession code: SwissProt P09919

```
-30  MAGPATQSPM KLMALQLLLW HSALWTVQEA
  1  TPLGPASSLP QSFLLKCLEQ VRKIQGDGAA LQEKLVSECA TYKLCHPEEL
 51  VLLGHSLGIP WAPLSSCPSQ ALQLAGCLSQ LHSGLFLYQG LLQALEGISP
101  ELGPTLDTLQ LDVADFATTI WQQMEELGMA PALQPTQGAM PAFASAFQRR
151  AGGVLVASHL QSFLEVSYRV LRHLAQP
```

Note that the sequence (VSE) in italics (amino acids 36–38) is spliced out in the 174-amino-acid variant.

Amino acid sequence for mouse G-CSF[7,8]

Accession code: SwissProt P09920

```
-30  MAQLSAQRRM KLMALQLLLW QSALWSGREA
  1  VPLVTVSALP PSLPLPRSFL LKSLEQVRKI QASGSVLLEQ LCATYKLCHP
 51  EELVLLGHSL GIPKASLSGC SSQALQQTQC LSQLHSGLCL YQGLLQALSG
101  ISPALAPTLD LLQLDVANFA TTIWQQMENL GVAPTVQPTQ SAMPAFTSAF
151  QRRAGGVLAI SYLQGFLETA RLALHHLA
```

THE G-CSF RECEPTOR

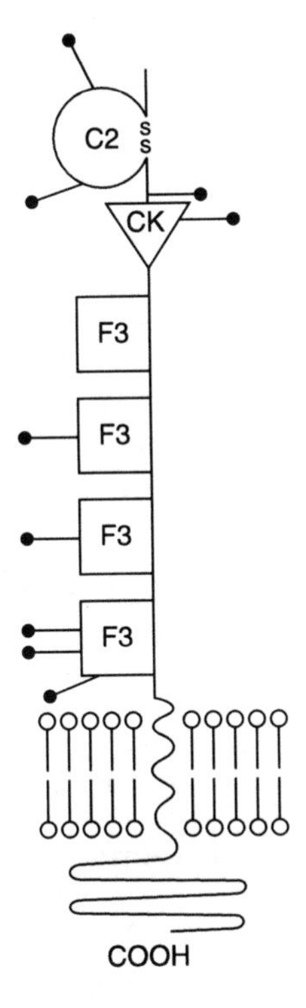

The G-CSF receptor is a hybrid structure containing an immunoglobulin domain, a haematopoietin domain and three FNIII domains[13,14]. The human receptor exists as two forms varying in their intracytoplasmic domains[13]. One form, 25.1, carries a C kinase phosphorylation site. A potential soluble form of the human receptor has been reported which has deleted the transmembrane region[15]. The only murine receptor reported resembles the 25.1 form[16]. Human and mouse receptors share 62.5% homology. The human receptor shares 46.3% sequence homology to the gp130 chain of the IL-6 receptor. Binding specificity for G-CSF resides in the haematopoietin domain, whereas the membrane proximal 57 amino acids are necessary for proliferative signal transduction and residues 57–96 are involved in G-CSF-mediated acute phase protein induction[14,17]. The receptor appears to confer high-affinity binding in transfected cells, possibly as a dimeric structure. A G-CSF receptor is present on platelets[19].

Distribution

Neutrophils, myeloid leukaemias, endothelium, platelets, placenta.

Physicochemical properties of the G-CSF receptor

Properties	Human	Mouse
Amino acids		
precursor	836 (783)[a]	837
mature	812 (759)[a]	812
M_r (K)		
predicted	90	90
expressed	150	95–125
Potential *N*-linked glycosylation sites	9	11
Affinity K_d (M)	$1–53 \times 10^{-10}$	$2–33 \times 10^{-10}$

[a] Truncated soluble form.

Signal transduction

G-CSF stimulation results in tyrosine phosphorylation of an M_r 56 000 protein in NFS60 cells, and tyrosine and serine phosphorylation of an M_r 75 000 protein. The receptor is thought to bind and mediate autophosphorylation of the Jak2 kinase[20].

Chromosomal location

The human receptor is on chromosome 1p35–p34.3.

Amino acid sequence for human G-CSF receptor[13]

Accession code: EMBL M59820, M38027, X55720, X55721

```
-24  MARLGNCSLT WAALIILLLP GSLE
  1  ECGHISVSAP IVHLGDPITA SCIIKQNCSH LDPEPQILWR LGAELQPGGR
 51  QQRLSDGTQE SIITLPHLNH TQAFLSCCLN WGNSLQILDQ VELRAGYPPA
101  IPHNLSCLMN LTTSSLICQW EPGPETHLPT SFTLKSFKSR GNCQTQGDSI
151  LDCVPKDGQS HCCIPRKHLL LYQNMGIWVQ AENALGTSMS PQLCLDPMDV
201  VKLEPPMLRT MDPSPEAAPP QAGCLQLCWE PWQPGLHINQ KCELRHKPQR
251  GEASWALVGP LPLEALQYEL CGLLPATAYT LQIRCIRWPL PGHWSDWSPS
301  LELRTTERAP TVRLDTWWRQ RQLDPRTVQL FWKPVPLEED SGRIQGYVVS
351  WRPSGQAGAI LPLCNTTELS CTFHLPSEAQ EVALVAYNSA GTSRPTPVVF
401  SESRGPALTR LHAMARDPHS LWVGWEPPNP WPQGYVIEWG LGPPSASNSN
451  KTWRMEQNGR ATGFLLKENI RPFQLYEIIV TPLYQDTMGP SQHVYAYSQE
501  MAPSHAPELH LKHIGKTWAQ LEWVPEPPEL GKSPLTHYTI FWTNAQNQSF
551  SAILNASSRG FVLHGLEPAS LYHIHLMAAS QAGATNSTVL TLMTLTPEGS
601  ELHIILGLFG LLLLLTCLCG TAWLCCSPNR KNPLWPSVPD PAHSSLGSWV
651  PTIMEEDAFQ LPGLGTPPIT KLTVLEEDEK KPVPWESHNS SETCGLPTLV
701  QTYVLQGDPR AVSTQPQSQS GTSDQVLYGQ LLGSPTSPGP GHYLRCDSTQ
751  PLLAGLTPSP KSYENLWFQA SPLGTLVTPA PSQEDDCVFG PLLNFPLLQG
801  IRVHGMEALG SF
```

Alternative C-terminus clone D7:

```
721  AGPPRRSAYF KDQIMLHPAP PNGLLCLFPI TSVL
```

Amino acid sequence for mouse G-CSF receptor[16]

Accession code: EMBL M32699

```
-25  MVGLGACTLT GVTLIFLLLP RSLES
  1  CGHIEISPPV VRLGDPVLAS CTISPNCSKL DQQAKILWRL QDEPIQPGDR
 51  QHHLPDGTQE SLITLPHLNY TQAFLFCLVP WEDSVQLLDQ AELHAGYPPA
101  SPSNLSCLMH LTTNSLVCQW EPGPETHLPT SFILKSFRSR ADCQYQGDTI
151  PDCVAKKRQN NCSIPRKNLL LYQYMAIWVQ AENMLGSSES PKLCLDPMDV
201  VKLEPPMLQA LDIGPDVVSH QPGCLWLSWK PWKPSEYMEQ ECELRYQPQL
251  KGANWTLVFH LPSSKDQFEL CGLHQAPVYT LQMRCIRSSL PGFWSPWSPG
301  LQLRPTMKAP TIRLDTWCQK KQLDPGTVSV QLFWKPTPLQ EDSGQIQGYL
351  LSWNSPDHQG QDIHLCNTTQ LSCIFLLPSE AQNVTLVAYN KAGTSSPTTV
401  VFLENEGPAV TGLHAMAQDL NTIWVDWEAP SLLPQGYLIE WEMSSPSYNN
451  SYKSWMIEPN GNITGILLKD NINPFQLYRI TVAPLYPGIV GPPVNVYTFA
501  GERAPPHAPA LHLKHVGTTW AQLEWVPEAP RLGMIPLTHY TIFWADAGDH
551  SFSVTLNISL HDFVLKHLEP ASLYHVYLMA TSRAGSTNST GLTLRTLDPS
601  DLNIFLGILC LVLLSTTCVV TWLCCKRRGK TSFWSDVPDP AHSSLSSWLP
651  TIMTEETFQL PSFWDSSVPS ITKITELEED KKPTHWDSES SGNGSLPALV
701  QAYVLQGDPR EISNQSQPPS RTGDQVLYGQ VLESPTSPGV MQYIRSDSTQ
751  PLLGGPTPSP KSYENIWFHS RPQETFVPQP PNQEDDCVFG PPFDFPLFQG
801  LQVHGVEEQG GF
```

References

[1] Nagata, S. (1990) In Peptide Growth Factors and their Receptors, Sporn, M.B. and Roberts, A.B. eds, Springer-Verlag, Heidelberg.
[2] Gabrilove, J.L. et al. (1988) N. Engl. J. Med. 318, 1414–1422.
[3] Bussolino, F. et al. (1989) Nature 337, 471–473.
[4] Souza, L.M. et al. (1986) Science 232, 61–65.
[5] Nagata, S. et al. (1986) Nature 319, 415–418.
[6] Nagata, S. et al. (1986) EMBO J. 5, 575–581.
[7] Tsuchiya, M. et al. (1986) Proc. Natl Acad. Sci USA 83, 7633–7637.
[8] Tsuchiya , M. et al. (1987) Eur. J. Biochem. 165, 7–12.
[9] Testa, N.G. et al. (1991) In Cytokines: A Practical Approach, Balkwill, F.A. ed., IRL Press Oxford, pp. 229–244.
[10] Hill, C.P. et al. (1993) Proc. Natl Acad. Sci. USA 90, 5167–5171.
[11] Diederichs, K. et al. (1991) Science 254, 1779–1782.
[12] Simmers, R.N. et al. (1987) Blood 70, 330–332.
[13] Larsen, A. et al. (1990) J. Exp. Med. 172, 1559–1570.
[14] Fukunaga, R. et al. (1991) EMBO J. 10, 2855–2865.
[15] Fukunaga, R. et al. (1990) Proc. Natl Acad. Sci. USA 87, 8702–8706.
[16] Fukunaga, R. et al. (1990) Cell 61, 341–350.
[17] Ziegler, S.F. et al. (1993) Mol. Cell. Biol. 13, 2384–2390.
[18] Tweardy, D.J. et al. (1992) Blood 79, 1148–1154.
[19] Shimoda, K. et al. (1993) J. Clin. Invest. 91, 1310–1313.
[20] Witthuhn, B.A. et al. (1993) Cell 74, 227–236.

Other names
None.

THE MOLECULE

Glial cell line-derived growth factor (GDNF) is a glycosylated disulfide-bonded homodimer which is distantly related to the TGFβ superfamily[1]. GDNF is a neurotrophic factor that promotes the survival of dopaminergic neurons, and has been implicated in their morphological and functional differentiation[1–3]. It has also been reported to have a role in the survival and phenotype expression of central noradrenergic and motor neurons and various peripheral sensory and sympathetic neurons[4]. Parkinson s disease is characterized by the progressive degeneration of the nigrostriatal dopaminergic system. Administration of GDNF in mouse models of Parkinson s disease resulted in significant recovery of this system with improvements in three key features of the disease: bradykinesia, rigidity and postural instability. Most recently a role for GDNF has been identified in the biochemical and behavioural adaptation to drugs of abuse[5]. GDNF is the founding member of the GDNF family which also includes neurturin (NTN), persephin (PSP), artemin[6] and enovin[7].

Crossreactivity
Not known.

Sources
GDNF PCR products have been observed in human striatum, hippocampus and spinal cord[2] and in newborn rat kidney, lung, bone, heart, liver, spleen, sciatic nerve and blood[3].

Bioassays
Promotes survival of dopaminergic neurons in midbrain cultures[1].

Physicochemical properties of GDNF

Property	Human	Mouse
pI[a]	9.44	9.44
Amino acids – precursor	211	211
– mature	134	134
M_r (K) – predicted	15.07	14.97
– expressed	?	?
Potential *N*-linked glycosylation sites	2	2
Disulfide bonds	4	4

[a] Predicted.

3-D structure
Not known.

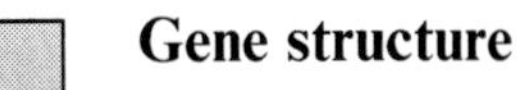

Gene structure

The gene coding for GDNF maps to chromosome 5p12–p13.1[8].

Amino acid sequence for human GDNF[1]

Accession code: SwissProt P39905

-19 MKLWDVVAVC LVLLHTASA
1 F*PLPAGKRPP EAPAEDRSLG RRRAPFALSS DSNMPEDYPD QFDDVMDFIQ*
51 *ATIKRLKR*SP DKQMAVLPRR ERNRQAAAAN PENSRGKGRR GQRGKNRGCV
101 LTAIHL**N**VTD LGLGYETKEE LIFRYCSGSC DAAETTYDKI LK**N**LSRNRRL
151 VSDKVGQACC RPIAFDDDLS FLDDNLVYHI LRKHSAKRCG CI

The propeptide sequence (20–77) shown in italics is removed to generate the mature protein. Disulfide bridges are formed between Cys118–179, 145–208, 149–210 by similarity and an interchain disulfide ridge is predicted using Cys178. A number of polymorphisms have also recently been identified in the GDNF gene sequence, with R→W at position 93, D→N at position 150 and T→S at position 154.

Amino acid sequence for mouse GDNF

Accession code: SwissProt P48540

-19 MKLWDVVAVC LVLLHTASA
1 *FPLPAGKRLL EAPAEDHSLG HRRVPFALTS DSNMPEDYPD QFDDVMDFIQ*
51 *ATIKRLKR*SP DKQAAALPRR ERNRQAAAAS PENSRGKGRR GQRGKNRGCV
101 LTAIHL**N**VTD LGLGYETKEE LIFRYCSGSC ESAETMYDKI LK**N**LSRSRRL
151 TSDKVGQACC RPVAFDDDLS FLDDNLVYHI LRKHSAKRCG CI

The propeptide sequence (20–77) shown in italics is removed to generate the mature protein. Disulfide bridges are formed between Cys118–179, 145–208, 149–210 by similarity and an interchain disulfide ridge is predicted using Cys178. Two isoforms of mouse GDNF are produced as a result of alternative splicing, residues 25–50 are missing in isoform 2 and amino acid residue 51 has an S→A substitution.

THE GDNF RECEPTOR

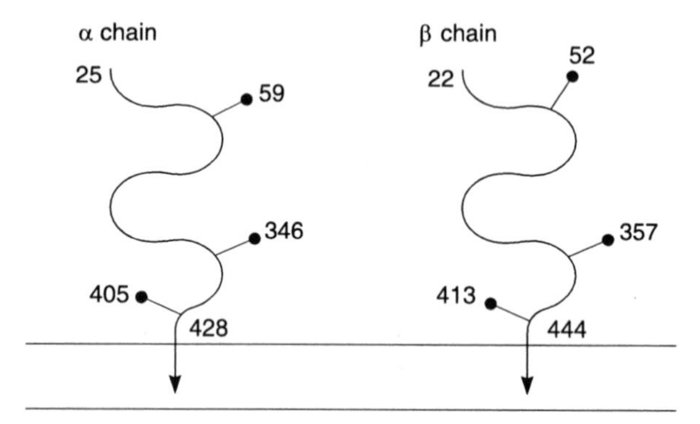

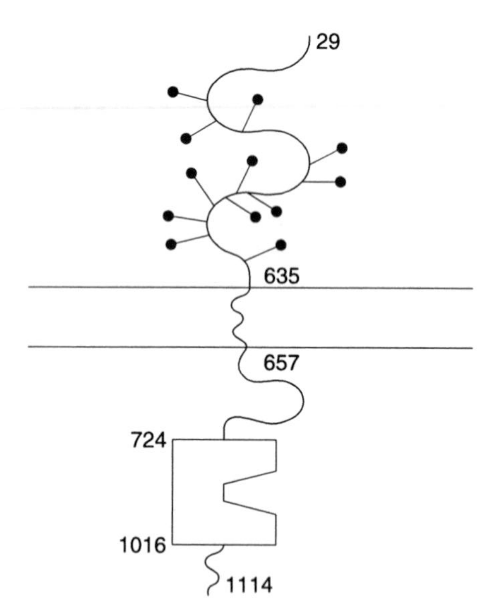

GDNF binds to and causes tyrosine autophosphorylation of the c-*ret* proto-oncogene product, Ret[4,9], a process mediated by GDNF receptors α[10] and β[11] (also known as TRNR1 and TRNR2 respectively[11]). GDNF receptor α is a glycosylphosphatidylinositol-linked cell surface protein sharing 48% homology with the β receptor[10,11]. Ret is found in a long (1114-bp) and a short (1072-bp) isoform; the long isoform has consensus sequences for binding the Shc PTB domain, and the short isoform has consensus sequences for the PTB and SH2 domains[12].

Distribution

GDNF receptor β and Ret are expressed in the adult brain and superior cervical and dorsal root ganglia neurons.

Physicochemical properties of GDNF receptors

Property	Alpha		Beta		Ret	
	Human	Mouse	Human	Mouse	Human	Mouse
Amino acids – precursor	464	468	464	463	1114	1115
– mature[a]	404 426	423	422	1086	1087	
M_r (K) – predicted	55.07	55.13	56.02	56.06	124.3	123.7
– expressed	?	?	?	?	?	?
Potential *N*-linked glycosylation sites	3	3	3	3	12	12
Affinity K_d (M)	?	?	?	?	?	?

[a] After removal of predicted signal peptide.

Signal transduction

GDNF induces Ret dimerization, resulting in tyrosine autophosphorylation; subsequently the Ret dimer recruits Shc and Grb2 proteins[12]. GDNF also activates the Ras-MAP kinase pathway in human neuroblastoma cells.

Chromosomal location

GDNF receptor β has been localized to the short arm of chromosome 8[11].

Amino acid sequence for human GDNF receptor α

Accession code: SwissProt P56159

-24 MFLATLYFAL PLLDLLLSAE VSGG
1 DRLDCVKASD QCLKEQSCST KYRTLRQCVA GKET**N**FSLAS GLEAKDECRS
51 AMEALKQKSL YNCRCKRGMK KEKNCLRIYW SMYQSLQGND LLEDSPYEPV
101 NSRLSDIFRV VPFISDVFQQ VEHIPKGNNC LDAAKACNLD DICKKYRSAY
151 ITPCTTSVSN DVCNRRKCHK ALRQFFDKVP AKHSYGMLFC SCRDIACTER
201 RRQTIVPVCS YEEREKPNCL LQDSCKTNYI CRSRLADFFT NCQPESRSVS
251 SCLKENYADC LLAYSGLIGT VMTPNYIDSS SLSVAPWCDC SNSGNDLEEC
301 LKFLNFFKDN TCLKNAIQAF G**N**GSDVTVWQ PAPPVQTTTA TTTTALRVKN
351 KPLGPAGSEN EIPTHVLPPC ANLQAQKLKS **N**VSGNTHLCI SNGNYEKEGL
401 GASSHITTKS MAAPPSCGLS PLLVLVVTAL STLLSLTETS

The propeptide sequence (429–464 of precursor protein) shown in italics is removed to generate the mature protein. The potential GPI anchor site is predicted at position 428.

Amino acid sequence for human GDNF receptor β

Accession code: SwissProt O00451

-21 MILANAFCLF FFLDETLRSL A
1 SPSSLQGPEL HGWRPPVDCV RANELCAAES **N**CSSRYRTLR QCLAGRDRNT
51 MLANKECQAA LEVLQESPLY DCRCKRGMKK ELQCLQIYWS IHLGLTEGEE
101 FYEASPYEPV TSRLSDIFRL ASIFSGTGAD PVVSAKSNHC LDAAKACNLN
151 DNCKKLRSSY ISICNREISP TERCNRRKCH KALRQFFDRV PSEYTYRMLF
201 CSCQDQACAE RRRQTILPSC SYEDKEKPNC LDLRGVCRTD HLCRSRLADF
251 HANCRASYQT VTSCPADNYQ ACLGSYAGMI GFDMTPNYVD SSPTGIVVSP
301 WCSCRGSGNM EEECEKFLRD FTENPCLRNA IQAFG**N**GTDV NVSPKGPSFQ
351 ATQAPRVEKT PSLPDDLSDS TSLGTSVITT CTSVQEQGLK ANNSKELSMC
401 FTELTTNIIP GS**N**KVIKPNS GPS*RARPSAA LTVLSVLMLK QAL*

The propeptide sequence (445–464 of precursor protein) shown in italics is removed to generate the mature protein. The potential GPI anchor site is predicted at position 444 of unprocessed sequence. A variant splice product lacking residues 14–146 of the unprocessed sequence represents the short isoform.

Amino acid sequence for mouse GDNF receptor α

Accession code: SwissProt P97785

-24 MFLATLYFVL PLLDLLMSAE VSGG
1 DRLDCVKASD QCLKEQSCST KYRTLRQCVA GKET**N**FSLTS GLEAKDECRS
51 AMEALKQKSL YNCRCKRGMK KEKNCLRIYW SMYQSLQGND LLEDSPYEPV
101 NSRLSDIFRA VPFISDVFQQ VEHISKGNNC LDAAKACNLD DTCKKYRSAY
151 ITPCTTSMSN EVCNRRKCHK ALRQFFDKVP AKHSYGMLFC SCRDVACTER
201 RRQTIVPVCS YEERERPNCL NLQDSCKTNY ICRSRLADFF TNCQPESRSV
251 SNCLKENYAD CLLAYSGLIG TVMTPNYIDS SSLSVAPWCD CSNSGNDLED
301 CLKFLNFFKD NTCLKNAIQA FG**N**GSDVTMW QPAPPVQTTT AMTTTAFRIK
351 NKPLGPAGSE NEIPTHVLPP CANLQAQKLK S**N**VSGSTHLC LSDNDYGKDG
401 LAGASSHITT KSMAAPPSCG LSSLPVMVFT ALAALLSVSL AETS

The propeptide sequence (431–468 of precursor protein) shown in italics is removed to generate the mature protein. The potential GPI anchor site is predicted at position 430 of unprocessed sequence.

Amino acid sequence for mouse GDNF receptor β

Accession code: SwissProt O09942

-21 MILANAFCLF FFLDETLRSL A
1 SPSSPQGSEL HGWRPQVDCV RANELCAAES **N**CSSRYRTLR QCLAGRDRNT
51 MLANKECQAA LEVLQESPLY DCRCKRGMKK ELQCLQIYWS IHLGLTEGEE
101 FYEASPYEPV TSRLSDIFRL ASIFSGTGAD PVVSAKSNHC LDAAKACNLN
151 DNCKKLRSSY ISICNREISP TERCNRRKCH KALRQFFDRV PSEYTYRMLF
201 CSCQDQACAE RRRQTILPSC SYEDKEKPNC LDLRSLCRTD HLCRSRLADF
251 HANCRASYRT ITSCPADNYQ ACLGSYAGMI GFDMTPNYVD SNPTGIVVSP
301 WCNCRGSGNM EEECEKFLKD FTENPCLRNA IQAFG**N**GTDV NMSPKGPTFS
351 ATQAPRVEKT PSLPDDLSDS TSLGTSVITT CTSIQEQGLK A**N**NSKELSMC
401 FTELTTNISP GS*KKVIKLYS* GSCRARLSTA LTALPLLMVT LA

The propeptide sequence (444–463 of precursor protein) shown in italics is removed to generate the mature protein. The potential GPI anchor site is predicted at position 443 of unprocessed sequence. A variant splice product lacking residues 14–146 of the unprocessed sequence represents the short isoform.

Amino acid sequence for human Ret

Accession code: SwissProt P07949

-28 MAKATSGAAG LRLLLLLLLP LLGKVALG
1 LYFSRDAYWE KLYVDQAAGT PLLYVHALRD APEEVPSFRL GQHLYGTYRT
51 RLHENNWICI QEDTGLLYL**N** RSLDHSSWEK LSVRNRGFPL LTVYLKVFLS
101 PTSLREGECQ WPGCARVYFS FF**N**TSFPACS SLKPRELCFP ETRPSFRIRE
151 NRPPGTFHQF RLLPVQFLCP **N**ISVAYRLLE GEGLPFRCAP DSLEVSTRWA
201 LDREQREKYE LVAVCTVHAG AREEVVMVPF PVTVYDEDDS APTFPAGVDT
251 ASAVVEFKRK EDTVVATLRV FDADVVPASG ELVRRYTSTL LPGDTWAQQT
301 FRVEHWP**N**ET SVQA**N**GSFVR ATVHDYRLVL NR**N**LSISE**N**R TMQLAVLV**N**D
351 SDFQGPGAGV LLLHF**N**VSVL PVSLHLPSTY SLSVSRRARR FAQIGKVCVE
401 NCQAFSGINV QYKLHSSGA**N** CSTLGVVTSA EDTSGILFV**N** DTKALRRPKC
451 AELHYMVVAT DQQTSRQAQA QLLVTVEGSY VAEEAGCPLS CAVSKRRLEC
501 EECGGLGSPT GRCEWRQGDG KGITR**N**FSTC SPSTKTCPDG HCDVVETQDI
551 NICPQDCLRG SIVGGHEPGE PRGIKAGYGT CNCFPEEEKC FCEPEDIQDP
601 LCDELCRTVI AAAVLFSFIV SVLLSAFCIH CYHKFAHKPP ISSAEMTFRR
651 PAQAFPVSYS SSGARRPSLD SMENQVSVDA FKILEDPKWE FPRKNLVLGK
701 TLGEGEFGKV VKATAFHLKG RAGYTTVAVK MLKENASPSE LRDLLSEFNV
751 LKQVNHPHVI KLYGACSQDG PLLLIVEYAK YGSLRGFLRE SRKVGPGYLG
801 SGGSRNSSSL DHPDERALTM GDLISFAWQI SQGMQYLAEM KLVHRDLAAR
851 NILVAEGRKM KISDFGLSRD VYEEDSYVKR SQGRIPVKWM AIESLFDHIY
901 TTQSDVWSFG VLLWEIVTLG GNPYPGIPPE RLFNLLKTGH RMERPDNCSE
951 EMYRLMLQCW KQEPDKRPVF ADISKDLEKM MVKRRDYLDL AASTPSDSLI
1001 YDDGLSEEET PLVDCNNAPL PRALPSTWIE NKLYGMSDPN WPGESPVPLT
1051 RADGTNTGFP RYPNDSVYAN WMLSPSAAKL MDTFDS

Conflicting sequences I→V at position 647 and S→P at position 904 of unprocessed sequence. A range of mutations occurr throughout the sequence (see SwissProt Database entry).

Amino acid sequence for mouse Ret

Accession code: SwissProt P35546

-28 MAKATSGAAG LGLKLILLLP LLGEAPLG
1 LYFSRDAYWE RLYVDQPAGT PLLYVHALRD APGEVPSFRL GQHLYGVYRT
51 RLHENDWIRI **N**ETTGLLYL**N** QSLDHSSWEQ LSIRNGGFPL LTIFLQVFLG
101 STAQREGECH WPGCTRVYFS FI**N**DTFP**N**CS SFKAQDLCIP ETAVSSRVRE
151 NXPPGTFYHF HMLPVQFLCP **N**ISVKYSLLG GDSLPFRCDP DCLEVSTRWA
201 LDRELREKYV LEALCIVAGP GANKETVTLS FPVTVYDEDD SAPTFSGGVG
251 TASAVVEFKR KEGTVVATLQ VFDADVVPAS GELVRRYTNT LLSGDSWAQQ
301 TFRVEHSPIE TLVQVN**N**NSV RATMHNYKLI L**N**RSLSISES RVLQLAVLV**N**
351 DSDFQGPGAG GILVLHF**N**VS VLPVTLNLPR AYSFPVNKRA RRYAQIGKVC
401 VENCQEFSGV SIQYKLQPSS I**N**CTALGVVT SPEDTSGTLF V**N**DTEALRRP
451 ECTKLQYTVV ATDRQTRRQT QASLVVTVEG TSITEEVGCP KSCAVNKRRP
501 ECEECGGLGS PTGRCEWRQG DGKGITR**N**FS TCSPSTRTCP DGHCDAVESR
551 DANICPQDCL RADIVGGHER GERQGIKAGY GICNCFPDEK KCFCEPEDSQ
601 GPLCDALCRT IITAALFSLI ISILLSIFCV CHHHKHGHKP PIASAEMTFC
651 RPAQGFPISY SSSGTRRPSL DSTENQVPVD SFKIPEDPKW EFPRKNLVLG
701 KTLGEGEFGK VVXATAFRLK GRAGYTTVAV KXLKENASQS ELRDLLSEFN
751 LLKQVNHPHV IKLYGACSQD GPLLLIVEYA KYGSLRGFLR DSRKIGPAYV
801 SGGGSRNSSS LDHPDERVLT MGDLISFAWQ ISRGMQYLAE MKLVHRDLAA
851 RNILVAEGRK MKISDFGLSR DVYEEDSYVK KSKGRIPVKW MAIESLFDHI
901 YTTQSDVWSF GVLLWEIVTL GGNPYPGIPP ERLFNLLKTG HRMERPDNCS
951 EEMYRLMLQC WKQEPDKRPV FADISKDLEK MMVKSRDYLD LAASTPSDSL
1001 LYDDGLSEEE TPLVDCNNAP LPRSLPSTWI ENKLYGMSDP NWPGESPVPL
1051 TRADGTSTGF PRYANDSVYA NWMVSPSAAK LMDTFDS

References

[1] Lin, L.F. et al. (1993) Science 260, 1130–1132.
[2] Springer, J.E. et al. (1994) Exp. Neurol. 127, 167–170.
[3] Suter-Crazzolara, C. and Unsicker, K. (1994) Neuroreport 5, 2486–2688.
[4] Trupp, M. et al. (1996) Nature 381, 785–789.
[5] Messer, C.J. et al. (2000) Neuron 26, 247–257.
[6] Baloh, R.H. et al. (1998) Neuron 21, 1291–1302.
[7] Masure, S. et al. (1999) Eur. J. Biochem. 266, 892–902.
[8] Schindelhauer, D. et al. (1995) Genomics 28, 605–607.
[9] Durbec, P. et al. (1996) Nature 381, 789–793.
[10] Jing, S. et al. (1996) Cell 85, 1113–1124.
[11] Baloh, R.H. et al. (1997) Neuron 18, 793–802.
[12] Ohiwa, M. et al. (1997) Biochem. Biophys. Res. Commun. 237, 747–751.

Other names
AITRL (activation-induced TNFR member ligand), TL6 and TNFSF18.

THE MOLECULE

Glucocorticoid-induced TNF family related gene ligand (GITRL) is a recently characterized member of the TNF superfamily subset which also contains 4-1BBL and CD27L. These have the ability to initiate and propel the ongoing immune response by providing strong costimulatory functions. In addition, interactions of GITRL with its receptor may also participate in the regulation of T-cell receptor-mediated cell death as transfection of GITR induces resistance to anti-CD3 monoclonal antibody-induced apoptosis[1]. Furthermore, cotransfection of AITRL and AITR in Jurkat T leukaemia cells inhibited antigen receptor-induced cell death, further supporting a role for this ligand receptor pair in the regulation of T lymphocyte survival in peripheral tissues[2]. The human orthologue is called AITRL (activation-inducible TNFR family member). AITRL has been found on endothelial cells, suggesting that AITRL–AITR interactions may function in activated T-cell trafficking[3].

Crossreactivity
Unknown.

Sources
AITRL has been shown to be expressed on endothelial cells[3].

Bioassays
Resistance to TCR/CD3-induced apoptosis of GITR-transfected T cells.

Physicochemical properties of GITRL

Property	Human	Mouse
pI	7.64[a]	?
Amino acids – precursor	177	?
– mature[b]	177	?
M_r (K) – predicted	20.3	?
– expressed	?	
Potential *N*-linked glycosylation sites	2	?
Disulfide bonds	?	?

[a] This represents the predicted pI of GITRL.
[b] Human GITRL lacks a signal peptide.

3-D structure
The crystal structure for GITRL is unknown.

Gene structure

The human GITRL gene mapped to chromosome 1q23, near the gene for the TNF homologue, Fas/CD95 ligand.

Amino acid sequence for human GITRL

Accession code: SPTREMBL O95852

```
  1 MPLSHSRTQG AQRSSWKLWL FCSIVMLLFL CSFSWLIFIF LQLETAKEPC
 51 MAKFGPLPSK WQMASSEPPC VNKVSDWKLE ILQNGLYLIY GQVAPNANYN
101 DVAPFEVRLY KNKDMIQTLT NKSKIQNVGG TYELHVGDTI DLIFNSEHQV
151 LKNNTYWGII LLANPQFIS
```

THE GITRL RECEPTOR

The GITR (glucocorticoid-induced TNFR family-related gene) was cloned originally from dexamethasone-treated murine T-cell hybridoma (3D0) cells by differential display analysis[1]. GITR is a type I transmembrane protein that belongs to the tumour necrosis factor/nerve growth factor receptor (TNF/NGFR) family, which is preferentially expressed on activated T lymphocytes and may function as a signalling molecule during T-cell development. GITR is a 228-amino-acid type I transmembrane protein characterized by three cysteine pseudorepeats in the extracellular domain and similar to CD27 and 4-1BB in the intracellular domain[1]. Furthermore, all three genes are closely linked on mouse chromosome 4, suggesting close functional roles. Expression of GITR appears to be activation dependent, as stimulation of T lymphocytes by anti-CD3 mAb, ConA or PMA/ionomycin readily upregulates its levels[1]. It is now known that GITR expression confers resistance to TCR/CD3-induced apoptosis of transfected T cells. However, these effects are not observed in response to other apoptosis-inducing stimuli such as Fas triggering, dexamethasone treatment and UV irradiation[1]. Recently, a human homologue of GITR was discovered. The human receptor is called AITR (activation-induced TNFR family member) or hGITR[2,3]. AITR has 55% amino acid identity with mouse GITR at the amino acid level[2]. In contrast to mouse GITR, AITR is not induced by dexamethasone treatment[3].

Distribution

AITR is expressed in lymph node and peripheral blood leukocytes and its expression is upregulated in human peripheral mononuclear cells mainly after stimulation with anti-CD3/CD28 monoclonal antibodies or PMA/ionomycin[3].

Physiochemical properties of the GITRL

Property	Human	Mouse
Amino acids – precursor	234	228
– mature[a]	209	209
M_r (K) – predicted	25	25.33
– expressed	?	?
Potential *N*-linked glycosylation sites		?
Affinity K_d (M)	?	?

[a] After removal of predicted signal peptide.

Signal transduction

Both mouse and human GITR (AITR) associate with Traf-1, -2 and -3 and induce NFκB activation via Traf-2.

Chromosomal location

The gene for mouse GITR maps to chromosome 4 (E region), where other TNF/NGFR members localize, including mouse 4-1BB and mouse OX40. The gene spans a 2543-bp region and consists of five exons (with a length ranging from 88 bp to 395 bp) and four introns (67–778 bp). The hGITR gene maps to chromosome 1p36.

Amino acid sequence for human GITR

Accession code: SPTREMBL O95851

-26 *MAQHGAMGAF RALCGLALLC ALSLGQ*
1 RPTGGPGCGP GRLLLGTGTD ARCCRVHTTR CCRDYPGEEC CSEWDCMCVQ
51 PEFHCGDPCC TTCRHHPCPP GQGVQSQGKF SFGFQCIDCA SGTFSGGHEG
101 HCKPWTDCTQ FGFLTVFPG**N** KTHNAVCVPG SPPAEPLGWL TVVLLAVAAC
151 VLLLTSAQLG LHIWQLRKTQ LLLEVPPSTE DARSCQFPEE ERGERSAEEK
201 GRLGDLWV

Amino acid sequence for mouse GITR

Accession code: SPTREMBL O35714

-19 *MGAWAMLYGV SMLCVLDLG*
1 QPSVVEEPGC GPGKVQNGSG NNTRCCSLYA PGKEDCPKER CICVTPEYHC
51 GDPQCKICKH YPCQPGQRVE SQGDIVFGFR CVACAMGTFS AGRDGHCRLW
101 TNCSQFGFLT MFPGNKTHNA VCIPEPLPTE QYGHLTVIFL VMAACIFFLT
151 TVQLGLHIWQ LRRQHMCPRE TQPFAEVQLS AEDACSFQFP EEERGEQTEE
201 KCHLGGRWP

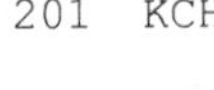

References

[1] Nocentini, G. et al. (1997) Proc. Natl Acad. Sci. USA 94, 6216–6221.
[2] Gurney, A.L. et al. (1999) Curr. Biol. 9, 215–218.
[3] Kwon, B. et al. (1999) J. Biol. Chem. 274, 6056–6061.

Other names
CSFα, pluripoietin-α.

THE MOLECULE

Granulocyte/macrophage colony-stimulating factor (GM-CSF) is a survival and growth factor for haematopoietic progenitor cells, a differentiation and activating factor for granulocytic and monocytic cells, and a growth factor for endothelial cells, erythroid cells, megakaryocytes and T cells[1–4].

Crossreactivity
There is about 56% homology between human and mouse GM-CSF and no cross-species activity.

Sources
T cells, macrophages, fibroblasts and endothelial cells.

Bioassays
There are no specific bioassays for GM-CSF. Bone marrow colony formation in agar or the AML-193 cell line can provide sensitive estimates of human GM-CSF activity. Mouse GM-CSF can be measured in bone marrow colony assays or using the DA-1 or FDC-P1 cell lines[5,6].

Physicochemical properties of GM-CSF

Property	Human	Mouse
pI	3.4–4.5	?
Amino acids – precursor	144	141
– mature	127	124
M_r (K) – predicted	16.3	16
– expressed	22	22
Potential *N*-linked glycosylation sites	2	2
Disulfide bonds[a]	2	2

[a] Disulfide bonds between Cys54–96 and 88–121 in human GM-CSF and 51–93 and 85–118 in mouse GM-CSF[7].

3-D structure
An X-ray crystal structure of human GM-CSF has been determined to 2.4 Å resolution. The molecule comprises a two-stranded antiparallel β-sheet with an open bundle of four α-helices, similar to IL-2, IL-4, G-CSF, M-CSF, IFNβ and growth hormone[8,9].

Gene structure[10,11]

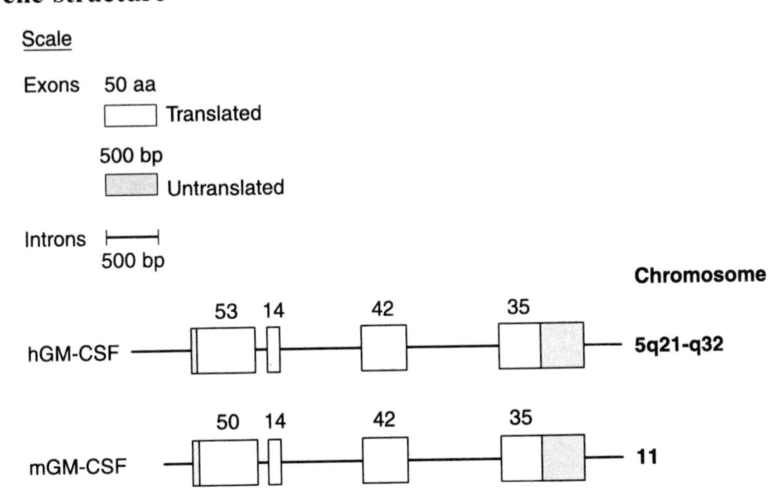

The genes for GM-CSF are located on human chromosome 5q21–q32 and contains four exons of 53, 14, 42 and 35 amino acids, while that of mouse GM-CSF is on chromosome 11 with a similar gene structure.

Amino acid sequence for human GM-CSF[12,13]

Accession code: SwissProt P04141

```
-17  MWLQSLLLLG TVACSIS
  1  APARSPSPST QPWEHVNAIQ EARRLLNLSR DTAAEMNETV EVISEMFDLQ
 51  EPTCLQTRLE LYKQGLRGSL TKLKGPLTMM ASHYKQHCPP TPETSCATQI
101  ITFESFKENL KDFLLVIPFD CWEPVQE
```

Conflicting sequence I→T at position 100.

Amino acid sequence for mouse GM-CSF[2,11,14]

Accession code: SwissProt P01587

```
-17  MWLQNLLFLG IVVYSLS
  1  APTRSPITVT RPWKHVEAIK EALNLLDDMP VTLNEEVEVV SNEFSFKKLT
 51  CVQTRLKIFE QGLRGNFTKL KGALNMTASY YQTYCPPTPE TDCETQVTTY
101  ADFIDSLKTF LTDIPFECKK PGQK
```

Conflicting sequence T→I at position 8, T→A at position 91, and Q→V/S at position 123.

THE GM-CSF RECEPTOR

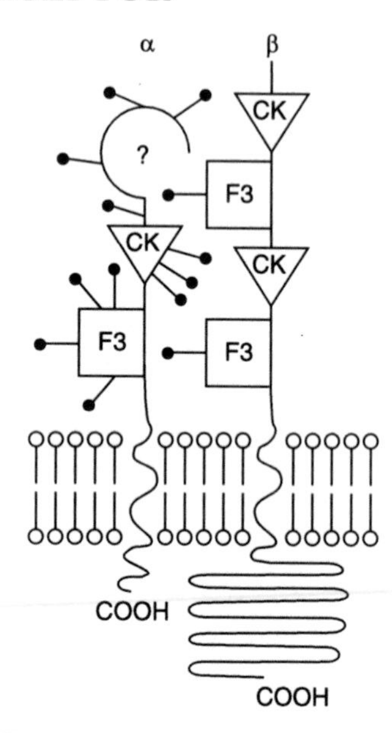

The high-affinity GM-CSF receptor is a complex of a low-affinity α-chain (CDw116) with a second affinity converting β-chain (human KH97 and murine AIC2B) which is also shared with the IL-3 and IL-5 receptor α-chains[15–18] (see IL-3 entry (page **51**)). Both the α- and β-chains are members of the haematopoietin family of receptors[19]. A potentially soluble form of the α-chain[20], and a splice variant with an altered intracytoplasmic domain of the α-chain have been described[21]. The intracytoplasmic domain of the β-chain is not required to form a high-affinity receptor[16].

Distribution

Granulocytes and monocytes and their precursors, endothelial cells, fibroblasts, Langerhans cells.

Physicochemical properties of the GM-CSF receptors

Property	Human α-chain	Human β-chain[a]
Amino acids – precursor	400	897
– mature	378	882
M_r (K) – predicted	46	95
– expressed	80	130
Potential *N*-linked glycosylation sites	11	3
Affinity K_d (M)[b]	3.23×10^{-9} 1.203×10^{-10}	0

[a] The β-chain is the same as the human IL-3 receptor KH97 and the murine IL-3 receptor AIC2B[17,18] (see IL-3 entry (page 51)).
[b] Low and high affinity sites.

Signal transduction

Ligand binding results in tyrosine phosphorylation of several intracellular proteins[19].The receptor is thought to bind and mediate phosphorylation of the Jak2 kinase[22].

Chromosomal location

The human α-chain is located on chromosome Xp22.3, Yp13.3[23] and the β-chain on chromosome 22.

Amino acid sequence for human GM-CSF receptor α-chain[15]

Accession code: SwissProt P15509

```
-22  MLLLVTSLLL CELPHPAFLL IP
  1  EKSDLRTVAP ASSLNVRFDS RTMNLSWDCQ ENTTFSKCFL TDKKNRVVEP
 51  RLSNNECSCT FREICLHEGV TFEVHVNTSQ RGFQQKLLYP NSGREGTAAQ
101  NFSCFIYNAD LMNCTWARGP TAPRDVQYFL YIRNSKRRRE IRCPYYIQDS
151  GTHVGCHLDN LSGLTSRNYF LVNGTSREIG IQFFDSLLDT KKIERFNPPS
201  NVTVRCNTTH CLVRWKQPRT YQKLSYLDFQ YQLDVHRKNT QPGTENLLIN
251  VSGDLENRYN FPSSEPRAKH SVKIRAADVR ILNWSSWSEA IEFGSDDGNL
301  GSVYIYVLLI VGTLVCGIVL GFLFKRFLRI QRLFPPVPQI KDKLNDNHEV
351  EDEIIWEEFT PEEGKGYREE VLTVKEIT
```

Amino acid sequence for human GM-CSF receptor β-chain

See IL-3 entry (page **51**).

Amino acid sequence for mouse GM-CSF receptor α-chain

Accession code: SPTREMBLE Q00941

```
-29  MTSSHAMNIT PLAQLALLFS TLLLPGTQA
  1  LLAPTTPDAG SALNLTFDPW TRTLTWACDT AAGNVTVTSC TVTSREAGIH
 51  RRVSPFGCRC WFRRMMALHH GVTLDVNGTV GGAAAHWRLS FVNESAAGSG
101  AENLTCEIRA ARFLSCAWRE GPAAPADVRY SLRVLNSTGH DVARCMADPG
151  DDVITQCIAN DLSLLGSEAY LVVTGRSGAG PVRFLDDVVA TKALERLGPP
201  RDVTASCNSS HCTVSWAPPS TWASLTARDF QFEVQWQSAE PGSTPRKVLV
251  VEETRLAFPS PAPHGGHKVK VRAGDTRMKH WGEWSPAHPL EAEDTRVPGA
301  LLYAVTACAV LLCALALGVT CRRFEVTRRL FPPIPGIRDK VSDDVRVNPE
351  TLRKDLLQP
```

References

[1] Wong, G.C. et al. (1985) Science 228, 810–815.
[2] Gough, N.M. et al. (1985) EMBO. J. 4, 645–653.
[3] Clarke, S. and Kamen, R. (1987) Science 236, 1229–1237.
[4] Groopman, J.E. et al. (1989) N. Engl. J. Med. 321, 1449–1459.
[5] Testa, N.G. et al. (1991) In Cytokines: A Practical Approach, Balkwill, F.A. ed., IRL Press, Oxford, pp. 229–234.
[6] Wadhwa, M. et al. (1991) In Cytokines: A Practical Approach, Balkwill, F.A. ed., IRL Press, Oxford, pp. 309–329.
[7] Schrimser, J.L. et al. (1987) Biochem. J. 247, 195–201.
[8] Reichert, P. et al. (1990) J. Biol. Chem. 265, 452–453.
[9] Diederichs, K. et al. (1991) Science 254, 1779–1782.
[10] Huebner, K. et al. (1985) Science 230, 1282–1285.
[11] Mitayake, S. et al. (1985) EMBO. J. 4, 2561–2568.
[12] Lee, F. et al. (1985) Proc. Natl Acad. Sci. USA 82, 4360–4364.
[13] Kaushansky, K. et al. (1986) Proc. Natl Acad. Sci. USA 83, 3101–3105.
[14] Delamater, J.F. et al. (1985) EMBO J. 4, 2575–2581.
[15] Gearing, D.P. et al. (1989) EMBO J. 8, 3667–3676.
[16] Hayashida, K. et al. (1990) Proc. Natl Acad. Sci. USA 87, 9655–9659.
[17] Kitamura, T. et al. (1991) Cell 66, 1165–1174.
[18] Gorman, D.M. et al. (1990) Proc. Natl Acad. Sci. USA 87, 5459–5463.
[19] Cosman, D. (1993) Cytokine 5, 95–106.
[20] Raines, M.S. et al. (1991) Proc. Natl Acad. Sci. USA 88, 8203–8207.
[21] Crosier, K.E. et al. (1991) Proc. Natl Acad. Sci. USA 88, 7744–7748.
[22] Witthuhn, B.A. et al. (1993) Cell 74, 227–236.
[23] Ashworth, A. and Kraft, A. (1990) Nucleic Acids Res. 18, 7178.

Other names

GROα is also known as MGSA, GROβ as MIP-2α and GROγ as MIP-2β. GRO/MGSA has also been known as NAP-3[1].

THE MOLECULES

The products of the GRO genes[2–4] are the human homologues of the mouse MIP-2 and KC genes[5,6]. They are members of the CXC family of chemokines. The molecules are chemoattractant and activating factors for neutrophils. MIP-2 is also a growth factor for myelopoietic cells and GRO/MGSA for fibroblasts and melanoma cells.

Crossreactivity

There is about 60% homology between human and mouse GRO proteins, 87% homology between the human proteins, and 63% homology between KC and MIP-2. Murine MIP-2 is active on human cells; GROα is active on rodent cells.

Sources

GROβ and γ are produced by cytokine- or LPS-activated monocytes; GROα/MGSA is made by activated monocytes, fibroblasts, epithelial and endothelial cells and KC by activated monocytes and fibroblasts.

Bioassays

There are no specific assays for these molecules. GRO and MIP-2 are chemotactic and activating factors for neutrophils. GRO is also a growth factor for fibroblasts and melanoma cells.

Physicochemical properties of GRO

	Human		Mouse		
Property	GROβ	GROγ	GROα/MGSA	MIP-2	KC
pI	9.7–9.9	?	?		
Amino acids – precursor	107	107	107	100	96
– mature	73	73	73	73	72
M_r (K) – predicted	11.4	11.4	11.3	10.6	10.2
– expressed	7.9	7.9	?	7.9	?
Potential *N*-linked glycosylation sites	0	0	0	0	0
Disulfide bonds[a]	2	2	2	2	2

[a] Disulfide bonds between Cys9–35 and 11–51 in each molecule.

3-D structure

Could be modelled on IL-8.

Gene structure

The genes for human GROs are located on chromosome 4q12–211.

Amino acid sequence for human GROβ[2]

Accession code: SwissProt P19875

-34 MARATLSAAP SNPRLLRVAL LLLLLVAASR RAAG

1 APLATELRCQ CLQTLQGIHL KNIQSVKVKS PGPHCAQTEV IATLKNGQKA

51 CLNPASPMVK KIIEKMLKNG KSN

Amino acid sequence for human GROγ[3]

Accession code: SwissProt P19876

-34 MAHATLSAAP SNPRLLRVAL LLLLLVAASR RAAG

1 ASVVTELRCQ CLQTLQGIHL KNIQSVNVRS PGPHCAQTEV IATLKNGKKA

51 CLNPASPMVQ KIIEKILNKG STN

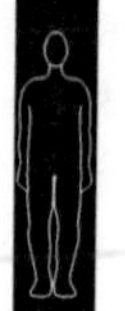

Amino acid sequence for human GROα/MGSA[4]

Accession code: SwissProt P09341

-34 MARAALSAAP SNPRLLRVAL LLLLLVAAGR RAAG

1 ASVATELRCQ CLQTLQGIHP KNIQSVNVKS PGPHCAQTEV IATLKNGRKA

51 CLNPASPIVK KIIEKMLNSD KSN

Amino acid sequence for mouse MIP-2[5]

Accession code: SwissProt P10889

-27 MAPPTCRLLS AALVLLLLLA TNHQATG

1 AVVASELRCQ CLKTLPRVDF KNIQSLSVTP PGPHCAQTEV IATLKGGQKV

51 CLDPEAPLVQ KIIQKILNKG KAN

Amino acid sequence for mouse KC[6]

Accession code: SwissProt P12850

-24 MIPATRSLLC AALLLLATSR LATG

1 APIANELRCQ CLQTMAGIHL KNIQSLKVIP SGPHCTQTEV IATLKNGREA

51 CLDPEAPLVQ KIVQKMLKGV PK

THE GRO FAMILY RECEPTOR, CXCR2

The GRO family have been shown to bind to the CXCR2 receptor with high affinity[7,9] (see IL-8 entry (page **80**) for full details).

References

[1] Oppenheim, J.J. et al. (1991) Annu. Rev. Immunol. 9, 617–648.
[2] Haskill, S. et al. (1990) Proc. Natl Acad. Sci. USA 87, 7732–7736.
[3] Sager, R. et al. (1991) Adv. Exp. Med. Biol. 305, 73–77.
[4] Aniscowicz, A. et al. (1987) Proc. Natl Acad. Sci. USA 84, 7188–7192.
[5] Tekamp-Olson, P. et al. (1990) J. Exp. Med. 172, 911–919.
[6] Oquendo, P. et al. (1989) J. Biol. Chem. 264, 4133–4137.
[7] LaRosa, G.J. et al. (1992) J. Biol. Chem. 267, 25402–25406.
[8] Walz, A. et al. (1991) J. Leucocyte Biol. 50, 279–286.
[9] Ahuja, S.K. et al. (1996) J. Biol. Chem. 271, 20545–20550

Other names
Somatotropin.

THE MOLECULE

Growth hormone (GH) is a secreted protein monomer belonging to a family of 22–25-kDa growth factors that includes prolactin, the placental lactogens, proliferins and somatolactin[1]. GH has many activities in addition to its somatogenic function, including metabolic effects especially on adipocytes. It is also a haematopoietic growth factor acting either directly on haematopoietic cells or indirectly by stimulating production of IGF-1[2]. GH receptors are expressed on lymphocytes and the cytokine has important immunoregulatory properties[2,3].

Crossreactivity
There is 66% amino acid sequence identity between human and mouse GH. Human GH binds to the mouse GH receptor.

Sources
Pituitary, placenta and white blood cells, including lymphocytes.

Bioassays
The standard bioassay for lactogenic hormones is proliferation of the rat Nb2 lymphoma cells[4]. Specificity should be determined with blocking antibodies.

Physicochemical properties of GH

Property	Human	Mouse
pI	5.1 (5.5)[a]	?
Amino acids – precursor	217	216
– mature[b]	191	190
M_r (K) – predicted		
– expressed	22–25 (20)[a]	22
Potential *N*-linked glycosylation sites	0	0
Disulfide bonds	2[c]	2[d]

[a] A 20 kDa short form produced by alternative splicing.
[b] After removal of signal peptide.
[c] Cys53–165 and 182–189.
[d] Cys52–163 and 180–188.

3-D structure
GH has a four α-helical bundle structure[5].

Gene structure[6]

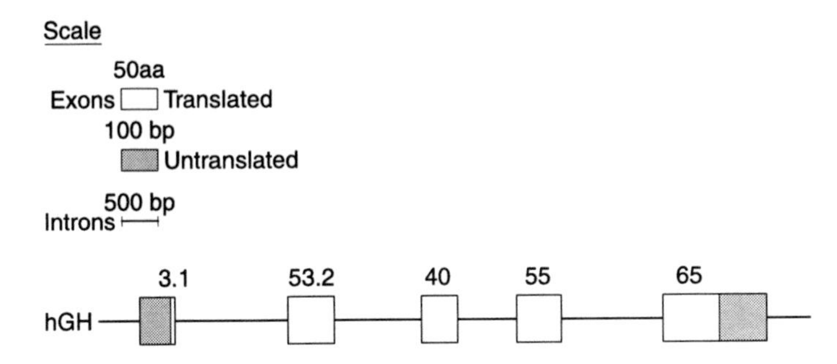

Gene for human pituitary GH is on 17q22–24.

Amino acid sequence for human GH[5]

Accession code: SwissProt P01241

```
-26  MATGSRTSLL LAFGLLCLPW LQEGSA
  1  FPTIPLSRLF DNAMLRAHRL HQLAFDTYQE FEEAYIPKEQ KYSFLQNPQT
 51  SLCFSESIPT PSNREETQQK SNLELLRISL LLIQSWLEPV QFLRSVFANS
101  LVYGASDSNV YDLLKDLEEG IQTLMGRLED GSPRTGQIFK QTYSKFDTNS
151  HNDDALLKNY GLLYCFRKDM DKVETFLRIV QCRSVEGSCG F
```

Amino acids Glu32–Gln46 missing in short version produced by alternative splicing at junction of exon 2. Conserved disulfide bonds between Cys53–165 and 182–189.

Amino acid sequence for mouse GH[7]

Accession code: SwissProt P06880

```
-26  MATDSRTSWL LTVSLLCLLW PQEASA
  1  FPAMPLSSLF SNAVLRAQHL HQLAADTYKE FERAYIPEGQ RYSIQNAQAA
 51  FCFSETIPAP TGKEEAQQRT DMELLRFSLL LIQSWLGPVQ FLSRIFTNSL
101  MFGTSDRVYE KLKDLEEGIQ ALMQELEDGS PRVGQILKQT YDKFDANMRS
151  DDALLKNYGL LSCFKKDLHK AETYLRVMKC RRFVESSCAF
```

Disulfide binds between Cys52–163 and 180–188.

THE GH RECEPTOR

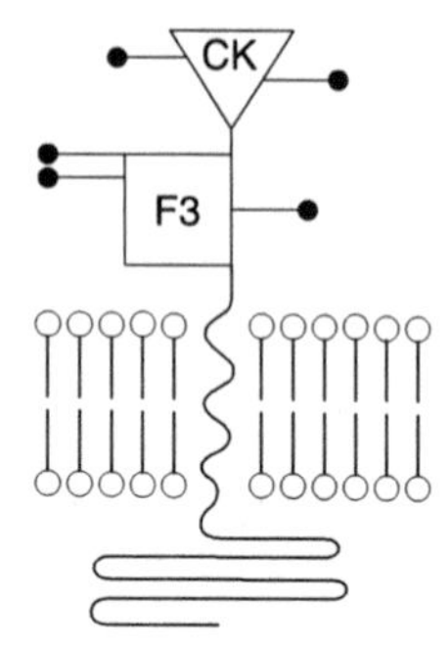

The GH receptors is a member of the CKR superfamily but does not possess the WSXWS motif characteristic of this receptor family[8]. The extracellular domain of the receptor (residues 1–246) produced by proteolysis of membrane hGHR occurs naturally in serum as a soluble GH-binding protein. Structural analysis has shown that one GH molecule binds to two GHR chains. There are two binding sites on opposite faces of the four α-helical GH molecule, each of which binds to essentially the same site on the receptor[9]. Mutations in the human GHR cause idiopathic short stature and Laron type dwarfism[10]. In the presence of physiological concentrations of zinc, hGH also binds to the PRL receptor but hPRL does not bind to the GH receptor[11].

Distribution

Liver, adipose tissue, intestine, brain, testes, heart, kidney, lung, pancreas, cartilage, skeletal muscle, corpus luteum and lymphocytes.

Physicochemical properties of the GH and PRL receptors

Property	Human	Mouse
Amino acids – precursor	638	650 (297)[a]
– mature[b]	620	626 (273)[a]
M_r (K) – predicted	70	72.8 (34)[a]
– expressed	120	102 (33,39)[a]
Potential *N*-linked glycosylation sites	5	4
Affinity K_d (M)	10^{-10}	?

[a] Short soluble form of the GHR.
[b] After removal of predicted signal peptide.

Signal transduction

Signalling depends on ligand-induced homodimerization which occurs sequentially[12]. Binding to one receptor chain through GH site 1 is required before binding through site 2. The cytoplasmic domain of the GH receptor has no intrinsic enzyme activity but signalling does involve tyrosine phosphorylation of several cytoplasmic proteins including the receptor chains themselves. Jak2 associates with the proline-rich conserved box 1 of the GHR in response to ligand activation, and STATS 1, 3 and 5 are activated.

Chromosomal location

The genes for growth hormone receptor are on human chromosome 5p13–14 and mouse chromosome 15.

Amino acid sequence for human GH receptor[13]

Accession code: EMBL XP_003896

```
-18  MDLWQLLLTL ALAGSSDA
  1  FSGSEATAAI LSRAPWSLQS VNPGLKTNSS KEPKFTKCRS PERETFSCHW
 51  TDEVHHGTKN LGPIQLFYTR RNTQEWTQEW KECPDYVSAG ENSCYFNSSF
101  TSIWIPYCIK LTSNGGTVDE KCFSVDEIVQ PDPPIALNWT LLNVSLTGIH
151  ADIQVRWEAP RNADIQKGWM VLEYELQYKE VNETKWKMMD PILTTSVPVY
201  SLKVDKEYEV RVRSKQRNSG NYGEFSEVLY VTLPQMSQFT CEEDFYFPWL
251  LIIIFGIFGL TVMLFVFLFS KQQRIKMLIL PPVPVPKIKG IDPDLLKEGK
301  LEEVNTILAI HDSYKPEFHS DDSWVEFIEL DIDEPDEKTE ESDTDRLLSS
351  DHEKSHSNLG VKDGDSGRTS CCEPDILETD FNANDIHEGT SEVAQPQRLK
401  GEADLLCLDQ KNQNNSPYHD ACPATQQPSV IQAEKNKPQP LPTEGAESTH
451  QAAHIQLSNP SSLSNIDFYA QVSDITPAGS VVLSPGQKNK AGMSQCDMHP
501  EMVSLCQENF LMDNAYFCEA DAKKCIPVAP HIKVESHIQP SLNQEDIYIT
551  TESLTTAAGR PGTGEHVPGS EMPVPDYTSI HIVQSPQGLI LNATALPLPD
601  KEFLSSCGYV STDQLNKIMP
```

Fibronectin type III domain 136–234. Disulfide bonds between Cys38–48, 83–94 and 108–122. Mutations in the receptor cause idiopathic short stature and Laron dwarfism[10].

Amino acid sequence for mouse GH receptor[14]

Accession code: EMBL NP_034414

```
-24  MDLCQVFLTL ALAVTSSTFS GSEA
  1  TPATLGKASP VLQRINPSLG TSSSGKPRFT KCRSPELETF SCYWTEGDNP
 51  DLKTPGSIQL YYAKRESQRQ AARIAHEWTQ EWKECPDYVS AGKNSCYFNS
101  SYTSIWIPYC IKLTTNGDLL DQKCFTVDEI VQPDPPIGLN WTLLNISLTG
151  IRGDIQVSWQ PPPNADVLKG WIILEYEIQY KEVNESKWKV MGPIWLTYCP
201  VYSLRMDKEH EVRVRSRQRS FEKYSEFSEV LRVIFPQTNI LEACEEDIQF
251  PWFLIIIFGI FGVAVMLFVV IFSKQQRIKM LILPPVPVPK IKGIDPDLLK
301  EGKLEEVNTI LGIHDNYKPD FYNDDSWVEF IELDIDEADV DEKTEGSDTD
351  RLLSNDHEKS AGILGAKDDD SGRTSCYDPD ILDTDFHTSD MCDGTLKFRQ
401  SQKLNMEADL LCLDQKNLKN LPYDASLGSL HPSITQTVEE NKPQPLLSSE
451  TEATHQLAST PMSNPTSLAN IDFYAQVSDI TPAGGDVLSP GQKIKAGIAQ
501  GNTQREVATP CQENYSMNSA YFCESDAKKC IAVARRMEAT SCIKPSFNQE
551  DIYITTESLT TTAQMSETAD IAPDAEMSVP DYTTVHTVQS PRGLILNATA
601  LPLPDKKNFP SSCGYVSTDQ LNKIMQ
```

Disulfide bonds between Cys32–42, 85–96 and 110–124 by similarity. Conflicting sequence T→A at position 1. Short soluble form has same sequence from position 1–246 followed by a short hydrophilic tail:

```
247  GTKSNSQHPH QEIDNHLYHQ LQRIRHP
```

References

[1] Goffin, V. et al. (1996) Endocr. Rev. 17, 385–410.
[2] Kooljman, R. et al. (1996) Adv. Immunol. 63, 377–454.
[3] Trivedi, H.N. et al. (1997) Hum. Immunol. 57, 69–79.
[4] Gout, P.W et al. (1980) Cancer Res. 40, 2433–2436.
[5] Ludin, C. et al. (1987) EMBO J. 6, 109–114.
[6] Chen, E.Y et al. (1989) Genomics 4, 479–497.
[7] Linzer, D.I.H. and Talamantes, F. (1985) J. Biol. Chem. 260, 9574–9579.
[8] Kelly, P.A et al. (1993) Rec. Progr. Horm. Res. 48, 123–164.
[9] De Vos, A.M. et al. (1992) Science 255, 306–312.
[10] Godowski, P.J. et al. (1989) Proc. Natl Acad. Sci. USA 86, 8083–8087.
[11] Somers, W. et al. (1994) Nature 372, 478–481.
[12] Goffin, V. and Kelly, P.A. (1996) Clin. Endocrinol. 45, 247–255.
[13] Leung, D.W et al. (1987) Nature 330, 537–543.
[14] Smith, W.C. et al. (1989) Endocrinology 3, 984–990.

Other names

HBEGF, diphtheria toxin receptor (DT-R).

THE MOLECULE

Heparin-binding epidermal growth factor (HB-EGF) is a 22-kDa growth factor, mitogenic for fibroblasts and smooth muscle (but not for endothelial cells) and also expressed in human macrophages, where it may be involved in macrophage-induced cellular proliferation[1]. HB-EGF is a more potent mitogen for smooth muscle than is EGF, and has been implicated in the migration and development of smooth muscle cells occurring in both the normal development of arterial walls and in the formation of atherosclerotic plaques[2]. HB-EGF binds the EGF receptor; concomitant binding of cell surface heparin sulfate proteoglycan appears to be required for optimal binding to the EGF receptor[3]. There are two forms of HB-EGF, a membrane-anchored precursor form (pro-HB-EGF) and the mature soluble form (sHB-EGF) cleaved from pro-HB-EGF, each with distinct biological activities[4]. The pro form is a juxtacrine growth and adhesion factor, acts as the receptor for diptheria toxin and is upregulated by binding of CD9 antigen[5,6].

Crossreactivity

Not known.

Sources

HB-EGF is synthesized in smooth muscle cells[7].

Bioassays

HB-EGF is mitogenic for BALB/c 3T3 cells, causes inhibition of ^{125}I-EGF binding to the EGF receptor and triggers autophosphorylation of the EGF receptor[8].

Physicochemical properties of HB-EGF

Property	Human	Mouse
pI[a]	9.70	9.34
Amino acids – precursor	208	208
– mature	86	86
M_r (K) – predicted	9.77	9.73
– expressed	22	?
Potential *N*-linked glycosylation sites	0	0
Disulfide bonds	3	3

[a] Predicted.

3-D structure

Not known.

Gene structure

The human HB-EGF gene has been localized to chromosome 5[9].

Amino acid sequence for human HB-EGF

Accession code: SwissProt Q99075

```
-19  MKLLPSVVLK LFLAAVLSA
  1  LVTGESLERL RRGLAAGTSN PDPPTVSTDQ LLPLGGGRDR KVRDLQEADL
 51  DLLRVTLSSK PQALATPNKE EHGKRKKKGK GLGKKRDPCL RKYKDFCIHG
101  ECKYVKELRA PSCICHPGYH GERCHGLSLP VENRLYTYDH TTILAVVAVV
151  LSSVCLLVIV GLLMFRYHRR GGYDVENEEK VKLGMTNSH
```

Propeptide in italics (20–62 or 72 or 73 or 76 or 81 of precursor). A C-terminal propeptide sequence is removed (149–208 of precursor) to generate soluble mature HB-EGF. Contains one EGF-like domain. Disulfide bridges between Cys108–121, 116–132 and 134–143 by similarity. One potential *O*-linked carbohydrate moiety at position 85 of precursor.

Amino acid sequence for mouse HB-EGF

Accession code: SwissProt Q06186

```
-23  MKLLPSVMLK LFLAAVLSAL VTG
  1  ESLERLRRGL AAATSNPDPP TGSTNQLLPT GGDRAQGVQD LEGTDLNLFK
 51  VAFSSKPQGL ATPSKERNGK KKKGKGLGK KRDPCLRKYK DYCIHGECRY
101  LQEFRTPSCK CLPGYHGHRC HGLTLPVENP LYTYDHTTVL AVVAVVLSSV
151  CLLVIVGLLM FRYHRRGGYD LESEEKVKLG VASSH
```

Propeptide in italics (24–62 of precursor). A C-terminal propeptide sequence is removed (149–208 of precursor) to generate soluble mature HB-EGF. The position of the transmembrane region is underlined. Contains one EGF like domain. Disulfide bridges between Cys108–121, 116–132 and 134–143 by similarity. One potential *O*-linked carbohydrate moiety at position 85 of precursor.

THE HB-EGF RECEPTOR

HB-EGF activates two subtypes of the EGF receptor HER1 (EGFR) and HER4/ErbB4[5]. See EGF entry (page **203**).

References

[1] Higashiyama, S. et al. (1991) Science 251, 936–939.
[2] Miyagawa, J. et al. (1995) J. Clin. Invest. 95, 404–411.
[3] Higashiyama, S. et al. (1993) J. Cell Biol. 122, 933–940.
[4] Iwamoto, R. et al. (1999) J. Biol. Chem. 274, 25906–25912.
[5] Raab, G. and Klagsbrun, M. (1997) Biochim. Biophys. Acta 1333, F179–F199.
[6] Sakuma, T. et al. (1997) J. Biochem. (Tokyo) 122, 474–480.
[7] Higashiyama, S. et al. (1994) Hormone Res. 42, 9–13.
[8] Higashiyama, S. et al. (1992) J. Biol. Chem. 267, 6205–6212.
[9] Fen, Z. et al. (1993) Biochemistry 32, 7932–7938.

Other names

Small inducible cytokine A14, chemokine CC-1/CC-3, HCC-1/HCC-3, NCC-2.

THE MOLECULE

Haemoinfiltrate CC chemokine 1 (HCC-1) has weak activity on human monocytes, inducing changes in intracellular calcium concentrations and enzyme release[1]. Mature HCC-1 shares 46% homology with MIP-1α and β. Unusually for chemokines, HCC-1 is not chemotactic for leukocytes[2].

Crossreactivity

Not known.

Sources

HCC-1 is constitutively secreted by tissues (spleen, liver, skeletal and heart muscle, gut and bone marrow) and found at high concentrations in plasma[1,2].

Bioassays

Enhances the proliferation of CD34+ myeloid progenitor cells[1].

Physicochemical properties of HCC-1

Property	Human	Mouse
pI[a]	8.72	?
Amino acids – precursor	93	?
– mature	74	?
M_r (K) – predicted	8.7	?
– expressed	8.7	?
Potential *N*-linked glycosylation sites	0	?
Disulfide bonds	2	?

[a] Predicted.

3-D structure

Not known.

Gene structure

The HCC-1 gene is localized to the human CC chemokine gene cluster on chromosome 17q11.2[3].

Amino acid sequence for human HCC-1

Accession code: SwissProt Q16627

```
-19  MKISVAAIPF FLLITIALG
  1  TKTESSSRGP YHPSECCFTY TTYKIPRQRI MDYYETNSQC SKPGIVFITK
 51  RGHSVCTNPS DKWVQDYIKD MKEN
```

Disulfide bridges between Cys35–59 and 36–75 by similarity. Splice variant HCC-3 at position 27 R→QTGGKPKVVKIQLKLVG.

THE HCC-1 RECEPTOR, CCR1

HCC-1 has been reported to specifically activate the CCR1 receptor[2]. The CCR1 receptor is a member of the G protein-coupled receptors family and induces increased intracellular calcium ion concentration. See MIP-1α entry (page **384**).

References

[1] Schulz-Knappe, P. et al. (1996) J. Exp. Med. 183, 295–299.
[2] Tsou, C.L. et al. (1998) J. Exp. Med. 188, 603–608.
[3] Nomiyama, H. et al. (1999) J. Interferon Cytokine Res. 19, 227–234.

Other names

Small inducible cytokine A16, IL-10 inducible chemokine, chemokine LEC, liver-expressed chemokine, monotactin-1 (MTN-1), chemokine CC-4, NCC-4, lymphocyte and monocyte chemoattractant, LMC, LCC-1.

THE MOLECULE

Haemoinfiltrate CC chemokine 4 (HCC-4) is a recently characterized beta (CC) chemokine which is chemotactic for monocytes[1]. HCC-4 also induces leukocyte and dendritic cell migration[1–3]. HCC-4 expression is markedly upregulated in response to IL-10, in contrast to the majority of chemokines, which are downregulated by IL-10[1]. HCC-4 has been reported to improve cross-talk between APCs and T cells[2].

Crossreactivity

Not known.

Sources

HCC-4 is mainly expressed in the liver, and also found in the spleen and thymus. HCC-4 is most highly expressed in LPS and IFNγ-activated monocytes, with weaker expression in natural killer cells and γδ T cells.

Bioassays

HCC-4 induces calcium ion flux in THP-1 cells previously desensitized by exposure to RANTES[1].

Physicochemical properties of HCC-4

Property	Human	Mouse
pI[a]	9.51	?
Amino acids – precursor	120	?
– mature	97	?
M_r (K) – predicted	11.2	?
– expressed	?	?
Potential *N*-linked glycosylation sites	0	?
Disulfide bonds	2	?

[a] Predicted.

3-D structure

Not known.

Gene structure

The HCC-4 gene has been localized to the human CC chemokine cluster on chromosome 17q11.2[4].

Amino acid sequence for human HCC-4

Accession code: SwissProt O15476

```
-23  MKVSEAALSL LVLILIITSA SRS
  1  QPKVPEWVNT PSTCCLKYYE KVLPRRLVVG YRKALNCHLP
 51  AIIFVTKRNR EVCTNPNDDW VQEYIKDPNL PLLPTRNLST
101  VKIITAK NGQPQLLNSQ
```

Disulfide bridges between Cys37–60 and 38–76 by similarity.

THE HCC-4 RECEPTORS

HCC-4 has been shown to interact with both CCR1 and CCR8 (K_d 2 nM)[3]. See MIP-1α entry (page **384**) for information on CCR1 and TARC (page **457**) for information on CCR8.

References

[1] Hedrick, J.A. et al. (1998) Blood 91, 4242–4247.
[2] Giovarelli, M. et al. (2000) J. Immunol. 164, 3200–3206.
[3] Howard, O.M. et al. Blood 96, 840–845.
[4] Tsou, C.L. et al. (1998) J. Exp. Med. 188, 603–608.

Other names

Scatter factor (SF) and hepatopoietin (HPTA).

THE MOLECULE

Hepatocyte growth factor (HGF) is a mesenchymally derived heparin-binding glycoprotein secreted as a single-chain, biologically inert propeptide[1–3]. HGF has potent motogenic, mitogenic and morphogenetic activities on epithelial cells *in vitro*. It is converted to the biologically active form by proteolytic cleavage into two subunits that associate to form the active molecule. The human converting enzyme has been cloned (SwissProt Q04756). The α-chain contains a hairpin loop of 27 amino acids at the N-terminus and four unique domains called kringles. The β-chain contains a serine protease-like structure but without protease activity. The kringle motif is an approximately 80-amino-acid double looped structure formed by three internal disulfide bridges and was first described for many enzymes involved in proteolysis which HGF resembles. An HGF-like protein (see MSP entry (page **392**)) has a similar structure and binds to the receptor tyrosine kinase Ron. HGF is a multifunctional cytokine with activity on a wide variety of different cells and tissues. HGF was originally described as an hepatocyte growth factor in regenerating liver. Its effects are now known to include induction of cell proliferation and motility, inhibition of cell growth, induction of morphogenesis, stimulation of T-cell adhesion to endothelial cells and migration, regulation of erythroid differentiation, and enhancement of neuron survival. HGF is also important in development. Mice lacking the HGF gene die *in utero* with severely impaired placentas and extensive loss of parenchymal cells in the liver[4,5].

Crossreactivity

There is crossreactivity between human and mouse HGF.

Sources

Various cells produce HGF, including polymorphonucleocytes, fibroblasts, epithelial cells, endothelial cells, Kupffer cells and fat-storing cells in the liver. HGF protein and high levels of mRNA have also been found in intestine, brain, thyroid, thymus and placenta. IL-1α and IL-1β both enhance HGF production.

Bioassays

Scatter response by MDCK cells at 12 hours. HGF units/ml are defined as the maximum dilution that will give a detectable response. One unit per ml is approximately equivalent to 0.2– 0.5 ng/ml of purified protein.

Physicochemical properties of HGF

Property	Human		Mouse	
	α-chain	β-chain	α-chain	β-chain
pI (calculated)	7.7	8.8	8.3	8.8
Amino acids – precursor[a]	728	728		
– mature[b]	463	234	463	233
M_r (K) – predicted	53.7	26.0	53.5	25.9
– expressed	60	30	60	30
Potential *N*-linked glycosylation sites	2	2	2	2
Disulfide bonds[c]	3	3	3	3

[a] Includes a 31 amino acid signal peptide.
[b] Propeptide is cleaved proteolytically to form two subunits that combine by a single disulfide bond to form the active heterodimer.
[c] Disulfide bonds between Cys39–65, Cys43–53 (human and mouse), and an interchain disulfide bond between Cys456–573 (human) and Cys456–575 (mouse).

3-D structure

Not known.

Gene structure[6]

Scale
50aa
Exons Translated
1 Kb
Untranslated
1 Kb
Introns

Chromosome

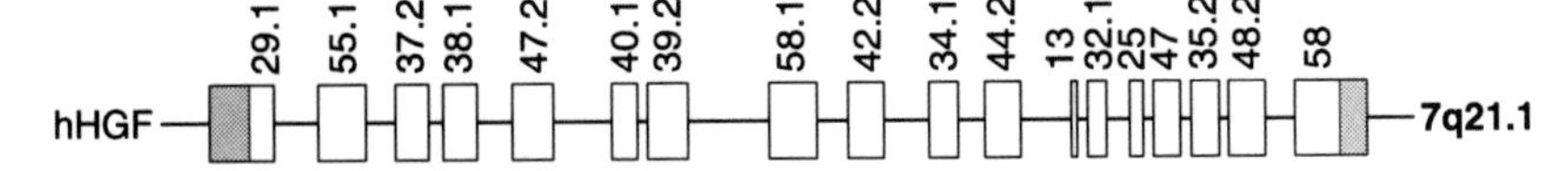

mouse - 5

The genes for hepatocyte growth factor are on human chromosome 7q21.1 and mouse chromosome 5.

Amino acid sequence for human HGF[7,6]

Accession code: SwissProt P14210

```
-31  MWVTKLLPAL LLQHVLLHLL LLPIAIPYAE G
  1  QRKRRNTIHE FKKSAKTTLI KIDPALKIKT KKVNTADQCA NRCTRNKGLP
 51  FTCKAFVFDK ARKQCLWFPF NSMSSGVKKE FGHEFDLYEN KDYIRNCIIG
101  KGRSYKGTVS ITKSGIKCQP WSSMIPHEHS FLPSSYRGKD LQENYCRNPR
151  GEEGGPWCFT SNPEVRYEVC DIPQCSEVEC MTCNGESYRG LMDHTESGKI
201  CQRWDHQTPH RHKFLPERYP DKGFDDNYCR NPDGQPRPWC YTLDPHTRWE
251  YCAIKTCADN TMNDTDVPLE TTECIQGQGE GYRGTVNTIW NGIPCQRWDS
301  QYPHEHDMTP ENFKCKDLRE NYCRNPDGSE SPWCFTTDPN IRVGYCSQIP
351  NCDMSHGQDC YRGNGKNYMG NLSQTRSGLT CSMWDKNMED LHRHIFWEPD
401  ASKLNENYCR NPDDDAHGPW CYTGNPLIPW DYCPISRCEG DTTPTIVNLD
451  HPVISCAKTK QLRVVNGIPT RTNIGWMVSL RYRNKHICGG SLIKESWVLT
501  ARQCFPSRDL KDYEAWLGIH DVHGRGDEKC KQVLNVSQLV YGPEGSDLVL
551  MKLARPAVLD DFVSTIDLPN YGCTIPEKTS CSVYGWGYTG LINYDGLLRV
601  AHLYIMGNEK CSQHHRGKVT LNESEICAGA EKIGSGPCEG DYGGPLVCEQ
651  HKMRMVLGVI VPGRGCAIPN RPGIFVRVAY YAKWIHKIIL TYKVPQS
```

Disulfide binds between Cys39–65, Cys43–53, and an interchain disulfide bond between Cys456–573. Propeptide is cleaved proteolytically to form two subunits (α-chain residues 1–463 and β-chain amino acids 464–697) that combine through a single disulfide bond to form the active heterodimer.

Amino acid sequence for mouse HGF[8]

Accession code: SwissProt Q08048

```
-31  MMWGTKLLPV LLLQHVLLHL LLLHVAIPYA EG
  1  QKKRRNTLHE FKKSAKTTLT KEDPLLKIKT KKVNSADECA NRCIRNRGFT
 51  FTCKAFVFDK SRKRCYWYPF NSMSSGVKKG FGHEFDLYEN KDYIRNCIIG
101  KGGSYKGTVS ITKSGIKCQP WNSMIPHEHS FLPSSYRGKD LQENYCRNPR
151  GEEGGPWCFT SNPEVRYEVC DIPQCSEVEC MTCNGESYRG PMDHTESGKT
201  CQRWDQQTPH RHKFLPERYP DKGFDDNYCR NPDGKPRPWC YTLDPDTPWE
251  YCAIKTCAHS AVNETDVPME TTECIQGQGE GYRGTSNTIW NGIPCQRWDS
301  QYPHKHDITP ENFKCKDLRE NYCRNPDGAE SPWCFTTDPN IRVGYCSQIP
351  KCDVSSGQDC YRGNGKNYMG NLSKTRSGLT CSMWDKNMED LHRHIFWEPD
401  ASKLNKNYCR NPDDDAHGPW CYTGNPLIPW DYCPISRCEG DTTPTIVNLD
451  HPVISCAKTK QLRVVNGIPT QTTVGWMVSL KYRNKHICGG SLIKESWVLT
501  ARQCFPARNK DLKDYEAWLG IHDVHERGEE KRKQILNISQ LVYGPEGSDL
551  VLLKLARPAI LDNFVSTIDL PSYGCTIPEK TTCSIYGWGY TGLINADGLL
601  RVAHLYIMGN EKCSQHHQGK VTLNESELCA GAEKIGSGPC EGDYGGPLIC
651  EQHKMRMVLG VIVPGRGCAI PNRPGIFVRV AYYAKWIHKV ILTYKL
```

Disulfide binds between Cys39–65, Cys43–53, and an interchain disulfide bond between Cys456–575. Propeptide is cleaved proteolytically to form two subunits (α-chain residues 1–463 and β-chain amino acids 463–696) that combine through a single disulfide bond to form the active heterodimer.

THE HGF RECEPTOR

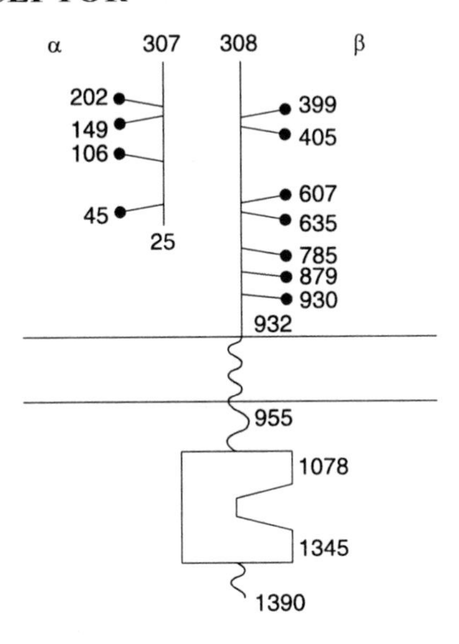

The HGF receptor is a transmembrane tyrosine kinase called Met, first identified as the proto-oncogene c-*met*. The mature receptor is a disulfide-linked heterodimer consisting of an entirely extracellular α-chain (M_r 50 000) and a transmembrane β-chain (M_r 145 000) that includes an intracellular kinase domain. Both chains are derived post-translationally by proteolytic cleavage of a single precursor molecule. Met belongs to a family of transmembrane receptors that includes Ron, the receptor for macrophage-stimulating protein (MSP), the orphan receptor Sea and four related proteins SEX, SEP, OCT and NOV[9]. Activation of the Met oncogene occurs by formation of a chimeric gene with TPR (translocated promoter region)[10]. HGF binding to Met and signal transduction is dependent on interactions between the N-terminal domain of HGF and heparan sulfate glycosaminoglycans (HSGAGs)[11].

Distribution

The HGF receptor is widely expressed on normal epithelium of most tissues, and on hepatocytes, keratinocytes, melanocytes, endothelial cells, microglia, neurons, haematopoietic cells and many tumour cell lines.

Physicochemical properties of the HGF receptor (Met)

Property	Human	Mouse
Amino acids – precursor	1390[a]	1379
– mature[b,c]	1366 (283 and 1083)[d]	1355 (282 and 1073)[d]
M_r (K) – predicted	153 (29.6 and 123.4)[d]	151 (29.3 and 121.8)[d]
– expressed	50 and 145[c]	50 and 145[c]
Potential *N*-linked glycosylation sites	11	10
Affinity K_d (M)[e]	approximately 2×10^{-10}	

[a] Alternative splicing can give rise to a 1408-amino-acid protein which differs by replacing Ser731 with STWWKEPLNI VSFLFCFAS[12].
[b] After removal of predicted signal peptide.
[c] The α- and β-chains of the receptor are obtained by proteolytic cleavage of the propeptide.
[d] α- and β-chains.
[e] HGF binding to heparan sulfate glycosaminoglycan is required for optimal binding to Met and signalling.

Signal transduction

Receptor tyrosine kinase that acts through a SH2 docking site to activate several signalling pathways including PI-3 kinase and Rac activation required for scattering, Ras-MAP required for growth and activation of STAT3 for tubulogenesis.[3,13]

Chromosomal location

The HGF receptor (c-*met*) is on human chromosome 7q31 and mouse chromosome 6.

Amino acid sequence for human HGF receptor (Met)[12,14,15]

Accession code: SwissProt P08581

```
-24  MKAPAVLAPG ILVLLFTLVQ RSNG
  1  ECKEALAKSE MNVNMKYQLP NFTAETPIQN VILHEHHIFL GATNYIYVLN
 51  EEDLQKVAEY KTGPVLEHPD CFPCQDCSSK ANLSGGVWKD NINMALVVDT
101  YYDDQLISCG SVNRGTCQRH VFPHNHTADI QSEVHCIFSP QIEEPSQCPD
151  CVVSALGAKV LSSVKDRFIN FFVGNTINSS YFPDHPLHSI SVRRLKETKD
201  GFMFLTDQSY IDVLPEFRDS YPIKYVHAFE SNNFIYFLTV QRETLDAQTF
251  HTRIIRFCSI NSGLHSYMEM PLECILTEKR KKRSTKKEVF NILQAAYVSK
301  PGAQLARQIG ASLNDDILFG VFAQSKPDSA EPMDRSAMCA FPIKYVNDFF
351  NKIVNKNNVR CLQHFYGPNH EHCFNRTLLR NSSGCEARRD EYRTEFTTAL
401  QRVDLFMGQF SEVLLTSIST FIKGDLTIAN LGTSEGRFMQ VVVSRSGPST
451  PHVNFLLDSH PVSPEVIVEH TLNQNGYTLV ITGKKITKIP LNGLGCRHFQ
501  SCSQCLSAPP FVQCGWCHDK CVRSEECLSG TWTQQICLPA IYKVFPNSAP
551  LEGGTRLTIC GWDFGFRRNN KFDLKKTRVL LGNESCTLTL SESTMNTLKC
601  TVGPAMNKHF NMSIIISNGH GTTQYSTFSY VDPVITSISP KYGPMAGGTL
651  LTLTGNYLNS GNSRHISIGG KTCTLKSVSN SILECYTPAQ TISTEFAVKL
701  KIDLANRETS IFSYREDPIV YEIHPTKSFI SGGSTITGVG KNLNSVSVPR
```

```
 751 MVINVHEAGR NFTVACQHRS NSEIICCTTP SLQQLNLQLP LKTKAFFMLD
 801 GILSKYFDLI YVHNPVFKPF EKPVMISMGN ENVLEIKGND IDPEAVKGEV
 851 LKVGNKSCEN IHLHSEAVLC TVPNDLLKLN SELNIEWKQA ISSTVLGKVI
 901 VQPDQNFTGL IAGVVSISTA LLLLLGFFLW LKKRKQIKDL GSELVRYDAR
 951 VHTPHLDRLV SARSVSPTTE MVSNESVDYR ATFPEDQFPN SSQNGSCRQV
1001 QYPLTDMSPI LTSGDSDISS PLLQNTVHID LSALNPELVQ AVQHVVIGPS
1051 SLIVHFNEVI GRGHFGCVYH GTLLDNDGKK IHCAVKSLNR ITDIGEVSQF
1101 LTEGIIMKDF SHPNVLSLLG ICLRSEGSPL VVLPYMKHGD LRNFIRNETH
1151 NPTVKDLIGF GLQVAKGMKY LASKKFVHRD LAARNCMLDE KFTVKVADFG
1201 LARDMYDKEY YSVHNKTGAK LPVKWMALES LQTQKFTTKS DVWSFGVVLW
1251 ELMTRGAPPY PDVNTFDITV YLLQGRRLLQ PEYCPDPLYE VMLKCWHPKA
1301 EMRPSFSELV SRISAIFSTF IGEHYVHVNA TYVNVKCVAP YPSLLSSEDN
1351 ADDEVDTRPA SFWETS
```

The propeptide is cleaved between residues 283 and 284 to form α- and β-chains of the receptor. Alternative splicing gives rise to a 1408-amino-acid protein which differs by replacing Ser731 with STWWKEPLNI VSFLFCFAS. Translocation breakpoint to form Met/TPR oncogene is between Glu1085 and Asp1086. See SwissProt entry for conflicting sequences. Tyr1332 is essential for signal transduction.

Amino acid sequence for mouse HGF receptor (Met)[16]

Accession code: SwissProt P16056

```
 -24 MKAPTVLAPG ILVLLLSLVQ RSHG
   1 ECKEALVKSE MNVNMKYQLP NFTAETPIQN VVLHGHHIYL GATNYIYVLN
  51 DKDLQKVSEF KTGPVLEHPD CLPCRDCSSK ANSSGGVWKD NINMALLVDT
 101 YYDDQLISCG SVNRGTCQRH VLPPDNSADI QSEVHCMFSP EEESGQCPDC
 151 VVSALGAKVL LSEKDRFINF FVGNTINSSY PPGYSLHSIS VRRLKETQDG
 201 FKFLTDQSYI DVLPEFLDSY PIKYIHAFES NHFIYFLTVQ KETLDAQTFH
 251 TRIIRFCSVD SGLHSYMEMP LECILTEKRR KRSTREEVFN ILQAAYVSKP
 301 GANLAKQIGA SPSDDILFGV FAQSKPDSAE PVNRSAVCAF PIKYVNDFFN
 351 KIVNKNNVRC LQHFYGPNHE HCFNRTLLRN SSGCEARSDE YRTEFTTALQ
 401 RVDLFMGRLN QVLLTSISTF IKGDLTIANL GTSEGRFMQV VLSRTAHLTP
 451 HVNFLLDSHP VSPEVIVEHP SNQNGYTLVV TGKKITKIPL NGLGCGHFQS
 501 CSQCLSAPYF IQCGWCHNQC VRFDECPSGT WTQEICLPAV YKVFPTSAPL
 551 EGGTVLTICG WDFGFRKNNK FDLRKTKVLL GNESCTLTLS ESTTNTLKCT
 601 VGPAMSEHFN VSVIISNSRE TTQYSAFSYV DPVITSISPR YGPQAGGTLL
 651 TLTGKYLNSG NSRHISIGGK TCTLKSVSDS ILECYTPAQT TSDEFPVKLK
 701 IDLANRETSS FSYREDPVVY EIHPTKSFIS GGSTITGIGK TLNSVSLPKL
 751 VIDVHEVGVN YTVACQHRSN SEIICCTTPS LKQLGLQLPL KTKAFFLLDG
 801 ILSKHFDLTY VHNPVFEPFE KPVMISMGNE NVVEIKGNNI DPEAVKGEVL
 851 KVGNQSCESL HWHSGAVLCT VPSDLLKLNS ELNIEWKQAV SSTVLGKVIV
 901 QPDQNFAGLI IGAVSISVVV LLLSGLFLWM RKRKHKDLGS ELVRYDARVH
 951 TPHLDRLVSA RSVSPTTEMV SNESVDYRAT FPEDQFPNSS QNGACRQVQY
1001 PLTDLSPILT SGDSDISSPL LQNTVHIDLS ALNPELVQAV QHVVIGPSSL
1051 IVHFNEVIGR GHFGCVYHGT LLDNDGKKIH CAVKSLNRIT DIEEVSQFLT
1101 EGIIMKDFSH PNVLSLLGIC LRSEGSPLVV LPYMKHGDLR NFIRNETHNP
1151 TVKDLIGFGL QVAKGMKYLA SKKFVHRDLA ARNCMLDEKF TVKVADFGLA
```

```
1201 RDMYDKEYYS VHNKTGAKLP VKWMALESLQ TQKFTTKSDV WSFGVLLWEL
1251 MTRGAPPYPD VNTFDITIYL LQGRRLLQPE YCPDALYEVM LKCWHPKAEM
1301 RPSFSELVSR ISSIFSTFIG EHYVHVNATY VNVKCVAPYP SLLPSQDNID
1351 GEGNT
```

The propeptide is cleaved between residues 282 and 283 to form the α- and β-chains of the receptor. Tyr1330 is essential for signal transduction. See SwissProt entry for conflicting sequences.

References

[1] Zarnegar, R. and Michalopoulos, G.K (1995) J. Cell Biol. 129, 1177–1180.
[2] Boros, P. and Miller, C.M. (1995) Lancet 345, 293–295.
[3] Jiang, W.G. and Hiscox, S. (1997) Histol. Histopathol. 12, 537–555.
[4] Schmidt, C. et al. (1995) Nature 373, 699–702.
[5] Uehara, Y. et al. (1995) Nature 373, 702–705.
[6] Seki, T. et al. (1991) Gene 102, 213–219.
[7] Miyazawa, K. et al. (1989) Biochem. Biophys. Res. Commun. 163, 967–973.
[8] Sasaki, M. et al. (1994) Biochem. Biophys. Res. Commun. 199, 772–779.
[9] Maestrini, E. et al. (1996) Proc. Natl Acad. Sci. USA 93, 674–678.
[10] Soman, N.R. et al. (1991) Proc. Natl Acad. Sci. USA 88, 4892–4896.
[11] Sakata, H. et al. (1997) J. Biol. Chem. 272, 9457–9463.
[12] Ponzetto, C. et al. (1991) Oncogene 6, 553–559.
[13] Boccaccio, C. et al. (1998) Nature 391, 285–288.
[14] Park, M. et al. (1987) Proc. Natl Acad. Sci. USA 84, 6379–6383.
[15] Duh, F.-M. et al. (1997) Oncogene 15, 1583–1586.
[16] Chan, A.M.L. et al. (1988) Oncogene 2, 593–599.

Other names

Scya1. A splice variant of TCA-3 is known as P500[1].

THE MOLECULES

I-309 is the human homologue of the murine TCA-3 gene[2,3]. Both molecules are members of the CC family of chemokine/intercrine cytokines[4]. I-309 is a monocyte chemoattractant[5]. It may play a role in Th2 lymphocyte recirculation *in vivo*[6].

Crossreactivity

There is about 42% homology between human I-309 and mouse TCA-3. I-309 is active on murine cells.

Sources

T lymphocytes, monocytes and mast cells[7].

Bioassays

Monocyte chemoattraction.

Physicochemical properties of I-309/TCA-3

Property	Human (I-309)	Mouse (TCA-3)
Amino acids – precursor	96	92
– mature	73	69
M_r (K) – predicted	8	8
– expressed	8–16	16
Potential *N*-linked glycosylation sites[a]	1	1
Disulfide bonds[b]	3	3

[a] The TCA-3 splice variant P500 lacks the *N*-linked glycosylation site in TCA-3.
[b] Putative.

3-D structure

Could be modelled on MCP-1.

Gene structure[8,9]

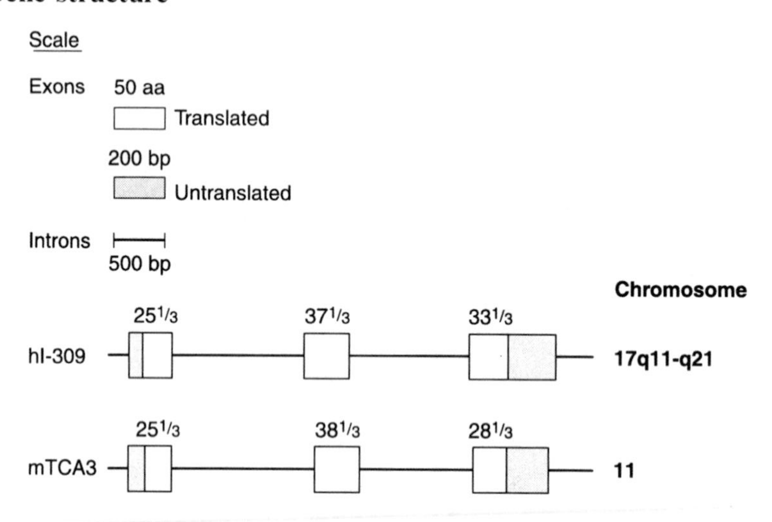

The gene for human I-309 is on chromosome 17q11–q21 and contains three exons of $25^{1/3}$, $37^{1/3}$ and $33^{1/3}$ amino acids. Murine TCA-3 is located on chromosome 11[9], and contains three exons of $25^{1/3}$, $38^{1/3}$ and $28^{1/3}$ amino acids.

Amino acid sequence for human I-309[9,10]

Accession code: SwissProt P22362

```
-23  MQIITTALVC LLLAGMWPED VDS
  1  KSMQVPFSRC CFSFAEQEIP LRAILCYRNT SSICSNEGLI FKLKRGKEAC
 51  ALDTVGWVQR HRKMLRHCPS KRK
```

Amino acid sequence for mouse TCA-3[3]

Accession code: SwissProt P10146

```
-23  MKPTAMALMC LLLAAVWIQD VDS
  1  KSMLTVSNSC CLNTLKKELP LKFIQCYRKM GSSCPDPPAV VFRLNKGRES
 51  CASTNKTWVQ NHLKKVNPC
```

In the TCA-3 splice variant P500, amino acids 42–69 are replaced with the sequence shown below[1]:

Accession code: SwissProt P14098

```
42   VRSSGVPGL TEAEKTVTDS SE
```

THE I-309/TCA-3 RECEPTOR, CCR8

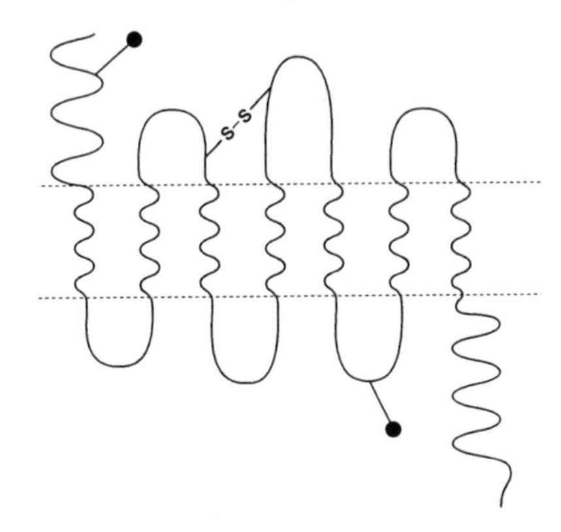

CCR8 (also known as TER-1, ChemR1, GPRCY6 or CKRL1) is expressed on T cells and NK cells, and also binds to TARC and MIP-1β[11–15]. CCR8 is also a receptor for viral chemokine antagonists including the Kaposi-associated herpesvirus (HHV-8) proteins viral MIP-1 and viral MIP-II[16] and the molluscum contagiosum poxvirus protein MC148[17]. See the TARC entry (page **457**).

Distribution

CCR8 is expressed on some Th2 cells and NK cell lines, weakly on IL-2-activated peripheral blood leukocytes, and in thymus and spleen[11–15].

Physicochemical properties of the CCR8 receptor

Property	Human	Mouse
Amino acids – precursor	355	353
– mature		
M_r (K) – predicted	40.8	40.6
– expressed	?	?
Potential *N*-linked glycosylation sites	1	2
Affinity K_d (M)	2×10^{-9}	2×10^{-9}

Signal transduction

I-309 causes a pertussis toxin-sensitive calcium flux following binding to cells transfected with human or murine CCR8 receptors[12,18].

Chromosomal location

The human CCR8 gene is on chromosome 3p21–24[5].

Amino acid sequence for human CCR8 receptor

Accession code: SwissProt P51685

```
  1  MDYTLDLSVT TVTDYYYPDI FSSPCDAELI QTNGKLLLAV FYCLLFVFSL
 51  LGNSLVILVL VVCKKLRSIT DVYLLNLALS DLLFVFSFPF QTYYLLDQWV
101  FGTVMCKVVS GFYYIGFYSS MFFITLMSVD RYLAVVHAVY ALKVRTIRMG
151  TTLCLAVWLT AIMATIPLLV FYQVASEDGV LQCYSFYNQQ TLKWKIFTNF
201  KMNILGLLIP FTIFMFCYIK ILHQLKRCQN HNKTKAIRLV LIVVIASLLF
251  WVPFNVVLFL TSLHSMHILD GCSISQQLTY ATHVTEIISF THCCVNPVIY
301  AFVGEKFKKH LSEIFQKSCS QIFNYLGRQM PRESCEKSSS CQQHSSRSSS
351  VDYIL
```

Disulfide bond links Cys106–183.

Amino acid sequence for mouse CCR8 receptor

Accession code: SwissProt P56484

```
  1  MDYTMEPNVT MTDYYPDFFT APCDAEFLLR GSMLYLAILY CVLFVLGLLG
 51  NSLVILVLVG CKKLRSITDI YLLNLAASDL LFVLSIPFQT HNLLDQWVFG
101  TAMCKVVSGL YYIGFFSSMF FITLMSVDRY LAIVHAVYAI KVRTASVGTA
151  LSLTVWLAAV TATIPLMVFY QVASEDGMLQ CFQFYEEQSL RWKLFTHFEI
201  NALGLLLPFA ILLFCYVRIL QQLRGCLNHN RTRAIKLVLT VVIVSLLFWV
251  PFNVALFLTS LHDLHILDGC ATRQRLALAI HVTEVISFTH CCVNPVIYAF
301  IGEKFKKHLM DVFQKSCSHI FLYLGRQMPV GALERQLSSN QRSSHSSTLD
351  DIL
```

Disulfide bond links Cys104–181.

References

[1] Brown, K.D. et al. (1989) J. Immunol. 142, 679–687.
[2] Wilson, S.D. et al. (1990) J. Immunol. 145, 2745–2744.
[3] Burd, P.R. et al. (1987) J. Immunol. 139, 3126–3131.
[4] Schall, T.J. (1991) Cytokine 3, 165–183.
[5] Miller, M.D. et al. (1992) Proc. Natl Acad. Sci. USA 89, 2950–2954.
[6] D Ambrosio, D. et al. (1988) J. Immunol. 161, 5111–5115.
[7] Burd, P.R. et al. (1989) J. Exp. Med. 170, 245–257.
[8] Wilson, S.D. et al. (1990) J. Exp. Med. 171, 1301–1314.
[9] Miller, M.D. et al. (1990) J. Immunol. 145, 2737–2744.
[10] Miller, M.D. et al. (1989) J. Immunol. 143, 2907–2916.
[11] Napolitano, M. et al. (1996) J. Immunol. 157, 2759–2763.
[12] Tiffany, H.L. et al. (1997) J. Exp. Med. 186, 165–170.
[13] Bernardini, G. et al. (1998) Eur. J. Immunol. 28, 582–588.
[14] Stuber-Roos, R. et al. (1997) J. Biol. Chem. 272, 17251–17254.
[15] Zingoni, A. et al. (1998) J. Immunol. 161, 547–551.
[16] Napolitano, M. and Santoni, A. (1999) Forum 9, 315–324.
[17] Luttichau, H.R. et al. (2000) J. Exp. Med. 191, 171–180.
[18] Goya, I. et al. (1998) J. Immunol. 160, 1975–1981.

Other names

IGF IA is also known as somatamedin C. IGF II is also known as somatamedin A and multiplication stimulating peptide.

THE MOLECULES

The insulin-like growth factors (IGFs) were first described as blood-borne factors that increased incorporation of sulfate into cartilage. They are now known to be pleiotropic cytokines that stimulate proliferation and survival of many cell types with important functions in development and growth[1,2]. They are synthesized throughout the body by most if not all tissues with a characteristic ontogeny of expression suggesting multiple and specific functions for the IGFs at defined stages of development. Both IGF I and IGF II can be detected readily in blood complexed to one or more of six different IGF binding proteins (IGFBP) that elevate blood levels, prolong their half-life and control their availability for receptor binding[3,4]. IGF may also promote survival of erythroid precursors and lymphocytes.

IGF I consists of A and B domains (homologous to the A and B chains of insulin) connected by a C peptide and an eight-amino-acid extension at the C-terminus termed the D domain. The E peptide cleaved from the propetide has an unknown function. The 70-amino-acid mature active IGF I protein is highly conserved in vertebrates and has 60% amino acid homology with IGF II and 40% amino acid homology with insulin. Specific binding of IGF I to its receptor involves the B domain and C peptide. The B domain is also important for binding to IGF-binding proteins[4].

IgF-II also contains A, B, C, and D domains. The A domain is critical for interactions with the IGF II receptor. The B domain is involved with interactions with IGF-binding proteins.

Crossreactivity

Not described.

Sources

Primary source is the liver but IGFs are produced by many different tissues at different times.

Bioassays

Most cultured cells proliferate in response to IGF, including fibroblasts, keratinocytes, muscles and neurons. Any responsive cell line can be used as a basis of a bioassay but a blocking antibody should be used to identify IGF as the growth factor. 3T3 fibroblasts are commonly used.

Physicochemical properties of IGF I and II

Property	Human			Mouse		
	IGF IA	IGF IB	IGF II	IGF IA	IGF IB	IGF II
pI (calculated)	7.8	7.8	6.5	8.3	8.3	5.4
Amino acids – precursor	153	195	180	127	133	180
– mature[a]	70[b,c]	70[b,c]	67	70[b,c]	70[b,c]	67
M_r (K) – predicted[d]	7.7	7.7	7.5	7.7	7.7	7.4
– expressed[e]	7.5	7.5	7.5	7.5	7.5	7.5
Potential *N*-linked glycosylation sites	1[f]	0	0[g]	2[f]	0	0
Disulfide bonds[h]	3	3	3	3	3	3

[a] After removal of signal peptide and N- and C-terminal propeptides.
[b] Removal of N- and C-terminal propeptides (see sequence) from IGF IA and IGF IB yields identical active IGF I[5,6].
[c] IGF IB and IGF IB prepropeptides formed by alternative splicing.
[d] For mature active IGF.
[e] Larger forms may be found in blood plasma with propeptides not removed
[f] On E peptide.
[g] IGF II is glycosylated on Thr75.
[h] Disulfide bonds given under sequence.

3-D structure

IGF I, IGF II and insulin have similar 3-D structures consisting of an α-helix followed by a turn and a strand in the B domain and two antiparallel α-helices in the A domain[7,8].

Gene structure[5]

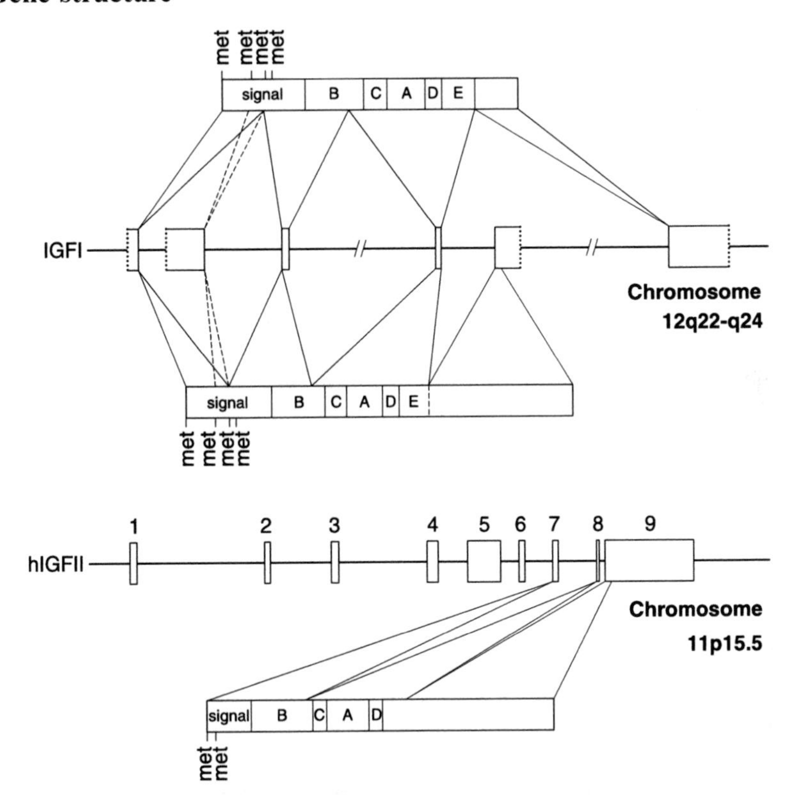

IGF IA and IGF IB are on human chromosome 12q22–q24.1 and mouse chromosome 10. The gene is complex and gives rise to multiple mRNA species from 1 to 7.5 kb that vary mostly at 3′ untranslated end. IGF II is on human chromosome 11p15.5 and mouse chromosome 7. The gene is also complex, giving rise to mRNA species of different lengths as a result of alternative exon use. The IGF II gene is paternally imprinted (in mice).

Amino acid sequence for human IGF IA[5]

Accession code: SwissProt P01343

```
-21  MGKISSLPTQ LFKCCFCDFL K
  1  VKMHTMSSSH LFYLALCLLT FTSSATAGPE TLCGAELVDA LQFVCGDRGF
 51  YFNKPTGYGS SSRRAPQTGI VDECCFRSCD LRRLEMYCAP LKPAKSARSV
101  RAQRHTDMPK TQKEVHLKNA SRGSAGNKNY RM
```

Propeptides in italics (1–27 and E peptide 98–132) are removed to give active IGF IA (28–97). B domain 28–56, C domain 57–68, A domain 69–89, D domain 90–97.

Disulfide bonds between Cys33–75, 45–88 and 74–79.

Amino acid sequence for human IGF IB[5,9]

Accession code: SwissProt P05019

```
-21   MGKISSLPTQ LFKCCFCDFL K
  1   VKMHTMSSSH LFYLALCLLT FTSSATAGPE TLCGAELVDA LQFVCGDRGF
 51   YFNKPTGYGS SSRRAPQTGI VDECCFRSCD LRRLEMYCAP LKPAKSARSV
101   RAQRHTDMPK TQKYQPPSTN KNTKSQRRKG WPKTHPGGEQ KEGTEASLQI
151   RGKKKEQRRE IGSRNAECRG KKGK
```

Propetides in italics (1–27 and E peptide 98–174)are removed to give active IGF IB (49–97). B domain 28–56, C domain 57–68, A domain 69–89, D domain 90–97.

Disulfide bonds between Cys33–75, 45–88 and 74–79

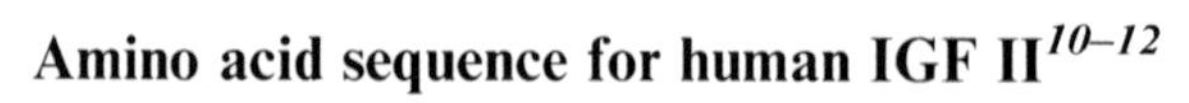

Amino acid sequence for human IGF II[10–12]

Accession code: SwissProt P01344

```
-24   MGIPMGKSML VLLTFLAFAS CCIA
  1   AYRPSETLCG GELVDTLQFV CGDRGFYFSR PASRVSRRSR GIVEECCFRS
 51   CDLALLETYC ATPAKSERDV STPPTVLPDN FPRYPVGKFF QYDTWKQSTQ
101   RLRRGLPALL RARRGHVLAK ELEAFREAKR HRPLIALPTQ DPAHGGAPPE
151   MASNRK
```

B domain 1–28, C domain 29–40, A domain 41–61, D domain 62–67. E peptide (in italics) 68–156. Disulfide bonds between Cys9–47, 21–60 and 46–51. Conflicting sequence I→M position −22, S→RLPG position 2953, RYPV→EIPL position 83–86, E→ELE position 123. Glycosylation on Thr75.

Amino acid sequence for mouse IGF IA[6]

Accession code: SwissProt P05017

```
-22   MSSSHLFYLA LCLLTFTSST TA
  1   GPETLCGAEL VDALQFVCGP RGFYFNKPTG YGSSIRRAPQ TGIVDECCFR
 51   SCDLRRLEMY CAPLKPTKAA RSIRAQRHTD MPKTQKEVHL KNTSRGSAGN
101   KTYRM
```

E propeptide 71–105 in italics is removed to give active IGF IA (1–70). B domain 1–29, C domain 30–41, A domain 42–62, D domain 63–70. Disulfide bonds between Cys6–48, 18–61 and 47–52.

Amino acid sequence for mouse IGF IB[6]

Accession code: SwissProt P05018

```
-22   MSSSHLFYLA LCLLTFTSST TA
  1   GPETLCGAEL VDALQFVCGP RGFYFNKPTG YGSSIRRAPQ TGIVDECCFR
 51   SCDLRRLEMY CAPLKPTKAA RSIRAQRHTD MPKTQKSPSL STNKKTKLQR
101   RRKGSTFEEH K
```

Sequence for mouse IGF IB is identical up to amino acid 76 where there is a 52-bp insertion in IGF IB due to alternative splicing. E propeptide 71–111 in italics is removed to give active IGF IB (1–70). B domain 1–29, C domain 30–41, A domain 42–62, D domain 63–70. Disulfide bonds between Cys6–48, 18–61 and 47–52.

Amino acid sequence for mouse IGF II[13,14]

Accession code: SwissProt P09535

```
-24  MGIPVGKSML VLLISLAFAL CCIA
  1  AYGPGETLCG GELVDTLQFV CSDRGFYFSR PSSRANRRSR GIVEECCFRS
 51  CDLALLETYC ATPAKSERDV STSQAVLPDD FPRYPVGKFF QYDTWRQSAG
101  RLRRGLPALL RARRGRMLAK ELKEFREAKR HRPLIVLPPK DPAHGGASSE
151  MSSNHQ
```

E propeptide 68–156 in italics is removed to give active IGF II (1–70). B domain 1–28, C domain 29–40, A domain 41–61, D domain 62–67. Disulfide bonds between Cys9–47, 21–60 and 46–51.

THE IGF RECEPTORS

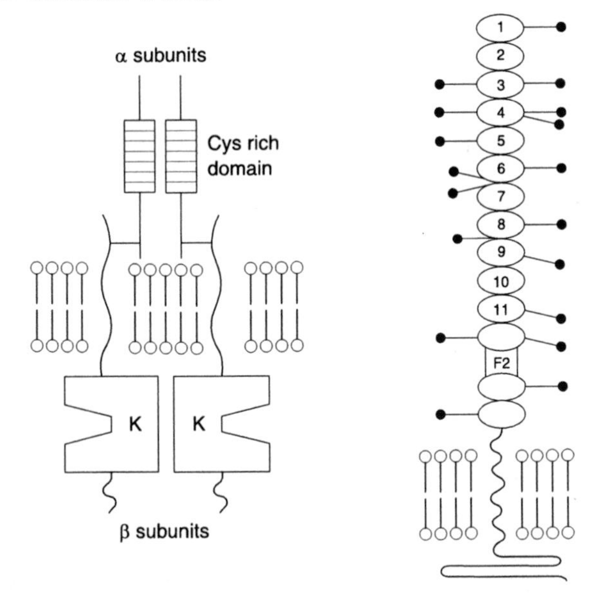

There are two known IGF receptors[1]. The type 1 receptor (IGF-IR) is a receptor tyrosine kinase and is responsible for signalling the mitogenic and probably all growth-promoting effects of both IGF I and IGF II[15]. It is a heterotetramer consisting of paired disulfide-linked α- and β-subunits that are derived from the same peptide precursor in much the same way as the HGF receptor. The α-subunits are entirely extracellular and bind IGF I > IGF II > > insulin. The β-subunit spans the membrane and includes a tyrosine kinase domain and three tyrosine residues at the C-terminus that are important for signalling. Disruption of the IGF-IR gene results in profound growth retardation during development. A receptor with structural homology to the IGF-IR called the insulin receptor-related receptor does not bind either insulin or IGF. The type II IGF receptor is identical to the mannose-6-phosphate receptor (IGF-IIR/MPR). Proteins containing mannose-6-phosphate and IGF-II are translocated by this receptor for degradation by lysosomes. The extracellular domain of the receptor has 15 repeating units of approximately 147 amino acids. The most highly conserved region within these repeating units consists of 13 amino acids with Cys residues at each end. It probably does not signal mitogenic or growth-promoting functions of the IGFs. Disruption of the gene for this receptor does not result in growth retardation but rather results in some fetal and placental overgrowth and cardiac hypertrophy, suggesting that this receptor is important for removing IGF.

Distribution

The IGF receptors are widely expressed in most cell types but their expression is regulated so that they are invariably expressed either by IGF-secreting cells or nearby cells. allowing localized autocrine and/or paracrine action.

Physicochemical properties of the IGF receptors

	IGF-IR	IGF-IIR	
Property	Human	Human	Mouse
Amino acids – precursor	1367	2491	2483
– mature[a,b]	1337 (710 and 627)[c]	2451	2448
M_r (K) – predicted[a,b]	151.9 (81 and 70.9)[c]	270	270
– expressed	(135 and 90)[d]		
Potential *N*-linked glycosylation sites	16	18	23
Affinity K_d (M)			
IGF I	$2–10 \times 10^{-10}$	$< 10^{-8}$	
IGF II	$0.2–5 \times 10^{-10}$	$0.2–7 \times 10^{-10}$	
Insulin	10^{-7}-10^{-6}	0	

[a] After removal of predicted signal peptide.
[b] The α- and β-chains of the receptor are obtained by proteolytic cleavage of the propeptide.
[c] α- and β-chains.

Signal transduction

The IGF-IR is a receptor tyrosine kinase[16]. Binding of IGFs to the receptor induces a conformational change and activation of the kinase with autophosphorylation of the β-subunit and activation of IRS-1 and IRS-2. Phosphorylation of Shc, PI-3 kinase and CRK also occurs. IRS-1/2 and Shc activate the Grb2/SOS pathway, leading to activation of Ras and the MAP kinase cascade. The IGF-IIR appears not to signal.

Chromosomal location

The IGF-IR is on human chromosome 15q25–q26 and mouse chromosome 7 and the IGF-IIR is on human chromosome 6q26 and mouse chromosome 17.

Amino acid sequence for human IGF-IR receptor[17,18]

Accession code: SwissProt P08069

```
 -30 MKSGSGGGSP TSLWGLLFLS AALSLWPTSG
   1 EICGPGIDIR NDYQQLKRLE NCTVIEGYLH ILLISKAEDY RSYRFPKLTV
  51 ITEYLLLFRV AGLESLGDLF PNLTVIRGWK LFYNYALVIF EMTNLKDIGL
 101 YNLRNITRGA IRIEKNADLC YLSTVDWSLI LDAVSNNYIV GNKPPKECGD
 151 LCPGTMEEKP MCEKTTINNE YNYRCWTTNR CQKMCPSTCG KRACTENNEC
 201 CHPECLGSCS APDNDTACVA CRHYYYAGVC VPACPPNTYR FEGWRCVDRD
 251 FCANILSAES SDSEGFVIHD GECMQECPSG FIRNGSQSMY CIPCEGPCPK
 301 VCEEEKKTKT IDSVTSAQML QGCTIFKGNL LINIRRGNNI ASELENFMGL
 351 IEVVTGYVKI RHSHALVSLS FLKNLRLILG EEQLEGNYSF YVLDNQNLQQ
 401 LWDWDHRNLT IKAGKMYFAF NPKLCVSEIY RMEEVTGTKG RQSKGDINTR
 451 NNGERASCES DVLHFTSTTT SKNRIIITWH RYRPPDYRDL ISFTVYYKEA
 501 PFKNVTEYDG QDACGSNSWN MVDVDLPPNK DVEPGILLHG LKPWTQYAVY
 551 VKAVTLTMVE NDHIRGAKSE ILYIRTNASV PSIPLDVLSA SNSSSQLIVK
 601 WNPPSLPNGN LSYYIVRWQR QPQDGYLYRH NYCSKDKIPI RKYADGTIDI
 651 EEVTENPKTE VCGGEKGPCC ACPKTEAEKQ AEKEEAEYRK VFENFLHNSI
 701 FVPRPERKRR DVMQVANTTM SSRSRNTTAA DTYNITDPEE LETEYPFFES
 751 RVDNKERTVI SNLRPFTLYR IDIHSCNHEA EKLGCSASNF VFARTMPAEG
 801 ADDIPGPVTW EPRPENSIFL KWPEPENPNG LILMYEIKYG SQVEDQRECV
 851 SRQEYRKYGG AKLNRLNPGN YTARIQATSL SGNGSWTDPV FFYVQAKTGY
 901 ENFIHLIIAL PVAVLLIVGG LVIMLYVFHR KRNNSRLGNG VLYASVNPEY
 951 FSAADVYVPD EWEVAREKIT MSRELGQGSF GMVYEGVAKG VVKDEPETRV
1001 AIKTVNEAAS MRERIEFLNE ASVMKEFNCH HVVRLLGVVS QGQPTLVIME
1051 LMTRGDLKSY LRSLRPEMEN NPVLAPPSLS KMIQMAGEIA DGMAYLNANK
1101 FVHRDLAARN CMVAEDFTVK IGDFGMTRDI YETDYYRKGG KGLLPVRWMS
1151 PESLKDGVFT TYSDVWSFGV VLWEIATLAE QPYQGLSNEQ VLRFVMEGGL
1201 LDKPDNCPDM LFELMRMCWQ YNPKMRPSFL EIISSIKEEM EPGFREVSFY
1251 YSEENKLPEP EELDLEPENM ESVPLDPSAS SSSLPLPDRH SGHKAENGPG
1301 PGVLVLRASF DERQPYAHMN GGRKNERALP LPQSSTC
```

α-chain amino acids 1–710 and β-chain 711–1337.

Amino acid sequence for human IGF-IIR receptor[19]

Accession code: SwissProt Q60751

(incomplete). The complete mouse IGF-IR sequence has not been reported.

Amino acid sequence for human IGF-IIR receptor[19]

Accession code: SwissProt P11717

```
 -40 MGAAAGRSPH LGPAPARRPQ RSLLLLQLLL LVAAPGSTQA
   1 QAAPFPELCS YTWEAVDTKN NVLYKINICG SVDIVQCGPS SAVCMHDLKT
  51 RTYHSVGDSV LRSATRSLLE FNTTVSCDQQ GTNHRVQSSI AFLCGKTLGT
 101 PEFVTATECV HYFEWRTTAA CKKDIFKANK EVPCYVFDEE LRKHDLNPLI
 151 KLSGAYLVDD SDPDTSLFIN VCRDIDTLRD PGSQLRACPP GTAACLVRGH
```

```
 201 QAFDVGQPRD GLKLVRKDRL VLSYVREEAG KLDFCDGHSP AVTITFVCPS
 251 ERREGTIPKL TAKSNCRYEI EWITEYACHR DYLESKTCSL SGEQQDVSID
 301 LTPLAQSGGS SYISDGKEYL FYLNVCGETE IQFCNKKQAA VCQVKKSDTS
 351 QVKAAGRYHN QTLRYSDGDL TLIYFGGDEC SSGFQRMSVI NFECNKTAGN
 401 DGKGTPVFTG EVDCTYFFTW DTEYACVKEK EDLLCGATDG KKRYDLSALV
 451 RHAEPEQNWE AVDGSQTETE KKHFFINICH RVLQEGKARG CPEDAAVCAV
 501 DKNGSKNLGK FISSPMKEKG NIQLSYSDGD DCGHGKKIKT NITLVCKPGD
 551 LESAPVLRTS GEGGCFYEFE WRTAAACVLS KTEGENCTVF DSQAGFSFDL
 601 SPLTKKNGAY KVETKKYDFY INVCGPVSVS PCQPDSGACQ VAKSDEKTWN
 651 LGLSNAKLSY YDGMIQLNYR GGTPYNNERH TPRATLITFL CDRDAGVGFP
 701 EYQEEDNSTY NFRWYTSYAC PEEPLECVVT DPSTLEQYDL SSLAKSEGGL
 751 GGNWYAMDNS GEHVTWRKYY INVCRPLNPV PGCNRYASAC QMKYEKDQGS
 801 FTEVVSISNL GMAKTGPVVE DSGSLLLEYV NGSACTTSDG RQTTYTTRIH
 851 LVCSRGRLNS HPIFSLNWEC VVSFLWNTEA ACPIQTTTDT DQACSIRDPN
 901 SGFVFNLNPL NSSQGYNVSG IGKIFMFNVC GTMPVCGTIL GKPASGCEAE
 951 TQTEELKNWK PARPVGIEKS LQLSTEGFIT LTYKGPLSAK GTADAFIVRF
1001 VCNDDVYSGP LKFLHQDIDS GQGIRNTYFE FETALACVPS PVDCQVTDLA
1051 GNEYDLTGLS TVRKPWTAVD TSVDGRKRTF YLSVCNPLPY IPGCQGSAVG
1101 SCLVSEGNSW NLGVVQMSPQ AAANGSLSIM YVNGDKCGNQ RFSTRITFEC
1151 AQISGSPAFQ LQDGCEYVFI WRTVEACPVV RVEGDNCEVK DPRHGNLYDL
1201 KPLGLNDTIV SAGEYTYYFR VCGKLSSDVC PTSDKSKVVS SCQEKREPQG
1251 FHKVAGLLTQ KLTYENGLLK MNFTGGDTCH KVYQRSTAIF FYCDRGTQRP
1301 VFLKETSDCS YLFEWRTQYA CPPFDLTECS FKDGAGNSFD LSSLSRYSDN
1351 WEAITGTGDP EHYLINVCKS LAPQAGTEPC PPEAAACLLG GSKPVNLGRV
1401 RDGPQWRDGI IVLKYVDGDL CPDGIRKKST TIRFTCSESQ VNSRPMFISA
1451 VEDCEYTFAW PTATACPMKS NEHDDCQVTN PSTGHLFDLS SLSGRAGFTA
1501 AYSEKGLVYM SICGENENCP PGVGACFGQT RISVGKANKR LRYVDQVLQL
1551 VYKDGSPCPS KSGLSYKSVI SFVCRPEAGP TNRPMLISLD KQTCTLFFSW
1601 HTPLACEQAT ECSVRNGSSI VDLSPLIHRT GGYEAYDESE DDASDTNPDF
1651 YINICQPLNP MHAVPCPAGA AVCKVPIDGP PIDIGRVAGP PILNPIANEI
1701 YLNFESSTPC LADKHFNYTS LIAFHCKRGV SMGTPKLLRT SECDFVFEWE
1751 TPVVCPDEVR MDGCTLTDEQ LLYSFNLSSL STSTFKVTRD SRTYSVGVCT
1801 FAVGPEQGGC KDGGVCLLSG TKGASFGRLQ SMKLDYRHQD EAVVLSYVNG
1851 DRCPPETDDG VPCVFPFIFN GKSYEECIIE SRAKLWCSTT ADYDRDHEWG
1901 FCRHSNSYRT SSIIFKCDED EDIGRPQVFS EVRGCDVTFE WKTKVVCPPK
1951 KLECKFVQKH KTYDLRLLSS LTGSWSLVHN GVSYYINLCQ KIYKGPLGCS
2001 ERASICRRTT TGDVQVLGLV HTQKLGVIGD KVVVTYSKGY PCGGNKTASS
2051 VIELTCTKTV GRPAFKRFDI DSCTYYFSWD SRAACAVKPQ EVQMVNGTIT
2101 NPINGKSFSL GDIYFKLFRA SGDMRTNGDN YLYEIQLSSI TSSRNPACSG
2151 ANICQVKPND QHFSRKVGTS DKTKYYLQDG DLDVVFASSS KCGKDKTKSV
2201 SSTIFFHCDP LVEDGIPEFS HETADCQYLF SWYTSAVCPL GVGFDSENPG
2251 DDGQMHKGLS ERSQAVGAVL SLLLVALTCC LLALLLYKKE RRETVISKLT
2301 TCCRRSSNVS YKYSKVNKEE ETDENETEWL MEEIQLPPPR QGKEGQENGH
2351 ITTKSVKALS SLHGDDQDSE DEVLTIPEVK VHSGRGAGAE SSHPVRNAQS
2401 NALQEREDDR VGLVRGEKAR KGKSSSAQQK TVSSTKLVSF HDDSDEDLLH
2451 I
```

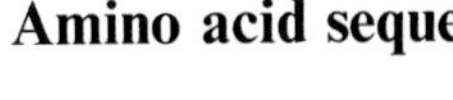

Amino acid sequence for mouse IGF-IIR receptor[20,21]

Accession code: SwissProt Q07113

-35 MRAVQLGPVP SGPRVALLPP LLLLLLLAAA GSAQA
1 QAVDLDALCS YTWEAVDSKN NAVYKINVCG NVGISSCGPT SAICMCDLKT
51 ENCRSVGDSL LRSSARSLLE F**N**TTMGCQPS DSQHRIQTSI TFLCGKTLGT
101 PEFVTATDCV HYFEWRTTAA CKKDIFKADK EVPCYAFDDK LQKHDLNPLI
151 KLNGGYLVDD SDPDTSLFIN VCRDIDSLRD PSTQLRVCPA GTAACLLKGN
201 QAFDVGRPKE GLKLLSKDRL VLTYVKEEGE KPDFCNGHSP AVTVTFVCPS
251 ERREGTIPKL TAKSNCRYEV EWITEYACHR DYLQSESCSL SSEQHDITID
301 LSPLAQYGGS PYVSDGREYT FFINVCGDTK VSLCNNKEAA VCQEKKADST
351 QVKIAGRHQ**N** QTLRYSDGDL TLIYSGGDEC SSGFQRMSVI NFEC**N**KTAGK
401 DGRGEPVFTG EVDCTYFFTW DTKYACIKEK EDLLCGAING KKRYDLSVLA
451 RHSESEQNWE AVDGSQAESE KYFFINVCHR VLQEGKARNC PEDAAVCAVD
501 K**N**GSKNLGKF VSSPTKEKGH IQLSYTDGDD CGSDKKIST**N** ITLVCKPGDL
551 ESAPVLRAAR SDGCFYEFEW HTAAACVLSK TEGE**N**CTVLD AQAGFSFDLS
601 LLTKKNGAYK VETEKYDFYI NVCGPVSMDP CQSNSGACQV AKSGKSWNLG
651 LSSTKLTYYD GMIQLSYR**N**G TPYNNEKHTP RATLITFLCD RDAGVGFPEY
701 QEED**N**STYNF RWYTSYACPE EPLECMVTDP SMMEQYDLSS LVKSEGGSGG
751 NWYAMENSRE HVTRRKYYLN VCRPLNPVPG CDRYASACQM KYENHEGSLA
801 ETVSISNLGV AKIGPVVEES GSLLLEYV**N**G SACTTSDGQL TTYSTRIHLV
851 CGRGFMNSHP IFTFNWECVV SFLWNTEAAC PIQTITETDQ ACSIRDPSSG
901 FVF**N**LSPL**N**D SAQGHVVLGI GKTFVFNICG AMPACGTVAG KPAYGCEAET
951 QIEDIKDLRP QRPVGMERSL QLSAEGFLTL TYKGSSPSDR GTAFIIRFIC
1001 NDDIYPGAPK FLHQDIDSTR GIRNTYFEFE TALACTPSLV DCQVTDPAGN
1051 EYDLSALSMV RKPWTAVDTS AYGKRRHFYL SVCNPLPYIP GCHGIALGSC
1101 MVSEDNSFNL GVVQISPQAT G**N**GSLSILYV NGDRCGDQRF STRIVFECAQ
1151 TSGSPMFQFV NNCEYVFVWR TVEACPVIRE EGDNCQVKDP RHGNLYDLKP
1201 LGL**N**DTIVSV GEYTYYLRVC GKLSSDVCSA HDGSKAVSSC QEKKGPQGFQ
1251 KVAGLLSQKL TFENGLLKM**N** YTGGDTCHKV YQRSTTIYFY CDRTTQKPVF
1301 LKETSDCSYM FEWRTQYACP PF**N**VTECSVQ DAAGNSIDLS SLSRYSDNWE
1351 AVTRTGATEH YLINVCKSLS PHAGTEPCPP EAAVCLL**N**GS KPVNLGKVRD
1401 GPQWTDGVTV LQYVDGDLCP DKIRRRSTII RFTCSDNQVN SRPLFISAVQ
1451 DCEYTFSWPT PSACPVKSNT HDDCQVTNPS TGHLFDLSSL SGRAGI**N**ASY
1501 SEKGLVFMSI CEENENCGPG VGACFGQTRI SVGKASKRLS YKDQVLQLVY
1551 E**N**GSPCPSLS DLRYKSVISF VCRPEAGPTN RPMLISLDKQ TCTLFFSWHT
1601 PLACEQATEC TVR**N**GSSIID LSPLIHRTGG YEAYDESEDD TSDTTPDFYI
1651 NICQPLNPMH GVPCPAGASV CKVPVDGPPI DIGRVTGPPI FNPVANEVYL
1701 NFESSTHCLA DRYM**N**YTSLI TFHCKRGVSM GTPKLIRTND CDFVFEWETP
1751 IVCPDEVKTQ GCAVTDEQLL YSF**N**LTSLST STFKVTRDAR TYSIGVCTAA
1801 AGLGQEGCKD GGVCLLSGNK GASFGRLASM QLDYRHQDEA VILSYVNGDP
1851 CPPETDDGEP CVFPFIYKGK SYDECVLEGR AKLWCSKTAN YDRDHEWGFC
1901 RQTNSYRMSA IIFTCDESED IGRPQVFSED RGCEVTFEWK TKVVCPPKKM
1951 ECKFVQKHKT YDLRLLSSLT GSWDFVHEGN SYFINLCQRV YKGPLDCSER
2001 ASICKKSATG QVQVLGLVHT QKLEVIDETV IVTYSKGYPC GG**N**KTASSVI
2051 ELTCAKTVGR PAFKRFDSVS CTYYFYWYSR AACAVRPQEV TMV**N**GTLTNP

```
2101 VTGKSFSLGE IYFKLFSASG DMRTNGDNYL YEIQLSSITS SSYPACAGAN
2151 ICQVKPNDQH FSRKVGTSDM TKYYVQDGDL DVVFTSSSKC GKDKTKSVSS
2201 TIFFHCDPLV KDGIPEFSHE TADCQYLFSW YTSAVCPLGV DFEDESAGPE
2251 YKGLSERSQA VGAVLSLLLV ALTGCLLALL LHKKERRETV INKLTSCCRR
2301 SSGVSYKYSK VSKEEETDEN ETEWLMEEIQ VPAPRLGKDG QENGHITTKA
2351 VKAEALSSLH GDDQDSEDEV LTVPEVKVHS GRGAEVESSQ PLRNPQRKVL
2401 KEREGERLGL VRGEKARKGK FRPGQRKPTA PAKLVSFHDD SDEDLLHI
```

References

[1] D'Ercole, A.J. (1996) Endocrin. Metab. Clin. North Am. 25, 573–590.
[2] Jones, J.I. and Clemmons, D.R. (1995) Endocr. Rev. 16, 3–34.
[3] Blakesley, V.A. et al. (1996) Cytokine Growth Factor Rev. 7, 153–159.
[4] Clemmons, D.R. (1997) Cytokine Growth Factor Rev. 8, 45–62.
[5] Rotwein, P. et al. (1986) J. Biol. Chem. 261, 4828–4832.
[6] Bell, G.I. et al. (1986) Nucleic Acids Res. 14, 7873–7882.
[7] Sato, A. et al. (1992) J. Biochem. 111, 529–536=.
[8] Terasawa, H. et al. (1994) EMBO J. 13, 5590–5597.
[9] Rotwein, P. (1986) Proc. Natl Acad. Sci. USA 83, 77–81.
[10] Dull, T.J. et al. (1984) Nature 310, 777–781.
[11] Bell, G.I. et al. (1984) Nature 310, 775–781.
[12] Shen, S.-J. et al. (1988) Proc. Natl Acad. Sci. USA 85, 1947–1951.
[13] Stempien, M.M. et al. (1986) DNA 5, 357–361.
[14] Sasaki, H. et al. (1996) DNA Res. 3, 331–335.
[15] Leroith, D. et al. (1995) Endocr. Rev. 16, 143–163.
[16] Leonard, E.J. et al. (1998) In Human Cytokines III, Aggarwal, B.B. ed., Blackwell, Oxford, pp. 235–265.
[17] Ullrich, A. et al. (1986) EMBO J. 5, 2503–2512.
[18] Abbott, A.M. et al. (1992) J. Biol. Chem. 267, 10759–10763.
[19] Morgan, D.O. et al. (1987) Nature 329, 301–307.
[20] Szebenyi, G. and Rotwein, P. (1994) Genomics 19, 120–129.
[21] Ludwig, T. et al. (1994) Gene 142, 311–312.

Other names

Type I interferon, leukocyte interferon, acid stable interferon, B cell interferon, lymphoblast interferon, Namalwa interferon, buffy coat interferon. IFNω is also known as IFNα (II).

THE MOLECULES

The α interferons including IFNω are a family of inducible secreted proteins, often produced in response to viral infection, which confer resistance to viruses on target cells, inhibit cell proliferation and regulate expression of MHC class I antigens[1–4]. IFNα and IFNβ bind to the same receptor and have similar biological properties. The IFNα family consists of 24 or more genes or pseudo-genes. There are two distinct families (IFNα and IFNω sometimes known as I and II) of human (and bovine) IFNα. Mature IFNα proteins are 166 amino acids long (one is 165 amino acids) whereas IFNω has 172 amino acids. The IFNα genes probably diverged 100 million years ago, prior to mammalian radiation and have about 30% homology with IFNβ. Various nomenclatures have been given to the IFNα proteins. The current accepted nomenclature with alternative names is given below.

Crossreactivity

There is about 40% homology between human and mouse IFNα subtypes and there is some species preference in their biological activities depending on the particular IFNα molecule[5].

Sources

Major producers of IFNα are lymphocytes, monocytes, macrophages and cell lines such as Namalwa and KG1.

Bioassays

Inhibition of cytopathic effect of EMCV (encephalomyocarditis virus), VSV (vesicular stomatitis virus) or SFV (Semliki forest virus) on Hep2/C (human epithelial) cell line for human IFN or L929 for mouse IFN. Inhibition of proliferation of Daudi human lymphoblastoid B cell line.

Physicochemical properties of IFNα and IFNω

Property	Human		Mouse
	IFNα	IFNω	IFNα
Number of isoforms	> 15	> 1	> 11
pI	5–6.5 for most subtypes		
Amino acids – precursor	188/189	195	189/190
– mature[a]	165/166	172	166/167
M_r (K) – predicted	19.2–19.7	20.1	19.1
– expressed[b]	16–27	?	16–27
Potential N-linked glycosylation sites[c]	0	1	1
Disulfide bonds[d]	2	2	2

[a] All human and murine IFNα have a signal peptide of 23 amino acids except for murine IFNα4, which has 24 amino acids.
[b] Apparent M_r on acrylamide gels. Some variation may be due to *O*-linked glycosylation.
[c] *N*-glycosylation site at position 78 is present on most murine IFNα sequences. The majority of human IFNα type I molecules are not glycosylated (IFNα14 has one potential *N*-linked site and at least three species have *O*-linked polysaccharides).
[d] Disulfide bonds between Cys1–99 and 29–139 are present in most human and murine IFNα, and are required for biological activity.

3-D structure

The 3-D structure for IFNα is similar to IFNβ. It consists of five helices, three of which are parallel to each other and the other two are antiparallel to the first three. The structure is a variant of the α-helix bundle, but with a new chain folding topology, common to all type I interferons[6].

Gene structure[7–9]

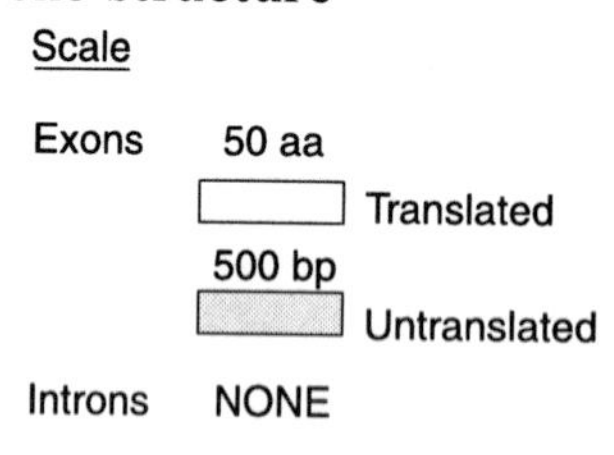

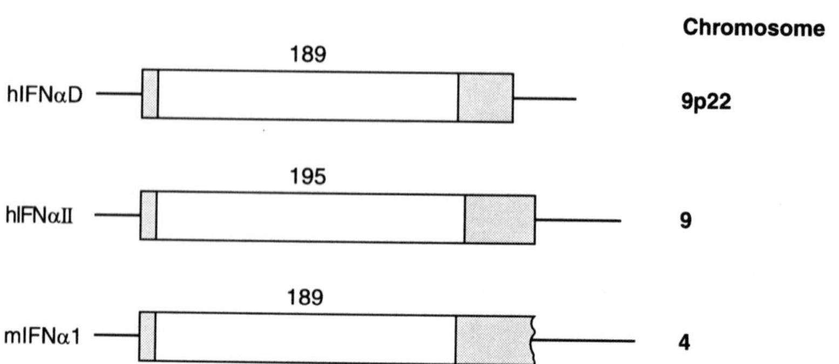

The human IFNαD gene is located on chromosome 9p22 and contains a single coding exon of 189 amino acids. The human IFNαIII gene is also on chromosome 9, contains a single coding exon of 195 amino acids. Mouse IFNa1 is located on chromosome 4, contains a single exon of 189 amino acids. The 3′ untranslated sequence of these genes is variable in length. It has not been determined for mouse IFNα.

More than 24 non-allelic genes or pseudo-genes for human IFNα/ω have been identified. Eighteen of these, including at least four pseudo-genes, are for IFNα, and six, of which at least five are pseudo-genes, are for IFNω also known as IFNα (II). The mature IFNα proteins are mostly 166 amino acids in length, although IFNα2 (IFNαA) has 165 amino acids. Mature IFNω has 172 amino acids. All have signal peptides of 23 amino acids. Sequences are given here for IFNα1 (IFNαD) as a typical IFNα, and for IFNω. For a comparison of human IFNα sequences see the table of SwissProt accession numbers below. An IFNα consensus sequence is given in ref. 10.

At least 12 nonallelic genes or pseudo-genes for mouse IFNα have been identified, 11 of which have been cloned and expressed as biologically active proteins. The mature proteins have 166 or 167 amino acids except for mouse IFNα4, which has a five codon deletion between codons 102 and 108. The example given here is of mouse IFNα1.

The names and SwissProt accession numbers of the human and mouse IFNα molecules are listed below:

IFNα (alternative name)	Primary accession number
Human	
IFNα1 (IFNαD, LEIFD)	P01562
IFNα2 (IFNαA, LEIFA)	P01563
IFNα4 (IFNα4B, IFNαM1, IFNα76)	P05014
IFNα5 (IFNαG, LEIFG, IFNα61)	P01569
IFNα6 (IFNαK, LEIFK, IFNα54)	P05013
IFNα7 (IFNαJ1, LEIFJ, IFNαJ)	P01567
IFNα8 (IFNαB2, LEIFB, IFNαB)	P32881
IFNα10 (IFNαC, LEIFC, IFNα6L)	P01566
IFNα14 (IFNαH, LEIFH, IFNα2H)	P01570
IFNα16 (IFNαWA)	P05015
IFNα17 (IFNα, IFNαT, IFNα88)	P01571
IFNα21 (IFNαF, LEIFF)	P01568
IFNω (IFNα II-1)	P05000
Mouse	
IFNα1	P01572
IFNα2	P01573
IFNα4	P07351
IFNα5	P07349
IFNα6	P07350
IFNα7	P06799
IFNα8	P17660
IFNα9	P09235

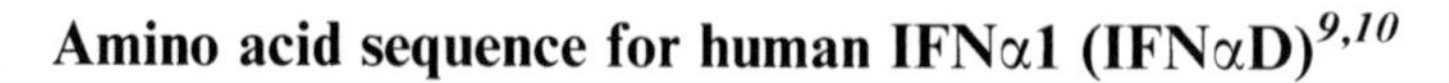

Amino acid sequence for human IFNα1 (IFNαD)[9,10]

Accession code: SwissProt P01562

```
-23   MASPFALLMV LVVLSCKSSC SLG
  1   CDLPETHSLD NRRTLMLLAQ MSRISPSSCL MDRHDFGFPQ EEFDGNQFQK
 51   APAISVLHEL IQQIFNLFTT KDSSAAWDED LLDKFCTELY QQLNDLEACV
101   MQEERVGETP LMNADSILAV KKYFRRITLY LTEKKYSPCA WEVVRAEIMR
151   SLSLSTNLQE RLRRKE
```

Conflicting sequence A→V at position 114[11]. Disulfide bonds between Cys1–99 and 29–139.

Amino acid sequence for human IFNω (IFNαII)[8]

Accession code: SwissProt P05000

```
-23   MALLFPLLAA LVMTSYSPVG SLG
  1   CDLPQNHGLL SRNTLVLLHQ MRRISPFLCL KDRRDFRFPQ EMVKGSQLQK
 51   AHVMSVLHEM LQQIFSLFHT ERSSAAWNMT LLDQLHTELH QQLQHLETCL
101   LQVVGEGESA GAISSPALTL RRYFQGIRVY LKEKKYSDCA WEVVRMEIMK
151   SLFLSTNMQE RLRSKDRDLG SS
```

Disulfide bonds between Cys1–99 and 29–139.

Amino acid sequence for mouse IFNα1[7]

Accession code: SwissProt P01572

```
-23   MARLCAFLMV LAVMSYWPTC SLG
  1   CDLPQTHNLR NKRALTLLVQ MRRLSPLSCL KDRKDFGFPQ EKVDAQQIKK
 51   AQAIPVLSEL TQQILNIFTS KDSSAAWNAT LLDSFCNDLH QQLNDLQGCL
101   MQQVGVQEFP LTQEDALLAV RKYFHRITVY LREKKHSPCA WEVVRAEVWR
151   ALSSSANVLG RLREEK
```

Disulfide bonds between Cys1–99 and 29–139.

THE IFNα RECEPTORS

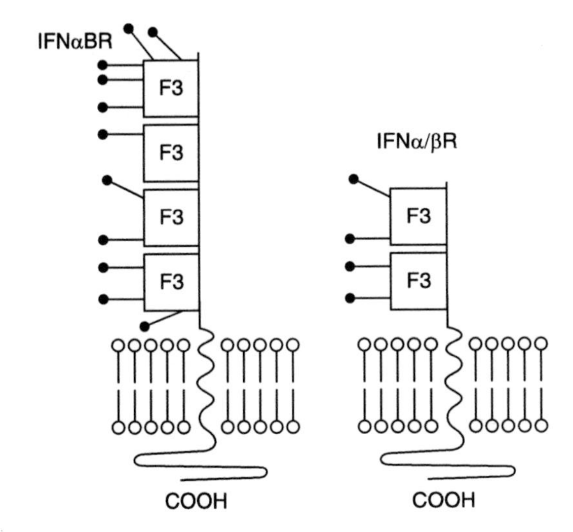

IFNα and IFNβ bind to a common receptor sometimes known as the IFNα/βR made up of two subunits (IFNAR-1 and IFNAR-2)[4,12]. Both are members of the class II cytokine receptor family which also includes the IFNγ receptor, IL-10 receptor and tissue factor[13–15]. The IFNAR-2 chain is also known as IFNRβ. It has three isoforms: two transmembrane forms and a soluble form produced by alternative splicing[12,16]. The receptor heterodimer appears to have two separate ligand-binding sites that are both necessary and sufficient to ensure that the different IFNs are recognized and can act selectively[12]. A soluble form of the IFNα/β receptor has been identified in human serum and urine[16,17]. A mouse IFNα receptor cDNA has also been cloned[18].

Distribution

The receptors are present on most cell types.

Physicochemical properties of the IFNα receptors

Property	Human		Mouse
	IFNAR-1	IFNAR-2	IFNAR-1
Amino acids – precursor	557	515[a]	590
– mature[b]	530	489 (239)[c]	564
M_r (K) – predicted	60.9	34.5	63.4
– expressed	95–110	95–100[d] (55)[e]	95–100
Potential *N*-linked glycosylation sites	11	5	8
Affinity K_d (M)[f]	10^{-9}–10^{-11}		

[a] Long form.
[b] After removal of predicted signal peptide.
[c] Soluble form in parenthesis.
[d] Probably dimer of two disulfide linked M_r 51 000 subunits.
[e] Short form in parenthesis.
[f] Receptor dimer.

Signal transduction

IFNAR-1 is constitutively associated with Tyk2 and IFNAR-2 with Jak1. The long and short forms of the IFNAR-2 subunit both interact with Jak1. The long form also activates the transcriptional factors STAT1 and STAT2. IFNAR-1 is also associated with SHP2 which is phosphorylated on binding of IFNα/β. IFNα binding to its receptor results in phosphorylation and activation of protein tyrosine kinases Tyk2 and Jak1 but not Jak2[19–21] which directly or indirectly tyrosine phosphorylate the cytoplasmic STAT (signal transducers and activators of transcription) proteins p84/91 and p113 which combine with p48 to form the transcription factor ISGF3 in the cytoplasm[22–27]. Activated ISGF3 moves to the nucleus and binds to the ISRE-responsive element on IFNα-responsive genes[28]. IFNα has also been shown to activate PLC-A2, releasing arachidonic acid and diacylglycerol (DAG)[29,30]. Activation of PKC (but not PKCα/β) in the absence of phospholipid turnover or elevation of intracellular calcium has also been described[30].

Chromosomal location

Both human IFNAR-1 and IFNAR-2 are located on chromosome 21q22.1. The mouse IFNAR receptor is on chromosome 16.

Amino acid sequence for human IFNAR-1[31]

Accession code: SwissProt P17181

```
-27  MMVVLLGATT LVLVAVGPWV LSAAAGG
  1  KNLKSPQKVE VDIIDDNFIL RWNRSDESVG NVTFSFDYQK TGMDNWIKLS
 51  GCQNITSTKC NFSSLKLNVY EEIKLRIRAE KENTSSWYEV DSFTPFRKAQ
101  IGPPEVHLEA EDKAIVIHIS PGTKDSVMWA LDGLSFTYSL LIWKNSSGVE
151  ERIENIYSRH KIYKLSPETT YCLKVKAALL TSWKIGVYSP VHCIKTTVEN
201  ELPPPENIEV SVQNQNYVLK WDYTYANMTF QVQWLHAFLK RNPGNHLYKW
251  KQIPDCENVK TTQCVFPQNV FQKGIYLLRV QASDGNNTSF WSEEIKFDTE
301  IQAFLLPPVF NIRSLSDSFH IYIGAPKQSG NTPVIQDYPL IYEIIFWENT
```

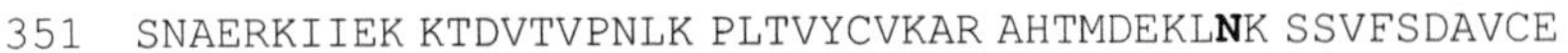

```
351  SNAERKIIEK KTDVTVPNLK PLTVYCVKAR AHTMDEKLNK SSVFSDAVCE
401  KTKPGNTSKI WLIVGICIAL FALPFVIYAA KVFLRCINYV FFPSLKPSSS
451  IDEYFSEQPL KNLLLSTSEE QIEKCFIIEN ISTIATVEET NQTDEDHKKY
501  SSQTSQDSGN YSNEDESESK TSEELQQDFV
```

The four pairs of Cys residues at positions 52–60; 172–193; 256–264 and 376–399 are conserved in the class II cytokine receptor family.

Amino acid sequence for human IFNAR-2[16,32–34]

Accession code: SwissProt P48551

```
-26  MLLSQNAFIF RSLNLVLMVY ISLVFG
  1  ISYDSPDYTD ESCTFKISLR NFRSILSWEL KNHSIVPTHY TLLYTIMSKP
 51  EDLKVVKNCA NTTRSFCDLT DEWRSTHEAY VTVLEGFSGN TTLFSCSHNF
101  WLAIDMSFEP PEFEIVGFTN HINVMVKFPS IVEEELQFDL SLVIEEQSEG
151  IVKKHKPEIK GNMSGNFTYI IDKLIPNTNY CVSVYLEHSD EQAVIKSPLK
201  CTLLPPGQES ESAESAKIGG IITVFLIALV LTSTIVTLKW IGYICLRNSL
251  PKVLNFHNFL AWPFPNLPPL EAMDMVEVIY INRKKKVWDY NYDDESDSDT
301  EAAPRTSGGG YTMHGLTVRP LGQASATSTE SQLIDPESEE EPDLPEVDVE
351  LPTMPKDSPQ QLELLSGPCE RRKSPLQDPF PEEDYSSTEG SGGRITFNVD
401  LNSVFLRVLD DEDSDDLEAP LMLSSHLEEM VDPEDPDNVQ SNHLLASGEG
451  TQPTFPSPSS EGLWSEDAPS DQSDTSESDV DLGDGYIMR
```

An alternative 4.5-kb mRNA transcript gives rise to a truncated soluble form. The pairs of Cys residues at positions 59–67 and 181–201 form disulfide bonds and are conserved in the class II cytokine receptor family.

Amino acid sequence for mouse IFNα (IFNAR-1) receptor[18]

Accession code: SwissProt P33896

```
-26  MLAVVGAAAL VLVAGAPWVL PSAAGG
  1  ENLKPPENID VYIIDDNYTL KWSSHGESMG SVTFSAEYRT KDEAKWLKVP
 51  ECQHTTTTKC EFSLLDTNVY IKTQFRVRAE EGNSTSSWNE VDPFIPFYTA
101  HMSPPEVRLE AEDKAILVHI SPPGQDGNMW ALEKPSFSYT IRIWQKSSSD
151  KKTINSTYYV EKIPELLPET TYCLEVKAIH PSLKKHSNYS TVQCISTTVA
201  NKMPVPGNLQ VDAQGKSYVL KWDYIASADV LFRAQWLPGY SKSSSGSHSD
251  KWKPIPTCAN VQTTHCVFSQ DTVYTGTFFL HVQASEGNHT SFWSEEKFID
301  SQKHILPPPP VITVTAMSDT LLVYVNCQDS TCDGLNYEII FWENTSNTKI
351  SMEKDGPEFT LKNLQPLTVY CVQARVLFRA LLNKTSNFSE KLCEKTRPGS
401  FSTIWIITGL GVVFFSVMVL YALRSVWKYL CHVCFPPLKP PRSIDEFFSE
451  PPSKNLVLLT AEEHTERCFI IENTDTVAVE VKHAPEEDLR KYSSQTSQDS
501  GNYSNEEEES VGTESGQAVL SKAPCGGPCS VPSPPGTLED GTCFLGNEKY
551  LQSPALRTEP ALLC
```

The four pairs of Cys residues at positions 52–60, 173–194, 258–266 and 371–393 are conserved in the class II cytokine receptor family.

References

[1] Pestka, S. and.Langer, J.A. (1987) Annu.Rev.Biochem. 56, 727–777.
[2] Balkwill, F.R. (1994) Cytokines in Cancer Therapy. Oxford University Press, Oxford.
[3] De Maeyer, E. and De Maeyer-Guignard, J. (1998) In The Cytokine Handbook, Thomson, A. ed., Academic Press, London, pp. 491–516.
[4] Uze, G. et al. (1995) J. Interferon Cytok. Res. 15, 3–26.
[5] Meager, A. (1994) Cytokines. Open University Press, Milton Keynes.
[6] Senda, T. et al. (1992) EMBO J. 11, 3193–3201.
[7] Shaw, G.D. et al. (1983) Nucleic Acids Res. 11, 555–573.
[8] Capon, D.J. et al. (1985) Mol. Cell Biol. 5, 768–779.
[9] Mantei, N. et al. (1980) Gene 10, 1–10.
[10] Zoon, K.C. (1987) Interferon 9, 1–12.
[11] Goeddel, D.V. et al. (1981) Nature 290, 20–26.
[12] Mogensen, K.E. et al. (1999) Interferon Cytok. Res. 19, 1069–1098.
[13] Bazan, J.F. (1990) Proc. Natl Acad. Sci. USA 87, 6934–6938.
[14] Aguet, M. (1991) Br. J. Haematol. 79 (suppl. 1), 6–8.
[15] Lutfalla, G. et al. (1992) J. Biol. Chem. 267, 2802–2809.
[16] Novick, D. et al. (1995) J. Leukoc. Biol. 57, 712–718.
[17] Novick, D. et al. (1992) FEBS Lett. 314, 445–448.
[18] Uze, G. et al. (1992) Proc. Natl Acad. Sci. USA 89, 4774–4778.
[19] Velazquez, L. et al. (1992) Cell 70, 313–322.
[20] Muller, M. et al. (1993) Nature 366, 129–135.
[21] Hunter, T. (1993) Nature 366, 114–116.
[22] Pellegrini, S. and Schindler, C. (1993) TIBS 18, 338–342.
[23] David, M. et al. (1993) J. Biol. Chem. 268, 6593–6599.
[24] Gutch, M.J. et al. (1992) Proc. Natl Acad. Sci. USA 89, 11411–11415.
[25] Marx, J. (1992) Science 257, 744–745.
[26] David, M. and Larner, A.C. (1992) Science 257, 813–815.
[27] Schindler, C. et al. (1992) Science 257, 809–813.
[28] Decker, T. et al. (1991) Mol. Cell Biol. 11, 5147–5153.
[29] Hannigan, G.E. and Williams, B.R.G. (1991) Science 251, 204–207.
[30] Pfeffer, L.M. et al. (1991) Proc. Natl Acad. Sci. USA 88, 7988–7992.
[31] Uze, G. et al. (1990) Cell 60, 225–234.
[32] Lutfalla, G. et al. (1995) EMBO J. 14, 5100–5108.
[33] Novick, D. et al. (1994) Cell 77, 391–400.
[34] Domanski, P. et al. (1995) J. Biol. Chem. 270, 21606–21611.

Other names

Type I interferon, fibroblast interferon, acid stable interferon, interferon-β1.

THE MOLECULE

Interferon β (IFNβ) is related to IFNα with about 30% amino acid sequence homology. IFNα and IFNβ share the same receptor and have very similar biological activities. Both confer resistance to viral infections on target cells, inhibit cell proliferation and regulate expression of MHC class I. Unlike IFNα, there is only one IFNβ gene in humans and in mice. The so called IFNβ2 is now known to be IL-6[1–4].

Crossreactivity

There is 48% homology between human and mouse IFNβ and no cross-species activity.

Sources

Fibroblasts and some epithelial cells.

Bioassays

See IFNα entry (page **311**).

Physicochemical properties of IFNβ

Property	Human	Mouse
pI	7.8–8.9	8–10
Amino acids – precursor	187	182
– mature[a]	166	161
M_r (K) – predicted	20	19.7
– expressed	20	26/35[b]
Potential *N*-linked glycosylation sites[b]	1	3
Disulfide bonds	1	0

[a] After removal of predicted signal peptide.
[b] Two forms of natural murine IFNβ with M_r of 26 000 and 35 000 due to differential glycosylation of the same protein have been reported.

3-D structure

IFNβ conisists of five helices. Three of these are parallel to each other and the other two are antiparallel to the first three. The structure is a variant of the α-helix bundle, but with a new chain-folding topology, common to all type I interferons[5].

Gene structure[6–8]

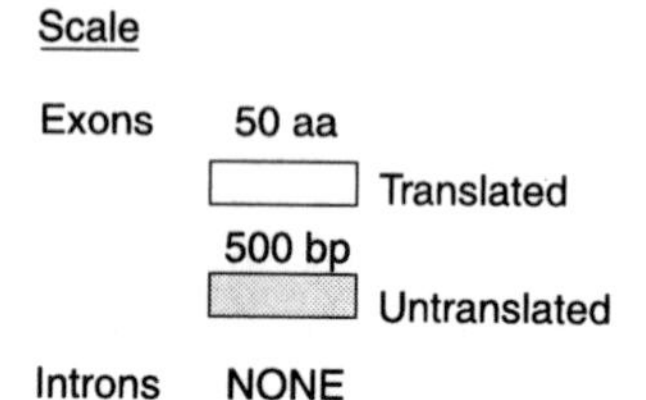

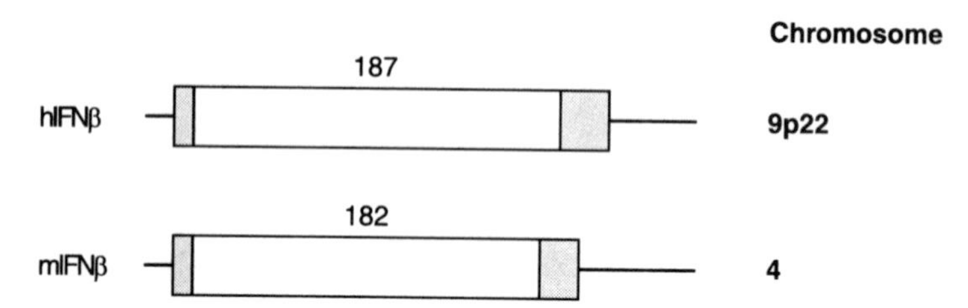

The gene for IFNβ is located on human chromosome 9p22 and contains a single exon of 187 amino acids, while the mouse gene is located on chromosome 4 also contains a single coding exon spanning 182 amino acids.

Amino acid sequence for human IFNβ[6,9]

Accession code: SwissProt P01574

```
-21  MTNKCLLQIA LLLCFSTTAL S
  1  MSYNLLGFLQ RSSNFQCQKL LWQLNGRLEY CLKDRMNFDI PEEIKQLQQF
 51  QKEDAALTIY EMLQNIFAIF RQDSSSTGWN ETIVENLLAN VYHQINHLKT
101  VLEEKLEKED FTRGKLMSSL HLKRYYGRIL HYLKAKEYSH CAWTIVRVEI
151  LRNFYFINRL TGYLRN
```

Disulfide bonds between Cys31–141. Variant C→Y at position 141 results in loss of disulfide bond and loss of antiviral activity.

Amino acid sequence for mouse IFNβ[10]

Accession code: SwissProt P01575

```
-21  MNNRWILHAA FLLCFSTTAL S
  1  INYKQLQLQE RTNIRKCQEL LEQLNGKINL TYRADFKIPM EMTEKMQKSY
 51  TAFAIQEMLQ NVFLVFRNNF SSTGWNETIV VRLLDELHQQ TVFLKTVLEE
101  KQEERLTWEM SSTALHLKSY YWRVQRYLKL MKYNSYAWMV VRAEIFRNFL
151  IIRRLTRNFQ N
```

THE IFNβ RECEPTOR

See The IFNα receptors under the IFNα and IFNω entry (page **311**).

References

[1] Pestka, S. and Langer, J.A. (1987) Annu. Rev. Biochem. 56, 727–777.
[2] Balkwill, F.R. (1994) Cytokines in Cancer Therapy. Oxford University Press, Oxford.
[3] De Maeyer, E. and De Maeyer-Guignard, J. (1998) In The Cytokine Handbook, Thomson, A. ed., Academic Press, London, pp. 491–516.
[4] Uze, G. et al. (1995) J. Interferon Cytok. Res. 15, 3–26.
[5] Senda, T. et al. (1992) EMBO J. 11, 3193–31201.
[6] Derynck, R. et al. (1980) Nature 285, 542–547.
[7] Ohno, S. and Taniguchi, T. (1981) Proc. Natl Acad. Sci. USA 78, 5305–5309.
[8] Kuga, T. et al. (1989) Nucleic Acids Res. 17, 3291.
[9] Taniguchi, T. et al. (1980) Gene 10, 11–15.
[10] Higashi, Y. et al. (1983) J. Biol. Chem. 258, 9522–9529.

Other names

Immune interferon, type II interferon, T cell interferon, macrophage-activating factor (MAF).

THE MOLECULE

Interferon-γ is a pleiotropic cytokine involved in the regulation of nearly all phases of immune and inflammatory responses, including the activation, growth and differentiation of T cells, B cells, macrophages, NK cells and other cell types such as endothelial cells and fibroblasts. It enhances MHC expression on antigen-presenting cells. IFNγ production is a hallmark of Th1 differentiation[1]. It also has weak antiviral and antiproliferative activity, and potentiates the antiviral and anti-tumour effects of IFNα/β[2–4]. Defects in the IFNγ activation pathway are associated with susceptibility to severe mycobacterial infection[5,6].

Crossreactivity

There is only 40% homology between human and mouse IFNγ and no significant cross-species activity.

Sources

CD8+ and CD4+ T cells, NK cells.

Bioassays

Mainly by activation (MHC expression) of monocytes or macrophages. See IFNα and IFNω entry (page **311**).

Physicochemical properties of IFNγ

Property	Human	Mouse
pI	7–9	5.5–6.0
Amino acids – precursor	166	155
– mature[a]	143	133
M_r (K) – predicted	17.1	15.9
– expressed[b]	40–70 (20, 25)	40–80
Potential *N*-linked glycosylation sites[c]	2	2
Disulfide bonds	0	0

[a] The predicted sequence after removal of a signal peptide of 20 amino acids for human IFNγ and 19 amino acids for murine IFNγ gives mature proteins of 146 and 136 amino acids respectively. However, sequencing of natural IFNγ has shown that the three N-terminal residues (Cys-Tyr-Cys) are removed during post translational modification yielding mature proteins of 143 and 133 amino acids[7].

[b] Two species (M_r 25 000 and 20 000) of monomeric human IFNγ have been identified on SDS–PAGE which differ only in the degree and sites of *N*-linked glycosylation. The higher molecular weight forms observed on gel filtration are due to dimerization and some multimerization.

[c] Glycosylation of human IFNγ at position 100 occurs only in the dimer.

3-D structure[8]

IFNγ is a homodimer formed by antiparallel association of two subunits. Each subunit has six α-helices held together by short nonhelical sequences. There are no β-sheets. The subunits have a flattened elliptical shape, whereas the overall structure of the dimer is globular.

Gene structure[9,10]

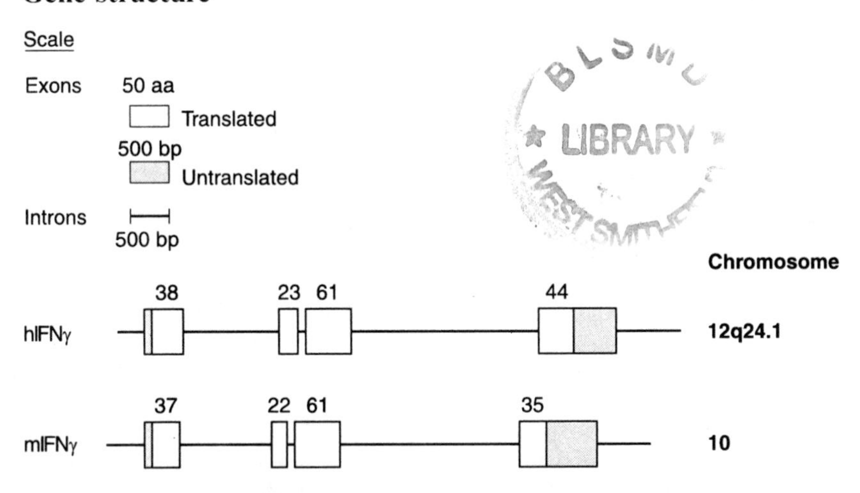

IFNγ has now been located on human chromosome 12q15[11] and on mouse chromosome 10 (telomeric) and consists of four exons.

Amino acid sequence for human IFNγ[12]

Accession code: SwissProt P01579

```
-20  MKYTSYILAF QLCIVLGSLG
  1  CYCQDPYVKE AENLKKYFNA GHSDVADNGT LFLGILKNWK EESDRKIMQS
 51  QIVSFYFKLF KNFKDDQSIQ KSVETIKEDM NVKFFNSNKK KRDDFEKLTN
101  YSVTDLNVQR KAIHELIQVM AELSPAAKTG KRKRSQMLFR GRRASQ
```

Variant sequences K→Q at position 9 and R→Q at position 140. The *N*-glycosylation at position 100 occurs only in the dimer. N-terminal amino acids Cys–Tyr–Cys are removed during post-translational modification.

Amino acid sequence for mouse IFNγ[10]

Accession code: SwissProt P01580

```
-19  MNATHCILAL QLFLMAVSG
  1  CYCHGTVIES LESLNNYFNS SGIDVEEKSL FLDIWRNWQK DGDMKILQSQ
 51  IISFYLRLFE VLKDNQAISN NISVIESHLI TTFFSNSKAK KDAFMSIAKF
101  EVNNPQVQRQ AFNELIRVVH QLLPESSLRK RKRSRC
```

N-Terminal amino acids Cys–Tyr–Cys are removed during post-translational modification.

THE IFNγ RECEPTOR

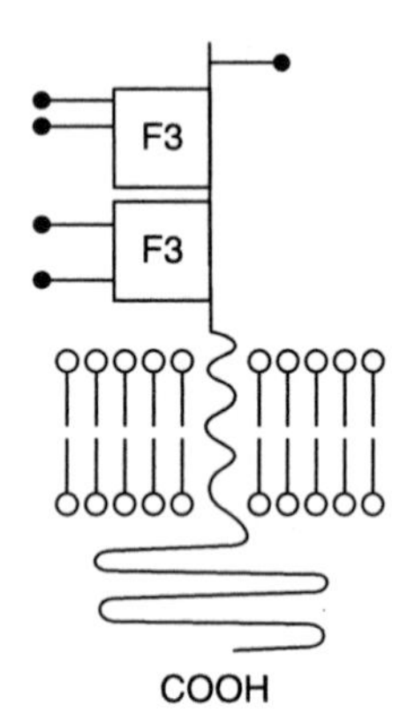

The IFNγ receptor is a complex of a high-affinity IFNγ-binding chain (CDw119) and a second accessory protein required for signal transduction[4,13]. The IFNγ binding subunit now known as the IGNγR α-chain is a transmembrane glycoprotein with a high content of serine and threonine in the intracytoplasmic domain. There is at least one disulfide bond which is essential for function. The receptor is a member of the class II cytokine receptor family which also includes the IFNα/β receptor, IL-10 receptor and tissue factor[14–16], and is characterized by an extracellular region of about 200 amino acids consisting of two homologous FNIII domains. The first of these has two conserved tryptophans and a pair of conserved cysteines whereas the second has a unique disulfide loop formed from the second pair of conserved cysteines, but no WSXWS motif characteristic of class I cytokine receptors[17]. Although the receptor α-chain binds IFNγ with high affinity, signal transduction requires a species specific accessory protein (IFNγR β-chain)[18–20] which associates with the extracellular domain of the receptor[21]. The accessory subunit is encoded by a gene on chromosome 21 and has been called accessory factor 1 (AF-1)[19] and the IFNγ receptor β-chain[20]. It also is a type I membrane-spanning glycoprotein and a member of the class II cytokine receptor family with two FNIII domains. The cloned human β-chain restores only MHC class I induction and not EMCV resistance by cells transfected with the IFNγ receptor, suggesting that there may be other components of the receptor that have not yet been identified[19]. Mutations in the IFNγR are associated with severe and sometimes lethal infections with mycobacteria[6,22].

Distribution

The IFNγ receptor is expressed on practically every cell in the body except for erythrocytes[23].

Physicochemical properties of the IFNγ receptor

Property	IFNγ R α-chain Human	IFNγ R α-chain Mouse	IFNγ R α-chain Human	IFNγ R α-chain Mouse
Amino acids – precursor	489	477	337	332
– mature[a]	472	455	310	314
M_r (K) – predicted	52.6	50.0	35.0	35.5
– expressed	90	90	62	62
Potential *N*-linked glycosylation sites	5	5	6	6
Affinity K_d (M)[b]	10^{-9}–10^{-10}	10^{-9}–10^{-10}		

[a] After removal of predicted signal peptide.
[b] For the receptor dimer. Most of the affinity for IFNγ can be attributed to the α-chain (10^{-8}–10^{-9} M).

Signal transduction[4,13]

IFNγ induces receptor dimerization and internalization. Receptor-mediated ligand internalization is not sufficient to induce a biological response. Signal transduction is mediated by phosphorylation and activation of Jak1 and Jak2 protein tyrosine kinases[24–26], and involves STAT1 activation and phosphorylation of the receptor[18,27,28]. The IFNγ and IFNα/β receptors share common signal transduction components including p91, which binds directly to the GAS response element after translocation to the nucleus[26,27,29].

Chromosomal location

The human receptor is on chromosome 6q12-q22 and the receptor accessory protein is on chromosome 21q22. The mouse receptor is on chromosome 10, and the accessory protein is on chromosome 16.

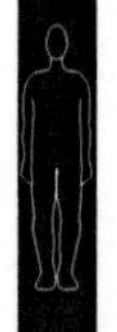

Amino acid sequence for human IFNγR α-chain[30]

Accession code: SwissProt P15260

-17 MALLFLLPLV MQGVSRA
1 EMGTADLGPS SVPTPT**N**VTI ESYNMNPIVY WEYQIMPQVP VFTVEVKNYG
51 VKNSEWIDAC I**N**ISHHYC**N**I SDHVGDPSNS LWVRVKARVG QKESAYAKSE
101 EFAVCRDGKI GPPKLDIRKE EKQIMIDIFH PSVFVNGDEQ EVDYDPETTC
151 YIRVYNVYVR M**N**GSEIQYKI LTQKEDDCDE IQCQLAIPVS SLNSQYCVSA
201 EGVLHVWGVT TEKSKEVCIT IF**N**SSIKGSL WIPVVAALLL FLVLSLVFIC
251 FYIKKINPLK EKSIILPKSL ISVVRSATLE TKPESKYVSL ITSYQPFSLE
301 KEVVCEEPLS PATVPGMHTE DNPGKVEHTE ELSSITEVVT TEENIPDVVP
351 GSHLTPIERE SSSPLSSNQS EPGSIALNSY HSRNCSESDH SRNGFDTDSS
401 CLESHSSLSD SEFPPNNKGE IKTEGQELIT VIKAPTSFGY DKPHVLVDLL
451 VDDSGKESLI GYRPTEDSKE FS

The two pairs of Cys residues at positions 60–68 and 197–218 are conserved in the class II cytokine receptor family.

Amino acid sequence for human IFNγR β-chain[19]

Accession code: SwissProt P38484

-27 MRPTLLWSLL LLLGVFAAAA AAPPDPL
1 SQLPAPQHPK IRLYNAEQVL SWEPVALS**N**S TRPVVYRVQF KYTDSKWFTA
51 DIMSIGV**N**CT QITATECDFT AASPSAGFPM DF**N**VTLRLRA ELGALHSAWV
101 TMPWFQHYR**N** VTVGPPENIE VTPGEGSLII RFSSPFDIAD TSTAFFCYYV
151 HYWEKGGIQQ VKGPFRSNSI SLDNLKPSRV YCLQVQAQLL W**N**KSNIFRVG
201 HLS**N**ISCYET MADASTELQQ VILISVGTFS LLSVLAGACF FLVLKYRGLI
251 KYWFHTPPSI PLQIEEYLKD PTQPILEALD KDSSPKDDVW DSVSIISFPE
301 KEQEDVLQTL

Cys59–67 and 147–207 are conserved in the IFNR family.

Amino acid sequence for mouse IFNγ receptor[31]

Accession code: SwissProt P15261

-22 MGPQAAAGRM ILLVVLMLSA KV
1 GSGALTSTED PEPPSVPVPT NVLIKSYNLN PVVCWEYQ**N**M SQTPIFTVQV
51 KVYSGSWTDS CT**N**ISDHCCN IYGQIMYPDV SAWARVKAKV GQKESDYARS
101 KEFLMCLKGK VGPPGLEIRR KKEEQLSVLV FHPEVVVNGE SQGTMFGDGS
151 TCYTFDYTVY VEH**N**RSGEIL HTKHTVEKEE C**N**ETLCEL**N**I SVSTLDSRYC
201 ISVDGISSFW QVRTEKSKDV CIPPFHDDRK DSIWILVVAP LTVFTVVILV
251 FAYWYTKKNS FKRKSIMLPK SLLSVVKSAT LETKPESKYS LVTPHQPAVL
301 ESETVICEEP LSTVTAPDSP EAAEQEELSK ETKALEAGGS TSAMTPDSPP
351 TPTQRRSFSL LSSNQSGPCS LTAYHSRNGS DSGLVGSGSS ISDLESLPNN
401 NSETKMAEHD PPPVRKAPMA SGYDKPHMLV DVLVDVGGKE SLMGYRLTGE
451 AQELS

The two pairs of Cys residues at positions 61–69 and 199–221 are conserved in the class II cytokine receptor family.

Amino acid sequence for mouse IFNγR β-chain[20]

Accession code: NCBI A49947

```
-18  MRPLPLWLPS LLLCGLGA
  1  AASSPDSFSQ LAAPLNPRLH LYNDEQILTW EPSPSSNDPR PVVYQVEYSF
 51  IDGSWHRLLE PNCTDITETK CDLTGGGRLK LFPHPFTVFL RVRAKRGNLT
101  SKWVGLEPFQ HYENVTVGPP KNISVTPGKG SLVIHFSPPF DVFHGATFQY
151  LVHYWEKSET QQEQVEGPFK SNSIVLGNLK PYRVYCLQTE AQLILKNKKI
201  RPHGLLSNVS CHETTANASA PLQQVILIPL GIFALLLGLT GACFTLFLKY
251  QSRVKYWFQA PPNIPEQIEE YLKDPDQFIL EVLDKDGSPK EDSWDSVSII
301  SSPEKERDDV LQTP
```

References

[1] Romagnani S. (1997) Immunol.Today 18, 263–268.
[2] Farrar, M.A. and Schreiber, R.D. (1993) Annu. Rev. Immunol. 11, 571–611.
[3] Billiau, A. (1996) Adv. Immunol. 62, 61–130.
[4] Stark, G.R. et al. (1998) Annu. Rev. Biochem. 67, 227–264.
[5] Ottenhoff, T.H.M. et al. (1998) Immunol. Today 19, 491–494.
[6] Newport, M. et al. (1996) N. Engl. J. Med. 335, 1941–1949.
[7] Rinderknecht, E. et al. (1984) J. Biol. Chem. 259, 6790–6797.
[8] Ealick, S.E. et al. (1991) Science 252, 698–252.
[9] Gray, P.W. and Goeddel, D.V. (1982) Nature 298, 859–863.
[10] Gray, P.W. and Goeddel, D.V. (1983) Proc. Natl Acad. Sci. USA 80, 5842–5846.
[11] Bureau, J.F. et al. (1995) Genomics 28, 109–112.
[12] Gray, P.W. et al. (1982) Nature 295, 503–508.
[13] Rodig, S. et al. (1998) Eur. Cytokine Netw. 9, 49–53.
[14] Bazan, J.F. (1990) Proc. Natl Acad. Sci. USA 87, 6934–6938.
[15] Aguet, M. (1991) Br. J. Haematol. 79 (suppl. 1), 6–8.
[16] Lutfalla, G. et al. (1992) J. Biol. Chem. 267, 2802–2809.
[17] Bazan, J.F. (1990) Cell 61, 753–754.
[18] Schreiber, R.D. et al. (1992) Int. J. Immunopharmacol. 14, 413–419.
[19] Soh, J. et al. (1994) Cell 76, 793–802.
[20] Hemmi, S. et al. (1994) Cell 76, 803–810.
[21] Gibbs, V.C. et al. (1991) Mol. Cell Biol. 11, 5860–5866.
[22] Jouanguy, E. et al. (1996) N. Engl. J. Med. 335, 1956–1961.
[23] Valente, G. et al. (1992) Eur. J. Immunol. 22, 2403–2412.
[24] Hunter, T. (1993) Nature 366, 114–116.
[25] Watling, D. et al. (1993) Nature 366, 166–170.
[26] Muller, M. et al. (1993) Nature 366, 129–135.
[27] Pellegrini, S. and Schindler, C. (1993) TIBS 18, 338–342.
[28] Kalina, U. et al. (1993) J. Virol. 67, 1702–1706.
[29] Loh, J.E. et al. (1992) EMBO J. 11, 1351–1363.
[30] Aguet, M. et al. (1988) Cell 55, 273–280.
[31] Gray, P.W. et al. (1989) Proc. Natl Acad. Sci. USA 86, 8497–8501.

Other names

Murine γIP-10 has been known as CRG-2 or C7[1,2].

THE MOLECULE

γIP-10 is a member of the non-ELR CXC family of chemokines[3–5]. It is a selective Th1 lymphocyte and monocyte chemoattractant and promotes lymphocyte adhesion to activated endothelial cells[6]. γIP-10 is also angiostatic, mitogenic for smooth muscle cells, and suppressive for haematopoietic cell proliferation[7,8]. γIP-10 has been detected in delayed-type reactions and leprosy lesions[9].

Crossreactivity

Human γIP-10 shares 67% identity with murine γIP-10 and is active on mouse cells[10].

Sources

Keratinocytes, monocytes, astrocytes, microglia, T lymphocytes, neutrophils, endothelial cells, fibroblasts. Induced by IFNγ stimulation (potentiated by IL-2), LPS and IFNβ[11–13].

Bioassays

Lymphocyte chemoattraction.

Physicochemical properties of γIP-10

Property	Human	Mouse
pI	10.8	>8
Amino acids – precursor	98	98
– mature	77	77
M_r (K) – predicted	10.9	10.8
– expressed	6 and 7	8
Potential *N*-linked glycosylation sites[a]	0	0

[a] One in signal sequence.

3-D structure

Could be modelled on IL-8.

Gene structure[14]

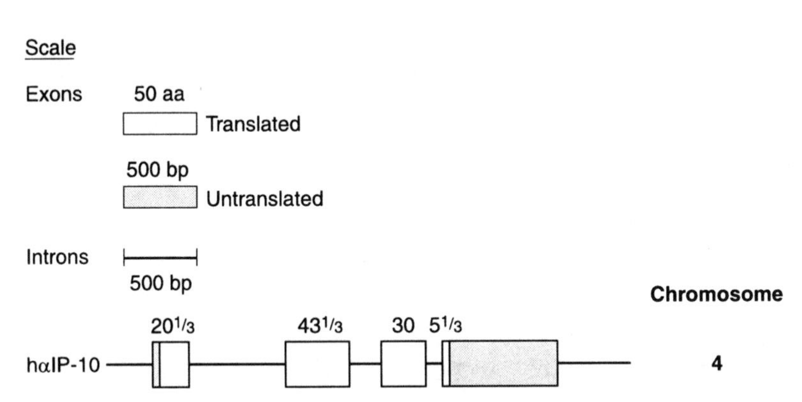

The gene for γIP-10 is located on human chromosome 4q2 and consists of four exons of $20^{1/3}$, $42^{1/3}$, 30 and $5^{1/3}$ amino acids.

Amino acid sequence for human γIP-10[3]

Accession code: SwissProt P02778

```
-21  MNQTAILICC LIFLTLSGIQ G
  1  VPLSRTVRCT CISISNQPVN PRSLEKLEII PASQFCPRVE IIATMKKKGE
 51  KRCLNPESKA IKNLLKAVSK EMSKRSP
```

Amino acid sequence for mouse γIP-10[1,2]

Accession code: SwissProt P17515

```
-21  MNPSAAVIFC LILLGLSGTQ G
  1  IPLARTVRCN CIHIDDGPVR MRAIGKLEII PASLSCPRVE IIATMKKNDE
 51  QRCLNPESKT IKNLMKAFSQ KRSKRAP
```

THE RECEPTOR, CXCR3

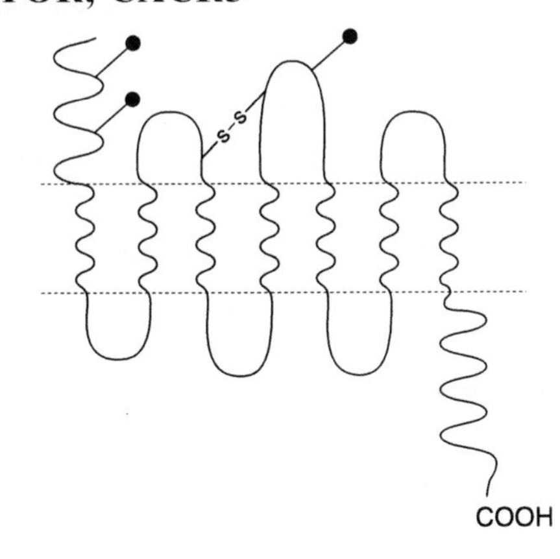

The γIP-10 receptor CXCR3, also known as GPR9, is a high-affinity functional receptor for γIP-10 mediating chemotaxis and adhesion of Th1 lymphocytes and NK cells[15–17]. CXCR3 is also receptor for three other CXC chemokines: I-TAC, MIG and 6Ckine[16,18,19]. Eotaxin may act as a natural receptor antagonist for CXCR3[20]. γIP-10 has also been shown to be an inverse agonist for the herpesvirus 8 G protein-coupled receptor (accession code P88966), acting to block the constituitive signalling mediated by the viral receptor[21]. MIG is inactive on this receptor.

Distribution

Expressed on some peripheral blood T and B lymphocytes, and NK cells, upregulated by IL-2 treatment. CXCR3 seems to be selectively expressed on Th1 and Th0 T cells[15,22].

Physicochemical properties of the γIP-10 receptor

Property	Human	Mouse
Amino acids	368	367
M_r (K) – predicted	40.7	41.0
Potential *N*-linked glycosylation sites	3	3
Affinity K_d (M)	$0.1–0.3 \times 10^{-9}$	?

Signal transduction

Binding of human γIP-10 to CXCR3 mediates a Ca flux with an EC_{50} of 10–20 nM, causes migration of T cells with an EC_{50} of about 5 nM, and promotes lymphocyte adhesion at subnanomolar concentrations[15,16,20]. Mouse CXCR3 ligation with γIP-10 causes a Ca flux at 1 μg/ml, and mediates T-cell migration at doses over 10 ng/ml[17].

Chromosomal location

The gene for CXCR3 is on human chromosome Xq13 and on the mouse X chromosome.

Amino acid sequence for human CXCR3 receptor

Accession code: SwissProt P49682

```
  1  MVLEVSDHQV LNDAEVAALL ENFSSSYDYG ENESDSCCTS PPCPQDFSLN
 51  FDRAFLPALY SLLFLLGLLG NGAVAAVLLS RRTALSSTDT FLLHLAVADT
101  LLVLTLPLWA VDAAVQWVFG SGLCKVAGAL FNINFYAGAL LLACISFDRY
151  LNIVHATQLY RRGPPARVTL TCLAVWGLCL LFALPDFIFL SAHHDERLNA
201  THCQYNFPQV GRTALRVLQL VAGFLLPLLV MAYCYAHILA VLLVSRGQRR
251  LRAMRLVVVV VVAFALCWTP YHLVVLVDIL MDLGALARNC GRESRVDVAK
301  SVTSGLGYMH CCLNPLLYAF VGVKFRERMW MLLLRLGCPN QRGLQRQPSS
351  SRRDSSWSET SEASYSGL
```

Disulfide bond links Cys124–203.

Amino acid sequence for mouse CXCR3 receptor

Accession code: SPTREMBLE Q9QWN6

```
  1  MYLEVSERQV LDASDFAFLL ENSTSPYDYG ENESDFSDSP PCPQDFSLNF
 51  DRTFLPALYS LLFLLGLLGN GAVAAVLLSQ RTALSSTDTF LLHLAVADVL
101  LVLTLPLWAV DAAVQWVFGP GLCKVAGALF NINFYAGAFL LACISFDRYL
151  SIVHATQIYR RDPRVRVALT CIVVWGLCLL FALPDFIYLS ANYDQRLNAT
201  HCQYNFPQVG RTALRVLQLV AGFLLPLLVM AYCYAHILAV LLVSRGQRRF
251  RAMRLVVVVV AAFAVCWTPY HLVVLVDILM DVGVLARNCG RESHVDVAKS
301  VTSGMGYMHC CLNPLLYAFV GVKFREQMWM LFTRLGRSDQ RGPQRQPSSS
351  RRESSWSET EASYLGL
```

Disulfide bond links Cys123–202.

References

[1] Ohmori, Y. et al. (1990) Biochem. Biophys. Res. Commun. 168, 1261–1267.
[2] Vanguri, P. and Farber, J.M. (1990) J. Biol. Chem. 265, 15047–15049.
[3] Luster, A.D. et al. (1985) Nature 315, 672–676.
[4] Luster, A.D. and Ravetch, J.V. (1987) J. Exp. Med. 166, 1084–1097.
[5] Luster, A.D. and Ravetch, J.V. (1987) Mol. Cell. Biol. 7, 3723–3731.
[6] Taub, D.D. et al. (1993) J. Exp. Med. 177, 1809–1814.
[7] Strieter, R.M. et al (1995) Shock 4, 155–160.
[8] Wang, X. et al. (2000) J. Biol. Chem. 271, 24286–24293.
[9] Kaplan, G. et al. (1987) J. Exp. Med. 166, 1098–1108.
[10] Luster, A.D. and Peder, P. (1993) J. Cell. Biol. 17B (suppl.), E547.
[11] Narumi, S. and Hamilton, T.A. (1991) J. Immunol. 146, 3038–3044.
[12] Narumi, S. et al. (1992) J. Leucoc. Biol. 52, 27–33.
[13] Ren, L.Q. et al. (1998) Brain Res. 58, 256–263.
[14] Luster, A.D. et al. (1987) Proc. Natl Acad. Sci. USA 84, 2868–2871.
[15] Loetscher, M. et al. (1998) Eur. J. Immunol. 28, 3696–3705.
[16] Loetscher, M. et al. (1996) J. Exp. Med. 184, 963–969.
[17] Tamaru, M. et al. (1998) Biochem. Biophys. Res. Commun. 251, 41–48.
[18] Soto, H. et al. (1998) Proc. Natl Acad. Sci. USA 95, 8205–8210.
[19] Cole, K.E. et al. (1998) J. Exp. Med. 187, 2009–2021.
[20] Weng, Y.M. et al. (1998) J. Biol. Chem. 273, 18288–18291.
[21] Geras-raker, E. et al. (1998) J. Exp. Med. 188, 405–408.
[22] Sallusto, F. et al. (1998) J. Exp. Med. 187, 875–883.

Other names

G1P2, UCRP (ubiquitin cross-reactive protein).

THE MOLECULE

Interferon-sensitive gene 15 (ISG-15) is a 15-kDa protein with homology to ubiquitin[1]. It is produced by IFNα/β-stimulated monocytes and T cells and induces IFNγ production by T cells but not NK cells[2,3]. ISG-15 stimulates proliferation of B cell-depleted lymphocytes including NK (CD56+) cells. It also augments NK killing of K562 and Daudi cell lines. The effect on NK cells is indirect and requires the presence of T cells[4].

Crossreactivity

There is no biological crossreactivity between human and mouse ISG-15.

Sources

IFNα/β-stimulated T cells and monocytes[5] as well as cell lines of T cell, B cell, monocyte, and epithelial origin[2].

Bioassays

Induction of IFNγ.

Physicochemical properties of ISG-15

Property	Human	Mouse
pI (calculated)	5.5	7.7
Amino acids – precursor	165	160
– mature[a]	156	
M_r (K) – predicted	17.9 (16.9)[b]	17.6
– expressed	15	
Potential *N*-linked glycosylation sites	0	1
Disulfide bonds	0	0

[a] Biologically active ISG-15 is derived by processing of the 17-kDa precursor that removes the N-terminal Met and a further eight amino acids from the C-terminus[6].
[b] Predicted M_r for ISG-17 precursor with ISG-15 in parenthesis.

3-D structure

Similar to ubiquitin, consisting mainly of antiparallel β-sheets.

Gene structure[3]

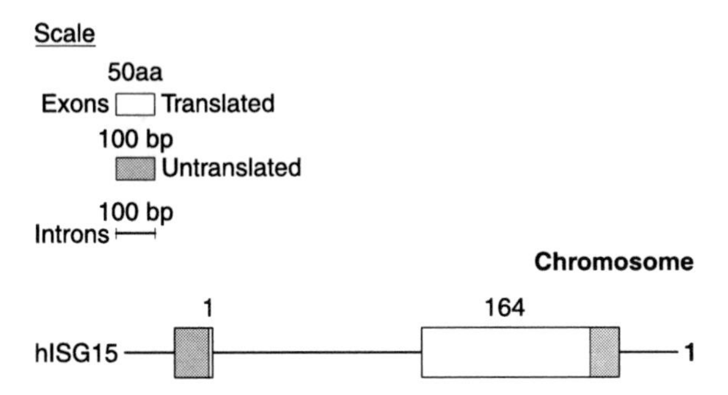

Human ISG-15 is on chromosome 1.

Amino acid sequence for human ISG-15[3,7]

Accession code: SwissProt P05161

```
  1  MGWDLTVKML AGNEFQVSLS SSMSVSELKA QITQNIGVHA FQQRLAVHPS
 51  GVALQDRVPL ASQGLGPGST VLLVVDKCDE PLSILVRNNK GRSSTYEVRL
101  TQTVAHLKQQ VSGLEGVQDD LFWLTFEGKP LEDQLPLGEY GLKPLSTVFM
151  NLRLRGGGTE PGGRS
```

ISG-15 is derived from interferon-induced 17-kDa protein 1–165[6]. Amino acids 1–78 and 79–165 are ubiquitin-like domains.

Amino acid sequence for mouse ISG-15

Accession code: SwissProt Q64339

```
  1  MAWDLKVKML GGNDFLVSVT NSMTVSELKK QIAQKIGVPA FQQRLAHQTA
 51  VLQDGLTLSS LGLGPSSTVM LVVQNCSEPL SILVRNERGH SNIYEVFLTQ
101  TVDTLKKKVS SGTSHEDQFW LSFEGRPMED KELLGEYGLK PQCTVIKHLR
151  LRGGGGDQCA
```

Amino acids 1–76 and 77–160 are ubiquitin-like domains.

THE ISG-15 RECEPTOR

No information.

References

[1] Haas, A.L. et al. (1987) J. Biol. Chem. 262, 11315–11323.
[2] D Cunha, J. et al. (1996) J. Immunol. 157, 4100–4108.
[3] Reich, N. et al. (1987) Proc. Natl Acad. Sci. USA 84, 6394–6398.
[4] D Cunha, J. et al. (1996) Proc. Natl Acad. Sci. USA 93, 211–215.
[5] Knight, E. and Cordova, B.J. (1991) Immunology 146, 2280–2284.
[6] Knight, E. et al. (1988) J. Biol. Chem. 263, 4520–4522.
[7] Blomstrom, D.C. et al. (1986) J. Biol. Chem. 261, 8811–8816.

Other names
B-R1.

THE MOLECULE

Interferon-inducible T-cell α chemokine (I-TAC) is a chemoattractant for activated T lymphocytes which is a member of the non-ELR subgroup of the CXC chemokines. It shares a receptor and has similar activities with two other interferon-inducible chemokines, γIP-10 and MIG. It has been proposed to play a role in attracting activated T cells to sites of Th1-mediated inflammation.

Crossreactivity
No murine homologue yet identified.

Sources
Astrocytes, fibrosarcoma cells, microglia and monocytes stimulated with IFNβ or IFNγ. mRNA was also found in peripheral blood leukocytes, pancreas, liver, thymus, spleen and lung[1,2].

Bioassays
Chemotaxis of T lymphocytes.

Physicochemical properties of I-TAC

Property	Human	Mouse
Amino acids – precursor	94	?
– mature	73	?
M_r (K) – predicted	8.0	?
– expressed	8.0	?
Potential *N*-linked glycosylation sites	0	?
Disulfide bonds	2	?

3-D structure
No information.

Gene structure

The gene for human I-TAC is on chromosome 4[1].

Amino acid sequence for human I-TAC[1]

Accession code: AF030514

```
-21  MSVKGMAIAL AVILCATVVQ G
  1  FPMFKRGRCL CIGPGVKAVK VADIEKASIM YPSNNCDKIE VIITLKENKG
 51  QRCLNPKSKQ ARLIIKKVER KNF
```

Potential splice variant mRNA species have been detected in liver and intestine[1].

THE RECEPTOR, CXCR3

See γIP-10 entry (page **328**). The CXCR3 receptor is shared with γIP-10 and MIG. I-TAC binds to CXCR3 with high affinity (K_d 0.3 nM), causing a Ca flux with an EC_{50} of 6 nM and mediates chemotaxis of activated T cells with a peak response at 10 nM[1]. I-TAC was more potent at stimulating a Ca flux than the other ligands, MIG or γIP-10.

References

[1] Cole, K.E. et al. (1998) J. Exp. Med. 187, 2009–2021.
[2] Rani, M.R.S. et al. (1996) J. Biol. Chem. 271, 22878–22884.

Other names
MIP-3α, SCYA20, Exodus-1

THE MOLECULE

Liver and activation related chemokine (LARC) is an inflammatory chemokine that is a potent lymphocyte and dendritic cell chemoattractant, a weak neutrophil chemoattractant, and that also inhibits the proliferation of myeloid progenitor cells[1–4]. LARC selectively promotes the adhesion of memory CD4+ T cells[5].

Crossreactivity
Human and murine LARC share 70% amino acid homology.

Sources
LARC mRNA is expressed in liver, lung, thymus, ileum and colon, inflamed tonsil and appendix, and in lung macrophages, endothelial cells, fibroblasts, dendritic cells and eosinophils[1–4]. LARC mRNA expression could be induced in U937 and melanoma cells by PMA treatment.

Bioassays
Chemotaxis of lymphocytes.

Physicochemical properties of LARC

Property	Human	Mouse
Amino acids – precursor	96	97
– mature	70	70
M_r (K) – predicted	10.8	10.8
– expressed	8	?
Potential *N*-linked glycosylation sites	0	0
Disulfide bonds	2	2

3-D structure
No information

Gene structure

The gene for human LARC is on chromosome 2q33-37[2].

Amino acid sequence for human LARC

Accession code: SwissProt P78556

```
-26  MCCTKSLLLA ALMSVLLLHL CGESEA
  1  ASNFDCCLGY TDRILHPKFI VGFTRQLANE GCDINAIIFH TKKKLSVCAN
 51  PKQTWVKYIV RLLSKKVKNM
```

The first residue A1 is absent from some EST sequences.

Amino acid sequence for mouse LARC

Accession code: SwissProt O89093

```
-27  MACGGKRLLF LALAWVLLAH LCSQAEA
  1  ASNYDCCLSY IQTPLPSRAI VGFTRQMADE ACDINAIIFH TKKRKSVCAD
 51  PKQNWVKRAV NLLSLRVKKM
```

THE RECEPTOR, CCR6

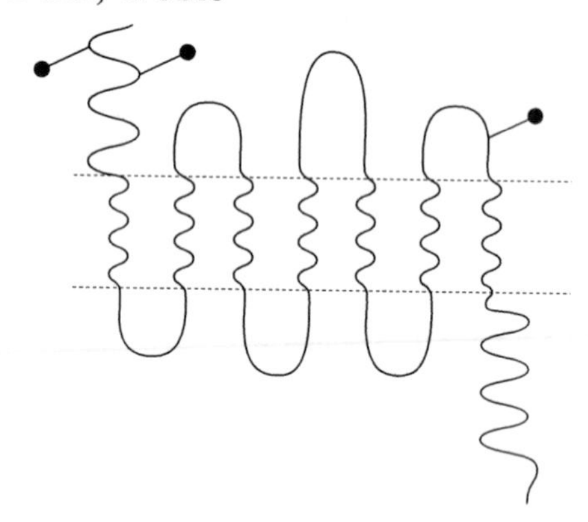

The LARC receptor, CCR6, is also known as GPR-CY4, CKRL-3, DRY-6 or STRL-22. CCR6 is a specific receptor binding LARC with a K_d of 0.9 nM, causing an IP_3-mediated calcium flux, and chemotactic response[4,6–9].

Distribution

CCR6 is expressed on CD4 and CD8+ T cells, B cells and CD34+ or lung dendritic cells, but is absent from other haematopoietic cells[6–8]. CCR6 expression was upregulated on T cells by IL-2 and downregulated by PHA[6]. The mRNA is expressed predominantly in mucosal and inflamed tissues.

Physicochemical properties of the LARC receptor

Property	Human	Mouse
Amino acids – precursor	374	?
– mature	?	?
M_r (K) – predicted	42 494	?
– expressed	?	
Potential *N*-linked glycosylation sites	2	?
Affinity K_d (M)	0.9×10^{-9}	?

Signal transduction

LARC binding causes an IP_3 dependent mobilization of calcium, which is inhibited by pertussis toxin and phospholipase C inhibitors. The EC_{50} of LARC for Ca mobilization was 50 nM and for chemotaxis was 12 nM, compared with the K_d of 0.9 nM. It has been suggested that for MCP-1 at least, a two-stage binding to chemokine receptors can occur, with an initial high-affinity binding followed by a second lower affinity interaction which mediates signalling[10].

Chromosomal location

The human gene is located on chromosome 6q26–27[11].

Amino acid sequence for human receptor

Accession code: SwissProt P51684

```
  1  MSGESMNFSD VFDSSEDYFV SVNTSYYSVD SEMLLCSLQE VRQFSRLFVP
 51  IAYSLICVFG LLGNILVVIT FAFYKKARSM TDVYLLNMAI ADILFVLTLP
101  FWAVSHATGA WVFSNATCKL LKGIYAINFN CGMLLLTCIS MDRYIAIVQA
151  TKSFRLRSRT LPRSKIICLV VWGLSVIISS STFVFNQKYN TQGSDVCEPK
201  YQTVSEPIRW KLLMLGLELL FGFFIPLMFM IFCYTFIVKT LVQAQNSKRH
251  KAIRVIIAVV LVFLACQIPH NMVLLVTAAN LGKMNRSCQS EKLIGYTKTV
301  TEVLAFLHCC LNPVLYAFIG QKFRNYFLKI LKDLWCVRRK YKSSGFSCAG
351  RYSENISRQT SETADNDNAS SFTM
```

Amino acid sequence for mouse receptor

No information.

References

[1] Rossi, D.L. et al. (1997) J. Immunol. 158, 1033–1036.
[2] Hieshima, K. et al. (1997) J. Biol. Chem. 272, 5846–5853.
[3] Hromas, R. et al. (1997) Blood 89, 3315–3322.
[4] Varona, R. et al. (1998) FEBS Lett. 440, 188–194.
[5] Campbell, J.J. et al. (1998) Science 381–384.
[6] Baba, M. et al. (1997) J. Biol. Chem. 272, 14893–14898.
[7] Power, C.A. et al. (1997) J. Exp. Med. 186, 825–835.
[8] Greaves, D.R. et al. (1997) J. Exp. Med. 186, 837–844.
[9] Liao, F. et al. (1997) Biochem. Biophys. Res. Commun. 236, 212–217.
[10] Monteclaro, F.S. and Charo, I.F. (1996) J. Biol. Chem. 271, 19084–19092.
[11] Liao, F. et al. (1997) Genomics 40, 175–180.

Other names

Obesity protein (OB).

THE MOLECULE

Leptin (Greek *leptos*: thin) is coded for by the obesity (*ob*) gene and is an important component in the complex processes that regulate food intake and body weight[1,2]. Mice expressing *ob/ob* mutations have five times the fat content of wild-type mice fed on the same diet and administration of leptin by injection or subcutaneous infusion results in a dose-dependent loss of body weight. The absence of leptin in *ob/ob* mice is also responsible for many of the abnormalities associated with starvation found in these animals, including decreased body temperature, hyperphagia, decreased energy expenditure, decreased loss of fertility and decreased immune function. Leptin preferentially stimulates proliferation of naïve (CD45RA+) human T cells and promotes Th1 immunity. The decreased immune function in starved mice can be restored by administration of leptin[3]. Leptin also acts directly on endothelial cells and stimulates angiogenesis[4].

Crossreactivity

Leptin from mouse and humans reduces food intake in rodents indicating some crossreactivity.

Sources

Leptin mRNA is found in white and brown adipose tissue, gastric epithelium and placenta.

Bioassays

There are no specific bioassays for leptin as yet but reduction in food intake after injection can be used as a crude measure.

Physicochemical properties of leptin

Property	Human	Mouse
pI (calculated)	5.71	5.2
Amino acids – precursor	167	167
– mature[a]	146	146
M_r (K) – predicted	16	16
– expressed	16	16
Potential *N*-linked glycosylation sites	0	0
Disulfide bonds	1	1

[a] After removal of signal peptide.

Disulfide bonds between Cys96–146 in human and mouse leptin by similarity.

3-D structure

Leptin has a four antiparallel α-helix structure and is a member of the short helix cytokine family[5].

Gene structure[6]

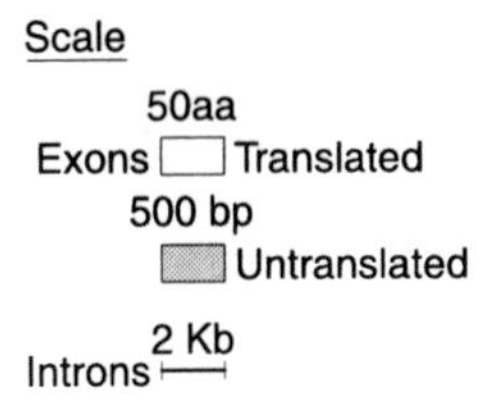

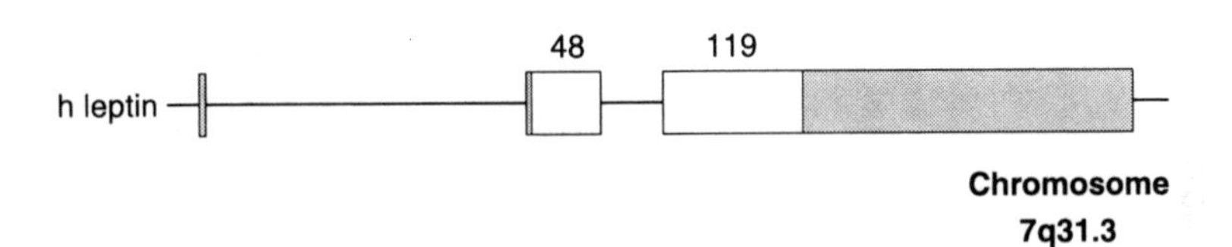

The leptin gene is on human chromosome 7q31.3 and mouse chromosome 6.

Amino acid sequence for human leptin[7]

Accession code: SwissProt P41159

```
-21  MHWGTLCGFL WLWPYLFYVQ A
  1  VPIQKVQDDT KTLIKTIVTR INDISHTQSV SSKQKVTGLD FIPGLHPILT
 51  LSKMDQTLAV YQQILTSMPS RNVIQISNDL ENLRDLLHVL AFSKSCHLPW
101  ASGLETLDSL GGVLEASGYS TEVVALSRLQ GSLQDMLWQL DLSPGC
```

Glu28 is missing in 30% of clones. Conflicting sequence Glu75→R.

Amino acid sequence for mouse leptin[7]

Accession code: SwissProt P41160

```
-21  MCWRPLCRFL WLWSYLSYVQ A
  1  VPIQKVQDDT KTLIKTIVTR INDISHTQSV SAKQRVTGLD FIPGLHPILS
 51  LSKMDQTLAV YQQVLTSLPS QNVLQIANDL ENLRDLLHLL AFSKSCSLPQ
151  TSGLQKPESL DGVLEASLYS TEVVALSRLQ GSLQDILQQL DVSPEC
```

Glu28 is missing in 30% of clones.Disulfide bonds between Cys117 and Cys167 of precursor.

THE LEPTIN RECEPTOR

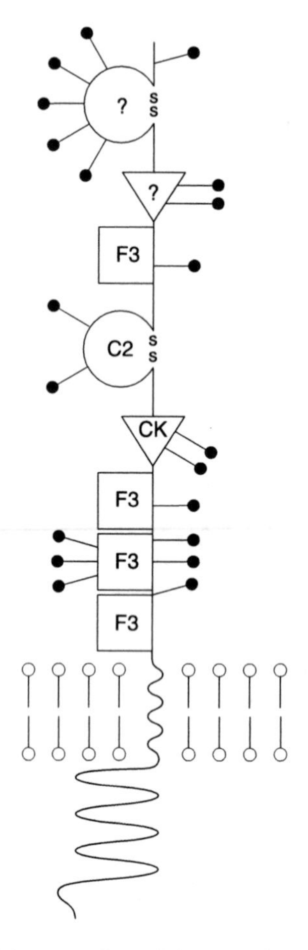

The leptin receptor (ObR) is a member of the cytokine receptor superfamily and is coded for by the diabetes gene *db*[7–9]. Mice with *db/db* mutation have the same obese phenotype as *ob/ob* mice. The *db* gene encodes at least five alternatively spliced forms of the receptor ObR-a–e in mice and four in humans[9–11]. The ObR b is the full-length receptor and is essential for the weight-reducing effects of leptin. ObR a is expressed in the choroid plexus and other tissues but its function is unknown. It has some weak signalling activity. The functions of the ObR-c, -d and -e forms are also unknown but may be involved in transport or form heterodimers with the other forms[1].

Distribution

ObR-b (the full-length form) is expressed at high levels in hypothalmic neurons, T cells and vascular endothelium. The ObR-a receptor is widely distributed.

Physicochemical properties of the leptin receptor

Property	Human	Mouse
Amino acids – precursor	1165	1162[a]
– mature[b]	?	?
M_r (K) – predicted	133	131
– expressed		
Potential *N*-linked glycosylation sites	20	16
Affinity K_d (M)	10^{-10}	7×10^{-10}

[a] Five forms of mouse leptin receptor (a–e) produced by alternative splicing have been described. The 1161-amino-acid variant given here is the longest form.
[b] Unknown signal peptide.

Signal transduction

Leptin binding to a ObR-b homodimer results in phosphorylation of Jak2, activation of STAT3, 5 and 6 (the fat STATs), phosphorylation of SHP-2 and expression of *fos*[2]. *In vivo* only STAT3 is activated.

Chromosomal location

The leptin receptor is found on human chromosome 1p31 and mouse chromosome 4.

Amino acid sequence for human leptin receptor (ObR)[11]

Accession code: SwissProt P48357

```
  1  MICQKFCVVL LHWEFIYVIT AFNLSYPITP WRFKLSCMPP NSTYDYFLLP
 51  AGLSKNTSNS NGHYETAVEP KFNSSGTHFS NLSKTTFHCC FRSEQDRNCS
101  LCADNIEGKT FVSTVNSLVF QQIDANWNIQ CWLKGDLKLF ICYVESLFKN
151  LFRNYNYKVH LLYVLPEVLE DSPLVPQKGS FQMVHCNCSV HECCECLVPV
201  PTAKLNDTLL MCLKITSGGV IFQSPLMSVQ PINMVKPDPP LGLHMEITDD
251  GNLKISWSSP PLVPFPLQYQ VKYSENSTTV IREADKIVSA TSLLVDSILP
301  GSSYEVQVRG KRLDGPGIWS DWSTPRVFTT QDVIYFPPKI LTSVGSNVSF
351  HCIYKKENKI VPSKEIVWWM NLAEKIPQSQ YDVVSDHVSK VTFFNLNETK
401  PRGKFTYDAV YCCNEHECHH RYAELYVIDV NINISCETDG YLTKMTCRWS
451  TSTIQSLAES TLQLRYHRSS LYCSDIPSIH PISEPKDCYL QSDGFYECIF
501  QPIFLLSGYT MWIRINHSLG SLDSPPTCVL PDSVVKPLPP SSVKAEITIN
551  IGLLKISWEK PVFPENNLQF QIRYGLSGKE VQWKMYEVYD AKSKSVSLPV
601  PDLCAVYAVQ VRCKRLDGLG YWSNWSNPAY TVVMDIKVPM RGPEFWRIIN
651  GDTMKKEKNV TLLWKPLMKN DSLCSVQRYV INHHTSCNGT WSEDVGNHTK
701  FTFLWTEQAH TVTVLAINSI GASVANFNLT FSWPMSKVNI VQSLSAYPLN
751  SSCVIVSWIL SPSDYKLMYF IIEWKNLNED GEIKWLRISS SVKKYYIHDH
801  FIPIEKYQFS LYPIFMEGVG KPKIINSFTQ DDIEKHQSDA GLYVIVPVII
851  SSSILLLGTL LISHQRMKKL FWEDVPNPKN CSWAQGLNFQ KPETFEHLFI
901  KHTASVTCGP LLLEPETISE DISVDTSWKN KDEMMPTTVV SLLSTTDLEK
951  GSVCISDQFN SVNFSEAEGT EVTYEAESQR QPFVKYATLI SNSKPSETGE
```

```
1001 EQGLINSSVT KCFSSKNSPL KDSFSNSSWE IEAQAFFILS DQHPNIISPH
1051 LTFSEGLDEL LKLEGNFPEE NNDKKSIYYL GVTSIKKRES GVLLTDKSRV
1101 SCPFPAPCLF TDIRVLQDSC SHFVENNINL GTSSKKTFAS YMPQFQTCST
1151 QTHKIMENKM CDLTV
```

Three alternative splicing variants have been described[10].

Amino acid sequence for mouse leptin receptor (ObR)[11]

Accession code: SwissProt P48356

```
   1 MMCQKFYVVL LHWEFLYVIA ALNLAYPISP WKFKLFCGPP NTTDDSFLSP
  51 AGAPNNASAL KGASEAIVEA KFNSSGIYVP ELSKTVFHCC FGNEQGQNCS
 101 ALTDNTEGKT LASVVKASVF RQLGVNWDIE CWMKGDLTLF ICHMEPLPKN
 151 PFKNYDSKVH LLYDLPEVID DSPLPPLKDS FQTVQCNCSL RGCECHVPVP
 201 RAKLNYALLM YLEITSAGVS FQSPLMSLQP MLVVKPDPPL GLHMEVTDDG
 251 NLKISWDSQT MAPFPLQYQV KYLENSTIVR EAAEIVSATS LLVDSVLPGS
 301 SYEVQVRSKR LDGSGVWSDW SSPQVFTTQD VVYFPPKILT SVGSNASFHC
 351 IYKNENQIIS SKQIVWWRNL AEKIPEIQYS IVSDRVSKVT FSNLKATRPR
 401 GKFTYDAVYC CNEQACHHRY AELYVIDVNI NISCETDGYL TKMTCRWSPS
 451 TIQSLVGSTV QLRYHRRSLY CPDSPSIHPT SEPKNCVLQR DGFYECVFQP
 501 IFLLSGYTMW IRINHSLGSL DSPPTCVLPD SVVKPLPPSN VKAEITVNTG
 551 LLKVSWEKPV FPENNLQFQI RYGLSGKEIQ WKTHEVFDAK SKSASLLVSD
 601 LCAVYVVQVR CRRLDGLGYW SNWSSPAYTL VMDVKVPMRG PEFWRKMDGD
 651 VTKKERNVTL LWKPLTKNDS LCSVRRYVVK HRTAHNGTWS EDVGNRTNLT
 701 FLWTEPAHTV TVLAVNSLGA SLVNFNLTFS WPMSKVSAVE SLSAYPLSSS
 751 CVILSWTLSP DDYSLLYLVI EWKILNEDDG MKWLRIPSNV KKFYIHDNFI
 801 PIEKYQFSLY PVFMEGVGKP KIINGFTKDA IDKQQNDAGL YVIVPIIISS
 851 CVLLLGTLLI SHQRMKKLFW DDVPNPKNCS WAQGLNFQKP ETFEHLFTKH
 901 AESVIFGPLL LEPEPISEEI SVDTAWKNKD EMVPAAMVSL LLTTPDPESS
 951 SICISDQCNS ANFSGSQSTQ VTCEDECQRQ PSVKYATLVS NDKLVETDEE
1001 QGFIHSPVSN CISSNHSPLR QSFSSSSWET EAQTFFLLSD QQPTMISPQL
1051 SFSGLDELLE LEGSFPEENH REKSVCYLGV TSVNRRESGV LLTGEAGILC
1101 TFPAQCLFSD IRILQERCSH FVENNLSLGT SGENFVPYMP QFQTCSTHSH
1151 KIMENKMCDL TV
```

Alternative splicing variants[10]:

A – PETFE→RTDTL at position 890–894 terminating at 894 (895–1162 missing).

B – full-length form sequence shown here.

C – PET→VTV at position 890–892 terminating at 892 (893–1162 missing)

D – PETFEHLFTKH→DISFHEVFIFR at position 890–900 terminating at 900 (901–1162 missing).

E – DNFIPIEK→GMCTVLFM at position 797–804 terminating at 804 (805–1162 missing) has no transmembrane domain and is released from the cell.

References

[1] Friedman, J.M. and Halaas, J.L. (1998) Nature 395, 763–770.

[2] Auwerx, J. (1998) Lancet 351, 737–742.

[3] Lord, G.M. et al. (1998) Nature 394, 897–901.

[4] Sierra-Honigmann, M.R. et al. (1998) Science 281, 1683–1686.

[5] Kline, A.D. et al. (1997) FEBS Lett. 407, 239–242.
[6] Gong, D.-W. et al. (1996) J. Biol. Chem. 271, 3971–3974.
[7] Zhang, Y. et al. (1994) Nature 372, 425–432.
[8] Chen, H. et al. (1996) Cell 84, 491–495.
[9] Lee, G.-H. et al. (1996) Nature 379, 632–635.
[10] Cioffi, J.A. et al. (1996) Nature Med. 2, 585–589.
[11] Tartaglia, L.A. et al. (1995) Cell 83, 1263–1271.

Other names

Human LIF has been known as human interleukin for DA cells (HILDA), melanoma-derived lipoprotein lipase inhibitor (MLPLI) and hepatocyte-stimulating factor III (HSF III). Mouse LIF has been known as differentiation-inhibiting factor (DIA), differentiation-retarding factor (DRF) and cholinergic nerve differentiation factor (CNDF).

THE MOLECULE

Leukaemia inhibitory factor (LIF) is produced by many different cell types and has pleiotropic actions[1]. LIF stimulates the differentiation of the macrophage cell line M1, and the proliferation of haematopoietic stem cells[2–4]. *In vivo*, it has profound effects on haematopoiesis, particularly in combination with other cytokines such as IL-3, causing increased platelet formation[5]. LIF allows embryonic stem cells to remain in an undifferentiated state and can maintain their proliferation in culture[6]. LIF stimulates synthesis of acute phase proteins by liver cells, increases bone resorption, stimulates differentiation of cholinergic nerves and loss of body fat[7–10].

Crossreactivity

There is 78% homology between human and mouse LIF[2,3]. Human LIF is active on mouse cells, mouse LIF is not active on human cells.

Sources

Multiple, including T cells, myelomonocytic lineages, fibroblasts, liver, heart and melanoma.

Bioassays

Human and murine LIF can be measured using the M1 murine leukaemia cell line assay[2].

Physicochemical properties of LIF

Property	Human	Mouse
pI	9	>9
Amino acids – precursor	202	203[a]
– mature	180	180
M_r (K) – predicted	20	20
– expressed	45	58
Potential *N*-linked glycosylation sites[b]	6	6
Disulfide bonds	3	3

[a] A second form of murine LIF has an alternative leader peptide which targets LIF to matrix[11].
[b] Disulfide bonds link Cys12–134, 18–131 and 60–163[12].

3-D structure

Predicted to be a four α-helical bundle structure similar to other haematopoietin cytokines[13].

Gene structure[14,15]

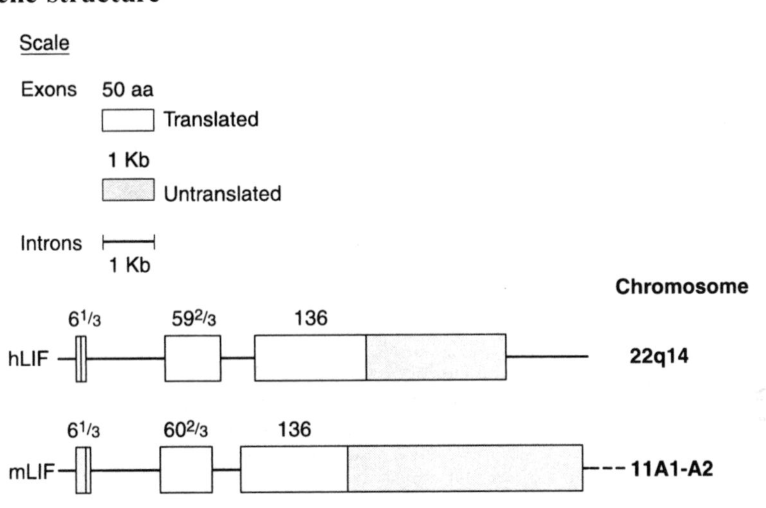

The gene for human LIF is on chromosome 22q14 and contains three exons of $6^{1/3}$, $59^{2/3}$ and 136 amino acids while that of murine LIF is located on chromosome 11A1–A2 also containing three exons of $6^{1/3}$, $60^{2/3}$ and 136 amino acids[15].

Amino acid sequence for human LIF[3,7,9]

Accession code: SwissProt P15018

```
-22  MKVLAAGVVP LLLVLHWKHG AG
  1  SPLPITPVNA TCAIRHPCHN NLMNQIRSQL AQLNGSANAL FILYYTAQGE
 51  PFPNNLDKLC GPNVTDFPPF HANGTEKAKL VELYRIVVYL GTSLGNITRD
101  QKILNPSALS LHSKLNATAD ILRGLLSNVL CRLCSKYHVG HVDVTYGPDT
151  SGKDVFQKKK LGCQLLGKYK QIIAVLAQAF
```

Amino acid sequence for mouse LIF[2]

Accession code: SwissProt P09056

```
-23  MKVLAAGIVP LLLLVLHWKH GAG
  1  SPLPITPVNA TCAIRHPCHG NLMNQIKNQL AQLNGSANAL FISYYTAQGE
 51  PFPNNVEKLC APNMTDFPSF HGNGTEKTKL VELYRMVAYL SASLTNITRD
101  QKVLNPTAVS LQVKLNATID VMRGLLSNVL CRLCNKYRVG HVDVPPVPDH
151  SDKEAFQRKK LGCQLLGTYK QVISVVVQAF
```

THE LIF RECEPTOR

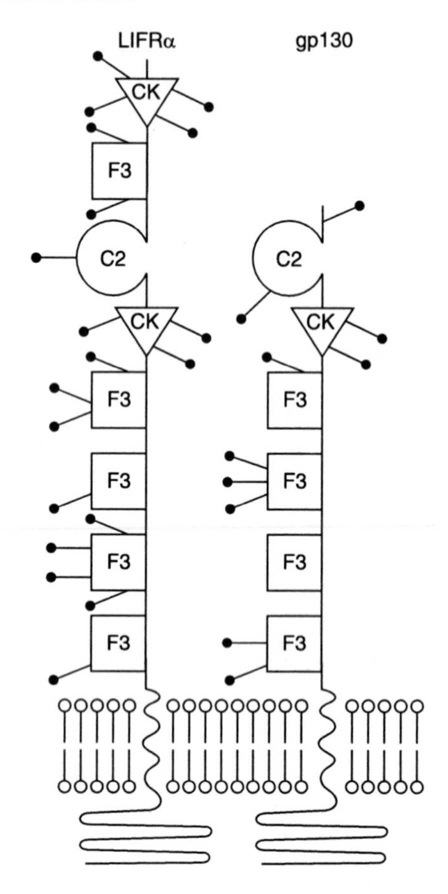

The high-affinity (K_d 10–200 × 10^{-12} M) LIF receptor is a complex of a low affinity (K_d 1–3 nM) receptor (LIFRα-chain) belonging to the cytokine receptor superfamily, and a second affinity conversion chain (gp130) (CDw130) which also forms part of the IL-6R complex[16] (see IL-6 entry (page **69**)). The high-affinity LIF receptor is also a high-affinity receptor for oncostatin M. The LIF receptor and gp130 also form part of the CNTF receptor complex[17]. Antibodies to gp130 block responses to LIF, IL-6, OSM and CNTF[18]. A soluble murine receptor has been described which shares 76% identity with the human receptor[16]. A membrane-bound murine LIFR also exists[19].

Distribution

Monocytes, liver, placenta, embryonic stem cells.

Physicochemical properties of the LIF receptor α-chain

Property	Human	Mouse
Amino acids – precursor	1097	719
– mature	1053	676
M_r (K) – predicted	111	76.2
– expressed	190	130
Potential *N*-linked glycosylation sites	19	13
Affinity K_d (M) – low	10^{-9}	$1–3 \times 10^{-9}$
– high	0.15×10^{-10}	$0.1–2.0 \times 10^{-10}$

Signal transduction

There is no evidence that the low-affinity receptor (LIFR α-chain) can transduce a signal. The C-terminal 110 amino acids can be removed from the receptor and it can still associate with the affinity converter. By analogy with the IL-6 receptor, the affinity converter is likely to be the signalling chain.

Chromosomal location

The human LIFR α-chain is on chromosome 5p12–13 and the mouse LIFR α-chain is on chromosome 15.

Amino acid sequence for human low-affinity LIF receptor[16]

Accession code: Genbank X61615

```
 -44 MMDIYVCLKR PSWMVDNKRM RTASNFQWLL STFILLYLMN QVNS
   1 QKKGAPHDLK CVTNNLQVWN CSWKAPSGTG RGTDYEVCIE NRSRSCYQLE
  51 KTSIKIPALS HGDYEITINS LHDFGSSTSK FTLNEQNVSL IPDTPEILNL
 101 SADFSTSTLY LKWNDRGSVF PHRSNVIWEI KVLRKESMEL VKLVTHNTTL
 151 NGKDTLHHWS WASDMPLECA IHFVEIRCYI DNLHFSGLEE WSDWSPVKNN
 201 SWIPDSQTKV FPQDKVILVG SDITFCCVSQ EKVLSALIGH TNCPLIHLDG
 251 ENVAIKIRNI SVSASSGTNV VFTTEDNIFG TVIFAGYPPD TPQQLNCETH
 301 DLKEIICSWN PGRVTALVGP RATSYTLVES FSGKYVRLKR AEAPTNESYQ
 351 LLFQMLPNQE IYNFTLNAHN PLGRSQSTIL VNITEKVYPH TPTSFKVKDI
 401 NSTAVKLSWH LPGNFAKINF LCEIEIKKSN SVQEQRNVTI QGVENSSYLV
 451 ALDKLNPYTL YTFRIRCSTE TFWKWSKWSN KKQHLTTEAS PSKGPDTWRE
 501 WSSDGKNLII YWKPLPINEA NGKILSYNVS CSSDEETQSL SEIPDPQHKA
 551 EIRLDKNDYI ISVVAKNSVG SSPPSKIASM EIPNDDLKIE QVVGMGKGIL
 601 LTWHYDPNMT CDYVIKWCNS SRSEPCLMDW RKVPSNSTET VIESDEFRPG
 651 IRYNFFLYGC RNQGYQLLRS MIGYIEELAP IVAPNFTVED TSADSILVKW
 701 EDIPVEELRG FLRGYLFYFG KGERDTSKMR VLESGRSDIK VKNITDISQK
 751 TLRIADLQGK TSYHLVLRAY TDGGVGPEKS MYVVTKENSV GLIIAILIPV
 801 AVAVIVGVVT SILCYRKREW IKETFYPDIP NPENCKALQF QKSVCEGSSA
 851 LKTLEMNPCT PNNVEVLETR SAFPKIEDTE IISPVAERPE DRSDAEPENH
 901 VVVSYCPPII EEEIPNPAAD EAGGTAQVIY IDVQSMYQPQ AKPEEEQEND
 951 PVGGAGYKPQ MHLPINSTVE DIAAEEDLDK TAGYRPQANV NTWNLVSPDS
1001 PRSIDSNSEI VSFGSPCSIN SRQFLIPPKD EDSPKSNGGG WSFTNFFQNK
1051 PND
```

Amino acid sequence for mouse low affinity LIF receptor[16]

Accession code: D26177, D17444

```
-43  MAAYSWWRQP SWMVDNKRSR MTPNLPWLLS ALTLLHLTMH ANG
  1  LKRGVQDLKC TTNNMRVWDC TWPAPLGVSP GTVKDICIKD RFHSCHPLET
 51  TNVKIPALSP GDHEVTINYL NGFQSKFTLN EKDVSLIPET PEILDLSADF
101  FTSSLLLKWN DRGSALPHPS NATWEIKVLQ NPRTEPVALV LLNTMLSGKD
151  TVQHWNWTSD LPLQCATHSV SIRWHIDSPH FSGYKEWSDW SPLKNISWIR
201  NTETNVFPQD KVVLAGSNMT ICCMSPTKVL SGQIGNTLRP LIHLYGQTVA
251  IHILNIPVSE NSGTNIIFIT DDDVYGTVVF AGYPPDVPQK LSCETHDLKE
301  IICSWNPGRI TGLVGPRNTE YTLFESISGK SAVFHRIEGL TNETYRLGVQ
351  MHPGQEIHNF TLTGRNPLGQ AQSAVVINVT ERVAPHDPTS LKVKDINSTV
401  VTFSWYLPGN FTKINLLCQI EICKANSKKE VRNATIRGAE DSTYHVAVDK
451  LNPYTAYTFR VRCSSKTFWK WSRWSDEKRH LTTEATPSKG PDTWREWSSD
501  GKNLIVYWKP LPINEANGKI LSYNVSCSLN EETQSVLEIF DPQHRAEIQL
551  SKNDYIISVV ARNSAGSSPP SKIASMEIPN DDITVEQAVG LGNRIFLTWR
601  HDPNMTCDYV IKWCNSSRSE PCLLDWRKVP SNSTETVIES DQFQPGVRYN
651  FYLYGCTNQG YQLLRSIIGY VEELEA
```

For the affinity converter (gp130) see IL-6 entry (page **69**).

References

[1] Gearing, D.P. (1991) Ann. N.Y. Acad. Sci. 628, 9–18.
[2] Gearing, D.P. et al. (1987) EMBO J. 6, 3995–4002.
[3] Gough, N.M. et al. (1988) Proc. Natl Acad. Sci. USA 85, 2623–2627.
[4] Escary, J.L. et al. (1993) Nature 363, 361–364.
[5] Waring, P. et al. (1993) Br. J. Haematol. 83, 80–87.
[6] Williams, R.L. et al. (1988) Nature 336, 684–687.
[7] Moreau, J.-F. et al. (1988) Nature 336, 690–692.
[8] Yamamori, T. et al. (1989) Science 246, 1412–1416.
[9] Baumann, H. and Wong, G.G. (1989) J. Immunol. 143, 1163–1167.
[10] Mori, M. et a.l (1989) Biochem. Biophys. Res. Commun. 160, 1085–1092.
[11] Rathjen, P.D. et al. (1990) Cell 62, 1105–1114.
[12] Nicola, N.A. et al. (1993) Biochem. Biophys. Res. Commun. 190, 20–26.
[13] Bazan, J.F. (1991) Neuron 7, 197- 208.
[14] Sutherland, G.R. et al. (1989) Leukemia 3, 9–13.
[15] Kola, I. et al. (1989) Growth Factors 2, 235–140.
[16] Gearing, D.P. et al. (1991) EMBO J. 10, 2839–2848.
[17] Gearing, D.P. et al. (1992) Science 255, 1434–1437.
[18] Taga, T. et al. (1993) Proc. Natl Acad. Sci. USA 89, 10998–11001.
[19] Gearing, D.P. et al. (1993) Adv. Immunol. 53, 31–58.

Other names
Herpesvirus entry mediator ligand (HVEM-L), LTγ and TNFSF-14.

THE MOLECULE

LIGHT is a relatively new member of the TNF family of membrane-anchored ligands. It was originally discovered on the surface of an activated T-cell hybridoma and shown to bind fusion proteins constituting the extracellular domain of herpesvirus entry mediator (HVEM) and LTβR[1]. LIGHT was subsequently cloned from a cDNA derived from activated peripheral blood lymphocytes based on its homology to FasL and its ability to bind HVEM. LIGHT has been shown to activate NFκB, stimulate the proliferation of lymphocytes and inhibit the growth of the adenocarcinoma cell line HT-29[2,3]. More recently, LIGHT has been implicated in apoptosis induction of human tumour cells[4]. LIGHT is closely related to LTα and LTβ as defined by amino acid sequence homologies and common receptor specificities. The name LIGHT is based on a number of its features – homology to lymphotoxins, inducible expression, competitive antagonism of HSV glycoprotein D binding to HVEM and its expression on T and B cells. A murine cDNA clone for a mouse homologue of LIGHT has recently been identified by signal sequence trap (SST) screening[3]. The ability of LIGHT to interefere with HSV entry suggests that LIGHT may function in antiviral defence mechanisms. In addition, LIGHT has been shown to function as a costimulatory molecule and is required for dendritic cell-mediated allogeneic T-cell responses[5].

Crossreactivity
Human and mouse LIGHT share 76% identity at the amino acid sequence level[3].

Sources
LIGHT mRNA is highly expressed in splenocytes, activated PBL, CD8+ tumour infiltrating lymphocytes, granulocytes, and monocytes[4]. Murine LIGHT transcripts have been detected in spleen, lung, and in heart[3].

Bioassays
Induction of apoptosis in HT-29 adenocarcinoma cells in presence of IFNγ using traditional growth inhibition assays. Induction of IFNγ secretion by peripheral blood lymphocytes and weak NFκB activation[2].

Physicochemical properties of LIGHT

Property	Human	Mouse
pI	8.49	?
Amino acids – precursor	240	239
– mature[a]	240	239
M_r (K) – predicted	26.351	?
– expressed	28–30	?
Potential *N*-linked glycosylation sites	1	?
Disulfide bonds	?	?

[a] Human LIGHT lacks a signal peptide.

3-D structure

A crystal structure of LIGHT is currently unavailable but is predicted to resemble that of TNF, LT and CD40L by sequence comparison.

Gene structure

LIGHT is localized on human chromosome 16p11.2. Fluorescence *in situ* hybridization analysis localized the murine LIGHT gene to chromosome 17 which is tightly linked with TNF, LTα and LTβ[3].

Amino acid sequence for human LIGHT

Accession code: SPTREMBLE O43557

```
  1  MEESVVRPSV FVVDGQTDIP FTRLGRSHRR QSCSVARVGL GLLLLLMGAG
 51  LAVQGWFLLQ LHWRLGEMVT RLPDGPAGSW EQLIQERRSH EVNPAAHLTG
101  ANSSLTGSGG PLLWETQLGL AFLRGLSYHD GALVVTKAGY YYIYSKVQLG
151  GVGCPLGLAS TITHGLYKRT PRYPEELELL VSQQSPCGRA TSSSRVWWDS
201  SFLGGVVHLE AGEEVVVRVL DERLVRLRDG TRSYFGAFMV
```

Human LIGHT lacks a signal peptide. The putative signal anchor (transmembrane region) is underlined.

THE LIGHT RECEPTORS, HVEM, LTβR and DcR3

LIGHT has been shown to bind three cell surface receptors: herpesvirus entry mediator type A (HVEM)[1], LTβR[4] and the decoy receptor, DcR3[6]. HVEM (also known as HveA) is a type I transmembrane receptor which is related to members of the TNF receptor superfamily[7]. HVEM binds a virus-encoded ligand, the envelope glycoprotein D of herpes simplex virus, and thereby facilitates virus entry which has been shown to correlate with immune suppression. LIGHT is the cellular ligand for HVEM, inhibits HSV entry into cells by competing with envelope glycoprotein D from HVEM[1]. The ability of HSV to block LIGHT binding may either mimic or block LIGHT-regulated T cell-mediated immune respnses. HVEM, also known as TR2, ATAR and TNFRSF14, is a 283-amino-acid protein, with a relatively short cytoplasmic domain[7]. Murine HVEM is 276 amino acids long and shares 44% amino acid identity with its human counterpart. The precise biological role of HVEM in normal cellular function is unclear but like LTβR, is speculated to participate as a regulator of lymphoid organogenesis and splenic architecture as well as NK-T cell differentiation, and due to its effects on NFκB and AP-1, it may also function in orchestrating lymphocyte differentiation and effector function. In addition to HVEM as a receptor, LIGHT also binds LTβR[1], but this binding is not inhibited by HSV glycoprotein D. DcR3/DR6 is a recently described member of the TNFR superfamily which binds LIGHT in addition to FasL[6] and thereby suppresses FasL and LIGHT-mediated cell death[6]. DcR3 is a secreted protein which is most closely related to osteoprotegerin.

Distribution

HVEM is broadly expressed, but is most prominent on cells of the immune system such as T cells and dendritic cells[8]. See FasL entry (page **225**) for information on DcR3.

Physicochemical properties of LIGHT receptors

Property	HVEM	
	Human	Mouse
Amino acids – precursor	283	276
– mature[a]	?	?
M_r (K) – predicted	30.42	?
– expressed	40[b]	?
Potential *N*-linked glycosylation sites	2	?
Affinity K_d (M)[c]	0.4×10^{-9}	?

[a] After removal of predicted signal peptide.
[b] The HVEM ectodomain (truncated at His200) under native conditions behaves as a dimer of ~40 kDa and treatment with reducing agents does not alter the dimeric structure, indicating that all cysteines are most likely involved in intramolecular disulfide bonds.
[c] As demonstrated by HVEM-Fc binding studies.

Signal transduction

Upon overexpression HVEM activates NFκB, through a Traf-mediated mechanism[2]. The cytoplasmic domain of HVEM binds Traf-2 and Traf-5, leading to activation of NFκB and AP-1 transcription factors. HVEM has also been shown to activate JNK/AP-1-dependent signalling pathways.

Chromosomal location

The HVEM gene maps to human chromosome 1p36.

Amino acid sequence for human HVEM

Accession code: SPTREMBLE Q92956

```
  1  MEPPGDWGPP PWRSTPRTDV LRLVLYLTFL GAPCYAPALP SCKEDEYPVG
 51  SECCPKCSPG YRVKEACGEL TGTVCEPCPP GTYIAHLNGL SKCLQCQMCD
101  PAMGLRASRN CSRTENAVCG CSPGHFCIVQ DGDHCAACRA YATSSPGQRV
151  QKGGTESQDT LCQNCPPGTF SPNGTLEECQ HQTKCSWLVT KAGAGTSSSH
201  WVWWFLSGSL VIVIVCSTVG LIICVKRRKP RGDVVKVIVS VQRKRQEAEG
251  EATVIEALQA PPDVTTVAVE ETIPSFTGRS PNH
```

References

1. Mauri, D.N. et al. (1998) Immunity 8, 21–30.
2. Harrop, J.A. et al. (1998) J. Biol. Chem. 273, 27548–27556.
3. Misawa, K. et al. (2000) Cytogenet. Cell. Genet. 89, 89–91.
4. Zhai, Y. et al. (1998) J. Clin. Invest. 102, 1142–1151.
5. Tamada, K. et al. (2000) J. Immunol. 164, 4105–4110.
6. Yu, K.Y. et al. (1999) J. Biol. Chem. 274, 13733–13736.
7. Kwon, B. et al. (1999) J. Biol. Chem. 274, 6056–6061.
8. Montgomery, R.I. et al. (1996) Cell 87, 427–436.

Other names
ATAC, SCM-1α.

THE MOLECULE

Lymphotactin (Lptn) is a chemokinetic and chemoattractant factor for T cells and NK cells. It is the prototype of the C or gamma class of chemokines which have a single disulfide bond in the mature protein[1–4].

Crossreactivity
Human Lptn is 60% homologous to mouse Lptn; human Lptn is active on murine cells.

Sources
Activated T cells (mainly CD8+), dendritic epidermal γδ T cells, NK cells, mast cells and basophils[6–8,11].

Bioassays
Chemokinesis of T cells.

Physicochemical properties of Lptn

Property	Human	Mouse
pI	11.3	?
Amino acids – precursor	114	114
– mature	93	93
M_r (K)	10.3	10.3
– expressed[a]	12, 15, 17, 19	13, 18
Potential *N*-linked glycosylation sites	0	0
Disulfide bonds[b]	1	1

[a] Lptn is *O*-glycosylated, leading to multiple M_r forms.
[b] Lptn only has a single disulfide bond unlike other chemokines.

3-D structure
No information.

Gene structure

The gene for human Lptn is on chromosome 1q23, the gene for mouse Lptn is on chromosome 1.

Amino acid sequence for human Lptn

Accession code: SwissProt P47992

```
-21  MRLLILALLG ICSLTAYIVE G
  1  VGSEVSDKRT CVSLTTQRLP VSRIKTYTIT EGSLRAVIFI TKRGLKVCAD
 51  PQATWVRDVV RSMDRKSNTR NNMIQTKPTG TQQSTNTAVT LTG
```

Lptn is probably *O*-glycosylated in the region of amino acids 76–92. The C-terminal 22 amino acids are necessary for function. A homologue of human Lptn called SCM-1β has been described with the N terminal sequence VGSEVHR replacing VGSEVSD.

Amino acid sequence for mouse Lptn

Accession code: SwissProt P47993

```
-21  MRLLLLTFLG VCCLTPWVVE G
  1  VGTEVLEESS CVNLQTQRLP VQKIKTYIIW EGAMRAVIFV TKRGLKICAD
 51  PEAKWVKAAI KTVDGRASTR KNMAETVPTG AQRSTSTAIT LTG
```

Lptn is probably *O*-glycosylated in the region of amino acids 76–92. A variant form with an I→V change at amino acid 89 has been described.

THE Lptn RECEPTOR, XCR1

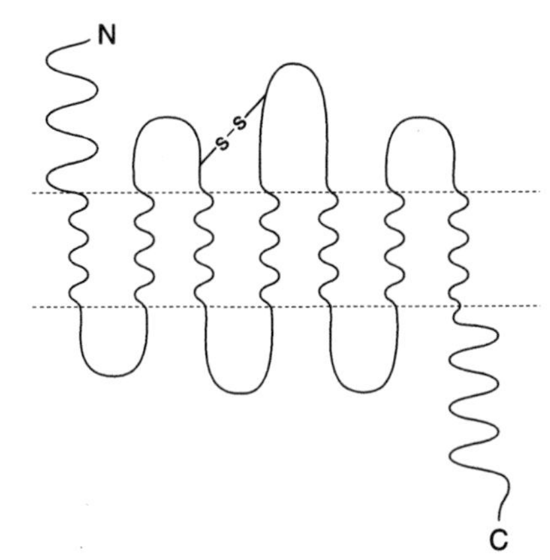

A specific human Lptn receptor designated XCR1(previously GPR5) has recently been cloned[12]. Pharmacological studies indicate that the Lptn receptor is linked to G_i and G_q and mediates a calcium flux in target cells[9,12]. Lptn binds the receptor with a K_d of 10 nm. It is unusual in that it does not bind to the Duffy chemokine receptor, but can bind to heparin[5,10].

Distribution

T cells, NK cells; mRNA is expressed in placenta.

Physicochemical properties of the Lptn receptor

Property	Human	Mouse
Amino acids	333	?
M_r (K) predicted	38 507	?
Potential *N*-linked glycosylation sites	0	?
Affinity K_d (M)	1×10^{-10}	?

Chromosomal location

The gene for the Lptn receptor is on human chromosome 3p21.3–p21.1[13].

Amino acid sequence for human Lptn receptor

Accession code: SwissProt P46094

```
  1 MESSGNPEST TFFYYDLQSQ PCENQAWVFA TLATTVLYCL VFLLSLVGNS
 51 LVLWVLVKYE SLESLTNIFI LNLCLSDLVF ACLLPVWISP YHWGWVLGDF
101 LCKLLNMIFS ISLYSSIFFL TIMTIHRYLS VVSPLSTLRV PTLRCRVLVT
151 MAVWVASILS SILDTIFHKV LSSGCDYSEL TWYLTSVYQH NLFFLLSLGI
201 ILFCYVEILR TLFRSRSKRR HRTVKLIFAI VVAYFLSWGP YNFTLFLQTL
251 FRTQIIRSCE AKQQLEYALL ICRNLAFSHC CFNPVLYVFV GVKFRTHLKH
301 VLRQFWFCRL QAPSPASIPH SPGAFAYEGA SFY
```

References

[1] Kelner, G.S. et al. (1994) Science 266, 1395–1399.
[2] Kennedy, J. et al. (1995) J. Immunol. 155, 203–209.
[3] Muller, S. et al. (1995) Eur. J. Immunol. 25, 1744–1748.
[4] Yoshida, T. et al. (1996) FEBS Lett. 395, 82–88.
[5] Szabo, M.C. et al. (1995) J. Biol. Chem. 270, 25348–25351.
[6] Hedrick, J.A. et al. (1997) J. Immunol. 158, 1533.
[7] Rumsaeng, V. et al. (1997) J. Immunol. 158, 1353–1360.
[8] Boismenu, R. et al. (1996) J. Immunol. 157, 985–992.
[9] Maghazachi, A.A. et al. (1997) FASEB J. 11, 765–774.
[10] Dorner, B. et al. (1997) J. Biol Chem. 272, 8817–8823.
[11] Hautamaa, D. et al. (1997) Cytokine 9, 375–382.
[12] Yoshida, T. et al. (1998) J. Biol. Chem. 273, 16551–16554.
[13] Heiber, M. et al. (1995) DNA Cell Biol. 14, 25–35.

Other names

Tumour necrosis factor β (TNFβ), cytotoxin, differentiation-inducing factor.

THE MOLECULE

Lymphotoxin (LT) has 35% homology with TNFα and binds to the same receptors. Like TNFα, it has a wide range of biological activities from killing of tumour cells *in vitro* to induction of gene expression and stimulation of fibroblast proliferation. It is also an important mediator of inflammation and immune function. Whereas TNFα can be expressed as a membrane protein attached by its long signal sequence, LT has a more conventional signal peptide and is secreted[1–3]. A membrane-bound molecule LTβ has been described. LTβ is a type II integral membrane protein that forms a heteromeric complex with LT on the cell surface[4,5]. The function of this complex is at present unknown.

Crossreactivity

There is 74% homology between human and mouse LT and significant cross-species activity.

Sources

LT is made mostly by activated T and B lymphocytes. There have also been some reports of its production by astrocytes.

Bioassays

The same bioassays for TNFα can be used. Cytotoxicity on murine L929 or WEHI 164 clone 13.

Physicochemical properties of LT

	LT (TNFβ)	LTβ	
Property	Human	Mouse	Human
pI	5.8	5.8	?
Amino acids – precursor	205	202	244
– mature[a]	171	169	244/240[b]
M_r (K) – predicted	18.7	18.6	25.4
– expressed[c]	25	25	33
N-linked glycosylation sites[d]	1	1	1
Disulfide bonds	0	0	0

[a] After removal of predicted signal peptide.

[b] LTβ is a type II integral membrane protein. N-terminal amino acid analysis has shown LTβ to begin with Gly at position 5. There is some evidence that translation is initiated by CTG coding for Leu at position 4 which is subsequently removed.

[c] Corresponds to the subunit secreted by the B lymphoblastoid cell line RPMI 1788. There is also a M_r 20 000 LT secreted by this line which lacks the 23 N-terminal amino acids and may be a breakdown product. The mature molecule is a trimer, except for the membrane-bound complex form[4].

[d] There is *N*-linked glycosylation and some *O*-linked glycosylation[6].

3-D structure

The structure of LT has been solved at a resolution of 1.9 Å[7]. It is very similar to TNFα.

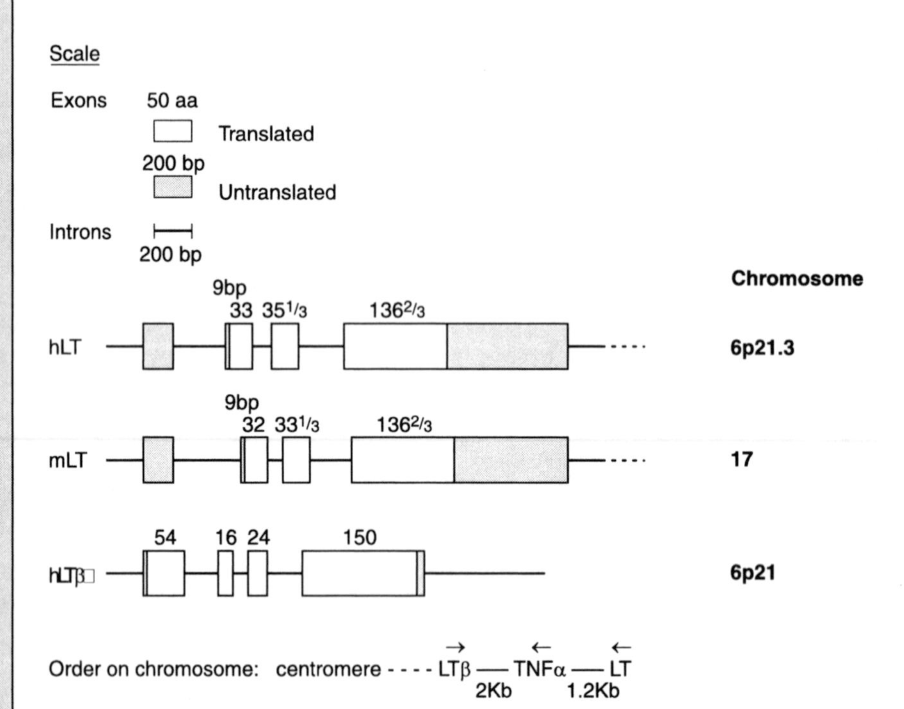

Gene structure[5,8–10]

The genes for LT are located on human chromosome 6p21.3 and mouse chromosome 17, while that of human LTβ is at position 6p21. The human LT gene contains three exons of 33, $35^{1/3}$ and $136^{2/3}$ amino acids while LTβ contains four exons of 54, 16, 24 and 150 amino acids. The murine LT gene also contains three exons of 32, $33^{1/3}$ and $136^{2/3}$ amino acids.

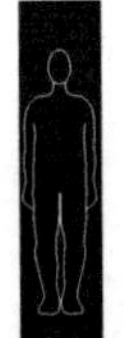

Amino acid sequence for human LT[11]

Accession code: SwissProt P01374

-34 MTPPERLFLP RVCGTTLHLL LLGLLLVLLP GAQG
1 LPGVGLTPSA AQTARQHPKM HLAHSTLKPA AHLIGDPSKQ NSLLWRANTD
51 RAFLQDGFSL SN**N**SLLVPTS GIYFVYSQVV FSGKAYSPKA TSSPLYLAHE
101 VQLFSSQYPF HVPLLSSQKM VYPGLQEPWL HSMYHGAAFQ LTQGDQLSTH
151 TDGIPHLVLS PSTVFFGAFA L

Variant sequence T→N at position 26.

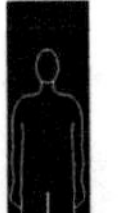

Amino acid sequence for human LTβ[5]

Accession code: GenEMBL L11015

1 MGALGLEGRG GRLQGRGSLL LAVAGATSLV TLLLAVPITV LAVLALVPQD
51 QGGLVTETAD PGAQAQQGLG FQKLPEEEPE TDLSPGLPAA HLIGAPLKGQ
101 GLGWETTKEQ AFLTSGTQFS DAEGLALPQD GLYYLYCLVG YRGRAPPGGG
151 DPQGRSVTLR SSLYRAGGAY GPGTPELLLE GAETVTPVLD PARRQGYGPL
201 WYTSVGFGGL VQLRRGERVY V**N**ISHPDMVD FARGKTFFGA VMVG

Translation may be initiated by CTG coding for Leu at position 4. Removal of the N-terminal amino acid then gives a mature protein begining with Gly at position 5.

Amino acid sequence for mouse LT[12]

Accession code: SwissProt P09225

-33 MTLLGRLHLL RVLGTPPVFL LGLLLALPLG AQG
1 LSGVRFSAAR TAHPLPQKHL THGILKPAAH LVGYPSKQNS LLWRASTDRA
51 FLRHGFSLS**N** NSLLIPTSGL YFVYSQVVFS GESCSPRAIP TPIYLAHEVQ
101 LFSSQYPFHV PLLSAQKSVY PGLQGPWVRS MYQGAVFLLS KGDQLSTHTD
151 GISHLHFSPS SVFFGAFAL

THE LT RECEPTORS

See under TNFα (page **474**). TNFα and LT bind the same receptor molecules.

References

[1] Paul, N.L. and Ruddle, N. H. (1988) Annu. Rev. Immunol. 6, 407–438.
[2] Ruddle, N.H. and Turetskaya, R.L. (1991) In The Cytokine Handbook, Thomson, A.W. ed., Academic Press, London, pp. 257–267.
[3] Ruddle, N.H. (1992) Curr. Opin. Immunol. 4, 327–332.
[4] Androlewicz, M.J. et al. (1992) J. Biol. Chem. 267, 2542–2547.
[5] Browning, J.L. et al. (1993) Cell 72, 847–856.

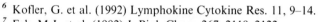

[6] Kofler, G. et al. (1992) Lymphokine Cytokine Res. 11, 9–14.
[7] Eck, M.J. et al. (1992) J. Biol. Chem. 267, 2119–2122.
[8] Nedwin, G.E. et al. (1985) Nucleic Acids Res. 13, 6361–6373.
[9] Semon, D. et al. (1987) Nucleic Acids Res. 15, 9083–9084
[10] Gray, P.W. et al. (1987) Nucleic Acids Res. 15, 3937–3937.
[11] Gray, P.W. et al. (1984) Nature 312, 721–724.
[12] Li, C.-B. et al. (1987) J. Immunol. 138, 4496–4501.

Other names

Human MCP-1 has been known as monocyte chemoattractant and activating factor (MCAF), human JE, lymphocyte-derived chemotactic factor (LDCF) and glioma-derived monocyte chemotactic factor (GDCF). Human MCP-2 has been known as HC14 or SCYA8. Human MCP-3 as been known as SCYA7. Human MCP-4 is also known as CK-BETA-10 NCC-1 and SCYA13. Mouse MCP-1 has been known as JE, and mouse MCP-3 as FIC, SCYA7 or MARC[1]. Murine MCP-5 has been known as SCYA12.

THE MOLECULES

Monocyte chemoattractant proteins 1–5 (MCP-1 to MCP-5) are all chemoattractant and activating factors for monocytes, basophils, T cells, NK cells and immature dendritic cells[2–6]. MCP-3, -4 and -5 are also eosinophil chemoattractant and activating factors[7,8]. There is currently no mouse homologue of MCP-4, nor a human homologue of MCP-5. The MCPs are thought to play key roles in chronic inflammatory and allergic disease.

Crossreactivity

There is about 55% homology between human and mouse MCP-1. MCPs are active across species.

Sources

MCPs are all produced by a wide range of tissues and cells *in vitro* and are upregulated by many proinflammatory stimuli such as IFNγ[9].

Bioassays

MCP activity is measured using Boyden chamber or agarose gel chemotaxis assays. Monocytes and basophils but not neutrophils or lymphocytes are affected. Half-maximal responses are seen with 0.5 ng/ml MCP-1. Histamine release from basophils is a sensitive measure of MCP levels.

Physicochemical properties of MCPs

Property	Human	Mouse
pI	9.7	10.5
Amino acids – precursor	99/99/100/98[a]	148/97/97/104
– mature	76/76/76/75	125[b]/74/74/72
M_r (K) – predicted	8.7	12.0/11.0/11.0/11.7
– expressed	8–18[c]	12–25[c]
Potential *N*-linked glycosylation sites	1/0/1/1[a]	1/0/1/0
Disulfide bonds[d]	2	2

[a] For MCP-1, -2 , -3 and -4 respectively in human and MCP-1, -2, -3 and -5 in the mouse.
[b] Mature mouse MCP-1 has a C-terminal extension of 49 amino acids.
[c] Variable *O*-linked glycosylation gives rise to the variation in molecular weight of human MCPs[10]. There is no *N*-linked glycosylation site in human or mouse MCP-2 or mouse MCP-5.
[d] Disulfide bonds link Cys11–36 and 12–52 in human and mouse MCP-1, a similar arrangement is predicted for MCP-2 to MCP-5.

3-D structure

A model of MCP-1 has been prepared, based on the solution structure of IL-8, indicating a dimeric structure consisting of two three-stranded Greek keys upon which lie two antiparallel α-helices[11]. Sequence differences between IL-8 and MCP-1 in surface residues suggest that biological specificity is determined by residues in the cleft between the two α-helices.

Gene structure[4,12]

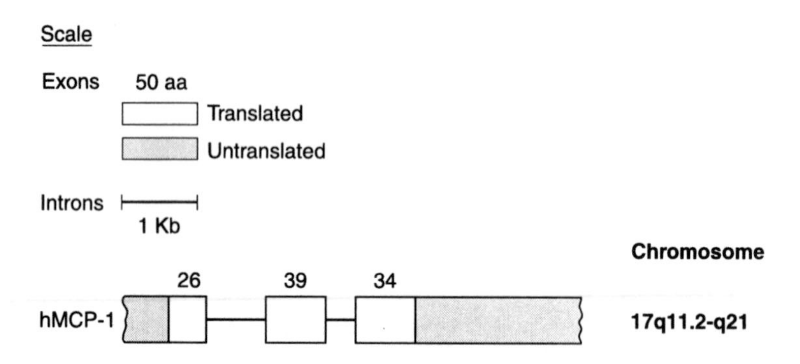

The genes for human MCP-1, -2, -3 and -4 are clustered on chromosome 17q11.2-q21.1, containing three exons of 26, 39 and 34 amino acids, and the mouse gene for MCP-1 is on the distal portion of chromosome 11[4,12].

Amino acid sequence for human MCP-1[13,14]

Accession code: SwissProt P13500

```
-23  MKVSAALLCL LLIAATFIPQ GLA
  1  QPDAINAPVT CCYNFTNRKI SVQRLASYRR ITSSKCPKEA VIFKTIVAKE
 51  ICADPKQKWV QDSMDHLDKQ TQTPKT
```

A proteolytically processed variant lacking five amino acids at the N-terminus has been described[15].

Amino acid sequence for mouse MCP-1[3]

Accession code: SwissProt P10148

```
-23  MQVPVMLLGL LFTVAGWSIH VLA
  1  QPDAVNAPLT CCYSFTSKMI PMSRLESYKR ITSSRCPKEA VVFVTKLKRE
 51  VCADPKKEWV QTYIKNLDRN QMRSEPTTLF KTASALRSSA PLNVKLTRKS
101  EANASTTFST TTSSTSVGVT SVTVN
```

Amino acid sequence for human MCP-2[5,16]

Accession code: SwissProt P78388

```
-23  MKVSAALLCL LLMAATFSPQ GLA
  1  QPDSVSIPIT CCFNVINRKI PIQRLESYTR ITNIQCPKEA VIFKTKRGKE
 51  VCADPKERWV RDSMKHLDQI FQNLKP
```

A K→Q variant at position 46 has been found.

Amino acid sequence for mouse MCP-2

Accession code: SwissProt Q9Z121

```
-23  MKIYAVLLCL LLIAVPVSPE KLT
  1  GPDKAPVTCC FHVLKLKIPL RVLKSYERIN NIQCPMEAVV FQTKQGMSLC
 51  VDPTQKWVSE YMEILDQKSQ ILQP
```

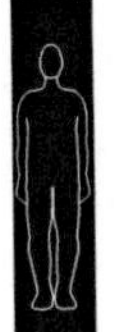

Amino acid sequence for human MCP-3[5]

Accession code: SwissProt P80098

```
-24  MKASAALLCL LLTAAAFSPQ GLA
  1  QPVGINTSTT CCYRFINKKI PKQRLESYRR TTSHSCPREA VIFKTKLDKE
 51  ICADPTQKWV QDFMKHLDKK TQTPKL
```

Amino acid sequence for mouse MCP-3[17–19]

Accession code: SwissProt Q03366

```
-23  MRISATLLCL LLIAAAFSIQ VWA
  1  QPDGPNASTC CYVKKQKIPK RNLKSYRRIT SSRCPWEAVI FKTKKGMEVC
 51  AEAHQKWVEE AIAYLDMKTP TPKP
```

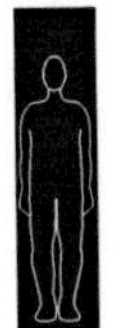

Amino acid sequence for human MCP-4[20–22]

Accession code: SwissProt Q99616

```
-23  MKVSAVLLCL LLMTAAFNPQ GLA
  1  QPDALNVPST CCFTFSSKKI SLQRLKSYVI TTSRCPQKAV IFRTKLGKEI
 51  CADPKEKWVQ NYMKHLGRKA HTLKT
```

Amino acid sequence for mouse MCP-5[7,8]

Accession code: SwissProt Q62401

```
-22  MKISTLLCLL LIATTISPQV LA
  1  GPDAVSTPVT CCYNVVKQKI HVRKLKSYRR ITSSQCPREA VIFRTILDKE
 51  ICADPKEKWV KNSINHLDKT SQTFILEPSC LG
```

THE MCP RECEPTORS, CCR1, CCR2, CCR3, CCR5, CCR11

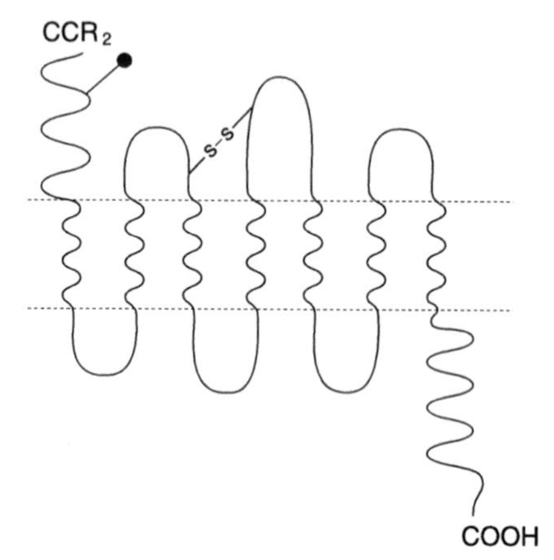

All of the five MCPs bind to the CCR2 receptors expressed on monocytes, basophils and memory T cells, forming the basis of their shared biological activity. In addition MCP-3 and MCP-4 also bind to CCR1 and to the eotaxin receptor CCR3, mediating their eosinophil effects. Recently the human MCPs have all been reported to bind to CCR5[23,24]. For the sequences for CCR1, CCR3 and CCR5 see MIP-1α (page **384**), eotaxin (page 213) and RANTES (page **444**).

Distribution

No information.

Signal transduction

MCPs cause pertussis toxin-sensitive calcium flux and chemotaxis in monocytes and the receptors are G protein-linked[7,25,26]. Although MCP-1 causes a Ca flux via CCR2b, it does not appear to do so via the CCR2a splice variant which differs only in the C-terminal intracellular tail[27].

Chromosomal location

The human gene for CCR2 is on chromosome 3p21 and the murine gene is on chromosome 9. The gene for human CCR11 is on chromosome 3q22.

Amino acid sequence for human MCP receptor CCR2a[28,29]

Accession code: SwissProt P41597

```
  1  MLSTSRSRFI RNTNESGEEV TTFFDYDYGA PCHKFDVKQI GAQLLPPLYS
 51  LVFIFGFVGN MLVVLILINC KKLKCLTDIY LLNLAISDLL FLITLPLWAH
101  SAANEWVFGN AMCKLFTGLY HIGYFGGIFF IILLTIDRYL AIVHAVFALK
151  ARTVTFGVVT SVITWLVAVF ASVPGIIFTK CQKEDSVYVC GPYFPRGWNN
201  FHTIMRNILG LVLPLLIMVI CYSGILKTLL RCRNEKKRHR AVRVIFTIMI
251  VYFLFWTPYN IVILLNTFQE FFGLSNCEST SQLDQATQVT ETLGMTHCCI
301  NPIIYAFVGE KFRSLFHIAL GCRIAPLQKP VCGGPGVRPG KNVKVTTQGL
351  LDGRGKGKSI GRAPEASLQD KEGA
```

Disulfide bond links Cys113–190.

Amino acid sequence for human MCP receptor CCR2b[28,29]

Accession code: SwissProt P41597

```
  1  MLSTSRSRFI RNTNESGEEV TTFFDYDYGA PCHKFDVKQI GAQLLPPLYS
 51  LVFIFGFVGN MLVVLILINC KKLKCLTDIY LLNLAISDLL FLITLPLWAH
101  SAANEWVFGN AMCKLFTGLY HIGYFGGIFF IILLTIDRYL AIVHAVFALK
151  ARTVTFGVVT SVITWLVAVF ASVPGIIFTK CQKEDSVYVC GPYFPRGWNN
201  FHTIMRNILG LVLPLLIMVI CYSGILKTLL RCRNEKKRHR AVRVIFTIMI
251  VYFLFWTPYN IVILLNTFQE FFGLSNCEST SQLDQATQVT ETLGMTHCCI
301  NPIIYAFVGE KFRRYLSVFF RKHITKRFCK QCPVFYRETV DGVTSTNTPS
351  TGEQEVSAGL
```

Disulfide bond links Cys113–190.

Amino acid sequence for mouse MCP receptor CCR2[30,31]

Accession code: SwissProt P51683

```
  1  MEDNNMLPQF IHGILSTSHS LFTRSIQELD EGATTPYDYD DGEPCHKTSV
 51  KQIGAWILPP LYSLVFIFGF VGNMLVIIIL IGCKKLKSMT DIYLLNLAIS
101  DLLFLLTLPF WAHYAANEWV FGNIMCKVFT GLYHIGYFGG IFFIILLTID
151  RYLAIVHAVF ALKARTVTFG VITSVVTWVV AVFASLPGII FTKSKQDDHH
201  YTCGPYFTQL WKNFQTIMRN ILSLILPLLV MVICYSGILH TLFRCRNEKK
251  RHRAVRLIFA IMIVYFLFWT PYNIVLFLTT FQESLGMSNC VIDKHLDQAM
301  QVTETLGMTH CCINPVIYAF VGEKFRRYLS IFFRKHIAKR LCKQCPVFYR
351  ETADRVSSTF TPSTGEQEVS VGL
```

Disulfide bond links Cys126–203.

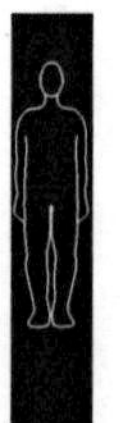

Amino acid sequence for human MCP receptor CCR11

Accession code: SwissProt Q9NPB9

```
  1 MALEQNQSTD YYYEENEMNG TYDYSQYELI CIKEDVREFA KVFLPVFLPI
 51 VFVIGLAGNS MVVAIYAYYK KQRTKTDVYI LNLAVADLLL LFTLPFWAVN
101 AVHGWVLGKI MCKITSALYT LNFVSGMQFL ACISIDRYVA VTKVPSQSGV
151 GKPCWIICFC VWMAAILLSI PQLVFYTVND NARCIPIFPR YLGTSMKALI
201 QMLEICIGFV VPFLIMGVCY FITARTLMKM PNIKISRPLK VLLTVVIVFI
251 VTQLPYNIVK FCRAIDIIYS LITSCNMSKR MDIAIQVTES IALFHSCLNP
301 ILYVFMGASF KNYVMKVAKK YGSWRRQRQS VEEFPFDSEG PTEPTSTFSI
```

References

[1] Kulmburg, P.A. et al. (1992) J. Exp. Med. 176, 1773–1778.
[2] Leonard, E.J. and Yoshimura, T. (1990) Immunol. Today 11, 97–101.
[3] Rollins, B.J. et al. (1988) Proc. Natl Acad. Sci. USA 85, 3738–3742.
[4] Schall, T.J. (1991) Cytokine 3, 165–183.
[5] Van Damme, J. et al. (1992) J. Exp. Med. 176, 59–65.
[6] Minty, A. et al. (1993) Eur. Cytok. Network 4, 99–110.
[7] Jia, G.Q. et al (1996). J. Exp. Med. 184:1939,
[8] Sarafi, M.Net al (1997) J. Exp. Med. 185:99–109).
[9] Cushing, S.D. et al. (1990) Proc. Natl Acad. Sci. USA 87, 5134–5138.
[10] Jiang, Y. et al. (1990) J. Biol. Chem. 265, 18318–18321.
[11] Gronenborn, A.M. and Clore G.M. (1991) Protein Eng. 4, 263–269.
[12] Wilson, S.D. et al. (1990) J. Exp. Med. 171, 1301–1314.
[13] Matsushima, K. et al. (1989) J. Exp. Med. 169, 1485–1490.
[14] Robinson, E.A. et al. (1989) Proc. Natl Acad. Sci. USA 86, 1850–1854.
[15] Decock, B. et al. (1990) Biochem. Biophys. Res. Commun. 167, 904–909.
[16] Van Coillie, E. et al. (1997) Genomics 40, 323–331.
[17] Kulmburg, P.A. et al. (1992) J. Exp. Med. 176, 1773–1778.
[18] Thirion, S. et al. (1994) Biochem. Biophys. Res. Commun. 201, 493–499.
[19] Heinrich, J.N. et al. (1993) Mol. Cell. Biol. 13, 2020–2030.
[20] Garcia-Zepeda, E.A. et al. (1996) J. Immunol. 157, 5613–5626.
[21] Uguccioni. M. et al. (1996) J. Exp. Med. 183, 2379–2384.
[22] Berkhout, T.A. et al. (1997) J. Biol. Chem. 272, 16404–16413.
[23] Schweikart, V.L. et al. (2000) J. Biol. Chem. 275, 9550–9556.
[24] Blanpain, C. et al. (1999) Blood 94, 1899–1905.
[25] Rollins, B.J. et al. (1991) Blood 78, 1112–1116.
[26] Yoshimura, T. and Leonard, E.J. (1990) J. Immunol. 145, 292–297.
[27] Sanders, S.K. et al. (2000) J. Immunol. 165, 4877–4883.
[28] Charo, I.F. et al. (1994) Proc. Natl Acad. Sci. USA 91, 2752–2756.
[29] Yamagami, S. et al. (1994) Biochem. Biophys. Res. Commun. 202, 1156–1162.
[30] Boring, L. et al. (1996) J. Biol. Chem. 271, 7551–7558.
[31] Kurihara, T et al. (1996) J. Biol. Chem. 271, 11603–11606.

Other names

Colony-stimulating factor 1 (CSF-1), macrophage and granulocyte inducer IM (MGI-IM).

THE MOLECULE

Macrophage colony-stimulating factor (M-CSF) is a survival, growth, differentiation and activating factor for macrophages and their progenitor cells[1,2]. M-CSF is a complex cytokine, which can be produced as integral cell surface or secreted protein variants[3–6].

Crossreactivity

There is about 82% homology between human and mouse M-CSF in the N-terminal 227 amino acid of the mature sequence, falling to 47% in the rest of the molecule[4]. Human M-CSF is active on murine cells but not vice versa[4].

Sources

Multiple, including lymphocytes, monocytes, fibroblasts, epithelial cells, endothelial cells, myoblasts and osteoblasts.

Bioassays

Macrophage colony formation from bone marrow in soft agar, or proliferation of the M-NFS60 murine cell line[7,8].

Physicochemical properties of M-CSF

Property	Human	Mouse
pI	3–5	?
Amino acids – precursor	554	552
– mature	522/406/224[a]	519
M_r (K) – predicted	60	61
– expressed[b]	45–90	45–90
Potential *N*-linked glycosylation sites	3	3
Disulfide bonds	7–9	7–9

[a] M-CSF is derived from three major mRNA species to yield 522 amino acid (M-CSFb), 406 amino acid (M-CSFg, lacking residues 332–447), and 224 amino acid (M-CSFa, lacking residues 150–447) polypeptides which share the same N- and C-termini.

[b] M-CSF is biologically active as a disulfide-linked homodimer[9]. The active homodimers are made up from N-terminal chains (amino acids 1–221) and are released from the cell membrane by proteolytic cleavage[1]. Residues 1–149 are required for biological activity.

3-D structure

The structure of a recombinant M-CSF residues 4–158 has been determined by X-ray crystallography to 2.5 Å[9]. The molecule is comprised of two bundles of four α-helices laid end to end. The connectivity of the helices is similar to GM-CSF, G-CSF, IL-4, IL-2 and growth hormone.

Gene structure[10,11]

Scale

Exons 50 aa
□ Translated

500 bp
▨ Untranslated

Introns ⊢⊣
2 Kb

Chromosome

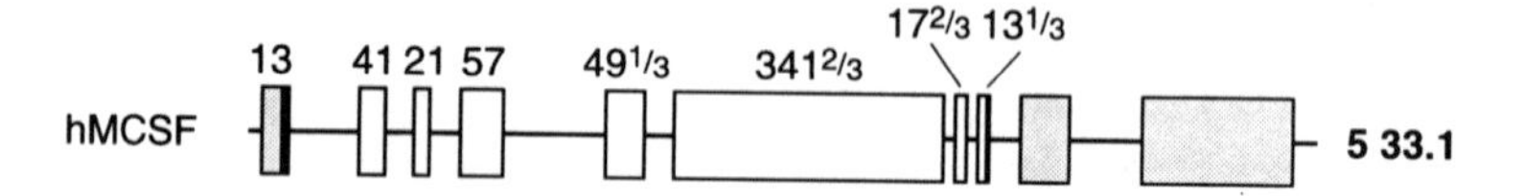

The gene for human M-CSF is located on chromosome 5q33.1 and consists of 10 exons (seven translated exons of 41, 21, 57, $49^{1/3}$, $341^{2/3}$, $17^{2/3}$ and $13^{1/3}$ amino acids). The first and last two exons are untranslated. The size of exon 6 can be $341^{2/3}$ or $43^{2/3}$ depending on which splice acceptor site is used[5]. Exon 10 (untranslated) gives an alternative 3' untranslated sequence[11]. The mouse M-CSF gene is on chromosome 11.

Amino acid sequence for human M-CSF[2,3,5]

Accession code: SwissProt P09603

-32 MTAPGAAGRC PPTTWLGSLL LLVCLLASRS IT
1 EEVSEYCSHM IGSGHLQSLQ RLIDSQMETS CQITFEFVDQ EQLKDPVCYL
51 KKAFLLVQDI MEDTMRFRDN TPNAIAIVQL QELSLRLKSC FTKDYEEHDK
101 ACVRTFYETP LQLLEKVKNV F**N**ETKNLLDK DWNIFSKNC**N** NSFAECSSQD
151 VVTKPDCNCL YPKAIPSSDP ASVSPHQPLA PSMAPVAGLT WEDSEGTEGS
201 SLLPGEQPLH TVDPGSAKQR P*PRSTCQSFE PPETPVVKDS TIGGSPQPRP*
251 *SVGAFNPGME DILDSAMGTN WVPEEASGEA SEIPVPQGTE LSPSRPGGGS*
301 *MQTEPARPSN FLSASSPLPA SAKGQQPADV TGTALPRVGP VRPTGQDW***N***H*
351 *TPQKTDHPSA LLRDPPEPGS PRISSLRPQG LSNPSTLSAQ PQLSRSHSSG*
401 *SVLPLGELEG RRSTRDRRSP AEPEGGPASE GAARPLPRFN SVPLTDTGHE*
451 *RQSEGSSSPQ LQESVFHLLV PSVILVLLAV GGLLFYRWRR RSHQEPQRAD*
501 *SPLEQPEGSP LTQDDRQVEL PV*

The biologically active soluble form of M-CSF (residues 1–221) is released by proteolytic cleavage from the membrane-bound propeptide (in italics).

Amino acid sequence for mouse M-CSF[4,12]

Accession code: SwissProt P07141

```
-33  MTARGRAGRC PSSTWLGSRL LLVCLLMSRS IAK
  1  EVSEHCSHMI GNGHLKVLQQ LIDSQMETSC QIAFEFVDQE QLDDPVCYLK
 51  KAFFLVQDII DETMRFKDNT PNANATERLQ ELSNNLNSCF TKDYEEQNKA
101  CVRTFHETPL QLLEKIKNFF NETKNLLEKD WNIFTKNCNN SFAKCSSRDV
151  VTKPDCNCLY PKATPSSDPA SASPHQPPAP SMAPLAGLAW DDSQRTEGSS
201  LLPSELPLRI EDAGSAKQRP PRSTCQTLES TEQPNHGDRL TEDSQPHPSA
251  GGPVPGVEDI LESSLGTNWV LEEASGEASE GFLTQEAKFS PSTPVGGSIQ
301  AETDRPRALS ASPFPKSTED QKPVDITDRP LTEVNPMRPI GQTQNNTPEK
351  TDGTSTLRED HQEPGSPHIA TPNPQRVSNS ATPVAQLLLP KSHSWGIVLP
401  LGELEGKRST RDRRSPAELE GGSASEGAAR PVARFNSIPL TDTGHVEQHE
451  GSSDPQIPES VFHLLVPGII LVLLTVGGLL FYKWKWRSHR DPQTLDSSVG
501  RPEDSSLTQD EDRQVELPV
```

THE M-CSF RECEPTOR

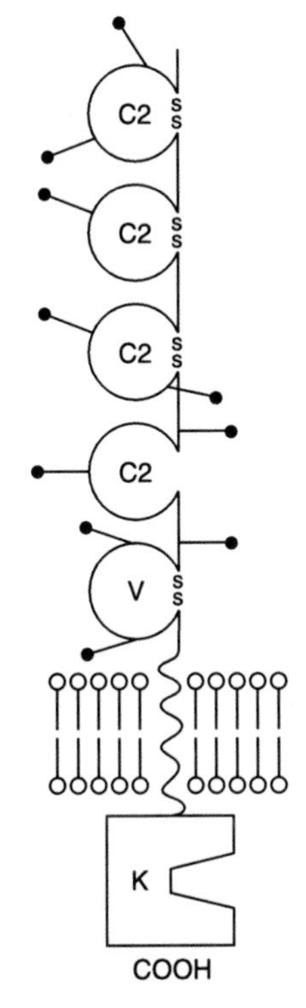

The M-CSF receptor (CD115) is identical to the product of the c-*fms* proto-oncogene and is related to the PDGF receptor, FGFb receptor and c-*kit* (the stem cell factor receptor)[13–17]. These receptors are all autophosphorylating tyrosine kinases[18]. The intracellular kinase domain contains a 70-amino-acid insert which is necessary for association with PI-3′ kinase. The receptor exists in a single affinity form.

Distribution

Macrophages and their progenitors, placental cells.

Physicochemical properties of the M-CSF receptor

Property	Human	Mouse
Amino acids – precursor	972	976
– mature	953	957
M_r (K) – predicted	106	109
– expressed	150	165
Potential *N*-linked glycosylation sites	11	9
Affinity K_d (M)	0.4×10^{-10}	$<2 \times 10^{-10}$

Signal transduction

M-CSF binding to its receptor stabilizes dimerization, leading to activation of the kinase which autophosphorylates the receptor, PI-3 kinase, and other cytoplasmic proteins[18]. The M-CSF receptor is also G protein-linked and stimulates protein kinase C translocation to the cell membrane[19].

Chromosomal location

The human gene is located on chromosome 5q33–q35[16].

Amino acid sequence for human M-CSF receptor[15]

Accession code: SwissProt P07333

```
-19  MGPGVLLLLL VATAWHGQG
  1  IPVIEPSVPE LVVKPGATVT LRCVGNGSVE WDGPASPHWT LYSDGSSSIL
 51  STNNATFQNT GTYRCTEPGD PLGGSAAIHL YVKDPARPWN VLAQEVVVFE
101  DQDALLPCLL TDPVLEAGVS LVRVRGRPLM RHTNYSFSPW HGFTIHRAKF
151  IQSQDYQCSA LMGGRKVMSI SIRLKVQKVI PGPPALTLVP AELVRIRGEA
201  AQIVCSASSV DVNFDVFLQH NNTKLAIPQQ SDFHNNRYQK VLTLNLDQVD
251  FQHAGNYSCV ASNVQGKHST SMFFRVVESA YLNLSSEQNL IQEVTVGEGL
301  NLKVMVEAYP GLQGFNWTYL GPFSDHQPEP KLANATTKDT YRHTFTLSLP
351  RLKPSEAGRY SFLARNPGGW RALTFELTLR YPPEVSVIWT FINGSGTLLC
401  AASGYPQPNV TWLQCSGHTD RCDEAQVLQV WDDPYPEVLS QEPFHKVTVQ
451  SLLTVETLEH NQTYECRAHN SVGSGSWAFI PISAGAHTHP PDEFLFTPVV
501  VACMSIMALL LLLLLLLYK YKQKPKYQVR WKIIESYEGN SYTFIDPTQL
551  PYNEKWEFPR NNLQFGKTLG AGAFGKVVEA TAFGLGKEDA VLKVAVKMLK
601  STAHADEKEA LMSELKIMSH LGQHENIVNL LGACTHGGPV LVITEYCCYG
651  DLLNFLRRKA EAMLGPSLSP GQDPEGGVDY KNIHLEKKYV RRDSGFSSQG
701  VDTYVEMRPV STSSNDSFSE QDLDKEDGRP LELRDLLHFS SQVAQGMAFL
751  ASKNCIHRDV AARNVLLTNG HVAKIGDFGL ARDIMNDSNY IVKGNARLPV
801  KWMAPESIFD CVYTVQSDVW SYGILLWEIF SLGLNPYPGI LVNSKFYKLV
851  KDGYQMAQPA FAPKNIYSIM QACWALEPTH RPTFQQICSF LQEQAQEDRR
901  ERDYTNLPSS SRSGGSGSSS SELEEESSSE HLTCCEQGDI AQPLLQPNNY
951  QFC
```

Tyr680, 689 and 790 are autophosphorylated.

Amino acid sequence for mouse M-CSF receptor[7]

Accession code: SwissProt P09581

-19 MELGPPLVLL LATVWHGQG
1 APVIEPSGPE LVVEPGETVT LRCVS**N**GSVE WDGPISPIWT LDPESPGSTL
51 TTS**N**ATFKNT GTYRCTELED PMAGSTTIHL YVKDPAHSWN LLAQEVTVVE
101 GQEAVLPCLI TDPALKDSVS LMREGGRQVL RKTVYFFSPW RGSIIRKAKV
151 LDSNTYVCKT MVNGRESTST GIWLKVNRVH PEPPQIKLEP SKLVRIRGEA
201 AQIVCSATNA EVGFNVILKR GDTKLEIPLN SDFQDNYYKK VRALSLNAVD
251 FQDAGIYSCV ASNDVGTRTA TMNFQVVESA YL**N**LTSEQSL LQEVSVGDSL
301 ILTVHADAYP SIQHY**N**WTYL GPFFEDQRKL EFITQRAIYR YTFKLFLNRV
351 KASEAGQYFL MAQNKAGWN**N** LTFELTLRYP PEVSVTWMPV **N**GSDVLFCDV
401 SGYPQPSVTW MECRGHTDRC DEAQALHLW**N** DTHPEVLSQK PFDKVIIQSQ
451 LPIGPLKH**N**M TYFCKTHNSV G**N**SSQYFRAV SLGQSKQLPD ESLFTPVVVA
501 CMSVMSLLVL LLLLLLYKYK QKPKYQVRWK IIERYEGNSY TFIDPTQLPY
551 NEKWEFPRNN LQFGKTLGAG AFGKVVEATA FGLGKEDAVL KVAVKMLKST
601 AHADEKEALM SELKIMSHLG QHENIVNLLG ACTHGGPVLV YTEYCCYGDH
651 LNFLRRKAEA MLGPSLSPGQ DSEGDSSYKN IHLEKKYVRR DSGFSSQGVD
701 TYVEMRPVST SSSDSFFKQD LDKEHSRPLE LWDLLHFSSQ VAQGMAFLAS
751 KNCIHRDVAA RNVLLTSGHV AKIGDFGLAR DIMNDSNYVV KGNALPVKWM
801 APESIFDCVI TVQSDVWSYG ILLWEIFSLG LNPYPGIHVN NKFYKLVKDG
851 YQMAQPVFAP KNIYSIMQSC WDLEPTRRPT FQQICFLLQE QARLERRDQD
901 YANLPSSGGS SGSDSGGGSS GGSSSEPEEE SSSEHLACCE PGDIAQPLLQ
951 PNNYQFC

Tyr678, 687 and 788 are autophosphorylated.

References

[1] Sherr, C.J. and Stanley, E.R. (1990) In Growth Factors, Differentiation Factors and Cytokines, Habenicht, A. ed., Springer-Verlag, Berlin.
[2] Kawasaki, E.S. et al. (1985) Science 230, 291–296.
[3] Wong, G.C. et al. (1987) Science 235, 1504–1508.
[4] Ladner, M.B. et al. (1988) Proc. Natl Acad. Sci. USA 85, 6706–6710.
[5] Ladner, M.B. et al. (1987) EMBO J. 6, 2693–2698.
[6] Cerreti, D.P. et al. (1988) Mol. Immunol. 25, 761–770.
[7] Testa, N.G. et al. (1991) In Cytokines: A Practical Approach, Balkwill, F.A. ed., IRL Press, Oxford, pp. 229–244.
[8] Wadhwa, M. et al. (1991) In Cytokines: A Practical Approach, Balkwill, F.A. ed., IRL Press, Oxford, pp. 309–329.
[9] Pandit, J. et al. (1992) Science 258, 1358–1362.
[10] Ladner, M.B. et al. (1987) EMBO J. 6, 2693–2698.
[11] Pettenati, M.J. et al. (1987) Proc. Natl Acad. Sci. USA 84, 2970–2974.
[12] Delamater, J.F. et al. (1987) Nucleic Acids Res. 15, 2389–2390.
[13] Sherr, C.J. (1990) Blood 75, 1–12.
[14] Sherr, C. et al. (1985) Cell 41, 665–676.
[15] Coussens, L. et al. (1986) Nature 320, 277–280.
[16] Hampe, A. et al. (1989) Oncogene Res. 4, 9–17.
[17] Rothwell, V.M. and Rohrschneider, L.R. (1987) Oncogene Res. 1, 311–324.

[18] Ullrich, A. and Schlessinger, J. (1990) Cell 61, 203–212.
[19] Vairo, G. and Hamilton, J.A. (1991) Immunol. Today 12, 362–369.

Other names

MDC is also known as stimulated T-cell chemoattractant protein 1 (STCP-1).

THE MOLECULE

Macrophage-derived chemokine is a chemoattractant for Th2 lymphocytes, monocyte-derived dendritic cells and IL-2 activated NK cells[1–4]. MDC is also a potent inhibitor of HIV infection of CD8+ T cells *in vitro*[6]. MDC is produced constituitively by B cells, macrophages and monocyte-derived dendritic cells and by activated monocytes, NK cells and CD4 T cells. Monocyte production of MDC is stimulated by the Th2 cytokines IL-4 and IL-13, and inhibited by the Th1 cytokine IFNγ[3,4]. MDC is one of the most highly expressed mRNA species in macrophages[5]. It shares most sequence homology (37%) with another CC chemokine, TARC, which also binds to a shared receptor CCR4[11].

Crossreactivity

Human MDC is active on mouse lymphocytes[2].

Sources

Macrophages, monocytes, dendritic cells, NK cells, T lymphocytes produce MDC, whilst mRNA is expressed in thymus, lymph nodes and appendix[1,2].

Bioassays

Chemotaxis of Th2 lymphocytes.

Physicochemical properties of MDC

Property	Human	Mouse
Amino acids – precursor	93	?
– mature	69	?
M_r (K) – predicted	10.6	?
– expressed	8	?
Potential *N*-linked glycosylation sites	0	?
Disulfide bonds	2	?

Purified natural MDC has lost the two N-terminal amino acids. This truncated form retains HIV inhibitory efficacy but has lost its chemoattractant ability.

3-D structure

No information.

Gene structure

The gene for human MDC is on chromosome 16q13[8].

Amino acid sequence for human MDC[6]

Accession code: SwissProt O00626

```
-24  MARLQTALLV VLVLLAVALQ ATEA
  1  GPYGANMEDS VCCRDYVRYR LPLRVVKHFY WTSDSCPRPG VVLLTFRDKE
 51  ICADPRVPWV KMILNKLSQ
```

The first two residues, GP, can be removed by proteolysis to give a new N-terminus of YGAN[6].

THE RECEPTOR, CCR4

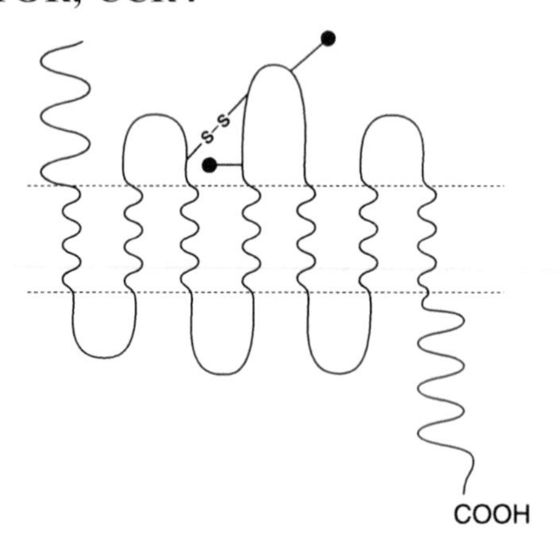

The MDC receptor, CCR4 (previously known as K5-5, CC CKR-4) is also shared with the related CC chemokine TARC[8–11]. CCR4 binds MDC with high affinity, mediating a Ca flux which is pertussis toxin sensitive. MDC is a more potent chemoattractant than TARC for lymphocte chemoattraction. The N-terminally truncated MDC shows impaired binding to CCR4, fails to cause chemotaxis, but retains HIV inhibitory properties[7]. This suggests that a second receptor for MDC exists which is utilized by HIV for infection.

Distribution

CCR4 is predominantly expressed on Th2 lymphocytes, but can also be found on basophils, megakaryocytes and platelets. mRNA expression is particularly strong in thymus[3,8–11].

Physicochemical properties of the CCR4 receptor

Property	Human	Mouse
Amino acids – precursor	360	360
– mature	?	?
M_r (K) – predicted	41.5	41.5
– expressed	?	?
Potential *N*-linked glycosylation sites	2	3
Affinity K_d (M)	0.2×10^{-9}	?

Signal transduction

MDC binds specifically to CCR4 with high affinity (0.18 nM), mediating a Ca flux with EC_{50} ranging between 0.1 and 1.0 nM[2,3,8].

Chromosomal location

No information.

Amino acid sequence for human CCR4 receptor

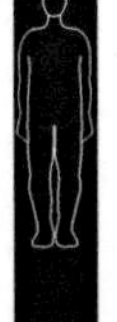

Accession code: SwissProt P51679

```
  1  MNPTDIADTT LDESIYSNYY LYESIPKPCT KEGIKAFGEL FLPPLYSLVF
 51  VFGLLGNSVV VLVLFKYKRL RSMTDVYLLN LAISDLLFVF SLPFWGYYAA
101  DQWVFGLGLC KMISWMYLVG FYSGIFFVML MSIDRYLAIV HAVFSLRART
151  LTYGVITSLA TWSVAVFASL PGFLFSTCYT ERNHTYCKTK YSLNSTTWKV
201  LSSLEINILG LVIPLGIMLF CYSMIIRTLQ HCKNEKKNKA VKMIFAVVVL
251  FLGFWTPYNI VLFLETLVEL EVLQDCTFER YLDYAIQATE TLAFVHCCLN
301  PIIYFFLGEK FRKYILQLFK TCRGLFVLCQ YCGLLQIYSA DTPSSSYTQS
351  TMDHDLHDAL
```

Disulfide bond links Cys110–187.

Amino acid sequence for mouse CCR4 receptor

Accession code: SwissProt P51680

```
  1  MNATEVTDTT QDETVYNSYY FYESMPKPCT KEGIKAFGEV FLPPLYSLVF
 51  LLGLFGNSVV VLVLFKYKRL KSMTDVYLLN LAISDLLFVL SLPFWGYYAA
101  DQWVFGLGLC KIVSWMYLVG FYSGIFFIML MSIDRYLAIV HAVFSLKART
151  LTYGVITSLI TWSVAVFASL PGLLFSTCYT EHNHTYCKTQ YSVNSTTWKV
201  LSSLEINVLG LLIPLGIMLF WYSMIIRTLQ HCKNEKKNRA VRMIFGVVVL
251  FLGFWTPYNV VLFLETLVEL EVLQDCTLER YLDYAIQATE TLGFIHCCLN
301  PVIYFFLGEK FRKYITQLFR TCRGPLVLCK HCDFLQVYSA DMSSSSYTQS
351  TVDHDFRDAL
```

Disulfide bond links Cys110–187.

References

[1] Godiska, R. et al. (1997) J. Exp. Med. 185, 1595–1604.
[2] Chang, M. et al. (1997) J. Biol. Chem. 272, 25229–25237.
[3] Andrew, D.P. et al. (1998) J. Immunol. 161, 5027–5038.
[4] Bonecchi, R. et al. (1998) Blood 92, 2668–2671.
[5] Chantry, D. et al. (1998) J. Leucocyte Biol. 64, 49–54.
[6] Pal, R. et al. (1998) Science 278, 695–698.
[7] Struyf, S. et al. (1998) J. Immunol. 161, 2672–2675.
[8] Imai, T. et al (1998) J. Biol. Chem. 273, 1764–1768.
[9] Power, C.A. et al. (1995) J. Biol. Chem. 270, 19495–19500.
[10] Hoogewerf, A.J. et al. (1996) Biochem. Biophys. Res. Commun. 218, 337–343.
[11] Imai, T. et al. (1997) J. Biol. Chem. 272, 15036–15402.

Other names
None.

THE MOLECULE

Mucosae-associated epithelial chemokine (MEC) is a CC chemokine that is a chemoattractant for T cells and eosinophils. It is produced by epithelial cells in mucosal sites such as salivary gland and colon. MEC shares 40% identity with CTACK.[1]

Crossreactivity
No murine homologue has been described.

Sources
Epithelial cells from bronchioles, salivary gland, colon and mammary gland.

Bioassays
Lymphocyte chemoattraction.

Physicochemical properties of MEC

Property	Human
pI	?
Amino acids – precursor	127
– mature	105
M_r (K) – predicted	?
– expressed	?
Potential *N*-linked glycosylation sites	1

3-D structure
No information.

Gene structure

The gene for MEC is located on human chromosome 5.

Amino acid sequence for human MEC

Accession code: AF266504

```
-22  MQQRGLAIVA LAVCAALHAS EA
  1  ILPIASSCCT EVSHHISRRL LERVNMCRIQ RADGDCDLAA VILHVKRRRI
 51  CVSPHNHTVK QWMKVQAAKK NGKGNVCHRK KHHGKRNSNR AHQGKHETYG
101  HKTPY
```

THE MEC RECEPTORS, CCR3 and CCR10

MEC is a chemoattractant for human CLA + T cells and for eosinophils via binding to the CCR3 or CCR10 receptors with an EC_{50} of 100–200 nM. For information on CCR3 see the eotaxin 1 entry (page **213**) and for CCR10 see the CTACK entry (page **200**).

References

[1] Pan, J. et al. (2000) J. Immunol. 165, 2943–2949.

Other names
Macrophage migration inhibition factor, glycosylation inhibition factor (GIF).

THE MOLECULE

Migration inhibition factor (MIF) is a highly atypical cytokine originally described to activate macrophages and also inhibit their migration[1,5,6]. MIF can be detected immunohistochemically in inflamed tissues, and in LPS-stimulated anterior pituitary cells acting as a mediator of lethal endotoxaemia[1,4]. MIF is inducible by glucocorticoids, and acts to overide the anti-inflammatory and immunosuppressive effects of glucocorticoids[7]. MIF has been shown to be ezymatically active, possessing tautomerase and oxidoreductase activity[8]. Glycosylation inhibition factor is a post-translationally modified form of MIF, lacking in MIF activity, that acts to inhibit glycosylation of IgE-binding factors and thereby reduces IgE synthesis[9].

Crossreactivity
Human MIF and GIF are active on murine cells. Murine MIF has 89% homology to human MIF.

Sources
MIF can be detected in most cell types, but it is readily produced by activated T cells, anterior pituitary cells, hepatocytes and possibly monocytes and endothelial cells[2,4].

Bioassays
Inhibition of macrophage migration from agar gel drops, or from capillary tubes[3]. The migration is also inhibited by IL-4 and IFNα. GIF activity is detected using a T-cell hybridoma assay for glycosylation of IgE-binding protein[9].

Physicochemical properties of MIF

Property	Human	Mouse
Amino acids – precursor	115	114
– mature	?	?
M_r (K) – predicted	12.3	12.3
– expressed	12.3, 12.4, 12.5	12.3
Potential *N*-linked glycosylation sites	2	2
Disulfide bonds	1	1

MIF produced by leukocytes exists as multimers ranging from 25 to 68 kDa. GIF is derived from MIF by post-translational modification cysteinylation of Cys60 and phosphorylation of Ser91[10]. The first Pro residue is critical for enzymatic activity of MIF.

3-D structure

The X-ray structure of MIF reveals a trimer of identical subunits, each monomer contains two antiparallel α-helices packed against a four-stranded β-sheet. The trimer forms a barrel with a central hole[11,12].

Gene structure

The gene for human MIF is located on chromosome 22q11.2[13] and for mouse MIF on chromosome 10[14].

Amino acid sequence for human MIF[2]

Accession code: SwissProt P14174

```
  1  MPMFIVNTNV PRASVPDGFL SELTQQLAQA TGKPPQYIAV HVVPDQLMAF
 51  GGSSEPCALC SLHSIGKIGG AQNRSYSKLL CGLLAERLRI SPDRVYINYY
101  DMNAASVGWN NSTFA
```

There is no obvious signal sequence. Conflicting sequence S→N at position 106.

Amino acid sequence for mouse MIF

Accession code: SwissProt P34884

```
  1  PMFIVNTNVP RASVPEGFLS ELTQQLAQAT GKPAQYIAVH VVPDQLMTFS
 51  GTNDPCALCS LHSIGKIGGA QNRNYSKLLC GLLSDRLHIS PDRVYINYYD
101  MNAANVGWNG STFA
```

THE MIF RECEPTORS

Although high-affinity cell surface binding of GIF has been demonstrated on T and NK cells[15], no specific receptors have been cloned. It has recently been shown that MIF produced intracellularly or taken up by cells acts by binding to, and enzymatically inactivating Jab-1, a co-activator of AP-1 transcription and promoter of degradation of p27^{kip1}[16].

References

[1] Malorny, U. et al. (1988) Clin. Exp. Immunol. 71, 164–170.
[2] Weiser, W.Y. et al. (1989) Proc. Natl Acad. Sci. USA 86, 7522–7526.
[3] Harrington, J.T. and Stastny, P. (1973) J. Immunol. 110, 752–759.
[4] Bernhagen, J. et al. (1993) Nature 365, 756–759.
[5] Cunha, F.Q. et al. (1993) J. Immunol. 150, 1908.
[6] Weiser, W.Y. et al. (1991) J. Immunol. 147, 2006.
[7] Calandra, T. et al. (1995) Nature 377, 68–71.
[8] Lubetsky, J.B. et al. (1999) Biochemistry 38, 7346–7354.
[9] Mikayama, T. et al. (1993) Proc. Natl Acad. Sci. USA 90, 10056–10060.
[10] Watarai, H. et al. (2000) Proc. Natl Acad. Sci. USA 97, 13251–13256.
[11] Sun, H.W. et al. (1996) Proc. Natl Acad. Sci. USA 93, 5191–5196.

[12] Kato, Y. et al. (1996) Proc. Natl Acad. Sci. USA 93, 3007–3010.
[13] Burdarf, M. et al. (1997) Genomics 39, 235–236.
[14] Mitchell, R. et al. (1995) J. Immunol. 154, 3863–3870.
[15] Sugie, K. et al. (1997) Proc. Natl Acad. Sci. USA 94, 5278–5283.
[16] Kleeman, R. et al. (2000) Nature 408, 211–216.

Other names
None.

THE MOLECULE

Monokine induced by interferon γ (MIG) is a non-ELR CXC chemokine which acts as a selective Th1 lymphocyte chemoattractant, causes lymphocyte adhesion to endothelial cells, is angiostatic, suppresses haematopoietic progenitor cell proliferation and mediates tumour regression *in vivo*[1–7].

Crossreactivity
Human MIG is active on murine cells[4,6].

Sources
Interferon γ stimulated monocytes, macrophages and endothelial cells[1–3].

Bioassays
Chemotaxis of IL-2-activated lymphocytes.

Physicochemical properties of MIG

Property	Human	Mouse
pI	11.1	?
Amino acids – precursor	125	126
– mature	103	105
M_r (K) – predicted	14.0	14.5
– expressed	?	14, 10, 8[10]
Potential *N*-linked glycosylation sites	0	1
Disulfide bonds	2	2

3-D structure
No information.

Gene structure

Human MIG is on chromosome 4q21.

Amino acid sequence for human MIG

Accession code: SwissProt Q07325

```
-22  MKKSGVLFLL GIILLVLIGV QG
  1  TPVVRKGRCS CISTNQGTIH LQSLKDLKQF APSPSCEKIE IIATLKNGVQ
 51  TCLNPDSADV KELIKKWEKQ VSQKKKQKNG KKHQKKKVLK VRKSQRSRQK
101  KTT
```

Amino acid sequence for mouse MIG

Accession code: SwissProt P18340

```
-22   MKSAVLFLLG IIFLEQCGVR G
  1   TLVIRNARCS CISTSRGTIH YKSLKDLKQF APSPNCNKTE IIATLKNGDQ
 51   TCLDPDSANV KKLMKEWEKK INQKKKQKRG KKHQKNMKNR KPKTPQSRRR
101   SRKTT
```

THE MIG RECEPTOR, CXCR3

MIG shares a receptor CXCR3 with two other chemokines, γIP-10 and I-TAC[9]. See γIP-10 entry (page 328). MIG binds to CXCR3 with high affinity, although precise quantitation of K_d was impossible due to high nonspecific binding to cell surface proteoglycans[9]. MIG causes a calcium flux with an EC_{50} of about 100 nM in IL-2-activated peripheral blood lymphocytes, but promotes adhesion to endothelial cells at subnanomolar concentrations[8].

References

1 Farber, J.M. et al. (1990) Proc. Natl Acad. Sci. USA 87, 5238–5242.
2 Farber, J.M. et al. (1993) Biochem. Biophys. Res. Commun. 192, 223–230
3 Liao, F. et al. (1995) J. Exp. Med. 182, 1301–1314.
4 Sgadari, C. et al. (1997) Blood 89, 2635–2643.
5 Strieter, R.M. et al. (1995) J. Biol. Chem. 270, 27348–27357.
6 Schwartz, G.N. et al. (1997) J. Immunol. 159, 895–904.
7 Sallusto, F et al. (1998) J. Exp. Med. 187, 875–883.
8 Piali, L. et al. (1998) Eur. J. Immunol. 28, 961–972.
9 Loetscher, M. et al. (1996) J. Exp. Med. 184, 963–969.
10 Tannenbaum, C.S. et al. (1998) J. Immunol. 161, 927–932.

Other names

Human MIP-1α has been known as Hu MIP-1α, pLD78, pAT464 and GOS19. Mouse MIP-1α has been known as TY5 and stem cell inhibitor.

THE MOLECULE

Macrophage inflammatory protein 1α (MIP-1α) is a monocyte-, neutrophil- and lymphocyte-derived CC chemokine family member with inflammatory and chemokinetic properties[1,2]. The principal functions of MIP-1α include cellular recruitment and trafficking, host inflammatory responses, immune regulation, bone remodelling and wound healing[3–7]. Most recently, an important link between MIP-1α and HIV-1 infection was revealed with the discovery that CCR5, a receptor for MIP-1α is utilized by macrophagetropic strains of HIV-1 as a co-receptor for entry into target cells. A second nonalleleic gene for MIP-1α in the human genome has been discovered, which shares 90% identity with MIP-1α and is referred to as MIP-1αP.

Crossreactivity

There is about 75% amino acid identity between human and mouse MIP-1α. Human and mouse MIP-1α are active on human and murine haematopoietic cells[8].

Sources

MIP-1α is not expressed constitutively in most cells (with some exceptions), but rather expression occurs as a result of celluolar activation either by cytokine stimulation or intracellular infection. Activated T cells, B cells, Langerhans cells, neutrophils and macrophages all produce MIP-1α[9,10,19].

Bioassays

There are two bioassays used to assess the activity of MIP-1α. First, inhibition of haematopoietic stem cell proliferation in the CFU-A assay[3]. The second involves traditional chemotaxis assays for monocytes or eosinophils.

Physicochemical properties of MIP-1α

Property	Human MIP-1α	Mouse MIP-1αP	
pI	4.6	4.77	4.6
Amino acids – precursor	92	93	92
– mature	70	70	69
M_r (K) – predicted	8.6	7.79	8
– expressed[a]	8–200	8–200	8–200
Potential *N*-linked glycosylation sites			
Disulfide bonds[b]	2	2	2

[a] MIP-1α probably exists as high-molecular-weight aggregates under physiological conditions.
[b] By homology with the structures of IL-8 and platelet factor 4, the disulfide bonds should link Cys11–51 and 12–35 (human) and 11–50 and 12–34 (mouse).

3-D structure

The crystal structure for MIP-1α is unknown but is predicted to resemble that of other members of the CC chemokine superfamily[5]. MIP-1α is thought to exist as a dimer or trimer in solution.

Gene structure[11,14]

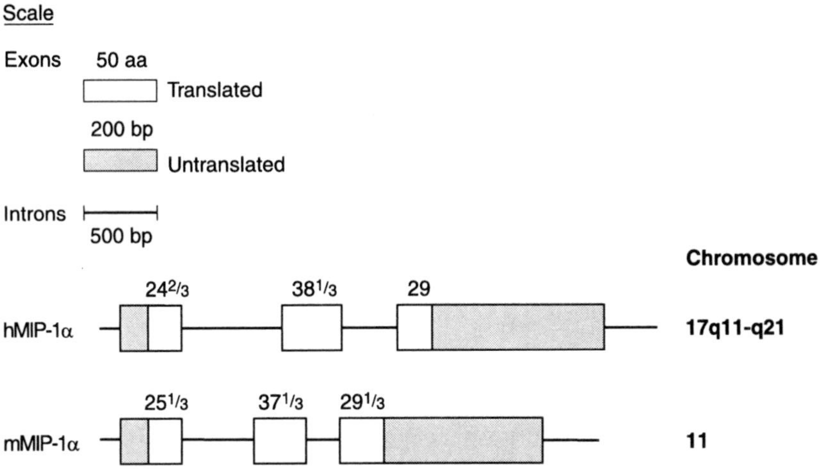

The genes for MIP-1α are located on human chromosome 17q11–q21 and mouse chromosome 11. Both genes contain three exons and two introns (exons $24^{2/3}$, $38^{1/3}$ and 29 amino acids in human and $25^{1/3}$, $37^{1/3}$ and $29^{1/3}$ in mouse). Human MIP-1α is 14 kbp from MIP-1β in a head-to-head configuration. A second nonallelic human MIP-1α gene which varies by five amino acids has also been identified.

Amino acid sequence for human MIP-1α[9,12,13]

Accession code: SwissProt P10147

```
-22  MQVSTAALAV LLCTMALCNQ FS
  1  ASLAADTPTA CCFSYTSRQI PQNFIADYFE TSSQCSKPGV IFLTKRSRQV
 51  CADPSEEWVQ KYVSDLELSA
```

Amino acid sequence for human MIP-1αP

Accession code: SwissProt P16619

```
-23  MQVSTAALAV LLCTMALCNQ VLS
  1  APLAADTPTA CCFSYTSRQI PQNFIADYFE TSSQCSKPSV IFLTKRGRQV
 51  CADPSEEWVQ KYVSDLELSA
```

Disulfide bridges between Cys34–58 and 35–74 by similarity.

Amino acid sequence for mouse MIP-1α[2,14]

Accession code: SwissProt P10855, P14096

```
-23  MKVSTTALAV LLCTMTLCNQ VFS
  1  APYGADTPTA CCFSYSRKIP RQFIVDYFET SSLCSQPGVI FLTKRNRQIC
 51  ADSKETWVQE YITDLELNA
```

Conflicting sequence F→L at position −2, V→A at position 39.

THE MIP-1α RECEPTORs, CCR1 AND CCR5

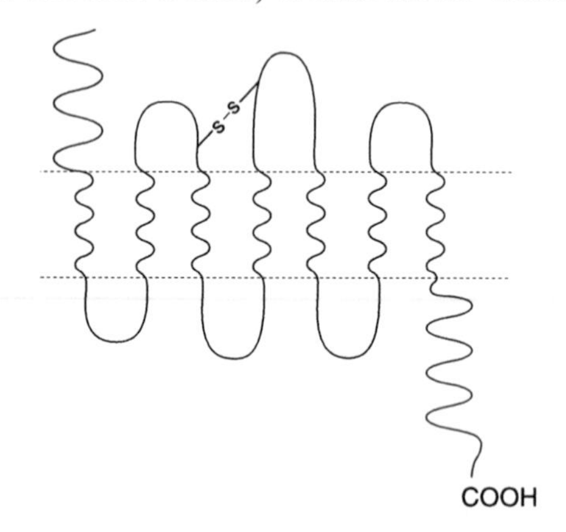

The two main receptors that mediate the biological activities of MIP-1α are CCR1 and CCR5, both of which are broadly expressed. CCR4 and CCR9/D6 have also been reported as receptors for MIP-1α, but their physiological relevance as MIP-1α receptors is questionable. CCR1 was originally cloned following a homology hybridization cloning approach from human leukocytes[15,18]. The cloned receptor expressed in human kidney cells binds MIP-1α (K_d 5 nM) and RANTES (K_d 468 nM) with high affinity and physiological concentrations of MIP-1α at 10 nm and RANTES at 100 nM trigger a significant calcium flux via the receptor. CCR1 has also been shown to bind MIP-1β and MCP-1, but with much lower affinity and show a poor response indicating an inability to initiate signalling. A gene product of cytomegalovirus, hcmv-us28, encodes a functional protein with homology to the MIP-1α receptor[15]. In addition CCR1 responds with high affinity to other chemokines such as HCC-1, MCP-3, MPIF-1 and leukotactin 1. Information on specific CCR1-mediated activities are unclear due to the lack of agonistic or antagonistic antibodies to CCR1 and the ability of CCR1 ligands to activate other chemokine receptors. For details of CCR5, see the RANTES entry (page **444**); for details of CCR9, see the TECK entry (page **460**); for details of CCR4, see the MDC entry (page **373**).

Distribution

CCR1 is widely expressed in different haematopoietic cells. CCR1 has been identified on human monocytes and lymphocytes, but not neutrophils. In contrast, mouse CCR1 is expressed on neutrophils, but the physiological basis for this difference is unclear at present. In addition the receptor is expressed on human dendritic cells and eosinophils[16,17].

Physicochemical properties of CCR1 receptors

Property	Human	Mouse
Amino acids – precursor	355	355
– mature	355	355
M_r (K) – predicted	41.2	40.9
– expressed	?	?
Potential *N*-linked glycosylation sites	1	0
Affinity K_d (M)	5×10^{-9}	?

Signal transduction

The CCR1 receptor is a member of the G protein-coupled receptors and induces increased intracellular calcium ion concentration. No associated or intrinsic kinases specific for CCR1 have as yet been demonstrated and the lack of CCR1-neutralizing antibodies have prevented CCR1-specific signalling pathways from being elucidated.

Chromosomal location

The gene for CCR1 is located on human chromosome 3p21.

Amino acid sequence for human CCR1[15]

Accession code: SwissProt P32246

```
  1  METPNTTEDY DTTTEFDYGD ATPCQKVNER AFGAQLLPPL YSLVFVIGLV
 51  GNILVVLVLV QYKRLKNMTS IYLLNLAISD LLFLFTLPFW IDYKLKDDWV
101  FGDAMCKILS GFYYTGLYSE IFFIILLTID RYLAIVHAVF ALRARTVTFG
151  VITSIIIWAL AILASMPGLY FSKTQWEFTH HTCSLHFPHE SLREWKLFQA
201  LKLNLFGLVL PLLVMIICYT GIIKILLRRP NEKKSKAVRL IFVIMIIFFL
251  FWTPYNLTIL ISVFQDFLFT HECEQSRHLD LAVQVTEVIA YTHCCVNPVI
301  YAFVGERFRK YLRQLFHRRV AVHLVKWLPF LSVDRLERVS STSPSTGEHE
351  LSAGF
```

Disulfide bridge between Cys106–183 by similarity. Conflict in sequence E→D at position 337.

Amino acid sequence for mouse CCR1

Accession code: SwissProt P51675

```
  1  MEISDFTEAY PTTTEFDYGD STPCQKTAVR AFGAGLLPPL YSLVFIIGVV
 51  GNVLMILVLM QHRRLQSMTS IYLFNLAVSD LVFLFTLPFW IDYKLKDDWI
101  FGDAMCKLLS GFYYLGLYSE IFFIILLTID RYLAIVHAVF ALRARTVTLG
151  IITSIITWAL AILASMPALY FFKAQWEFTH RTCSPHFPYK SLKQWKRFQA
201  LKLNLLGLIL PLLVMIICYA GIIRILLRRP SEKKVKAVRL IFAITLLFFL
251  LWTPYNLSVF VSAFQDVLFT NQCEQSKHLD LAMQVTEVIA YTHCCVNPII
301  YVFVGERFWK YLRQLFQRHV AIPLAKWLPF LSVDQLERTS SISPSTGEHE
351  LSAGF
```

Disulfide bridge between Cys106–183 by similarity. Conflict in sequence M→V at position 55 of precursor.

References

[1] Wolpe, S.D. et al. (1988) J. Exp. Med. 167, 570–581.
[2] Davatelis, G. et al. (1988) J. Exp. Med. 167, 1939–1944.
[3] Graham, G.J. et al. (1990) Nature 344, 442–445.
[4] Lord, B.I. et al. (1992) Blood 79, 2605–2609.
[5] Clements, J.C. et al. (1992) Cytokine 4, 76–82.
[6] Davatelis, G. et al. (1989) Science 243, 1066–1068.
[7] Schall, T.J. (1991) Cytokine 3, 165–183.
[8] Broxmeyer, H.E. et al. (1989) J. Exp. Med. 170, 1583–1594.
[9] Obaru, K. et al. (1986) J. Biochem. 99, 885–894.
[10] Yamamura, Y. et al. (1989) J. Clin. Invest. 84, 1707–1712.
[11] Wilson, S.D. et al. (1990) J. Exp. Med. 171, 1301–1314.
[12] Irving, S.G. et al. (1990) Nucleic Acids Res. 18, 3261–3270.
[13] Nakao, M. et al. (1990) Mol. Cell Biol. 10, 3646–3658.
[14] Grove, M. et al. (1990) Nucleic Acids Res. 18, 5561.
[15] Neote, K. et al. (1993) Cell 72, 415–425.
[16] Oh, K.O. et al. (1991) J. Immunol. 147, 2978–2983.
[17] Rot, A. et al. (1992). J. Exp. Med. 176, 1489–1495.
[18] Gao, J.-L. et al. (1993) J. Exp. Med. 177, 1421–1427.
[19] Kasama, T. et al. (1993) J. Exp. Med 178, 63–72.

Other names

Human MIP-1β has been known as Hu Mip-1β, ACT-2, pAT744, hH400, hSISα, G26, HC21, MAD-5 and HIMAP. Mouse MIP-1β has been known as MIP-1β, H400 and SISγ.

THE MOLECULE

Macrophage inflammatory protein 1β is a monocyte- and lymphocyte-derived chemokine of the CC superfamily which is structurally and functionally related to MIP-1α and may function in concert with it. MIP-1β was origionally cloned from a T-cell library and was shown to be chemotactic for monocytes and more recently T lymphocytes (Th1-type T lymphocytes), NK cells and dendritic cells. Its range of biological activities is now known to extend beyond the realm of chemotaxis. MIP-1β provides an important signal for T-cell activation resulting in enhanced proliferation, IL-2 secretion and cell surface IL-2 receptor expression. In addition to its many shared functions with MIP-1α, it exhibits specific functional differences; unlike MIP-1α it is chemotactic for neutrophils. MIP-1β has stimulatory effects on myelopoietic cell growth while MIP-1α has inhibitory effects. More recently an antiviral role for MIP-1β, like that observed for MIP-1α, in HIV-1 infection has been observed. MIP-1β binding to its receptor CCR5 interferes with viral use of CCR5 for cell entry[1–5].

Crossreactivity

There is about 70% homology between human and mouse MIP-1β[4]. Human and mouse MIP-1β are active on human and murine haematopoietic cells.

Sources

T cells, B cells and macrophages[5,7].

Bioassays

Human MIP-1β is a chemoattractant for memory T cells. Murine MIP-1β enhances haematopoietic colony formation in the presence of GM-CSF. MIP-1β may antagonize the effects of MIP-1β[8].

Physicochemical properties of MIP-1β

Property	Human	Mouse
pI	4.6	4.6
Amino acids – precursor	92	92
– mature	69	69
M_r (K) – predicted	8.6	8
– expressed	7.8	8–100
Potential *N*-linked glycosylation sites[a]	0	0
Disulfide bonds[b]	2	2

[a] There may be some *O*-linked glycosylation.
[b] By homology with the structures of IL-8 and platelet factor 4, the disulfide bonds should link Cys11–35 and 12–51 in human and mouse MIP-1β.

3-D structure

High-resolution NMR solution structure demonstrates that MIP-1β exists as a dimer stabilized by hydrogen bonding and hydrophobic interactions. The main secondary structure elements are composed of a triple-stranded antiparallel β-sheet arranged in a Greek key, on top of which lies an α-helix.

Gene structure[9–11]

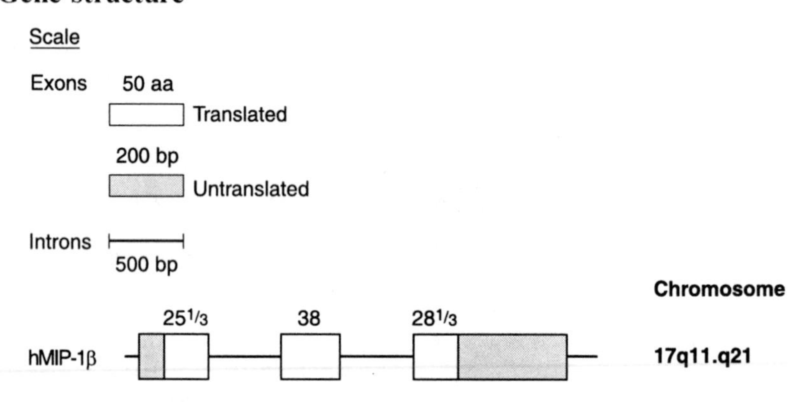

There are multiple nonallelic genes for human MIP-1β located on chromosome 17 in bands q11–q21. Like all members of the CC chemokine family, the gene consists of three exons which contain $25^{1/3}$, 38 and $28^{1/3}$ amino acids. The genes for murine MIP-1β are located on the distal portion of chromosome 11[9–11]. Both are closely linked to MIP-1α. Human MIP-1β is 14 kb from MIP-1α in a head-to-head configuration.

Amino acid sequence for human MIP1β[5,7]

Accession code: SwissProt P13236

```
-23  MKLCVTVLSL LMLVAAFCSP ALS
  1  APMGSDPPTA CCFSYTARKL PRNFVVDYYE TSSLCSQPAV VFQTKRSKQV
 51  CADPSESWVQ EYVYDLELN
```

Conflicting sequence L→I at position −14, P→L at position −4, KLPR→EASS/H, at position 19–22, S→I at position 32, S→G at position 47 and S→T at position 57.

Amino acid sequence for mouse MIP-1β[1]

Accession code: SwissProt P14097

```
-23  MKLCVSALSL LLLVAAFCAP GFS
  1  APMGSDPPTS CCFSYTSRQL HRSFVMDYYE TSSLCSKPAV VFLTKRGRQI
 51  CPNPSQPWVT EYMSHLELN
```

Conflicting sequence C→A at position 51, Q→E at position 56 and L→D at position 66.

THE MIP-1β RECEPTOR, CCR5

At present, CCR5 is accepted as the major MIP-1β-binding protein on cells[12]. CCR5 binds MIP-1β with high affinity (K_d 1.56 nM). CCR9 and CCR8 have also been reported to bind MIP-1β, but whether these interactions are physiologically relevant remains to be determined. MIP-1β also binds with very low affinity to the cloned MIP-1α receptor, CCR1, but does not initiate signalling. Murine MIP-1α and β compete with human MIP-1β for binding to the human receptor[13]. CCR5 has also been shown to be an important co-receptor for HIV infection, with numerous polymorphisms in the coding and noncoding regions of the gene affecting HIV infection. For details of CCR5, see the RANTES entry (page **444**).

References

1 Sherry, B. et al. (1988) J. Exp. Med. 168, 2251–2259.
2 Wolpe, S.D. et al. (1988) J. Exp Med 167, 570–581.
3 Broxmeyer, H.E. et al. (1989) J. Exp. Med. 170, 1583–1594.
4 Schall, T.J. (1991) Cytokine 3, 165–183.
5 Proudfoot, A.E. et al. (1999) Biochem. Pharmacol. 57, 451–463.
6 Lipes, M.A. et al. (1988) Proc. Natl Acad. Sci. USA 85, 9704–9708.
7 Zipfel, P.F. et al. (1989) J. Immunol. 142, 1582–1590.
8 Fahey, T.J. et al. (1992) J. Immunol. 148, 2764–2769.
9 Irving, S.G. et al. (1990) Nucleic Acids Res. 18, 3261–3270.
10 Wilson, S.D. et al. (1990) J. Exp. Med. 171, 1301–1314.
11 Napolitano, M. et al. (1990) J. Biol. Chem. 266, 17531–17536.
12 Wu, L. et al. (1997) J. Exp. Med. 186, 1373–1381.
13 Napolitano, M. et al. (1990) J. Exp. Med. 172, 285–289.

Other names

HGF-like protein (HGF1).

THE MOLECULE

Macrophage-stimulating protein (MSP) is secreted by the liver as a single-chain, biologically inactive propeptide that is activated at local tissue sites by proteolytic cleavage, yielding the active disulfide linked α/β-chain heterodimer[1,2]. MSP is structurally related to HGF with four conserved triple disulfide loops known as kringles. The MSP receptor is the transmembrane tyrosine kinase RON (CD136). MSP stimulates motility of resident peritoneal macrophages but not exudate macrophages or blood monocytes which do not express the receptor. MSP also stimulates keratinocytes and may play a role in wound healing.

Crossreactivity

Some.

Sources

Liver.

Bioassays

Proliferation of BK-1 murine keratinocyte cell line or PC12 neuroendocrine cell line.

Physicochemical properties of MSP

	Human		Mouse	
Property	α-chain	β-chain	α-chain	β-chain
PI (calculated)	7.9	7.4		
Amino acids				
– precursor[a]	711 (680)	716 (685)		
– mature[b]	453	227	457	228
M_r (K)				
– predicted	51.9	25.0	52.1	25.0
– expressed[c]	78	78		
Potential *N*-linked glycosylation sites	2	1	3	1
Disulfide bonds[d]	20	20		

[a] Prepropeptide (after removal of predicted signal peptide in parenthesis).
[b] Propeptide is cleaved proteolytically to form two subunits that combine by a single disulfide bond to form the active heterodimer.
[c] Precursor is 78 kDa.
[d] Potential – includes one interchain disulfide bond between Cys437–557 (human) and Cys446–562 (mouse) and 12 disulfide bonds in four kringle domains.

3-D structure

Structurally related to plasminogen and HGF with four disulfide-linked kringle loops. The β-subunit has homology to the catalytic domain of serine proteases but has no enzyme activity due to substitution of two critical amino acids in the active site.

Gene structure[3]

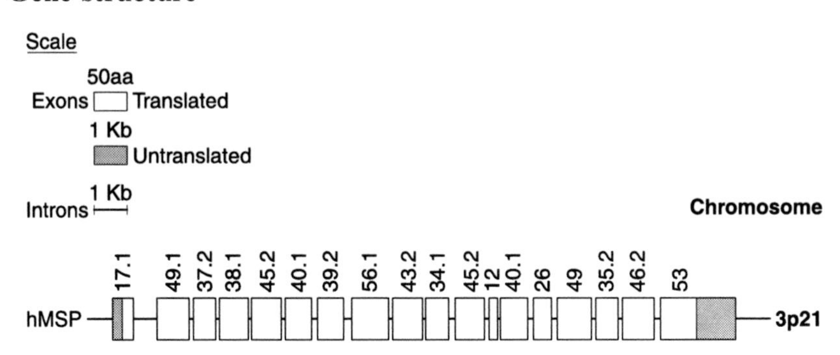

Human MSP consists of 18 exons on chromosome 3p21, mouse MSP is on chromosome 9.

Amino acid sequence for human MSP[4]

Accession code: SwissProt P26927

-31 MGWLPLLLLL TQYLGVPGQR SPLNDFQVLR G

1 TELQHLLHAV VPGPWQEDVA DAEECAGRCG PLMDCRAFHY **N**VSSHGCQLL

51 PWTQHSPHTR LRRSGRCDLF QKKDYVRTCI MNNGVGYRGT MATTVGGLPC

101 QAWSHKFPND HKYTPTLRNG LEENFCRNPD GDPGGPWCYT TDPAVRFQSC

151 GIKSCREAAC VWCNGEEYRG AVDRTESGRE CQRWDLQHPH QHPFEPGKFL

201 DQGLDDNYCR NPDGSERPWC YTTDPQIERE FCDLPRCGSE AQPRQEATTV

251 SCFRGKGEGY RGTA**N**TTTAG VPCQRWDAQI PHQHRFTPEK YACKDLRENF

301 CRNPDGSEAP WCFTLRPGMR AAFCYQIRRC TDDVRPQDCY HGAGEQYRGT

351 VSKTRKGVQC QRWSAETPHK PQFTFTSEPH AQLEENFCRN PDGDSHGPWC

401 YTMDPRTPFD YCALRRCADD QPPSILDPPD QVQFEKCGKR VDRLDQRRSK

451 LRVVGGHPGN SPWTVSLRNR QGQHFCGGSL VKEQWILTAR QCFSSCHMPL

501 TGYEVWLGTL FQNPQHGEPS LQRVPVAKMV CGPSGSQLVL LKLERSVTLN

551 QRVALICLPP EWYVVPPGTK CEIAGWGETK GTG**N**DTVLNV ALLNVISNQE

601 CNIKHRGRVR ESEMCTEGLL APVGACEGDY GGPLACFTHN CWVLEGIIIP

651 NRVCARSRWP AVFTRVSVFV DWIHKVMRLG

Propeptide is cleaved proteolytically to form two subunits (α-chain residues 1–453 and β-chain amino acids 454–680) that combine through a single disulfide bond between Cys437–557 to form the active heterodimer.

Amino acid sequence for mouse MSP[5]

Accession code: SwissProt P26928

```
-31  MGWLPLLLLL VQCSRALGPR SPLNDFQLFR G
  1  TELRNLLHTA VPGPWQEDVA DAEECARRCG PLLDCRAFHY NMSSHGCQLL
 51  PWTQHSLHTQ LYHSSLCHLF QKKDYVRTCI MDNGVSYRGT VARTAGGLPC
101  QAWSRRFPND HKYTPTPKNG LEENFCRNPD GDPRGPWCYT TNRSVRFQSC
151  GIKTCREAVC VLCNGEDYRG EVDVTESGRE CQRWDLQHPH SHPFQPEKFL
201  DKDLKDNYCR NPDGSERPWC YTTDPNVERE FCDLPSCGPN LPPTVKGSKS
251  QRRNKGKALN CFRGKGEDYR GTTNTTSAGV PCQRWDAQSP HQHRFVPEKY
301  ACKDLRENFC RNPDGSEAPW CFTSRPGLRM AFCHQIPRCT EELVPEGCYH
351  GSGEQYRGSV SKTRKGVQCQ HWSSETPHKP QFTPTSAPQA GLEANFCRNP
401  DGDSHGPWCY TLDPDILFDY CALQRCDDDQ PPSILDPPDQ VVFEKCGKRV
451  DKSNKLRVVG GHPGNSPWTV SLRNRQGQHF CGGSLVKEQW VLTARQCIWS
501  CHEPLTGYEV WLGTINQNPQ PGEANLQRVP VAKAVCGPAG SQLVLLKLER
551  PVILNHHVAL ICLPPEQYVV PPGTKCEIAG WGESIGTSNN TVLHVASMNV
601  ISNQECNTKY RGHIQESEIC TQGLVVPVGA CEGDYGGPLA CYTHDCWVLQ
651  GLIIPNRVCA RPRWPAIFTR VSVFVDWINK VMQLE
```

Propeptide is cleaved proteolytically to form two subunits (α-chain residues 1–457 and β-chain amino acids 458–685) that combine through a single disulfide bond to form the active heterodimer. Iinterchain disulfide bond between Cys477–562.

THE MSP RECEPTOR (RON)

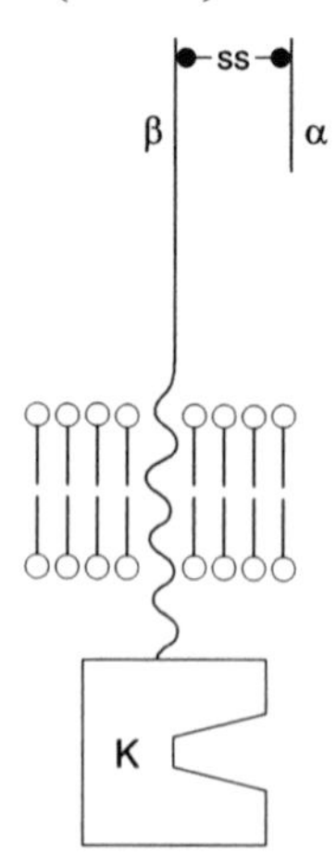

The MSP receptor RON (CD136) belongs to a family of transmembrane receptor kinases that includes the receptor for HGF (c-Met), the orphan receptor Sea and four related proteins SEX, SEP, OCT and NOV[2,6–8]. The mature receptor is a disulfide-linked heterodimer consisting of an entirely extracellular α-chain (M_r 50 000) and a transmembrane β-chain (M_r 145 000) that includes an intracellular kinase domain. Both chains are derived post-translationally by proteolytic cleavage of a single precursor molecule. MSP binds to its receptor through the β-chain[9].

Distribution

Mainly epithelial cells such as skin, kidney, lung, liver, intestine and some haematopoietic cell lines.

Physicochemical properties of the MSP receptor

Property	Human
Amino acids – precursor	1400[a]
– mature[b,c]	1376 (280 and 1090)[d]
M_r (K) – predicted	
– expressed	180 (40 and 150)[c]
Potential *N*-linked glycosylation sites	8
Affinity K_d (M)	10^{-9}–10^{-10}

[a] Alternative splicing gives rise to a truncated protein with 49 amino acids deleted from the extracellular domain of the β-chain.
[b] After removal of predicted signal peptide.
[c] The α- and β-chains of the receptor are obtained by proteolytic cleavage of the propeptide.
[d] α- and β-chains.

Signal transduction

Receptor dimerization and autophosphorylation. Receptor tyrosine kinase acts through a SH2 docking site to activate several signalling pathways including PI-3 kinase and Ras.

Chromosomal location

3p21.3.

Amino acid sequence for human MSP receptor RON[7]

Accession code: SwissProt Q04912

```
-24  MELLPPLPQS FLLLLLLPAK PAAG
  1  EDWQCPRTPY AASRDFDVKY VVPSFSAGGL VQAMVTYEGD RNESAVFVAI
 51  RNRLHVLGPD LKSVQSLATG PAGDPGCQTC AACGPGPHGP PGDTDTKVLV
101  LDPALPALVS CGSSLQGRCF LHDLEPQGTA VHLAAPACLF SAHHNRPDDC
151  PDCVASPLGT RVTVVEQGQA SYFYVASSLD AAVAGSFSPR SVSIRRLKAD
201  ASGFAPGFVA LSVLPKHLVS YSIEYVHSFH TGAFVYFLTV QPASVTDDPS
251  ALHTRLARLS ATEPELGDYR ELVLDCRFAP KRRRRGAPEG GQPYPVLQVA
301  HSAPVGAQLA TELSIAEGQE VLFGVFVTGK DGGPGVGPNS VVCAFPIDLL
351  DTLIDEGVER CCESPVHPGL RRGLDFFQSP SFCPNPPGLE ALSPNTSCRH
401  FPLLVSSSFS RVDLFNGLLG PVQVTALYVT RLDNVTVAHM GTMDGRILQV
451  ELVRSLNYLL YVSNFSLGDS GQPVQRDVSR LGDHLLFASG DQVFQVPIRG
501  PGCRHFLTCG RCLRAWHFMG CGWCGNMCGQ QKECPGSWQQ DHCPPKLTEF
551  HPHSGPLRGS TRLTLCGSNF YLHPSGLVPE GTHQVTVGQS PCRPLPKDSS
```

```
601  KLRPVPRKDF VEEFECELEP LGTQAVGPTN VSLTVTNMPP GKHFRVDGTS
651  VLRGFSFMEP VLIAVQPLFG PRAGGTCLTL EGQSLSVGTS RAVLVNGTEC
701  LLARVSEGQL LCATPPGATV ASVPLSLQVG GAQVPGSWTF QYREDPVVLS
751  ISPNCGYINS HITICGQHLT SAWHLVLSFH DGLRAVESRC ERQLPEQQLC
801  RLPEYVVRDP QGWVAGNLSA RGDGAAGFTL PGFRFLPPPH PPSANLVPLK
851  PEEHAIKFEY IGLGAVADCV GINVTVGGES CQHEFRGDMV VCPLPPSLQL
901  GQDGAPLQVC VDGECHILGR VVRPGPDGVP QSTLLGILLP LLLLVAALAT
951  ALVFSYWWRR KQLVLPPNLN DLASLDQTAG ATPLPILYSG SDYRSGLALP
1001 AIDGLDSTTC VHGASFSDSE DESCVPLLRK ESIQLRDLDS ALLAEVKDVL
1051 IPHERVVTHS DRVIGKGHFG VVYHGEYIDQ AQNRIQCAIK SLSRITEMQQ
1101 VEAFLREGLL MRGLNHPNVL ALIGIMLPPE GLPHVLLPYM CHGDLLQFIR
1151 SPQRNPTVKD LISFGLQVAR GMEYLAEQKF VHRDLAARNC MLDESFTVKV
1201 ADFGLARDIL DREYYSVQQH RHARLPVKWM ALESLQTYRF TTKSDVWSFG
1251 VLLWELLTRG APPYRHIDPF DLTHFLAQGR RLPQPEYCPD SLYQVMQQCW
1301 EADPAVRPTF RVLVGEVEQI VSALLGDHYV QLPATYMNLG PSTSHEMNVR
1351 PEQPQFSPMP GNVRRPRPLS EPPRPT
```

The propeptide is cleaved to form the α-chain (residues 1–280) and and β-chain (residues 286–1376) of the receptor (by similarity). Tyr1329 is autophosphorylated. Variant sequence Q→R at position 298.

References

[1] Leonard, E.J. et al. (1998) In Human Cytokines III, Aggarwal, B.B. ed., Blackwell, Oxford, pp. 235-265.

[2] Leonard, E.J. (1997) Ciba Foundation Symp. 212, 183–191.

[3] Han, S. et al. (1991) Biochemistry 30, 9768–9780.

[4] Yoshimura, T. et al. (1993) J. Biol. Chem. 268, 15461–15468.

[5] Friezner Degen, S.J. et al. (1991) Biochemistry 30, 9781–9791.

[6] Maestrini, E. et al. (1996) Proc. Natl Acad. Sci. USA 93, 674–678.

[7] Ronsin, C. et al. (1993) Oncogene 8, 1195–1202.

[8] Wang, M.-H. et al. (1994) Science 266, 117–119.

[9] Wang, M.-H. et al. (1997) J. Biol. Chem. 272, 16999–17004.

Other names
β-NGF.

THE MOLECULE

Nerve growth factor (NGF) enhances the survival, growth and neurotransmitter biosynthesis of certain sympathetic and sensory neurons[1–5]. NGF is a neurotrophic factor[6,7] and plays a key role in establishing cutaneous innervation and regulating neuropeptide expression[8]. It is also a growth and differentiation factor for B lymphocytes and promotes B-cell survival[9–11]. NGF has about 50% homology with BDNF, NT-3 and NT-4/5.

Crossreactivity
There is 90% sequence homology between human and mouse NGF and complete cross-species reactivity.

Sources
Submaxillary gland of male mice, prostate, brain and nervous system, and lymphocytes.

Bioassays
Survival and outgrowth of neurites from embryo chicken dorsal root ganglia. Differentiation of PC12 cell line.

Physicochemical properties of NGF

Property	Human	Mouse
pI	~10	~10
Amino acids – precursor	241	241/307/313[a]
– mature[b]	120	118
M_r (K) – predicted	13.5	13.3
– expressed[c]	26	26 (130)
Potential *N*-linked glycosylation sites[d]	3	3
Disulfide bonds[e]	3	3

[a] There are three forms of murine NGF precursor derived from alternative mRNA splicing: two high-molecular-weight forms (M_r 33 900 and 33 800) of 313 and 307 amino acids respectively, and a low-molecular-weight form (M_r 27 000) of 241 amino acids (see sequence and gene diagram).
[b] After removal of propeptide. The C-terminal dipeptide is also removed from mouse NGF.
[c] Biologically active NGF is a homodimer. Murine NGF is also found in the submaxillary gland as a high-molecular-weight complex (M_r 130 000) with α- and γ-NGF which belong to the kallikrein family of serine proteases. These are involved in the processing of the β-NGF precursor to form active NGF by removal of an N-terminal prepropeptide and a C-terminal dipeptide[3]. The β- and γ-NGF kallikreins are however lacking in other species, and processing of human precursor NGF does not include removal of the C-terminal dipeptide.
[d] Includes two sites on propeptide. The one site on mature NGF is not normally glycosylated.
[e] Disulfide bonds are required for biological activity. All the Cys residues are strictly conserved between species and between NGF, BDNF, NT-3 and NT-4.

3-D structure

Functional NGF is a noncovalently linked parallel homodimer. Crystal structure of the NGF dimer has been resolved at 2.3 Å. It consists of three antiparallel pairs of β-strands together forming a flat surface through which the two subunits associate. There are four loops which contain many of the variable residues between the different NGF-related molecules, and may determine receptor specificity. Clusters of positively charged side-chains may interact with the acidic low-affinity NGFR[12].

Gene structure[13]

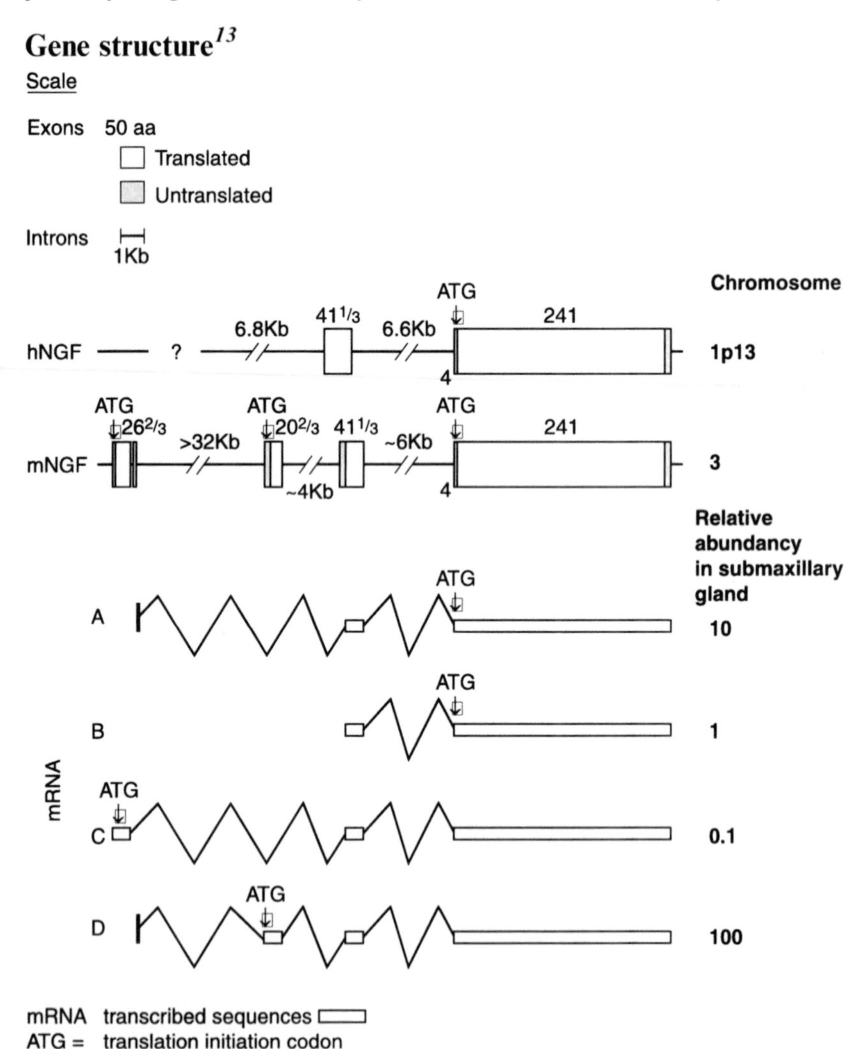

The gene for NGF is located on human chromosome 1p13 and mouse chromosome 3. Four species of murine NGF precursor mRNA produced by alternative splicing and independent promoter elements are shown as A–D above[13,14]. These give rise to three different precursor molecules. The abundance of the four mRNA species varies in different sites of production. The A form is the most abundant in submaxillary

gland whereas the short forms are more abundant in other sites. The 5′ end of the human gene has not yet been sequenced.

Amino acid sequence for human NGF[15]

Accession code: SwissProt P01138

-18 MSMLFYTLIT AFLIGIQA
1 *EPHSESNVPA GHTIPQVHWT KLQHSLDTAL RRARSAPAAA IAARVAGQTR*
51 ***N****ITVDPRLFK KRRLRSPRVL FSTQPPREAA DTQDLDFEVG GAAPF***N***RTHR*
101 *SKR*SSSHPIF HRGEFSVCDS VSVWVGDKTT ATDIKGKEVM VLGEVNI**N**NS
151 VFKQYFFETK CRDPNPVDSG CRGIDSKHWN SYCTTTHTFV KALTMDGKQA
201 AWRFIRIDTA CVCVLSRKAV RRA

Mature human NGF is formed by removal of a predicted signal peptide and a propeptide (1–103, in italics). Disulfide bonds between Cys118–183, 161–211 and 171–213 by similarity.

Amino acid sequence for mouse NGF[13,16]

Accession code: SwissProt P01139

(A B)
-18 MSMLFYTLIT AFLIGVQA
1 *EPYTDSNVPE GDSVPEAHWT KLQHSLDTAL RRARSAPTAP IAARVTGQTR*
51 ***N****ITVDPRLFK KRRLHSPRVL FSTQPPPTSS DTLDLDFQAH GTIPF***N***RTHR*
101 *SKR*SSTHPVF HMGEFSVCDS VSVWVGDKTT ATDIKGKEVT VLAEVNI**N**NS
151 VFRQYFFETK CRASNPVESG CRGIDSKHWN SYCTTTHTFV KALTTDEKQA
201 AWRFIRIDTA CVCVLSRKAT R*RG*

Mature murine NGF is formed by removal of a predicted signal peptide, an N-terminal propeptide (1–103, in italics), and a C-terminal Arg–Gly dipeptide (in italics). Disulfide bonds between Cys118–183, 161–211, and 171–213 by similarity.

Three murine NGF precursors derived from alternative mRNA splicing have been identified. The short form shown above with a total of 241 amino acids is derived from mRNA transcripts A and B shown on the gene diagram. Two other longer forms with additional N-terminal sequences (shown below) are derived from mRNA transcripts C and D. The letters A–D in bold indicate the translated peptides from the four mRNA species.

(C)
MSHQLGSYPS LVPRTLTTRT PGSSHSRVLA CGRAVQGAGW HAGPKLTSVS
GPNKGFAKDA AFYTGRSEVH SV

(D)
MLCLKPVKLG SLEVGHGQHG GVLACGRAVQ GAGWHAGPKL TSVSGPNKGF
AKDAAFYTGR SEVHSV

THE NGF RECEPTORS

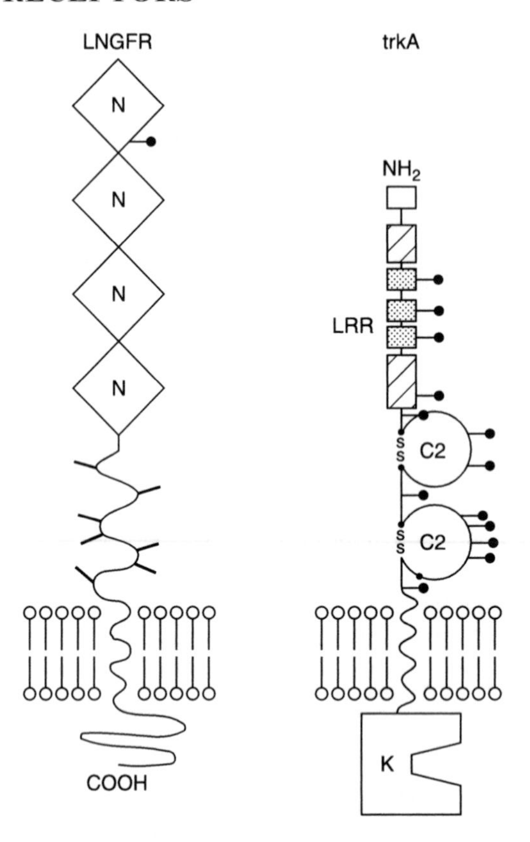

NGF binds to both low-affinity (K_d 10^{-9} M) and high-affinity (K_d 10^{-11} m) cell surface receptors[17–19]. The low-affinity NGF receptor (LNGFR), also known as p75 LNGFR or fast NGF receptor, belongs to the NGFR/TNFR superfamily with four Cys-rich repeats in the extracellular domain. It is also a low-affinity receptor (K_d 10^{-9} M) for the NGF-related neurotrophins, BDNF, NT-3 and NT-4/5. The LNGFR does not mediate the neurotrophic effects of the NGF family[20,21] and it is expressed in non-neuronal cells that do not respond to any of the known members of the NGF family of neurotrophins[22]. Mice lacking the LNGFR gene are viable, and for the most part exhibit normal neuronal development except in certain sensory terminals. Expression of LNGFR has been shown to induce apoptosis which can be inhibited by either NGF or antibody binding to the receptor[23,24]. The p75 LNGFR activates NFκB[25] and interacts in a ligand-dependent manner with RhoA, a known regulator of actin assembly[26].

The high-affinity receptors (K_d 10^{-11} M) mediate NGF neurotrophic activity and NGF activation of B cells[10,27,28]. At least one component of the high-affinity receptor is the product of the *trk* proto-oncogene gp140^{trk} (trkA). Transfection of trkA into 3T3 cells gives rise to low-affinity receptors (K_d 10^{-9} M) and a small proportion of high-affinity receptors (K_d 10^{-10} M). It has been proposed that the high-affinity receptors result from receptor dimerization (or oligomerization). trkA can mediate NGF signal transduction in the absence of LNGFR but the LNGFR may modify trkA signalling[29–31].

TrkA is a member of the trk family of tyrosine kinases which includes trkB, a receptor for BDNF and NT-3, and trkC, a receptor for NT-3 but not NGF or BDNF[32–34]. This family of receptors is characterized by a novel extracellular domain made up of an N-terminal leucine-rich region (LRR) composed of three tandem leucine-rich motifs flanked by two distinct cysteine-rich clusters, followed by two Ig-SF C2 set domains[35]. The LRR region is thought to be involved in cell adhesion and the Ig-SF domains may be important for interaction with other cell surface proteins. Fusion products of trkA with tropomyosin, L7a ribosomal protein, and the EGF receptor signal peptide have been identified as oncogenes[36].

Distribution

NGF receptors are expressed on sensory and sympathetic neurons, non-neuronal cells derived from the neural crest including melanocytes, and Schwann cells, and on mast cells, B lymphocytes, monocytes and the PC12 cell line[5,22,37–39].

Physicochemical properties of the NGF receptors

Property	LNGF receptor		trkA	
	Human	Rat	Human	Rat
Amino acids – precursor	427	425	790	799
– mature[a]	399	396	758	765
M_r (K) – predicted	42.5	42.5	83.6	84.4
– expressed	70–75	83	140	150
Potential *N*-linked glycosylation sites[b]	1	2	13	11
Affinity K_d (M)[c]	10^{-9}	10^{-9}–10^{-10}	10^{-9} and 10^{-10}	10^{-9}

[a] After removal of predicted signal peptide.

[b] There is some *O*-linked glycosylation of LNGFR on Ser/Thr-rich region proximal to the transmembrane domain.

[c] Rat and mouse trkA expressed in 3T3 cells have a K_d of approximately 10^{-9} M[18,40,41]. Human trkA expressed in 3T3 cells gives rise to two populations of receptors, one with a K_d of about 10^{-9} M and a second minor population with a K_d of about 10^{-10} M[29]. The higher affinity receptors may be dimers or oligomers[30].

Signal transduction

TrkA has intrinsic protein tyrosine kinase activity[18,42]. Binding of NGF induces tyrosine phosphorylation of several cellular proteins including PLC-γ1[43,44], PI-3 kinase[45] and autophosphorylation of trkA[43]. Association with the Erk1 Ser/Thr kinase has also been reported[46]. Activation of PLC-γ1 by phosphorylation results in hydrolysis of PIP_2 releasing IP_3 and DAG[44,47]. There is evidence for ligand-induced receptor dimerization[30]. The low-affinity receptor is not required for growth and differentiation of neurons in response to NGF[31]. Expression of LNGFR has been shown to induce neuronal cell death by apoptosis which can be prevented by either NGF or monoclonal antibody binding to the receptor[24]. It also modulates signalling through trkA[17]. The p75 LNGFR activates NFκB[25] and interacts in a ligand-dependent manner with RhoA, a known regulator of actin assembly[26].

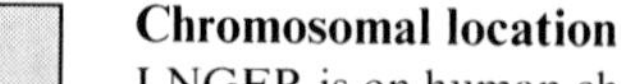

Chromosomal location

LNGFR is on human chromosome 17q21–q22 and trkA is on human chromosome 1q23–q31.

Amino acid sequence for human low-affinity NGF receptor (LNGFR)[20]

Accession code: SwissProt P08138

-28 MGAGATGRAM DGPRLLLLLL LGVSLGGA
1 KEACPTGLYT HSGECCKACN LGEGVAQPCG A**N**QTVCEPCL DSVTFSDVVS
51 ATEPCKPCTE CVGLQSMSAP CVEADDAVCR CAYGYYQDET TGRCEACRVC
101 EAGSGLVFSC QDKQNTVCEE CPDGTYSDEA NHVDPCLPCT VCEDTERQLR
151 ECTRWADAEC EEIPGRWITR STPPEGSDST APSTQEPEAP PEQDLIASTV
201 AGVVTTVMGS SQPVVTRGTT DNLIPVYCSI LAAVVVGLVA YIAFKRWNSC
251 KQNKQGANSR PVNQTPPPEG EKLHSDSGIS VDSQSLHDQQ PHTQTASGQA
301 LKGDGGLYSS LPPAKREEVE KLLNGSAGDT WRHLAGELGY QPEHIDSFTH
351 EACPVRALLA SWATQDSATL DALLAALRRI QRADLVESLC SESTATSPV

Cys-rich repeats: 3–37, 38–79, 80–119, 120–161. Ser/Thr-rich segment 169–220.

Amino acid sequence for human high-affinity NGF receptor (trkA)[36,48]

Accession code: SwissProt P04629, P08119

-32 MLRGGRRGQL GWHSWAAGPG SLLAWLILAS AG
1 AAPCPDACCP HGSSGLRCTR DGALDSLHHL PGAE**N**LTELY IENQQHLQHL
51 ELRDLRGLGE LR**N**LTIVKSG LRFVAPDAFH FTPRLSRL**N**L SFNALESLSW
101 KTVQGLSLQE LVLSGNPLHC SCALRWLQRW EEEGLGGVPE QKLQCHGQGP
151 LAHMP**N**ASCG VPTLKVQVP**N** ASVDVGDDVL LRCQVEGRGL EQAGWILTEL
201 EQSATVMKSG GLPSLGLTLA **N**VTSDLNRK**N** LTCWAENDVG RAEVSVQV**N**V
251 SFPASVQLHT AVEMHHWCIP FSVDGQPAPS LRWLF**N**GSVL **N**ETSFIFTEF
301 LEPAA**N**ETVR HGCLRLNQPT HVNNG**N**YTLL AANPFGQASA SIMAAFMDNP
351 FEFNPEDPIP DT**N**STSGDPV EKKDETPFGV SVAVGLAVFA CLFLSTLLLV
401 LNKCGRRNKF GINRPAVLAP EDGLAMSLHF MTLGGSSLSP TEGKGSGLQG
451 HIIENPQYFS DACVHHIKRR DIVLKWELGE GAFGKVFLAE CHNLLPEQDK
501 MLVAVKALKE ASESARQDFQ REAELLTMLQ HQHIVRFFGV CTEGRPLLMV
551 FEYMRHGDLN RFLRSHGPDA KLLAGGEDVA PGPLGLGQLL AVASQVAAGM
601 VYLAGLHFVH RDLATRNCLV GQGLVVKIGD FGMSRDIYST DYYRVGGRTM
651 LPIRWMPPES ILYRKFTTES DVWSFGVVLW EIFTYGKQPW YQLSNTEAID
701 CITQGRELER PRACPPEVYA IMRGCWQREP QQRHSIKDVH ARLQALAQAP
751 PVYLDVLG

Tyr at position 642 is site of autophosphorylation. Translocation breakpoint forming oncogene between Pro and Asp at position 360–361.

Amino acid sequence for rat low-affinity NGF receptor (LNGFR)[21]

(Mouse LNGFT sequence has not been published)

Accession code: SwissProt P07174

-29 MRRAGAACSA MDRLRLLLLL ILGVSSGGA
1 KETCSTGLYT HSGECCKACN LGEGVAQPCG A**N**QTVCEPCL D**N**VTFSDVVS
51 ATEPCKPCTE CLGLQSMSAP CVEADDAVCR CAYGYYQDEE TGHCEACSVC
101 EVGSGLVFSC QDKQNTVCEE CPEGTYSDEA NHVDPCLPCT VCEDTERQLR
151 ECTPWADAEC EEIPGRWIPR STPPEGSDST APSTQEPEVP PEQDLVPSTV
201 ADMVTTVMGS SQPVVTRGTT DNLIPVYCSI LAAVVVGLVA YIAFKRWNSC
251 KQNKQGANSR PVNQTPPPEG EKLHSDSGIS VDSQSLHDQQ THTQTASGQA
301 LKGDGNLYSS LPLTKREEVE KLLNGDTWRH LAGELGYQPE HIDSFTHEAC
351 PVRALLASWG AQDSATLDAL LAALRRIQRA DIVESLCSES TATSPV

Cys-rich repeats: 3–37, 38–79, 80–119, 120–161. Ser/Thr-rich domain 169–220.

Amino acid sequence for rat high-affinity NGF receptor (trkA)[49]

(Mouse trkA sequence has not been published)

Accession code: SwissProt P04629, P08119

-34 MLRGQRHGQL GWHRPAAGLG GLVTSLMLAC ACAA
1 SCRETCCPVG PSGLRCTRAG TLNTLRGLRG AG**N**LTELYVE NQRDLQRLEF
51 EDLQGLGELR SLTIVKSGLR FVAPDAFHFT PRLSHL**N**LSS NALESLSWKT
101 VQGLSLQDLT LSGNPLHCSC ALLWLQRWEQ EDLCGVYTQK LQGSGSGDQF
151 LPLGH**N**NSCG VPSVKIQMP**N** DSVEVGDDVF LQCQVEGQAL QQADWILTEL
201 EGTATMKKSG DLPSLGLTLV **N**VTSDLNKK**N** VTCWAENDVG RAEVSVQVSV
252 SFPASVHLGK AVEQHHWCIP FSVDGQPAPS LRWFF**N**GSVL **N**ETSFIFTQF
301 LESALT**N**ETM RHGCLRLNQP THVNNG**N**YTL LAANPYGQAA ASIMAAFMDN
351 PFEFNPEDPI PVSFSPVDT**N** STSRDPVEKK DETPFGVSVA VGLAVSAALF
400 LSALLLVLNK CGQRSKFGIN RPAVLAPEDG LAMSLHFMTL GGSSLSPTEG
451 KGSGLQGHIM ENPQYFSDTC VHHIKRQDII LKWELGEGAF GKVFLAECYN
501 LLNDQDKMLV AVKALKETSE NARQDFHREA ELLTMLQHQH IVRFFGVCTE
551 GGPLLMVFEY MRHGDLNRFL RSHGPDAKLL AGGEDVAPGP LGLGQLLAVA
601 SQVAAGMVYL ASLHFVHRDL ATRNCLVGQG LVVKIGDFGM SRDIYSTDYY
651 RVGGRTMLPI RWMPPESILY RKFSTESDVW SFGVVLWEIF TYGKQPWYQL
701 SNTEAIECIT QGRELERPRA CPPDVYAIMR GCWQREPQQR LSMKDVHARL
751 QALAQAPPSY LDVLG

References

[1] Levi-Montalcini, R. (1987) Science 237, 1154–1162.
[2] Levi-Montalcini, R. (1998) Neuroreport 9, R71-R83.
[3] Fahnestock, M. (1991) Curr. Topics Microbiol. Immunol. 165, 1–26.
[4] Bradshaw, R.A. et al. (1993) TIBS 18, 48–52.
[5] Ebendal, T.J. (1992) Neurosci. Res.32, 461–470.
[6] Snider, W.D. (1994) Cell 77, 627–638.
[7] Johnson, J. and Oppenheim, R. (1994) Curr. Biol. 4, 662–665.

[8] Davies, A.M. (2000) Curr. Biol. 10, R374–R376.
[9] Otten, U. et al. (1989) Proc. Natl Acad. Sci. USA 86, 10059–10063.
[10] Melamed, I. et al. (1996) Eur. J. Immunol. 26, 1985–1992.
[11] Torcia, M. et al. (1996) Cell 85, 345–356.
[12] McDonald, N.Q. et al. (1991) Nature 354, 411–414.
[13] Selby, M.J. et al. (1987) Mol. Cell Biol. 7, 3057–3064.
[14] Edwards, R.H. et al. (1986) Nature 319, 784–787.
[15] Ullrich, A. et al. (1983) Nature 303, 821–825.
[16] Scott, J. et al. (1983) Nature 302, 538–540.
[17] Frade, J.M. and Barde, Y.A. (1998) BioEssays 20, 137–145.
[18] Meakin, SO,.Shooter EM. Trends. Neurosci. 1992;15:323–31.
[19] Barker, P.A. and Murphy, R.A. (1992) Mol. Cell Biochem. 110, 1–15.
[20] Johnson, D. et al. (1986) Cell 47, 545–554.
[21] Radeke, M.J. et al. (1987) Nature 325, 593–597.
[22] Thomson, T.M. et al. (1988) J. Exp. Cell Res. 174, 533–539.
[23] Frade, J.M. et al. (1996) Nature 383, 166–168.
[24] Rabizadeh, S. et al. (1993) Science 261, 345–348.
[25] Carter, B.D. et al. (1996) Science 272, 542–545.
[26] Davies, A.M. (2000) Curr. Biol. 10, R198–R200.
[27] Barbacid, M. (1995) Ann. N.Y. Acad. Sci. 766, 442–458.
[28] Glass, D.J. and Yancopoulos, G.D. (1993) Trends Cell Biol. 3, 262–268.
[29] Klein, R. et al. (1991) Cell 65, 189–197.
[30] Jing, S. et al. (1992) Neuron 9, 1067–1079.
[31] Cordon-Cardo, C. et al. (1991) Cell 66, 173–183.
[32] Soppet, D. et al. (1991) Cell 65, 895–903.
[33] Squinto, S.P. et al. (1991) Cell 65, 885–893.
[34] Lamballe, F. et al. (1991) Cell 66, 967–979.
[35] Schneider, R. and Schweiger, M. (1991) Oncogene 6, 1807–1811.
[36] Barbacid, M. et al. (1991) Biochim. Biophys. Acta 1072, 115–127.
[37] Brodie, C. and Gelfand, E.W. (1992) J. Immunol. 148, 3492–3497.
[38] Ehrhard, P.B. et al. (1993) Proc. Natl Acad. Sci. USA 90, 5423–5427.
[39] Ernfors, P. et al. (1988) Neuron 1, 983–996.
[40] Hempstead, B.L. et al. (1991) Nature 350, 678–683.
[41] Kaplan, D.R. et al. (1991) Science 252, 554–558.
[42] Kaplan, D.R. and Miller, F.D. (1997) Curr. Opin. Cell Biol. 9, 213–221.
[43] Ohmichi, M. et al. (1991) Biochem. Biophys. Res. Commun. 179, 217–223.
[44] Vetter, M.L. et al. (1991) Proc. Natl Acad. Sci. USA 88, 5650–5654.
[45] Ohmichi, M. et al. (1992) Neuron 9, 769–777.
[46] Loeb, D.M. et al. (1992) Neuron 9, 1053–1065.
[47] Kaplan, D.R. et al. (1991) Nature 350, 158–160.
[48] Martin-Zanca, D. et al. (1989) Mol. Cell Biol. 9, 24–33.
[49] Meakin, S.O. et al. (1992) Proc. Natl Acad. Sci. USA 89, 2374–2378.

Other names
Hippocampus-derived neurotrophic factor (HDNF), nerve growth factor 2 (NGF-2).

THE MOLECULE

Neurotrophin 3 (NT-3) is a neurotrophic factor important in the development and maintenance of the vertebrate nervous system. *In vitro*, it promotes the survival and outgrowth of neural crest-derived sensory and sympathetic neurons. It is similar to NGF and NT-3 but with different neuronal specificities[1–4].

Crossreactivity
The amino acid sequences of mature NT-3 from human, mouse and rat are identical. There is complete cross-species reactivity.

Sources
To date, NT-3 protein has not been isolated from natural sources. NT-3 mRNA is present at high levels in the hippocampus and other sites in the fetal brain and at lower levels in adult brain. High levels are also present in the ovary. Transcripts have also been found in all adult tissues tested, including skin, spleen, thymus, liver, muscle, lung, intestine and kidney[2].

Bioassays
Survival and outgrowth of neurites from embryo chicken dorsal root ganglia. Survival and differentiation of trkC-transfected cell lines.

Physicochemical properties of NT-3

Property	Human	Mouse
pI	9.5	9.5
Amino acids – precursor	257	258
– mature[a]	119	119
M_r (K) – predicted[a]	13.6	13.6
– expressed[b]	27	27
Potential *N*-linked glycosylation sites[c]	0	0
Disulfide bonds[d]	3	3

[a] After removal of prepropeptide.
[b] Homodimer.
[c] There is one *N*-linked glycosylation site in propeptide.
[d] Conserved between NGF, BDNF NT-3 and NT-4.

3-D structure
Has 60% β-sheet. Exists as tightly linked homodimer. Crystal structure probably very similar to NGF.

Gene structure[5]

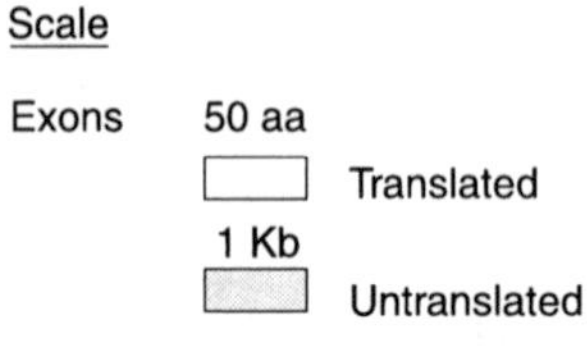

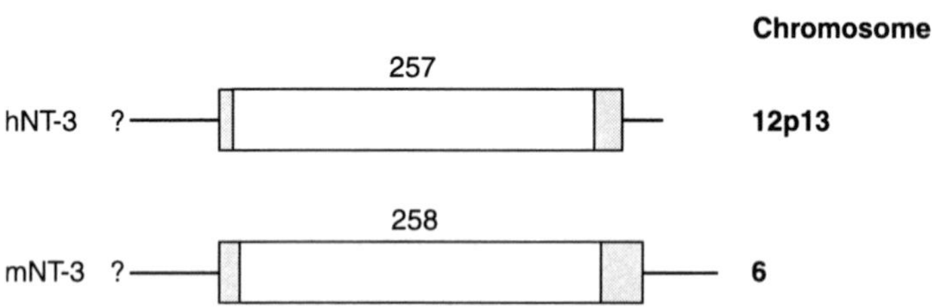

The gene for NT-3 is located on human chromosome 12p13 and consists of a single coding exon of 257 amino acids. The gene for murine NT-3 consists of a single 258 amino acid exon and is located on chromosome 6. There is evidence for more than one RNA species arising by alternative splicing and ATG initiation sites derived from additional upstream exons similar to mouse NGF.

Amino acid sequence for human NT-3[5,6]

Accession code: SwissProt P20783

-16 MSILFYVIFL AYLRGI
1 *QGNNMDQRSL PEDSLNSLII KLIQADILKN KLSKQMVDVK ENYQSTLPKA*
51 *EAPREPERGG PAKSAFQPVI AMDTELLRQQ RRYNSPRVLL SDSTPLEPPP*
101 *LYLMEDYVGS PVVA*__N__*RTSRR KRYAEHKSHR GEYSVCDSES LWVTDKSSAI*
151 DIRGHQVTVL GEIKTGNSPV KQYFYETRCK EARPVKNGCR GIDDKHWNSQ
201 CKTSQTYVRA LTSENNKLVG WRWIRIDTSC VCALSRKIGR T

Mature human NT-3 is formed by removal of a predicted signal peptide and a propeptide (1–122, in italics). Other precursor forms with extended N-terminal sequences similar to murine NGF may also exist. Disulfide bonds between Cys136–201, 179–230 and 189–232 by similarity.

Amino acid sequence for mouse NT-3[7]

Accession code: SwissProt P20181

-16 MSILFYVIFL AYLRGI
1 *QGNSMDQRSL PEDSLNSLII KLIQADILKN KLSKQMVDVK ENYQSTLPKA*
51 *EAPREPEQGE ATRSEFQPMI ATDTELLRQQ RRYNSPRVLL SDSTPLEPPP*
101 *LYLMEDYVGN PVVA*__N__*RTSPR RKRYAEHKSH RGEYSVCDSE SLWVTDKSSA*
151 IDIRGHQVTV LGEIKTGNSP VKQYFYETRC KEARPVKNGC RGIDDKHWNS
201 QCKTSQTYVR ALTSENNKLV GWRWIRIDTS CVCALSRKIG RT

Mature murine NT-3 is formed by removal of a predicted signal peptide and a propeptide (1–123, in italics). Other precursor forms with extended N-terminal sequences similar to NGF may also exist. Disulfide bonds between Cys137–202, 180–231 and 190–233 by similarity.

THE NT-3 RECEPTORS

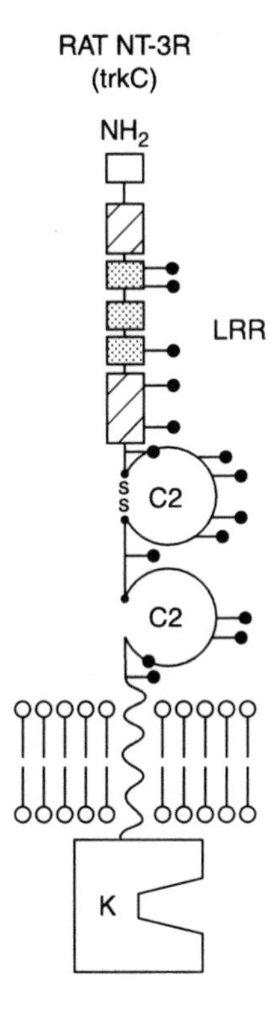

NT-3 binds to both low-affinity (K_d 10^{-9} M) and high-affinity (K_d 10^{-11} M) cell surface receptors[8–11]. The low-affinity receptor is LNGFR, which also binds the other members of the NGF neurotrophin family, NGF, BDNF, NT-3 and NT-4[12] (see NGF entry (page **397**)). Functional high-affinity NT-3 receptors (K_d 10^{-11} M) have been identified as gp145^{trkC} (trkC), which is a member of the trk family of protein tyrosine kinases[8,13–16]. The characteristics of this family of receptors are described in the entry for NGF receptors. The *trkC* locus is complex and encodes at least four distinct polypeptides by variable splicing, of which three are full-length receptor tyrosine kinases that differ by novel amino acid insertions in the kinase domain. The different isoforms do not have identical signalling properties. Isoform b is expressed in a relatively large amount in the adult brain compared with fetal brain[15–18]. TrkA and trkB are also functional receptors for NT-3 in certain *in vitro* systems[14,19–22]. In contrast, trkC does not bind NGF, BDNF or NT-4[13,23]. The LNGFR is not required for high-affinity binding of NT-3 nor for signal transduction but may modify trkC signalling[10,20].

Distribution

TrkC mRNA is expressed in pyramidal cells of the hippocampus, dentate gyrus, cortex and some sites in the cerebellum. *In situ* hybridization experiments have shown it to be present in multiple structures of the peripheral nervous system. Expressed in brain by Northern blot studies, but not in lung, heart, intestine, kidney, spleen, liver, testes or muscle. Possibly in ovaries.

Physicochemical properties of the NT-3 receptor (trkC)

Property	Human	Mouse	Rat
Amino acids – precursor	839	854	825
– mature[a,b]	808	833	794
M_r (K) – predicted	91.0	94.2	89.3
– expressed	145	145	145
Potential *N*-linked glycosylation sites[c]	13	5	14
Affinity K_d (M)	10^{-10}–10^{-11}	10^{-10}–10^{-11}	10^{-11}

[a] After removal of predicted signal peptide.
[b] At least four splice variants have been identified.
[c] Glycosylation is required for Trk to reach the cell surface and is required for signal transduction.

Signal transduction

Signalling requires ligand-induced dimerization. TrkC has intrinsic protein tyrosine kinase activity. NT-3 binding induces a complex set of phosphorylation events including receptor autophosphorylation, phosphorylation of PLCγ1 resulting in PIP_2 hydrolysis with release of IP_3 and DAG, activation of SHC and PI-3 kinase[24–26]. LNGFR is not required for signal transduction but may modify trkC activation[9,10,14,20].

Chromosomal location

trkC is on human chromosome 15q25 and mouse chromosome 7.

Amino acid sequence for human NT-3 receptor (trkC)[27,28]

Accession code: SwissProt Q16288

-31 MDVSLCPAKC SFWRIFLLGS VWLDYVGSVL A
1 CPANCVCSKT EINCRRPDDG NLFPLLEGQD SGNSNGNANI **N**ITDISR**N**IT
51 SIHIENWRSL HTLNAVDMEL YTGLQKLTIK NSGLRSIQPR AFAKNPHLRY
101 I**N**LSSNRLTT LSWQLFQTLS LRELQLEQNF F**N**CSCDIRWM QLWQEQGEAK
151 LNSQNLYCIN ADGSQLPLFR M**N**ISQCDLPE ISVSHV**N**LTV REGDNAVITC
201 **N**GSGSPLPDV DWIVTGLQSI NTHQTNL**N**WT NVHAI**N**LTLV **N**VTSEDNGFT
251 LTCIAENVVG MS**N**ASVALTV YYPPRVVSLE EPELRLEHCI EFVVRGNPPP
301 TLHWLHNGQP LRESKIIHVE YYQEGEISEG CLLFNKPTHY NNG**N**YTLIAK
351 NPLGTA**N**QTI NGHFLKEPFP ESTDNFILFD EVSPTPPITV THKPEEDTFG
401 VSIAVGLAAF ACVLLVVLFV MINKYGRRSK FGMKGPVAVI SGEEDSASPL
451 HHINHGITTP SSLDAGPDTV VIGMTRIPVI ENPQYFRQGH NCHKPDTYVQ
501 HIKRRDIVLK RELGEGAFGK VFLAECYNLS PTKDKMLVAV KALKDPTLAA
551 RKDFQREAEL LTNLQHEHIV KFYGVCGDGD PLIMVFEYMK HGDLNKFLRA
601 HGPDAMILVD GQPRQAKGEL GLSQMLHIAS QIASGMVYLA SQHFVHRDLA
651 TRNCLVGANL LVKIGDFGMS RDVYSTDYYR LFNPSGNDFC IWCEVGGHTM
701 LPIRWMPPES IMYRKFTTES DVWSFGVILW EIFTYGKQPW FQLSNTEVIE
751 CITQGRVLER PRVCPKEVYD VMLGCWQREP QQRLNIKEIY KILHALGKAT
801 PIYLDILG

At least four splice variants are known (see SwissProt entry). Tyr678 is site of autophosphorylation.

Amino acid sequence for mouse NT-3 receptor (trkC)[29,30]

Accession code: SwissProt Q62371

-21 MIPIPRMPLV LLLLLLILGS A
1 KAQVNPAICR YPLGMSGGHI PDEDITASSQ WSESTAAKYG RLDSEEGDGA
51 WCPEIPVQPD DLKEFLQIDL RTLHFITLVG TQGRHAGGHG IEFAPMYKI**N**
101 YSRDGSRWIS WRNRHGKQVL DGNSNPYDVF LKDLEPPIVA RFVRLIPVTD
151 HSMNVCMRVE LYGCVWLDGL VSYNAPAGQQ FVLPGGSIIY L**N**DSVYDGAV
201 GYSMTEGLGQ LTDGVSGLDD FTQTHEYHVW PGYDYVGWR**N** ESATNGFIEI
251 MFEFDRIR**N**F TTMKVHCNNM FAKGVKIFKE VQCYFRSEAS EWEPTAVYFP
301 LVLDDVNPSA RFVTVPLHHR MASAIKCQYH FADTWMMFSE ITFQSDAAMY
351 **N**NSGALPTSP MAPTTYDPML KVDDSNTRIL IGCLVAIIFI LLAIIVIILW
401 RQFWQKMLEK ASRRMLDDEM TVSLSLPSES SMFNNNRSSS PSEQESNSTY
451 DRIFPLRPDY QEPSRLIRKL PEFAPGEEES GCSGVVKPAQ PNGPEGVPHY
501 AEADIVNLQG VTGGNTYCVP AVTMDLLSGK DVAVEEFPRK LLAFKEKLGE
551 GQFGEVHLCE VEGMEKFKDK DFALDVSANQ PVLVAVKMLR ADANKNARND
601 FLKEIKIMSR LKDPNIIRLL AVCITEDPLC MITEYMENGD LNQFLSRHEP
651 LSSCSSDATV SYANLKFMAT QIASGMKYLS SLNFVHRDLA TRNCLVGKNY
701 TIKIADFGMS RNLYSGDYYR IQGRAVLPIR WMSWESILLG KFTTASDVWA
751 FGVTLWETFT FCQEQPYSQL SDEQVIENTG EFFRDQGRQI YLPQPALCPD
801 SVYKLMLSCW RRETKHRPSF QEIHLLLLQQ GAE

Tyr718 is site of autophosphorylation.

Amino acid sequence for rat NT-3 receptor (trkC)[31]

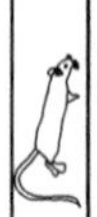

Accession code: SwissProt P24786

```
-31  MDVSLCPAKC SFWRIFLLGS VWLDYVGSVL A
  1  CPANCVCSKT EINCRRPDDG NLFPLLEGQD SGNSNGNASI NITDISRNIT
 51  SIHIENWRGL HTLNAVDMEL YTGLQKLTIK NSGLRNIQPR AFAKNPHLRY
101  INLSSNRLTT LSWQLFQTLS LRELRLEQNF FNCSCDIRWM QLWQEQGEAR
151  LDSQSLYCIS ADGSQLPLFR MNISQCDLPE ISVSHVNLTV REGDNAVITC
201  NGSGSPLPDV DWIVTGLQSI NTHQTNLNWT NVHAINLTLV NVTSEDNGFT
251  LTCIAENVVG MSNASVALTV YYPPRVVSLV EPEVRLEHCI EFVVRGNPTP
301  TLHWLYNGQP LRESKIIHMD YYQEGEVSEG CLLFNKPTHY NNGNYTLIAK
351  NALGTANQTI NGHFLKEPFP ESTDFFDFES DASPTPPITV THKPEEDTFG
401  VSIAVGLAAF ACVLLVVLFI MINKYGRRSK FGMKGPVAVI SGEEDSASPL
451  HHINHGITTP SSLDAGPDTV VIGMTRIPVI ENPQYFRQGH NCHKPDTYVQ
501  HIKRRDIVLK RELGEGAFGK VFLAECYNLS PTKDKMLVAV KALKDPTLAA
551  RKDFQREAEL LTNLQHEHIV KFYGVCGDGD PLIMVFEYMK HGDLNKFLRA
601  HGPDAMILVD GQPRQAKGEL GLSQMLHIAS QIASGMVYLA SQHFVHRDLA
651  TRNCLVGANL LVKIGDFGMS RDVYSTDYYR VGGHTMLPIR WMPPESIMYR
701  KFTTESDVWS FGVILWEIFT YGKQPWFQLS NTEVIECITQ GRVLERPRVC
751  PKEVYDVMLG CWQREPQQRL NIKEIYKILH ALGKATPIYL DILG
```

Tyr678 is site of autophosphorylation.

References

[1] Yancopoulos, G.D. et al. (1990) Cold Spring Harbor Symp. Quant. Biol. LV, 371–379.
[2] Maisonpierre, P.C. et al. (1990) Science 247, 1446–1451.
[3] Chalazonitis, A. (1996) Mol. Neurobiol. 12, 39–53.
[4] Lu, B. and Figurov, A. (1997) Rev. Neurosci. 8, 1–12.
[5] Maisonpierre, P.C. et al. (1991) Genomics 10, 558–568.
[6] Jones, K.R. and Reichardt, L.F. (1990) Proc. Natl Acad. Sci. USA 87, 8060–8064.
[7] Hohn, A. et al. (1990) Nature 344, 339–341.
[8] Meakin, S.O. and Shooter, E.M. (1992) Trends Neurosci. 15, 323–331.
[9] Barker, P.A. and Murphy, R.A. (1992) Mol. Cell Biochem. 110, 1–15.
[10] Frade, J.M. and Barde, Y.A. (1998) BioEssays 20, 137–145.
[11] Ip, N.Y. and Yancopoulos, G.D. (1994) Ann. Neurol. 35 (Suppl), S13–S16.
[12] Rodriguez-Tebar, A. et al. (1992) EMBO J. 11, 917–922.
[13] Lamballe, F. et al. (1991) Cell 66, 967–979.
[14] Cordon-Cardo, C. et al. (1991) Cell 66, 173–183.
[15] Rodriguez-Tebar, A. et al. (1991) Phil. Trans. R. Soc. Lond. B Biol. Sci. 331, 255–258.
[16] Barbacid, M. (1995) Ann. N.Y. Acad. Sci. 766, 442–458.
[17] Tsoulfas, P. et al. (1993) Neuron 10, 975–990.
[18] Valenzuela, D.M. et al. (1993) Neuron 10, 963–974.
[19] Klein, R. et al. (1991) Cell 66, 395–403.
[20] Glass, D.J. et al. (1991) Cell 66, 405–413.
[21] Soppet, D. et al. (1991) Cell 65, 895–903.
[22] Squinto, S.P. et al. (1991) Cell 65, 885–893.

[23] Klein, R. et al. (1992) Neuron 8, 947–956.
[24] Widmer, H.R. et al. (1992) J. Neurochem. 59, 2113–2124.
[25] Kaplan, D.R. and Miller, F.D. (1997) Curr. Opin. Cell Biol. 9, 213–221.
[26] Friedman, W.J. and Greene, L.A. (1999) Exp. Cell Res. 253, 131–142.
[27] Shelton, D.L. et al. (1995) J. Neurosci. 15, 477–491.
[28] McGregor, L.M. et al. (1994) Genomics 22, 267–272.
[29] Lai, C. and Lemke, G. (1994) Oncogene 9, 877–883.
[30] Karn, T. et al. (1993) Oncogene 8, 3433–3440.
[31] Merlio, J.P.et al. (1992) Neuroscience 51, 513–532.

Other names

Mammalian NT-4 has also been called NT-5 and NT4/5 to distinguish it from *Xenopus* NT-4.

THE MOLECULE

Neurotrophin 4 (NT-4) is a neurotrophic factor which supports the survival and outgrowth of sensory neurons from embryonic chicken dorsal root ganglia, but has no significant effect on embryonic day 8 sympathetic ganglia[1]. It has a strong survival/proliferative action on NIH 3T3 cells expressing trkB but little activity on 3T3 cells expressing trkA. NT-4 was originally identified in *Xenopus* and viper, and was subsequently identified in rat and humans.

Crossreactivity

The mature human and rat NT-4 sequences are 95% identical. Human, rat and *Xenopus* NT-4 are all active on neurites from chicken embryo dorsal root ganglia.

Sources

High levels of NT-4 mRNA have been found in prostate with lower levels in thymus, placenta and skeletal muscle.

Bioassays

Survival and outgrowth of neurites from embryo chicken dorsal root ganglia. Survival and proliferation on NIH 3T3 cells transfected with *trkB*.

Physicochemical properties of NT-4

Property	Human	Rat
Amino acids – precursor	210	209
– mature[a]	130	130
M_r (K) – predicted	13.9	13.9
– expressed[b]	14	14
Potential *N*-linked glycosylation sites[c]	0	0
Disulfide bonds[d]	3	3

[a] After removal of prepropeptide.
[b] May also be expressed as a homodimer like the other neurotrophins.
[c] There is one *N*-linked glycosylation site in the propeptide.
[d] Conserved between NGF, BDNF, NT-3 and NT-4.

3-D structure

Not determined but likely to be very similar to NGF.

Gene structure[1,2]

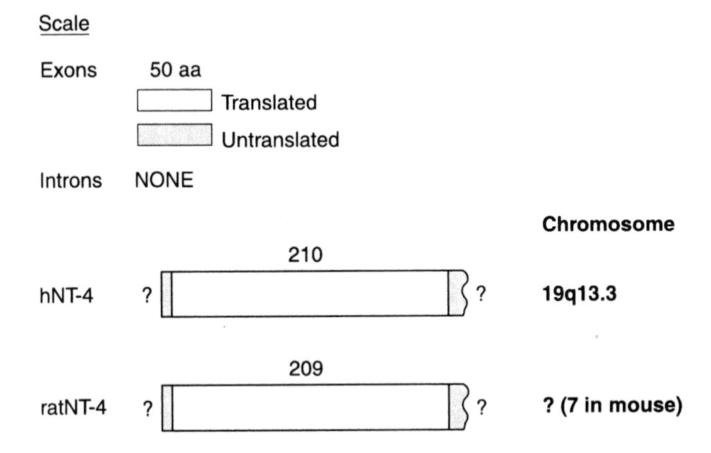

The gene for NT-4 is located on human chromosome 19q13.3. The human and mouse genes consist of a single coding exon of 210 and 209 amino acids respectively. There may be more than one RNA species arising by alternative splicing and additional ATG initiation sites[1]. Three other partial ORFs that may encode additional members of this acidic protein family have been found on the same region of human chromosome 19[2].

Amino acid sequence for human NT-4[1]

Accession code: SwissProt P34130

```
-19  MLPLPSCSLP ILLLFLLPS
  1  VPIESQPPPS TLPPFLAPEW DLLSPRVVLS RGAPAGPPLL FLLEAGAFRE
 51  SAGAPANRSR RGVSETAPAS RRGELAVCDA VSGWVTDRRT AVDLRGREVE
101  VLGEVPAAGG SPLRQYFFET RCKADNAEEG GPGAGGGGCR GVDRRHWVSE
151  CKAKQSYVRA LTADAQGRVG WRWIRIDTAC VCTLLSRTGR A
```

Mature human NT-4 is formed by removal of a predicted signal peptide and a propeptide (1–61, in italics). Other precursor forms with extended N-terminal sequences similar to NGF may also exist. Disulfide bonds between Cys78–151, 122–180, and 139–182 by similarity.

Amino acid sequence for rat NT-4[1]

(Mouse NT-4 has been cloned and the untranslated cDNA is available on NCBI AA271974.)

Accession code: SwissProt P34131

```
-18  MLPRHSCSLL LFLLLLPS
  1  VPMEPQPPSS TLPPFLAPEW DLLSPRVALS RGTPAGPPLL FLLEAGAYGE
 51  PAGAPANRSR RGVSETAPAS RRGELAVCDA VSGWVTDRRT AVDLRGREVE
101  VLGEVPAAGG SPLRQYFFET RCKAESAGEG GPGVGGGGCR GVDRRHWLSE
151  CKAKQSYVRA LTADSQGRVG WRWIRIDTAC VCTLLSRTGR A
```

Mature rat NT-4 is formed by removal of a predicted signal peptide and a propeptide (1–61, in italics). Other precursor forms with extended N-terminal sequences similar to NGF may also exist. Disulfide bonds between Cys78–151, 122–180 and 139–182 by similarity.

THE NT-4 RECEPTORS

NT-4 binds to LNGFR[3] (see NGF entry (page **397**)), and to trkB[4] (see BDNF entry (page **160**)).

References

[1] Ip, N.Y. et al. (1992) Proc. Natl Acad. Sci. USA 89, 3060–3064.
[2] Berkemeier, L.R. et al. (1992) Somat. Cell. Mol. Genet. 18, 233–245.
[3] Rodriguez-Tebar, A. et al. (1992) EMBO J. 11, 917–922.
[4] Klein, R. et al. (1992) Neuron 8, 947–956.

Other names

A number of abbreviations are in common use, such as OSM, OM, Onco M and ONC. OSM is the preferred abbreviation.

THE MOLECULE

Oncostatin M (OSM) is a pleiotropic cytokine which participates in the regulation of cell growth and differentiation during haematopoiesis, neurogenesis and osteogenesis. OSM has been shown to inhibit the growth of some tumour and normal cell lines, stimulates fibroblast, smooth muscle and Kaposi's sarcoma cell proliferation, cytokine release from endothelial cells and low-density lipoprotein receptor expression on hepatoma cells[1–6]. It shares many structural and biological properties with LIF, IL-6 and CNTF[6,7].

Crossreactivity

Human and mouse OSM share 48% amino acid identity. The most conserved motif is within the C-terminal helix and contains the site of interaction with the OSM receptor. OSM also shares significant sequence and structural homology with LIF and other members of the IL-6 cytokine family. Human OSM is active on mouse cells.

Sources

Activated T cells and monocytes, Kaposi s sarcoma cells[2,4].

Bioassays

The biological activity of human OSM has been determined in a number of assays including the growth inhibition of human melanoma A375 cells[1,2], differentiation of human Th1 or murine M1 cells. It is important to note that the assays on mouse cells are observed using human OSM, which interacts with the murine LIF receptor and not the murine OSM receptor (see discussion below on OSM receptors).

Physicochemical properties of OSM

Property	Human	Mouse
pI[a]	10.71	11.78
Amino acids – precursor	252	263
– mature[b]	196	239
M_r (K) – predicted	28.48	30.113
– expressed	28–32	?
Potential *N*-linked glycosylation sites	2	3
Disulfide bonds[c]	2	0

[a] Predicted.

[b] The mature human peptide is formed after removal of the predicted signal peptide and propeptide (shown in italics). In mouse OSM, the C-terminal domain is 26 amino acid residues longer than that of human OSM, whcih must be removed to generate a biologically active mouse OSM protein.

[c] Cys6 and 127, and Cys49 and 167 form disulfide bonds[8].

3-D structure

The crystal structure of OSM is unknown, but based on its comparison with LIF and on NMR structural data it is predicted that OSM forms a four α-helical bundle structure similar to that determined for LIF, CNTF, IL-6, G-CSF and IL-11[8–10].

Gene structure[11,12]

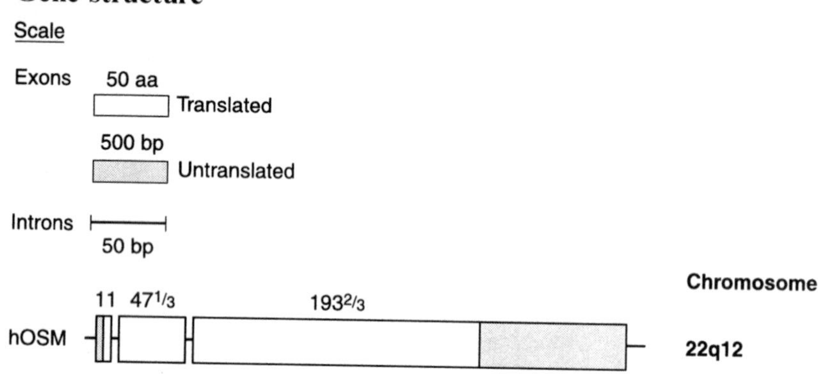

The gene encoding OSM is located on human chromosome 22q12[13] and murine chromosome 11[14] within 20 kbp from the gene encoding LIF. Both LIF and OSM genes are organized in a similar manner with three exons and two introns. The first exon of OSM is very short with 11 amino acids, the other two exons code for 48 and 193 amino acids respectively. The LIF and OSM genes are derived from a common ancestor.

Amino acid sequence for human OSM[13]

Accession code: SwissProt P13725

```
-25  MGVLLTQRTL LSLVLALLFP SMASM
  1  AAIGSCSKEY RVLLGQLQKQ TDLMQDTSRL LDPYIRIQGL DVPKLREHCR
 51  ERPGAFPSEE TLRGLGRRGF LQTLNATLGC VLHRLADLEQ RLPKAQDLER
101  SGLNIEDLEK LQMARPNILG LRNNIYCMAQ LLDNSDTAEP TKAGRGASQP
151  PTPTPASDAF QRKLEGCRFL HGYHRFMHSV GRVFSKWGES PNRSRRHSPH
201  QALRKGVRRT RPSRKGKRLM TRGQLPR
```

The OSM gene encodes a pre-pro OSM precursor which is processed at the N- and C-termini. Mature OSM is formed by removal of a predicted signal peptide (23–25 amino acid residues). An additional cleavage site occurs after a pair of basic residues (propeptide sequence shown in italics 235–252). Disulfide bonds between Cys31–152 and 74–192, by similarity.

Amino acid sequence for mouse OSM[14]

Accession code: SwissProt P53347

```
-24   MQTRLLRTLL SLTLSLLILS MALA
  1   NRGCSNSSSQ LLSQLQNQAN LTGNTESLLE PYIRLQNLNT PDLRAACTQH
 51   SVAFPSEDTL RQLSKPHFLS TVYTTLDRVL YQLDALRQKF LKTPAFPKLD
101   SARHNILGIR NNVFCMARLL NHSLEIPEPT QTDSGASRST TTPDVFNTKI
151   GSCGFLWGYH RFMGSVGRVF REWDDGSTRS RRQSPLRARR KGTRRIRVRH
201   KGTRRIRVRR KGTRRIWVRR KGSRKIRPSR STQSPTTRA
```

Mature OSM is formed by removal of a predicted signal peptide. Disulfide bonds between Cys28–139 and 71–177, by similarity.

THE OSM RECEPTOR

High- and low-affinity receptors for OSM have been detected in a wide variety of cell types. The low-affinity receptor for OSM has been identified as gp130, a molecule that forms a component of the IL-6 and LIF receptor complex (see IL-6 entry (page **69**) and LIF entry (page **346**)). Receptor binding in this family of receptors is characterized by low-affinity binding to an α receptor which is converted to a high-affinity complex due to a second unrelated β receptor. gp130 is the high-affinity converter for LIF and IL-6Rs, but can bind OSM directly. Hence, gp130 is often referred to as the low-affinity OSMR, hence OSMRα. Antibodies to gp130 block responses to OSM, LIF, IL-6 and CNTF[15–18]. gp130 however, is insufficient to confer OSM responsiveness in permissive cells and requires the low-affinity LIFR to convert gp130 into a high-affinity converter for OSM. In humans, two different high-affinity heterodimeric receptor complexes have been identified. Human OSM signals through the LIF/OSM shared receptor, composed of gp130 and the LIFR[19] as described above, and the OSM-specific receptor which binds OSM uniquely, consisting of OSMRα (gp130) and recently described high-affinity converting receptor subunit (OSMRβ)[20]. Human OSMR β-chain is a 979-amino-acid transmembrane protein that belongs to the haematopoietin receptor family, which also includes gp130, LIFR (gp190) and G-CSFR. OSMRβ shows considerable homology to LIFRα (32%) and to gp130 (21%), except that OSMRβ lacks the cysteine containing module of the N-terminal haematopoietin domain which is present in LIFR[20]. More recently, mouse OSMRβ has been described[21]. Interestingly, in mouse cells, only mouse OSM is capable of activating the mouse OSM receptor. Human OSM acts via gp130/LIFR complexes, suggesting that all previous studies with human OSM in mouse models did not elucidate the biology of OSM, but rather reflected that of LIF.

Distribution

OSMRβ mRNA is detected in mouse heart, brain, lung, spleen, lung, liver, skeletal muscle and kidney tissue.

Physicochemical properties of the OSM receptor β chain

Property	Human	Mouse
Amino acids – precursor	979	971
– mature[a]	952	?
M_r (K) – predicted	110.5	?
– expressed	180	180
Potential *N*-linked glycosylation sites	?	?
Affinity K_d (M)	$3–1000 \times 10^{-12}$	?

[a] After removal of predicted signal peptide.

Signal transduction

Involves a tyrosine kinase in hepatoma, endothelial and smooth muscle cells[8,22]. OSM, but not IL-6 or LIF, induces tyrosine phosphorylation of the Shc isoforms p52 and p66 and their association with Grb2. Concomitantly, OSM turns out to be a stronger activator of Erk1/2 MAP kinases. Shc is recruited to the OSM receptor (OSMR), but not to gp130[23].

Chromosomal location

Not known.

Amino acid sequence for human OSMR β-chain[20]

Accession code: SPTREMBL Q99650

```
-27  MALFAVFQTT FFLTLLSLRT YQSEVLA
  1  ERLPLTPVSL KVSTNSTRQS LHLQWTVHNL PYHQELKMVF QIQISRIETS
 51  NVIWVGNYST TVKWNQVLHW SWESELPLEC ATHFVRIKSL VDDAKFPEPN
101  FWSNWSSWEE VSVQDSTGQD ILFVFPKDKL VEEGTNVTIC YVSRNIQNNV
151  SCYLEGKQIH GEQLDPHVTA FNLNSVPFIR NKGTNIYCEA SQGNVSEGMK
201  GIVLFVSKVL EEPKDFSCET EDFKTLHCTW DPGTDTALGW SKQPSQSYTL
251  FESFSGEKKL CTHKNWCNWQ ITQDSQETYN FTLIAENYLR KRSVNILFNL
301  THRVYLMNPF SVNFENVNAT NAIMTWKVHS IRNNFTYLCQ IELHGEGKMM
351  QYNVSIKVNG EYFLSELEPA TEYMARVRCA DASHFWKWSE WSGQNFTTLE
401  AAPSEAPDVW RIVSLEPGNH TVTLFWKPLS KLHANGKILF YNVVVENLDK
451  PSSVEEERIA GTEGGFSLSW KPQPGDVIGY VVDWCDHTQD VLGDFQWKNV
501  GPNTTSTVIS TDAFRPGVRY DFRIYGLSTK RIACLLEKKT GYSQELAPSD
551  NPHVLVDTLT SHSFTLSWKD YSTESQPGFI QGYHVYLKSK ARQCHPRFEK
601  AVLSDGSECC KYKIDNPEEK ALIVDNLKPE SFYEFFITPF TSAGEGPSAT
651  FTKVTTPDEH SSMLIHILLP MVFCVLLIMV MCYLKSQWIK ETCYPDIPDP
701  YKSSILSLIK FKENPHLIIM NVSDCIPDAI EVVSKPEGTK IQFLGTRKSL
751  TETELTKPNY LYLLPTEKNH SGPGPCICFE NLTYNQAASD SGSCGHVPVS
801  PKAPSMLGLM TSPENVLKAL EKNYMNSLGE IPAGETSLNY VSQLASPMFG
851  DKDSLPTNPV EAPHCSEYKM QMAVSLRLAL PPPTENSSLS SITLLDPGEH
901  YC
```

References

[1] Malik, N. et al. (1989) Mol. Cell Biol. 9, 2847–2853.
[2] Zarling, J.M. et al. (1986) Proc. Natl Acad. Sci. USA 83, 9739–9743.
[3] Miles, S.A. et al. (1992) Science 255, 1432–1434.
[4] Radka, S.F. et al. (1993) J. Immunol. 150, 5195–5201.
[5] Brown, T.J. et al. (1991) J. Immunol. 147, 2175–2180.
[6] Bruce, A.G. et al. (1992) J. Immunol. 149, 1271–1275.
[7] Rose, T.M. and Bruce, A.G. (1991) Proc. Natl Acad. Sci. USA 88, 8641–8645.
[8] Bruce, A.G. et al. (1992) Progr. Growth Factor Res. 4, 157–170.
[9] Bazan, J.F. (1991) Neuron 7, 197–208.
[10] Nicola, N.A. et al. (1993) Biochem. Biophys. Res. Commun. 190, 20–26.
[11] Giovannini, M. et al. (1993) Cytogenet. Cell Genet. 62, 32–34.
[12] Giovannini, M. et al. (1993) Cytogenet. Cell Genet. 64, 240–244.
[13] Rose, T.M. et al. (1993) Genomics 17, 136–140.
[14] Yoshimura, A. et al. (1996) EMBO J. 15, 1055–1063.
[15] Watanabe, D. et al. (1996) Eur. J. Neurosci. 8, 1630–1640.
[16] Koshimizu, U. et al. (1996) Development 122, 1235–1242.
[17] Zhang, X.G. et al. (1994) J. Exp. Med. 179, 1337–1342.
[18] Taga, T. (1992) Nippon Rinsho 50, 1802–1810.
[19] Gearing, D.P. and Bruce, A.G. (1992) New Biol. 4, 61–65.
[20] Mosley, B. et al. (1996) J. Biol. Chem. 271, 32635–32643.
[21] Lindberg, R.A. et al. (1998) Mol. Cell Biol. 18, 3357–3367.
[22] Grove, R.I. et al. (1993) Proc. Natl Acad. Sci. USA 90, 823–827.
[23] Hermanns, H.M. et al. (2000) J. Biol. Chem. 275, 40742–40748.

Other names
CD134L, gp34, TXGP-1 and TNFSF4.

THE MOLECULE

OX-40 ligand (OX-40L) was originally identified in a search for molecules specifically expressed in HTLV-I-infected T cells as a 34-kDa glycoprotein regulated by the HTLV-I protein Tax[1]. In an independent expression cloning study using OX-40-Fc constructs as molecular probes, isolation of a cDNA revealed the identity of this 34-kDa protein to be the ligand for OX-40. Accumulating evidence now suggests that OX-40L interactions with its cognate receptor act after initial activation events to prolong clonal expansion and enhance effector cytokine secretion, and may be involved in promoting long-lived primary CD4 responses. The human ligand is a 32-kDa, 183-amino-acid residue glysosylated polypeptide that consists of a 21-amino-acid cytoplasmic domain, a 23-amino-acid transmembrane domain and a 139-amino-acid residue extracellular domain[2]. When compared with the extracellular region of TNFα, OX-40L shares only 15% amino acid sequence identity, but it is included in the TNF superfamily on the basis of secondary and tertiary structure[2]. Consistent with other TNF-SF members, OX-40L is reported to exist as a trimer. In addition, the HTLV-I regulation of gp34 suggests a possible connection between virally induced pathogenesis and the OX-40 system, however the significance of OX-40/OX-40L upregulation by HTLV-I infection for the pathology of the virus remains to be clarified.

Crossreactivity

Human OX-40L is 46% identical to mouse OX-40L at the amino acid level. Mouse OX-40L is active in humans but human OX-40L is inactive in mice.

Sources

OX-40L, like OX-40 shows limited expression. Currently only activated CD4+, CD8+ T cells, B cells and vascular endothelial cells have been reported to express this cytokine[1,3–6].

Bioassays

The cell-bound recombinant ligands are biologically active, costimulating T-cell proliferation and cytokine production[1].

Physicochemical properties of OX-40L

Property	Human	Mouse
pI	6.95[a]	6.09
Amino acids – precursor	183	198
– mature[b]	183	198
M_r (K) – predicted	21.05	22.254
– expressed	34	34
Potential *N*-linked glycosylation sites	4	1
Disulfide bonds	0	0

[a] This represents the predicted pI's of OX-40L.
[b] Human and mouse OX-40L lack signal peptides.

3-D structure

The crystal structure for OX-40L is unknown, however the predicted structures are similar to, but more compact than, those of other ligands of the TNF family[1].

Gene structure

Mapping of the mouse OX-40L gene revealed tight linkage to *gld*, the FasL gene, on chromosome 1. Human OX-40L (gp34) maps to a homologous region in the human genome, 1q25[1].

Amino acid sequence for human OX-40L

Accession code: SwissProt P23510

```
  1  MERVQPLEEN VGNAARPRFE RNKLLLVASV IQGLGLLLCF TYICLHFSAL
 51  QVSHRYPRIQ SIKVQFTEYK KEKGFILTSQ KEDEIMKVQN NSVIINCDGF
101  YLISLKGYFS QEVNISLHYQ KDEEPLFQLK KVRSVNSLMV ASLTYKDKVY
151  LNVTTDNTSL DDFHVNGGEL ILIHQNPGEF CVL
```

Human OX-40L lacks a signal peptide. The putative signal anchor (transmembrane region) is underlined.

Amino acid sequence for mouse OX-40L

Accession code: SwissProt P43488

```
  1  MEGEGVQPLD ENLENGSRPR FKWKKTLRLV VSGIKGAGML LCFIYVCLQL
 51  SSSPAKDPPI QRLRGAVTRC EDGQLFISSY KNEYQTMEVQ NNSVVIKCDG
101  LYIIYLKGSF FQEVKIDLHF REDHNPISIP MLNDGRRIVF TVVASLAFKD
151  KVYLTVNAPD TLCEHLQIND GELIVVQLTP GYCAPEGSYH STVNQVPL
```

Mouse OX-40L lacks a signal peptide. The putative signal anchor (transmembrane region) is underlined.

THE OX-40L RECEPTOR

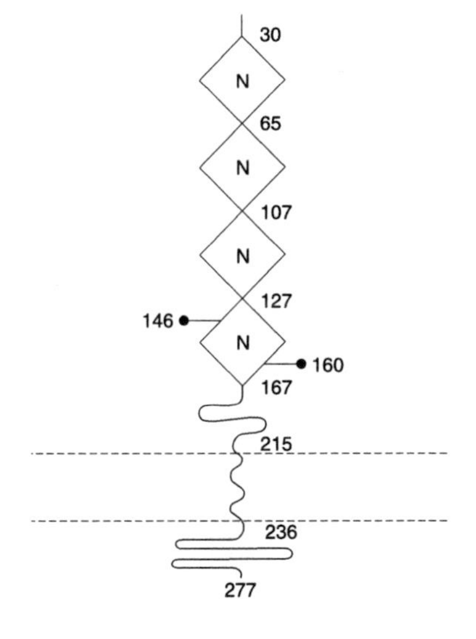

The OX-40L receptor (OX-40R) is a cell surface glycoprotein of the tumour necrosis factor receptor family that is expressed primarily on activated CD4+ T cells[7]. Engagement of OX-40R by the OX-40 ligand (OX-40L) is known to costimulate the production of cytokines by activated T lymphocytes[8] and to rescue effector T cells from activation-induced cell death. OX-40-deficient mice have greatly diminished T helper cell responses. Accumulating evidence now favours a crucial role for OX-40L/OX-40 interactions in the development of Th2 cells *in vivo*, possibly through T cell/DC interactions in the draining lymph nodes. Human OX-40 is a 48-kDa type I (N-terminal extracellular) transmembrane glycoprotein. The mature molecule is a 250-amino-acid residue polypeptide that consists of a 188-amino-acid residue extracellular region, a 26-amino-acid transmembrane segment and a 36-amino-acid residue cytoplasmic domain. In the extracellular domain there is an approximately 60% amino acid sequence identity from human to mouse. There is also marked species crossreactivity in this system.

Distribution

OX-40 is expressed primarily on activated CD4 T cells.

Physicochemical properties of OX-40L

Property	Human	Mouse
Amino acids – precursor	277	272
– mature[a]	249	253
M_r (K) – predicted	29.34	30.153
– expressed	48	48
Potential *N*-linked glycosylation sites	2	1
Affinity K_d (M)	?	?

[a] After removal of predicted signal peptide.

Signal transduction

Signal transduction processes initiated following OX-40L-OX-40 interactions are as yet unclear, but like other members of this ligand/receptor superfamily, OX-40 signalling involves members of the Traf family of signal transducers, namely, Traf-2, Traf-3 and Traf-5, which lead to NFκB activation[9,10].

Chromosomal location

Unknown.

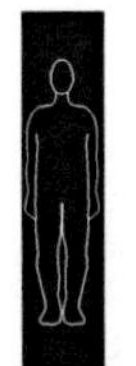

Amino acid sequence for human OX-40

Accession code: SwissProt P43489

```
-19  MCVGARRLGR GPCAALLL
  1  LGLGLSTVTG LHCVGDTYPS NDRCCHECRP GNGMVSRCSR SQNTVCRPCG
 51  PGFYNDVVSS KPCKPCTWCN LRSGSERKQL CTATQDTVCR CRAGTQPLDS
101  YKPGVDCAPC PPGHFSPGDN QACKPWTNCT LAGKHTLQPA SNSSDAICED
151  RDPPATQPQE TQGPPARPIT VQPTEAWPRT SQGPSTRPVE VPGGRAVAAI
201  LGLGLVLGLL GPLAILLALY LLRRDQRLPP DAHKPPGGGS FRTPIQEEQA
251  DAHSTLAKI
```

Amino acid sequence for mouse OX-40

Accession code: SwissProt P47741

```
-19  MYVWVQQPTA LLLLALTLG
  1  VTARRLNCVK HTYPSGHKCC RECQPGHGMV SRCDHTRDTL CHPCETGFYN
 51  EAVNYDTCKQ CTQCNHRSGS ELKQNCTPTQ DTVCRCRPGT QPRQDSGYKL
101  GVDCVPCPPG HFSPGNNQAC KPWTNCTLSG KQTRHPASDS LDAVCEDRSL
151  LATLLWETQR PTFRPTTVQS TTVWPRTSEL PSPPTLVTPE GPAFAVLLGL
201  GLGLLAPLTV LLALYLLRKA WRLPNTPKPC WGNSFRTPIQ EEHTDAHFTL
251  AKI
```

Conflicting sequence at position 15 A→G.

References

[1] Baum, P.R. et al. (1994) EMBO J. 13, 3992–4001.
[2] Godfrey, W.R. et al. (1994) J. Exp. Med. 180, 757–762.
[3] Calderhead, D.M. et al. (1993) J. Immunol. 151, 5261–5271.
[4] Stuber, E. and Strober, W. (1996) J. Exp. Med. 183, 979–989.
[5] Stuber, E. et al. (1996) J. Exp. Med. 183, 693–698.
[6] Imura, A. et al. (1996) J. Exp. Med. 183, 2185–2195.
[7] Watts, T.H. and DeBenedette, M.A. (1999) Curr. Opin. Immunol. 11, 286–293.
[8] Kaleeba, J.A. et al. (1998) Int. Immunol. 10, 453–461.
[9] Arch, R.H. et al. (1998) Genes Dev. 12, 2821–2830.
[10] Arch, R.H. and Thompson, C.B. (1998) Mol. Cell Biol. 18, 558–565.

Other names
CTAP-III is also known as low-affinity platelet factor 4 and CXCL7[1].

THE MOLECULES

Platelet basic protein (PBP) is a CXC chemokine produced by platelets[2,3] which is sequentially proteolytically processed at the N-terminus to form the active mediators CTAP-III (a chondrocyte mitogen[1]), β-thromboglobulin (βTG)[4,5] (a fibroblast chemoattractant) and NAP-2 (a neutrophil chemoattractant[6,7]). NAP-2 is produced by cleavage with cathepsin G[8]. CTAP-III and NAP-2 are both heparan sulfate-degrading enzymes[9].

Crossreactivity
There is no known murine homologue of human PBP.

Sources
Platelets.

Bioassays
CTAP-III: mitogenesis for fibroblasts[10]; βTG: chemotaxis for fibroblasts[11]; NAP-2: chemotaxis of neutrophils[12]. PBP has no known biological activity[5].

Physicochemical properties of CTAP-III, βTG and NAP-2

Property	Human	Human	Human
	CTAP-III	βTG	NAP-2
Amino acids – precursor	128	128	128
– mature	85	81	69
M_r (K) – predicted	9.2	8.8	7.3
– expressed	7.5	7	6
Potential *N*-linked glycosylation sites	0	0	0
Disulfide bonds	2	2	2

Seven-amino-acid C-terminal truncated forms of NAP-2 and CTAP-III have been described with up to 5-fold enhanced potency[13].

3-D structure
NAP-2 monomer consists of a triple-stranded antiparallel β-sheet in a Greek key motif and a C-terminal helix, similar to PF4 and IL-8[14,15]. The association state of CTAP-III has been studied by NMR; under physiological conditions the active form is a monomer[16].

Gene structure for human PBP

The gene for PBP is located on chromosome 4q12–21[17].

Amino acid sequence for human PBP[1,2]

Accession code: SwissProt P02775

```
-34  MSLRLDTTPS CNSARPLHAL QVLLLLSLLL TALA
  1  SSTKGQTKRN LAKGKEESLD SDLYAELRCM CIKTTSGIHP KNIQSLEVIG
 51  KGTHCNQVEV IATLKDGRKI CLDPDAPRIK KIVQKKLAGD ESAD
```

The N-terminal amino acids for mature CTAP-III, βTG, and NAP-2 are Asn10, Lys15 and Arg28 respectively. The natural form of NAP-2 truncates at L87.

THE RECEPTOR CXCR1 and CXCR2

NAP-2 binds selectively to the CXCR2 receptor with a 2.8 nM affinity and mediates chemotaxis at 10–100 ng/ml. It also binds to the CXCR1 receptor but with lower affinity (9 nM) and mediates chemotaxis at 1000–3000 ng/ml. In contrast, IL-8 binds to both CXCR1 and CXCR2 with similar potency[18,19]. For descriptions of both receptors, see the IL-8 entry (page **80**).

References

1 Caster, C.W. et al. (1983) Proc. Natl Acad. Sci. USA 80, 765–769.
2 Wenger, R.H. et al. (1989) Blood 73, 1498–1503.
3 Brandt E et al. (2000) J. Leukoc. Biol. 67, 471–478.
4 Beggs, G.S. et al. (1978) Biochemistry 17, 1739–1744.
5 Holt, J.C. et al. (1986) Biochemistry 25, 1988–1996.
6 Walz, A. et al. (1989) J. Exp. Med. 170, 1745–1750.
7 Walz, A. and Baggiolini, M. (1990) J. Exp. Med. 171, 449–454.
8 Cohen, A.B. et al. (1992) Am. J. Physiol. 263, 1249–1256.
9 Hoogerwerf AJ et al. (1995) J. Biol. Chem. 270:3268–3277.
10 Mullenbach, G.T. et al. (1986) J. Biol. Chem. 261, 719–722.
11 Senior, R.M. et al. (1983) J. Cell. Biol. 96, 382–385.
12 Walz, A. et al. (1989) J. Exp. Med. 170, 1745–1750.
13 Ehlert, J.E. et al. (1998) J. Immunol. 161, 4975–4982.
14 Mayo, R.T. et al. (1994) Biochem. J. 304, 371–376.
15 Malkowski, M.G. et al. (1995) J. Biol. Chem. 270, 7077–7087.
16 Mayo, K.H. (1991) Biochemistry 30, 925–934.
17 Modi, W.S. et al. (1990) Hum. Genet. 84, 185–187.
18 Ben-Baruch, A. et al. (1997) Cytokine 9, 37–45.
19 Ludwig, A. et al. (1997) Blood 90, 4588–4597.

Other names

Osteosarcoma-derived growth factor (ODGF), glioma-derived growth factor (GDGF).

THE MOLECULES

Platelet-derived growth factor (PDGF) is a mitogen for connective tissue cells and glial cells. It plays an important role in wound healing, and may act as an autocrine and/or paracrine growth factor for some malignant cells[1–3]. It is also a chemoattractant for fibroblasts, smooth muscle cells, monocytes and neutrophils. Functional PDGF is secreted as a dimer of disulfide-linked A and B chains (PDGF-AA, PDGF-BB or PDGF-AB). All three forms are produced naturally. The mature A and B chains have 60% homology and eight conserved Cys residues in each chain. The A-chain occurs in two variants arising from alternative splicing in which the three C-terminal amino acids in the short form are replaced by 18 different amino acids derived from exon 6 in the long form[4,5]. The gene coding for human PDGF B-chain is the c-*sis* proto-oncogene[1,6].

Crossreactivity

There is greater than 90% homology between human and mouse mature PDGF-A and complete crossreactivity.

Sources

Platelets (human 70% AB, 20% BB, 10% AA), placental cytotrophoblasts, macrophages, endothelial cells, megakaryocytes, fibroblasts, vascular smooth muscle cells, glial cells, type I astrocytes, myoblasts, kidney epithelial cells mesangial cells and many different tumour cells.

Bioassays

Mitogenic activity on fibroblasts.

Physicochemical properties of PDGF

Property	PDGF-A		PDGF-B	
	Human	Mouse	Human	Mouse
pI	10.2	10	10.2	10
Amino acids				
– precursor	211/196[b]	196[b]	241	241
– mature[a,b]	125/110[b]	110[b]	109	160
M_r (K)				
– predicted[b]	14.3/12.5	12.5	12.3	12
– expressed[c]	14–18	16	16	16
Potential *N*-linked glycosylation sites	1	1	0[d]	0[d]
Disulfide bonds[e]	3	3	3	3

[a] After removal of signal peptide and propeptide (see sequences). Precursors are proteolytically cleaved after dimerization.
[b] Two forms of human PDGF-A derived from alternative mRNA splicing have been identified (see sequence). Long and short forms of murine PDGF-A from alternative splicing have also been found, but mRNA for the short form only has been cloned.
[c] For monomeric PDGF. M_r of dimers is 30 000–32 000.
[d] There is one site on the propeptide.
[e] Dimeric PDGF has two additional interchain disulfide bonds (see sequence).

3-D structure

Biologically active PDGF is an antiparallel disulfide-linked dimer (AA, AB or BB). The crystal structure of the human PDGF BB isoform has been determined to a resolution of 3 Å[7]. The polypeptide chain is folded into two highly twisted antiparallel pairs of β-strands with an unusual knotted arrangement of three intramolecular disulfide bonds. A cluster of three surface loops at each end of the dimer may form the receptor recognition sites.

Gene structure[8,9]

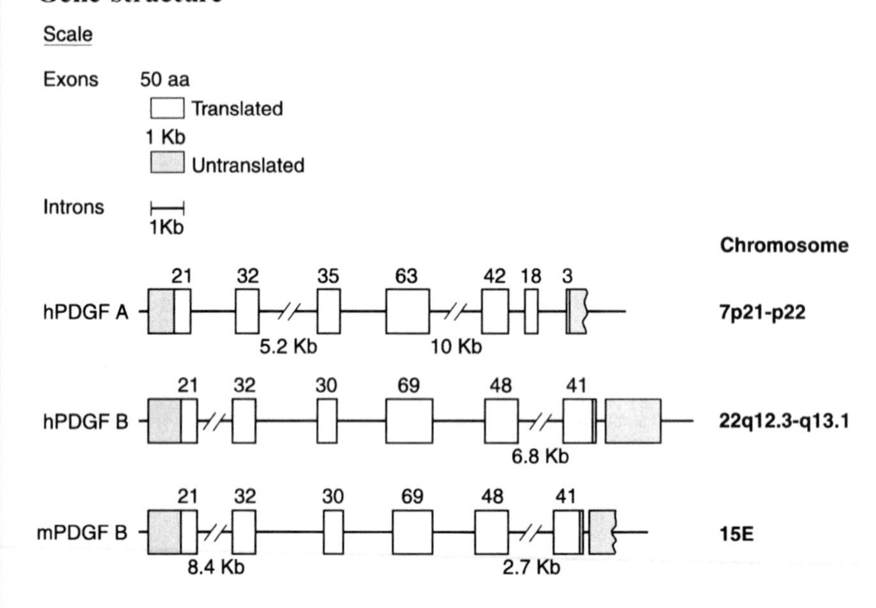

The gene for human PDGFA (six coding exons of 21, 32, 35, 63, 42, 18 and 3 amino acids) is located on chromosome 7p21–22 while that of PDGFB is on chromosome 22q12.3–q13.1 (six coding exons of 21, 32, 30, 69, 48 and 41 amino acids). Human chromosome 7p21–p22 is the site of a defined subset of mitogen-induced fragile sites, many of which coincide with well-known oncogenes and the breakpoint in myelodysplastic syndrome. The gene for mouse PDGF-A is located on chromosome 5, which is closely linked to the gene for the IL-2R β-chain. The gene for mouse PDGFB is located on chromosome 15E and consists of six coding exons of 21, 32, 30, 69, 48 and 41 amino acids).

Amino acid sequence for human PDGF-A (long form[10] and short form[4,5,11,12])

Accession code: SwissProt P04085

```
-20  MRTLACLLLL GCGYLAHVLA
  1  EEAEIPREVI ERLARSQIHS IRDLQRLLEI DSVGSEDSLD TSLRAHGVHA
 51  TKHVPEKRPL PIRRKRSIEE AVPAVCKTRT VIYEIPRSQV DPTSANFLIW
101  PPCVEVKRCT GCCNTSSVKC QPSRVHHRSV KVAKVEYVRK KPKLKEVQVR
151  LEEHLECACA TTSLNPDYRE EDTGRPRESG KKRKRKRLKP T
```

Mature PDGF-A is formed by removal of signal peptide and propeptide (1–66, in italics). The short form of human PDGF-A is derived from alternative splicing and terminates with the sequence DVR replacing GRP at position 174–176[4,5]. Intrachain disulfide bonds between Cys76–120, 109–157, 113–159, and interchain disulfide bonds between Cys103–103, 112–112 by similarity.

Amino acid sequence for human PDGF-B (c-*sis* proto-oncogene)[8,13–15]

Accession code: SwissProt P01127

```
-20  MNRCWALFLS LCCYLRLVSA
  1  EGDPIPEELY EMLSDHSIRS FDDLQRLLHG DPGEEDGAEL DLNMTRSHSG
 51  GELESLARGR RSLGSLTIAE PAMIAECKTR TEVFEISRRL IDRTNANFLV
101  WPPCVEVQRC SGCCNNRNVQ CRPTQVQLRP VQVRKIEIVR KKPIFKKATV
151  TLEDHLACKC ETVAAARPVT RSPGGSQEQR AKTPQTRVTI RTVRVRRPPK
201  GKHRKFKHTH DKTALKETLG A
```

Mature PDGF-B is formed by removal of a signal peptide and both N-terminal (1–61) and C-terminal (171–221) propeptides (in italics). Conflicting sequence E→R at position 1, T→E at position 81, E→C at position 85, and S→C at position 87. Intrachain disulfide bonds between Cys77–121, 110–158, 114–160, and interchain disulfide bonds between Cys104–104, 113–113.

Amino acid sequence for mouse PDGF-A (short form)[16]

Accession code: SwissProt P20033

```
-20  MRTWACLLLL GCGYLAHALA
  1  EEAEIPRELI ERLARSQIHS IRDLQRLLEI DSVGAEDALE TSLRAHGSHA
 51  INHVPEKRPV PIRRKRSIEE AIPAVCKTRT VIYEIPRSQV DPTSANFLIW
101  PPCVEVKRCT GCCNTSSVKC QPSRVHHRSV KVAKVEYVRK KPKLKEVQVR
151  LEEHLECACA TSNLNPDHRE EETDVR
```

Mature PDGF-A is formed by removal of a signal peptide and a propetide (1–66, in italics). Different forms of mouse PDGF-A are produced by alternative splicing. The sequence given here is for the short form. The C-terminal sequence for the long form has not yet been determined. Intrachain disulfide bonds between Cys76–120, 109–157, 113–159, and interchain disulfide bonds between Cys103–103, 112–112 by similarity.

Amino acid sequence for mouse PDGF-B[9]

Accession code: SwissProt P31240

```
-20  MNRCWALFLP LCCYLRLVSA
  1  EGDPIPEELY EMLSDHSIRS FDDLQRLLHR DSVDEDGAEL DLNMTRAHSG
 51  VELESSSRGR RSLGSLAAAE PAVIAECKTR TEVFQISRNL IDRTNANFLV
101  WPPCVEVQRC SGCCNNRNVQ CRASQVQMRP VQVRKIEIVR KKPIFKKATV
151  TLEDHLACKC ETIVTPRPVT RSPGTSREQR AKTPQARVTI RTVRIRRPPK
201  GKHRKFKHTH DKAALKETLG A
```

Mature mouse PDGF-B is formed by removal of a signal peptide and both N-terminal (1–61) and C-terminal (171–221) propeptides (in italics). Intrachain disulfide bonds between Cys77–121, 110–158, 114–160, and interchain disulfide bonds between Cys104–104, 113–113 by similarity.

THE PDGF RECEPTORS

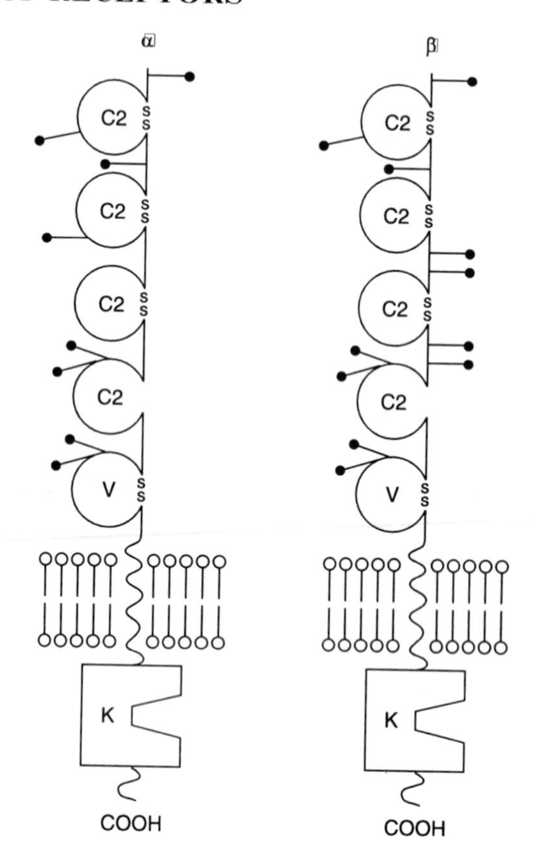

PDGF receptors (α and β) are single transmembrane glycoproteins with an intracellular tyrosine kinase domain split by an inserted sequence of about 100 amino acids[2,3]. They are structurally related to the M-CSF receptor (c-*fms*) and the SCF receptor (c-*kit*). Binding of divalent PDGF (AA, AB or BB) induces receptor dimerization with three possible configurations (αα, αβ, ββ). The PDGF receptor α-subunit binds both PDGF-A and -B chains, whereas the receptor β-subunit binds only PDGF-B chains. This specificity predicts that PDGF-AA binds only to PDGF receptor αα dimers, PDGF-AB binds to receptor αα and αβ dimers, and PDGF-BB binds to all three possible configurations (PDGF receptor αα, αβ, ββ)[17–19]. Murine and human α- and β-receptors have the same isoform specificity. Both receptors stimulate mitogenic responses, but only the β-receptor can induce chemotaxis and actin reorganization to form circular membrane ruffles. A soluble form of the PDGF-α receptor has been described[20].

Distribution

Widely distributed on cells of mesenchymal origin including fibroblasts, smooth muscle cells, glial cells and chondrocytes. Also on synovia in patients with rheumatoid arthritis and uterine endometrial cells. There are two examples of cells which express one receptor only: O-2A glial progenitor cells of the rat optic nerve express only α-receptors, and capillary endothelial cells of the brain express only β-receptors. PDGF receptors have also been detected on the SMS-SB and NALM6 pre-B cell lines.

Physicochemical properties of the PDGF receptors

	α-receptor		β-receptor	
Property	Human	Mouse	Human	Mouse
Amino acids – precursor	1089	1089	1106	1098
– mature[a]	1066	1065	1074	1067
M_r (K) – predicted	120.3	120.1	120.6	119.6
– expressed	170	?	180	180
Potential *N*-linked glycosylation sites	8	10	11	11
Affinity K_d (M) – PDGF-AB	10^{-10}	?	0	?
– PDGF-AA	23×10^{-10}	?	53×10^{-9}	?
– PDGF-BB	5×10^{-10}	?	5×10^{-10}	?

[a] After removal of predicted signal peptide.

Signal transduction

The PDGF receptor belongs to subclass III of receptor tyrosine kinases that includes SCFR (c-*kit*) and M-CSFR. PDGF binding induces receptor dimerization and autophosphorylation. Detailed studies of the human β-receptor have identified six autophosphorylation sites (Tyr708, Tyr719 and Tyr739 in the kinase insert, Tyr825 in the second part of the kinase domain, and Tyr977 and Tyr989 in the C-terminal tail) (reviewed in ref. 2). PLCγ, PI-3′ kinase, and GTPase-activating protein (GAP) bind through their SH2 domains to specific phosphotyrosine-containing motifs on the PDGF receptor and are activated (phosphorylated). Tyr708 and Tyr719 mediate binding of PI-3' kinase, and Tyr739 mediates binding of GAP[21,22] whereas Tyr977 and Tyr989 mediate binding of PLCγ[23]. The PDGF α- and β-receptors activate both common and distinct signal transduction pathways[24].

Chromosomal location

Human PDGFR-α is on chromosome 4q11–q12 and PDGFR-β is on chromosome 5q33–q35. The mouse β-receptor is on chromosome 18, closely linked to the M-CSFR.

Amino acid sequence for human PDGFR-α[15,25]

Accession code: SwissProt P16234

-23 MGTSHPAFLV LGCLLTGLSL ILC
1 QLSLPSILPN ENEKVVQL**N**S SFSLRCFGES EVSWQYPMSE EESSDVEIRN
51 EE**N**NSGLFVT VLEVSSASAA HTGLYTCYY**N** HTQTEENELE GRHIYIYVPD
101 PDVAFVPLGM TDYLVIVEDD DSAIIPCRTT DPETPVTLHN SEGVVPASYD
151 SRQGF**N**GTFT VGPYICEATV KGKKFQTIPF NVYALKATSE LDLEMEALKT
201 VYKSGETIVV TCAVFNNEVV DLQWTYPGEV KGKGITMLEE IKVPSIKLVY
251 TLTVPEATVK DSGDYECAAR QATREVKEMK KVTISVHEKG FIEIKPTFSQ
301 LEAVNLHEVK HFVVEVRAYP PPRISWLKN**N** LTLIE**N**LTEI TTDVEKIQEI
351 RYRSKLKLIR AKEEDSGHYT IVAQNEDAVK SYTFELLTQV PSSILDLVDD
401 HHGSTGGQTV RCTAEGTPLP DIEWMICKDI KKCN**N**ETSWT ILAN**N**VSNII
451 TEIHSRDRST VEGRVTFAKV EETIAVRCLA KNLLGAENRE LKLVAPTLRS
501 ELTVAAAVLV LLVIVIISLI VLVVIWKQKP RYEIRWRVIE SISPDGHEYI
551 YVDPMQLPYD SRWEFPRDGL VLGRVLGSGA FGKVVEGTAY GLSRSQPVMK
601 VAVKMLKPTA RSSEKQALMS ELKIMTHLGP HLNIVNLLGA CTKSGPIYII
651 TEYCFYGDLV NYLHKNRDSF LSHHPEKPKK ELDIFGLNPA DESTRSYVIL
701 SFENNGDYMD MKQADTTQYV PMLERKEVSK YSDIQRSLYD RPASYKKKSM
751 LDSEVKNLLS DDNSEGLTLL DLLSFTYQVA RGMEFLASKN CVHRDLAARN
801 VLLAQGKIVK ICDFGLARDI MHDSNYVSKG STFLPVKWMA PESIFDNLYT
851 TLSDVWSYGI LLWEIFSLGG TPYPGMMVDS TFYNKIKSGY RMAKPDHATS
901 EVYEIMVKCW NSEPEKRPSF YHLSEIVENL LPGQYKKSYE KIHLDFLKSD
951 HPAVARMRVD SDNAYIGVTY KNEEDKLKDW EGGLDEQRLS ADSGYIIPLP
1001 DIDPVPEEED LGKRNRHSSQ TSEESAIETG SSSSTFIKRE DETIEDIDMM
1051 DDIGIDSSDL VEDSFL

The extracellular Cys residues form disulfide bonds in Ig-SF domains and are conserved in both PDGF receptors and other members of the split tyrosine kinase family (c-*kit*, M-CSFR). Tyr708, 719, 739, 826, 970 and 995 are autophosphorylated (by similarity with the β-receptor) and all but Tyr826 have been shown to be involved in binding to SH2 domains of signalling proteins (see above under signal transduction)[2].

Amino acid sequence for human PDGFR-β[26,27]

Accession code: SwissProt P09619

-32 MRLPGAMPAL ALKGELLLLS LLLLLEPQIS QG
1 LVVTPPGPEL VL**N**VSSTFVL TCSGSAPVVW ERMSQEPPQE MAKAQDGTFS
51 SVLTLT**N**LTG LDTGEYFCTH **N**DSRGLETDE RKRLYIFVPD PTVGFLPNDA
101 EELFIFLTEI TEITIPCRVT DPQLVVTLHE KKGDVALPVP YDHQRGFSGI
151 FEDRSYICKT TIGDREVDSD AYYVYRLQVS SI**N**VSVNAVQ TVVRQGE**N**IT

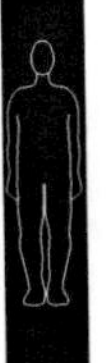

201 LMCIVIGNEV VNFEWTYPRK ESGRLVEPVT DFLLDMPYHI RSILHIPSAE
251 LEDSGTYTC**N** VTESVNDHQD EKAI**N**ITVVE SGYVRLLGEV GTLQFAELHR
301 SRTLQVVFEA YPPPTVLWFK D**N**RTLGDSSA GEIALSTR**N**V SETRYVSELT
351 LVRVKVAEAG HYTMRAFHED AEVQLSFQLQ INVPVRVLEL SESHPDSGEQ
401 TVRCRGRGMP QPNIIWSACR DLKRCPRELP PTLLG**N**SSEE ESQLET**N**VTY
451 WEEEQEFEVV STLRLQHVDR PLSVRCTLRN AVGQDTQEVI VVPHSLPFKV
501 VVISAILALV VLTIISLIIL IMLWQKKPRY EIRWKVIESV SSDGHEYIYV
551 DPMQLPYDST WELPRDQLVL GRTLGSGAFG QVVEATAHGL SHSQATMKVA
601 VKMLKSTARS SEKQALMSEL KIMSHLGPHL NVVNLLGACT KGGPIYIITE
651 YCRYGDLVDY LHRNKHTFLQ HHSDKRRPPS AELYSNALPV GLPLPSHVSL
701 TGESDGGYMD MSKDESVDYV PMLDMKGDVK YADIESSNYM APYDNYVPSA
751 PERTCRATLI NESPVLSYMD LVGFSYQVAN GMEFLASKNC VHRDLAARNV
801 LICEGKLVKI CDFGLARDIM RDSNYISKGS TFLPLKWMAP ESIFNSLYTT
851 LSDVWSFGIL LWEIFTLGGT PYPELPMNEQ FYNAIKRGYR MAQPAHASDE
901 IYEIMQKCWE EKFEIRPPFS QLVLLLERLL GEGYKKKYQQ VDEEFLRSDH
951 PAILRSQARL PGFHGLRSPL DTSSVLYTAV QPNEGDNDYI IPLPDPKPEV
1001 ADEGPLEGSP SLASSTLNEV NTSSTISCDS PLEPQDEPEP EPQLELQVEP
1051 EPELEQLPDS GCPAPRAEAE DSFL

The extracellular Cys residues form disulfide bonds in Ig-SF domains and are conserved in both PDGF receptors and other members of the split tyrosine kinase family (c-*kit*, M-CSFR). Tyr708, 719, 739, 825, 977 and 989 are autophosphorylated and all but Tyr825 have been shown to be involved in binding to SH2 domains of signalling proteins (see above under signal transduction)[2]. Conflicting amino acid sequence E→D at position 209.

Amino acid sequence for mouse PDGFR-α[2]

Accession code: SwissProt P26618

-24 MGTSHQVFLV LSCLLTGPGL ISCQ
1 LLLPSILPNE NEKIVQL**N**SS FSLRCVGESE VSWQHPMSEE EDPNVEIRSE
51 E**N**NSGLFVTV LEVV**N**ASAAH TGWYTCYY**N**H TQTDESEIEG RHIYIYVPDP
101 DMAFVPLGMT DSLVIVEEDD SAIIPCRTTD PETQVTLHNN GRLVPASYDS
151 RQGF**N**GTFSV GPYICEAAVK GRTFKTSAFN VYALKATSEL NLEMDARQTV
201 YKAGETIVVT CAVFNNEVVD LQWTYPGGVR NKGITMLEEI KLPSIKVVYT
251 LTVPKATVKD SGEYECAARQ ATKEVKEMKR VTISVHEKGF VEIEPTFSQL
301 EPVNLHEVRE FVVEVQAYPT PRISWLKD**N**L TLIE**N**LTEIT TDVQKSQETR
351 YQSKLKLIRA KEEDSGHYTI IVQNEDDVKS YTFELSTLVP ASILDLVDDH
401 HGSGGGQTVR CTAEEGPLPE IDWMICKHIK KCN**N**DTSWTV LAS**N**VSNIIT
451 ELPRRGRSTV EGRVSFAKVE ETIAVRCLAK N**N**LSVVAREL KLVAPTLRSE
501 LTVAAAVLVL LVIVIVSLIV LVVIWKQKPR YEIRWRVIES ISPDGHEYIY
551 VDPMQLPYDS RWEFPRDGLV LGRILGSGAF GKVVEGTAYG LSRSQPVMKV
601 AVKMLKPTAR SSEKQALMSE LKIMTHLGPH LNIVNLLGAC TKSGPIYIIT
651 EYCFYGDLVN YLHKNRDSFM SQHPEKPKKD LDIFGLNPAD ESTRSYVILS
701 FENNGDYMDM KQDDTTQYVP MLERKEVSKY SDIQRSLYDR PASYKKKSML
751 DSEVKNLLSD DDSEGLTLLD LLSFTYQVAR GMEFLASKNC VHRDLAARNV

```
801   LLAQGKIVKI CDFGLARDIM HDSNDVSKGS TFLPVKWMAP ESIFDNLYTT
851   LSDVWSYGIL LWEIFSLGGT PYPGMMVDST FYNKIKSGYR MAKPDHATSE
901   VYEIMVQCWN SDPEKRPSFY HLSEILENLL PGQYKKSYEK IHLDFLKSDH
951   PAVARMRVDS DNAYIGVTYK NEEDKLKDWE GGLDEQRLSA DSGYIIPLPD
1001  IDPVPEEEDL GKRNRHSSQT SEESAIETGS SSSTFIKRED ETIEDIDMMD
1051  DIGIDSSDLV EDSFL
```

The extracellular Cys residues form disulfide bonds in Ig-SF domains and are conserved in both PDGF receptors and other members of the split tyrosine kinase family (c-*kit*, M-CSFR). Tyr707, 718, 738, 969 and 994 are autophosphorylated (by similarity with the human β-receptor) and are involved in binding to SH2 domains of signalling proteins (see above under signal transduction)[2].

Amino acid sequence for mouse PDGFR-β[28]

Accession code: SwissProt P05622

```
-31   MGLPGVIPAL VLRGQLLLSV LWLLGPQTSR G
  1   LVITPPGPEF VLNISSTFVL TCSGSAPVMW EQMSQVPWQE AAMNQDGTFS
 51   SVLTLTNVTG GDTGEYFCVY NNSLGPELSE RKRIYIFVPD PTMGFLPMDS
101   EDLFIFVTDV TETTIPCRVT DPQLEVTLHE KKVDIPLHVP YDHQRGFTGT
151   FEDKTYICKT TIGDREVDSD TYYVYSLQVS SINVSVNAVQ TVVRQGESIT
201   IRCIVMGNDV VNFQWTYPRM KSGRLVEPVT DYLFGVPSRI GSILHIPTAE
251   LSDSGTYTCN VSVSVNDHGD EKAINISVIE NGYVRLLETL GDVEIAELHR
301   SRTLRVVFEA YPMPSVLWLK DNRTLGDSGA GELVLSTRNM SETRYVSELI
351   LVRVKVSEAG YYTMRAFHED DEVQLSFKLQ VNVPVRVLEL SESHPANGEQ
401   TIRCRGRGMP QPNVTWSTCR DLKRCPRKLS PTPLGNSSKE ESQLETNVTF
451   WEEDQEYEVV STLRLRHVDQ PLSVRCMLQN SMGGDSQEVT VVPHSLPFKV
501   VVISAILALV VLTVISLIIL IMLWQKKPRY EIRWKVIESV SSDGHEYIYV
551   DPVQLPYDST WELPRDQLVL GRTLGSGAFG QVVEATAHGL SHSQATMKVA
601   VKMLKSTARS SEKQALMSEL KIMSHLGPHL NVVNLLGACT KGGPIYIITE
651   YCRYGDLVDY LHRNKHTFLQ RHSNKHCPPS AELYSNALPV GFSLPSHLNL
701   TGESDGGYMD MSKDESIDYV PMLDMKGDIK YADIESPSYM APYDNYVPSA
751   PERTYRATLI NDSPVLSYTD LVGFSYQVAN GMDFLASKNC VHRDLAARNV
801   LICEGKLVKI CDFGLARDIM RDSNYISKGS TYLPLKWMAP ESIFNSLYTT
851   LSDVWSFGIL LWEIFTLGGT PYPELPMNDQ FYNAIKRGYR MAQPAHASDE
901   IYEIMQKCWE EKFETRPPFS QLVLLLERLL GEGYKKKYQQ VDEEFLRSDH
951   PAILRSQARF PGIHSLRSPL DTSSVLYTAV QPNESDNDYI IPLPDPKPDV
1001  ADEGLPEGSP SLASSTLNEV NTSSTISCDS PLELQEEPQQ AEPEAQLEQP
1051  QDSGCPGPLA EAEDSFL
```

The extracellular Cys residues form disulfide bonds in Ig-SF domains and are conserved in both PDGF receptors and other members of the split tyrosine kinase family (c-*kit*, M-CSFR). Tyr708, 719, 739, 825, 977 and 989 are autophosphorylated (by similarity with the human β-receptor) and all but Tyr825 are involved in binding to SH2 domains of signalling proteins (see above under signal transduction)[2].

References

[1] Ross, R. et al (1986) Cell 46, 155–169.
[2] Heldin, C.-H. (1992) EMBO J. 11, 4251–4259.
[3] Raines, E.W. et al. (1991) In Peptide Growth Factors and their Receptors I, Sporn, M.B. and Roberts, A.B. eds, Springer-Verlag, New York, pp. 173–262.
[4] Collins, T. et al. (1987) Nature 328, 621–624.
[5] Tong, B.D. et al. (1987) Nature 328, 619–621.
[6] Heldin, C.-H. and Westermark, B. (1989) Br. Med. Bull. 45, 453–464.
[7] Oefner, C. et al. (1992) EMBO J. 11, 3921–3926.
[8] Rao, C.D. et al. (1986) Proc. Natl Acad. Sci. USA 83, 2392–2396.
[9] Bonthron, D.T. et al. (1991) Genomics 10, 287–292.
[10] Betsholtz, C. et al. (1986) Nature 320, 695–699.
[11] Bonthron, D.T. et al. (1988) Proc. Natl Acad. Sci. USA 85, 1492–1496.
[12] Rorsman, F. et al. (1988) Mol. Cell. Biol. 8, 571–577.
[13] Chiu, I.-M. et al. (1984) Cell 37, 123–129.
[14] Josephs, S.F. et al. (1984) Science 225, 636- 639.
[15] Collins, T. et al. (1985) Nature 316, 748–750.
[16] Mercola, M. et al. (1990) Dev. Biol. 138, 114–122.
[17] Seifert, R.A. et al. (1989) J. Biol. Chem. 264, 8771–8778.
[18] Hart, C.E. and Bowen-Pope, D.F. (1990) J. Invest. Dermatol. 94, 53S-57S.
[19] Westermark, B. et al. (1989) Prog. Growth Factor Res. 1, 253–266.
[20] Tiesman, J. and Hart, C.E. (1993) J. Biol. Chem. 268, 9621- 9628.
[21] Fantl, W.J. et al. (1992) Cell 69, 413–423.
[22] Kashishian, A. et al. (1992) EMBO J. 11, 1373–1382.
[23] Ronnstrand, L. et al. (1992) EMBO J. 11, 3911–3919.
[24] Eriksson, A. et al. (1992) EMBO J. 11, 543–550.
[25] Claesson-Welsh, L. et al. (1989) Proc. Natl Acad. Sci. USA 86, 4917–4921.
[26] Gronwald, R.G.K. et al. (1988) Proc. Natl Acad. Sci. USA 85, 3435–3439.
[27] Claesson-Welsh, L. et al. (1988) Mol. Cell Biol. 8, 3476–3486.
[28] Yarden, Y. et al. (1986) Nature 323, 226- 232.

Other names
None.

THE MOLECULES

Platelet factor 4 (PF-4) is an alpha granule protein released by platelets with wide biological activities. PF-4 is normally bound to a 53-kDa carrier proteoglycan[1]. It binds to and neutralizes heparin, is a potent anti-angiogenic factor *in vivo*, it can suppress or enhance haematopoietic progenitor cell survival, activates basophils and mast cells to release histamine, stimulates neutrophil adherence to endothelium, is chemotactic for fibroblasts, and promotes monocyte survival[2–7].

Crossreactivity
There is no murine homologue of PF-4.

Sources
Platelets.

Bioassays
Neutrophil chemoattraction.

Physicochemical properties of PF-4

Property	Human
Amino acids	
– precursor	101
– mature	70
M_r (K)	
– predicted	10.8
– expressed	?
Potential *N*-linked glycosylation sites	0
Disulfide bonds	2

[a] Disulfide bonds link Cys41–67 and 43–83.

3-D structure
PF-4 exists as a homo-tetramer. Each monomer consists of an extended loop, three strands of antiparallel β-sheet arranged in a Greek key motif and an α-helix[8,9].

Gene structure

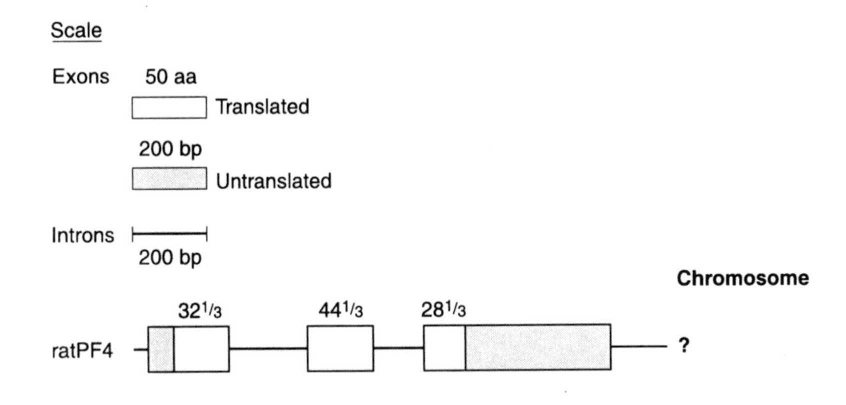

The gene for human PF-4 is on chromosome 4q12-q13[10].

Amino acid sequence for human PF-4[10–13]

Accession code: SwissProt P02776

```
-31  MSSAAGFCAS RPGLLFLGLL LLPLVVAFAS A
  1  EAEEDGDLQC LCVKTTSQVR PRHITSLEVI KAGPHCPTAQ LIATLKNGRK
 51  ICLDLQAPLY KKIIKKLLES
```

THE PF-4 RECEPTOR

Unlike most other chemokines, no specific receptor for PF-4 has yet been cloned. However, tetrameric PF-4 has been shown to bind to 23-kDa chondroitin sulfate type glycosaminoglycans on neutrophils.

References

[1] Huang, J.S. et al. (1982) J. Biol. Chem. 257, 11546–11550.
[2] Ravid, K. et al. (1991) Mol. Cell Biol. 11, 6116–6127.
[3] Scheuerer, B. et al. (2000) Blood 95, 1158–1166.
[4] Petersen, F. et al. (1999) Blood 94, 4020–4028.
[5] Lecomte-Raclet, L. et al. (1998) Blood 91, 2772–2780.
[6] Han, Z.C. et al. (1997) Blood 89, 2328–2335.
[7] Brandt, E. et al. (2000) J. Leukoc. Biol. 67, 471–478.
[8] Walz, D.A. et al. (1977) Thromb. Res. 11, 893–898.
[9] Zhang, X. et al. (1994) Biochemistry 33, 8361–8366.

[10] Eisman, R. et al. (1990) Blood 76, 336–344.
[11] Poncz, M. et al. (1987) Blood 69, 219–223.
[12] Hermodson, M. et al. (1977) J. Biol. Chem. 252, 6276–6279.
[13] Deuel, T.F. et al. (1977) Proc. Natl Acad. Sci. USA 74, 2256–2258.

Other names
None.

THE MOLECULE

Prolactin (PRL) is a secreted protein monomer of about 22–25 kDa and is a member of a family of growth factors that includes growth hormone (GH), the placental lactogens, proliferins and somatolactin[1]. More than 85 biological properties of PRL have been described that include activities in reproduction and lactation, water and salt balance, growth and morphogenesis, metabolism, behaviour, and effects on the ectoderm and skin. Prolactin is also an haematopoietic growth factor, either by direct action on haematopoietic cells or indirectly by stimulating production of IGF-1[2]. Receptors for PRL are expressed on lymphocytes and, like GH, prolactin has important immunoregulatory properties[2].

Crossreactivity
There is 61% amino acid sequence identity between human and mouse PRL.

Sources
Pituitary, uterus, decidua, mammary gland, brain and lymphocytes.

Bioassays
The standard bioassay for lactogenic hormones is proliferation of the rat Nb2 lymphoma cells[3]. Specificity should be determined with blocking antibodies.

Physicochemical properties of PRL

Property	Human	Mouse
pI	?[a]	?[a]
Amino acids – precursor	227	226
– mature[b]	219	197
M_r (K) – predicted		
– expressed	22	22
Potential *N*-linked glycosylation sites	1	0
Disulfide bonds	3[c]	3[d]

[a] pI for rat PRL is 5.1–5.3.
[b] After removal of signal peptide.
[c] Cys4–11, 58–174 and 191–199.
[d] Cys4–9, 56–172 and 189–197.

3-D structure
PRL is a four α-helical bundle cytokine.

Gene structure[4]

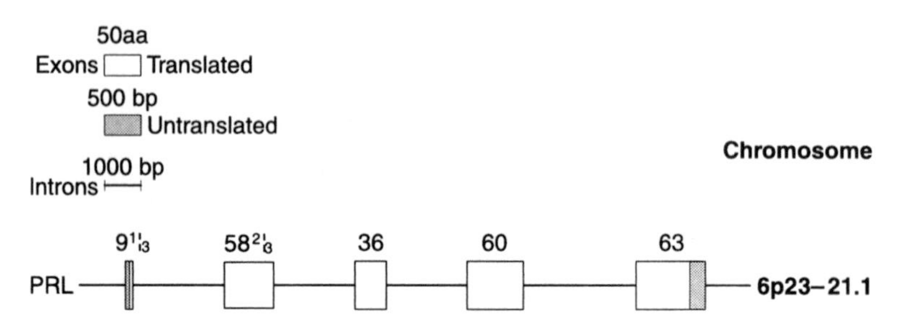

The gene for human prolactin is on chromosome 6p23–21.1.

Amino acid sequence for human PRL[4,5]

Accession code: SwissProt P01236

```
-28   MNIKGSPWKG SLLLLLVSNL LLCQSVAP
  1   LPICPGGAAR CQVTLRDLFD RAVVLSHYIH NLSSEMFSEF DKRYTHGRGF
 51   ITKAINSCHT SSLATPEDKE QAQQMNQKDF LSLIVSILRS WNEPLYHLVT
101   EVRGMQEAPE AILSKAVEIE EQTKRLLEGM ELIVSQVHPE TKENEIYPVW
151   SGLPSLQMAD EESRLSAYYN LLHCLRRDSH KIDNYLKLLK CRIIHNNNC
```

Disulfide binds between Cys4–11, 58–174 and 191–199. Conflicting sequence SL→VS at position 82–83, VS→L at position 85–86, E→Q at position 120, N→D at position 144, ES→SE at position 162–163 and D→H at position 178.

Amino acid sequence for mouse PRL[6]

Accession code: SwissProt P06879

```
-31   MT
-29   MNSQGSAQKA GTLLLLLISN LLFCQNVQP
  1   LPICSAGDCQ TSLRELFDRV VILSHYIHTL YTDMFIEFDK QYVQDREFMV
 51   KVINDCPTSS LATPEDKEQA LKVPPEVLLN LILSLVQSSS DPLFQLITGV
101   GGIQEAPEYI LSRAKEIEEQ NKQLLEGVEK IISQAYPEAK GNGIYFVWSQ
151   LPSLQGVDEE SKILSLRNTI RCLRRDSHKV DNFLKVLRCQ IAHQNNC
```

Disulfide bonds Cys4–9, Cys56–172 and Cys189–197.
−31 MT is an alternative initiation site[7]. Conflicting sequence D→H at position 176.

THE PRL RECEPTORS

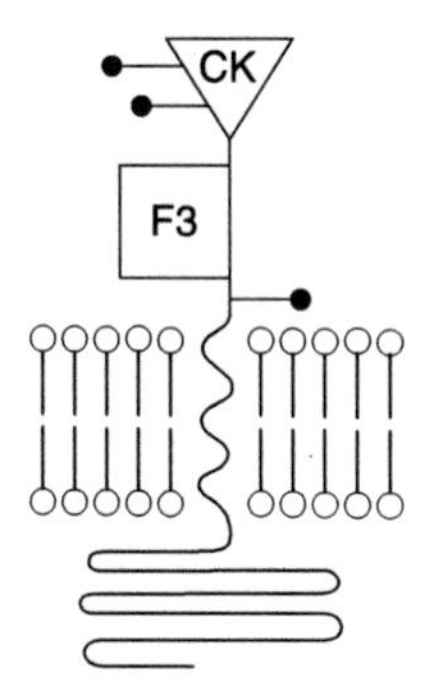

The PRL receptor belongs to the CKR superfamily[8–10]. Long and short forms of mouse PRL receptor have been described which have common signal sequences, extracellular domains and transmembrane domains but which differ markedly at the C-terminus after a short identical sequence in the cytoplasmic domain, probably as a result of alternative splicing. The different PRL receptors are coupled to distinct signal transduction pathways accounting for the different biological activities of PRL[8]. Soluble PRL-binding protein with a sequence identical to the extracellular domain of the membrane-bound receptor has also been identified[11]. Human PRL does not bind to the GH receptor but human GH binds to both the GH and PRL receptors[12].

Distribution

PRL receptor expression is very widespread, including liver, intestine, brain, testis, kidney, pancreas, mammary gland, ovary, uterus, corpus luteum, placenta, prostate, adrenal, eye and lymphocytes.

Physicochemical properties of the PRL receptors

Property	Human	Mouse – long form
Amino acids – precursor[a]	622	608 (303, 292)
– mature[b]	598	589 (284, 273)
M_r (K) – predicted		(32.7, 31.4)
– expressed	?	?[c] (42, 42)
Potential *N*-linked glycosylation sites	3	3
Affinity K_d (M)	5×10^{-10}	?

[a] Short form(s) of receptor in parenthesis.
[b] After removal of predicted signal peptide.
[c] Long form of rat PRL receptor is 82 kDa and short form is 40 kDa.

Signal transduction

PRL signalling is thought to occur by ligand-induced homodimerization but the evidence is not as unequivocal as it is for GH[8]. The cytoplasmic domain of the PRL receptor has no intrinsic enzyme activity but signalling does involve tyrosine

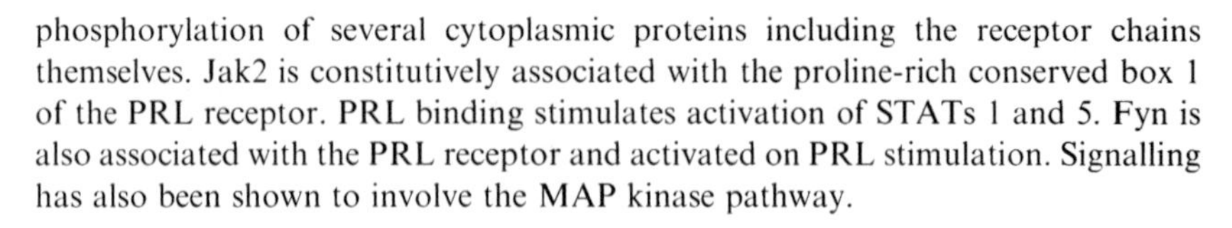

phosphorylation of several cytoplasmic proteins including the receptor chains themselves. Jak2 is constitutively associated with the proline-rich conserved box 1 of the PRL receptor. PRL binding stimulates activation of STATs 1 and 5. Fyn is also associated with the PRL receptor and activated on PRL stimulation. Signalling has also been shown to involve the MAP kinase pathway.

Chromosomal location

Human PRLR is on 5p13–14 and mouse PRLR is on chromosome 15.

Amino acid sequence for human PRL receptor[13]

Accession code: SwissProt P16471

```
-24  MKENVASATV FTLLLFLNTC LLNG
  1  QLPPGKPEIF KCRSPNKETF TCWWRPGTDG GLPTNYSLTY HREGETLMHE
 51  CPDYITGGPN SCHFGKQYTS MWRTYIMMVN ATNQMGSSFS DELYVDVTYI
101  VQPDPPLELA VEVKQPEDRK PYLWIKWSPP TLIDLKTGWF TLLYEIRLKP
151  EKAAEWEIHF AGQQTEFKIL SLHPGQKYLV QVRCKPDHGY WSAWSPATFI
201  QIPSDFTMND TTVWISVAVL SAVICLIIVW AVALKGYSMV TCIFPPVPGP
251  KIKGFDAHLL EKGKSEELLS ALGCQDFPPT SDYEDLLVEY LEVDDSEDQH
301  LMSVHSKEHP SQGMKPTYLD PDTDSGRGSC DSPSLLSEKC EEPQANPSTF
351  YDPEVIEKPE NPETTHTWDP QCISMEGKIP YFHAGGSKCS TWPLPQPSQH
401  NPRSSYHNIT DVCELAVGPA GAPATLLNEA GKDALKSSQT IKSREEGKAT
451  QQREVESFHS ETDQDTPWLL PQEKTPFGSA KPLDYVEIHK VNKDGALSLL
501  PKQRENSGKP KKPGTPENNK EYAKVSGVMD NNILVLVPDP HAKNVACFEE
551  SAKEAPPSLE QNQAEKALAN FTATSSKCRL QLGGLDYLDP ACFTHSFH
```

Disulfide bonds between Cys12–22 and 51–62. Fibronectin type III domains 1–98 and 99–203.

Amino acid sequence for mouse PRL receptor (long form)[14]

Accession code: SwissProt Q08501

```
-19  MSSALAYMLL VLSISLLNG
  1  QSPPGKPEIH KCRSPDKETF TCWWNPGSDG GLPTNYSLTY SKEGEKNTYE
 51  CPDYKTSGPN SCFFSKQYTS IWKIYIITVN ATNEMGSSTS DPLYVDVTYI
101  VEPEPPRNLT LEVKQLKDKK TYLWVKWLPP TITDVKTGWF TMEYEIRLKS
151  EEADEWEIHF TGHQTQFKVF DLYPGQKYLV QTRCKPDHGY WSRWGQEKSI
201  EIPNDFTLKD TTVWIIVAVL SAVICLIMVW AVALKGYSMM TCIFPPVPGP
251  KIKGFDTHLL EKGKSEELLS ALGCQDFPPT SDCEDLLVEF LEVDDNEDER
301  LMPSHSKEYP GQGVKPTHLD PDSDSGHGSY DSHSLLSEKC EEPQAYPPAF
351  HIPEITEKPE NPEANIPPTP NPQNNTPNCH TDTSKSTTWP LPPGQHTRRS
401  PYHSIADVCK LAGSPGDTLD SFLDKAEENV LKLSEDAGEE EVAVQEGAKS
451  FPSDKQNTSW PPLQEKGPIV YAKPPDYVEI HKVNKDGVLS LLPKQRENHQ
501  TENPGVPETS KEYAKVSGVT DNNILVLVPD SRAQNTALLE ESAKKVPPSL
551  EQNQSEKDLA SFTATSSNCR LQLGRLDYLD PTCFMHSFH
```

Disulfide bonds between Cys12–22, and 51–62 (by similarity). Fibronectin type III domains 1–98 and 100–203.

Two short forms of the mouse PLR receptor have been described[10] which are identical to the long form up to residue 261 including the signal peptide, extracellular domain and transmembrane domain. The sequences of these short forms are given from amino acid 251 with the divergent C-terminal sequences shown in italics.

Amino acid sequence for mouse PRL receptor (short form) 1[10]

Accession code: SwissProt P15212

251 KIKGFDTHLL E*LWCSILQLT SLVKIPTTEF LCDL*

Amino acid sequence for mouse PRL receptor (short form) 2[10]

Accession code: SwissProt P15213

251 KIKGFDTHLL E*VHNKEQLEN YVY*

References

1 Goffin, V. et al. (1996) Endocr. Rev. 17, 385–410.
2 Kooljman, R. et al. (1996) Adv. Immunol. 63, 377–454.
3 Gout, P.W. et al. (1980) Cancer Res. 40, 2433–2436.
4 Truong, A.T. et al. (1984) EMBO J. 3, 429–437.
5 Cooke, N.E. et al. (1981) J. Biol. Chem. 256, 4007–4016.
6 Linzer, D.I.H. and Talamantes, F. (1985) J. Biol. Chem. 260, 9574–9579.
7 Harigaya, T. et al. (1986) Biochim. Biophys. Acta 868, 30–38.
8 Goffin, V. and Kelly, P.A. (1996) Clin. Endocrinol. 45, 247–255.
9 Kelly, P.A. et al. (1993) Rec. Prog. Horm. Res. 48, 123–164.
10 Davis, J.A. and Linzer, D.I.H. (1989) Mol. Endocrinol. 3, 674–680.
11 Fuh, G. and Wells, J.A. (1995) J. Biol. Chem. 270, 13133–13137.
12 Somers, W. et al. (1994) Nature 372, 478–481.
13 Boutin, J.-M. et al. (1989) Mol. Endocrinol. 3, 1455–1461.
14 Moore, R.C. and Oka, T. (1993) Gene 134, 263–265.

Other names
Human RANTES has been known as human sisδ[1].

THE MOLECULE

RANTES is a member of the CC family of chemokines[2]. It is chemotactic for monocytes and memory T helper cells[3–5], as well as eosinophils[6]. RANTES also causes the release of histamine from human basophils[7], and activates eosinophils[8]. RANTES is an inhibitor of M-trophic HIV infection. RANTES, like MIP-1α and MIP-1β forms high-molecular-weight aggregates in solution, the dimeric form and aggregates having different biological activities; the dimeric RANTES mediates classical GPCR effects of activation and chemoattraction and also inhibts HIV infection, the aggregated RANTES causes T-cell and neutrophil activation and potentiates HIV infection[12].

Crossreactivity
There is 85% homology between human and murine RANTES[9]. The murine protein is active on human cells and vice versa.

Sources
RANTES is produced by T lymphocytes and macrophages, but is unusual in exhibiting a reduction in mRNA levels on activation of T cells[5].

Bioassays
Chemotaxis of monocytes, memory T cells[4], or eosinophils[6]. Release of histamine from mast cells[7].

Physicochemical properties of RANTES

Property	Human	Mouse
pI	9.5	9.0
Amino acids – precursor	91	91
– mature	68, 66[a]	68
M_r (K) – predicted	8	8
– expressed	8	8
Potential *N*-linked glycosylation sites	0[b]	0
Disulfide bonds	2	2

[a] Dipeptidylpeptidase IV truncated form of RANTES lacking residues 1–2 retains CCR5 binding but loses CCR1 activity[23].
[b] Possibly some *O*-linked glycosylation

3-D structure
Each monomer consists of a three-stranded antiparallel β-sheet in a Greek key motif with a C-terminal helix packed across the sheet, an arrangement similar to the monomeric structure of other members of this chemokine family (IL-8, PF-4, MGSA/Groα and MIP-1β). Overall, the RANTES dimer resembles that previously reported for MIP-1β[13,14].

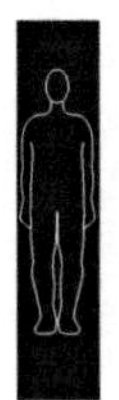

Amino acid sequence for human RANTES[3]

Accession code: SwissProt P13501

```
-23  MKVSAARLAV ILIATALCAP ASA
  1  SPYSSDTTPC CFAYIARPLP RAHIKEYFYT SGKCSNPAVV FVTRKNRQVC
 51  ANPEKKWVRE YINSLEMS
```

Amino acid sequence for mouse RANTES[9]

Accession code: SwissProt P30882

```
-23  MKISAAALTI ILTAAALCTP APA
  1  SPYGSDTTPC CFAYLSLELP RAHVKEYFYT SSKCSNLAVV FVTRRNRQVC
 51  ANPEKKWVQE YINYLEMS
```

THE RANTES RECEPTORS CCR1, CCR3, CCR5 and CCR9

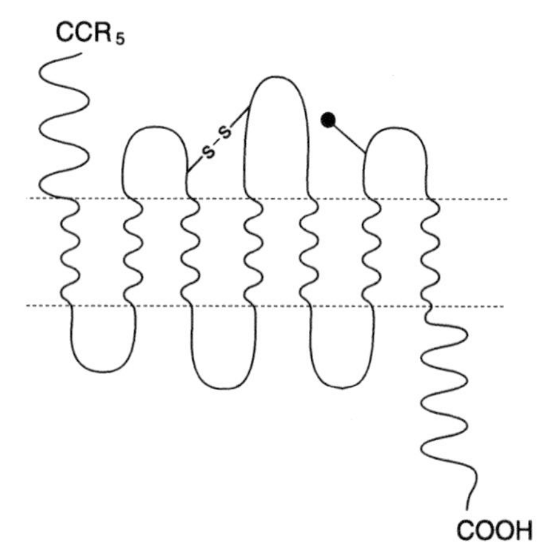

RANTES binds to the CCR1, CCR3, CCR5 and CCR9 receptors, mediating chemotaxis and activation of the various cell types bearing these receptors, lymphocytes, monocytes, eosinophils and basophils. CCR5 has also been shown to be an important co-receptor for HIV infection, with numerous polymorphisms in the coding and noncoding regions of the gene affecting HIV infection[16,24]. For a full description of the CCR1 receptor see the MIP-1α entry (page **384**), for CCR3 see the eotaxin 1 entry (page **213**) and for CCR9 see the TECK entry (page **460**).

Distribution

CCR5 is predominantly expressed on monocytes and T cells[18–22].

Physicochemical properties of the CCR5 RANTES receptor

Property	Human	Mouse
Amino acids – precursor	352	354
– mature	?	?
M_r (K) – predicted	40.5	40.9
– expressed	?	?
Potential *N*-linked glycosylation sites	1	1
Affinity K_d (M)	?	?

Signal transduction

RANTES causes a calcium flux and chemotaxis at low nanomolar concentrations, at high concentrations >1 μM the aggregated form causes a tyrosine kinase signal activating lymphocytes and neutrophils although it is not clear how this is mediated[8,11,17].

Chromosomal location

The human CCR5 genes are on chromosome 3p21[18] and the mouse gene is on chromosome 9[19].

Amino acid sequence for human RANTES receptor CCR5[18,20,21]

Accession code: SwissProt P51681

```
  1  MDYQVSSPIY DINYYTSEPC QKINVKQIAA RLLPPLYSLV FIFGFVGNML
 51  VILILINCKR LKSMTDIYLL NLAISDLFFL LTVPFWAHYA AAQWDFGNTM
101  CQLLTGLYFI GFFSGIFFII LLTIDRYLAV VHAVFALKAR TVTFGVVTSV
151  ITWVVAVFAS LPGIIFTRSQ KEGLHYTCSS HFPYSQYQFW KNFQTLKIVI
201  LGLVLPLLVM VICYSGILKT LLRCRNEKKR HRAVRLIFTI MIVYFLFWAP
251  YNIVLLLNTF QEFFGLNNCS SSNRLDQAMQ VTETLGMTHC CINPIIYAFV
301  GEKFRNYLLV FFQKHIAKRF CKCCSIFQQE APERASSVYT RSTGEQEISV
351  GL
```

Disulfide bonds link Cys101–178. Multiple polymorphisms – see database entry.

Amino acid sequence for mouse RANTES receptor CCR5[19,22]

Accession code: SwissProt P51682

```
  1  MDFQGSVPTY IYDIDYGMSA PCQKINVKQI AAQLLPPLYS LVFIFGFVGN
 51  MMVFLILISC KKLKSVTDIY LLNLAISDLL FLLTLPFWAH YAANEWIFGN
101  IMCKVFTGVY HIGYFGGIFF IILLTIDRYL AIVHAVFALK VRTVNFGVIT
151  SVVTWVVAVF ASLPEIIFTR SQKEGFHYTC SPHFPHTQYH FWKSFQTLKM
201  VILSLILPLL VMIICYSGIL HTLFRCRNEK KRHRAVRLIF AIMIVYFLFW
251  TPYNIVLLLT TFQEFFGLNN CSSSNRLDQA MQATETLGMT HCCLNPVIYA
301  FVGEKFRSYL SVFFRKHIVK RFCKRCSIFQ QDNPDRVSSV YTRSTGEHEV
351  STGL
```

Disulfide bonds link Cys103–180. Multiple polymorphisms – see database entry.

References

[1] Brown, K.D. et al. (1989). J. Immunol. 142, 679–687.
[2] Oppenheim, J.J. et al. (1991) Annu. Rev. Immunol 9, 617–648.
[3] Schall, T.J. et al. (1988) J. Immunol. 141, 1018–1025.
[4] Schall, T.J. et al. (1990) Nature 347, 669–671.
[5] Schall, T.J. (1991) Cytokine 3, 165–183.
[6] Kamayoshi, Y. et al. (1992) J. Exp. Med. 196, 187–192.
[7] Kuna, P. et al. (1992) J. Immunol. 149, 636–642.
[8] Rot, A. et al. (1992). J. Exp. Med. 176, 1489–1495.
[9] Schall, T.J. et al. (1992) Eur. J. Immunol. 22, 1477–1481.
[10] Donlon, T.A. et al. (1990) Genomics 6, 548–553.
[11] Neote, K. et al. (1993) Cell 72, 415–425.
[12] Czaplewski, L.G. et al. (1999) J. Biol. Chem. 274, 16077–16084.
[13] Chung, C.-W. et al. (1995) Biochemistry 34, 9307–9314.
[14] Skelton, N.J. et al. (1995) Biochemistry 34, 5329–5342.
[15] Danoff, T.M. et al. (1994) J. Immunol. 152, 1182–1189.
[16] Deng, H. et al. (1996) Nature 381, 661–666.
[17] Appay, V. et al. (1999) J. Biol. Chem. 274, 27505–27512.
[18] Raport, C.J. et al. (1996) J. Biol. Chem. 271, 17161–17166.
[19] Boring, L. et al. (1996) J. Biol. Chem. 271, 7551–7558.
[20] Samson, M. et al. (1996) Biochemistry 35, 3362–3367.
[21] Combadiere, C. et al. (1996) J. Leukoc. Biol. 60, 147–152.
[22] Meyer, A. et al. (1996) J. Biol. Chem. 271, 14445–14451.
[23] Oravecz, T. et al. (1997) J. Exp. Med. 186, 1865–1872.
[24] Dragic, T. et al. (1996) Nature 381, 667–673.

Other names

Mast cell growth factor (MGF), kit ligand (KL), steel factor (SLF).

THE MOLECULE

Stem cell factor (SCF) is involved in the development of haematopoietic, gonadal and pigment cell lineages. It has a very wide range of activities with direct effects on myeloid and lymphoid cell development and powerful synergistic effects with other growth factors such as GM-CSF, IL-7 and erythropoietin. SCF is encoded by the steel (Sl) locus of the mouse and is the ligand for the c-*kit* proto-oncogene[1,2]. Alternative mRNA splicing gives rise to two forms of SCF, both of which have a transmembrane domain and are inserted into the cell membrane. The larger form contains a peptide cleavage site and is processed to yield secreted SCF[3,4]. Both membrane-bound and soluble forms are biologically active.

Crossreactivity

There is 81% homology between human and mouse SCF. Human SCF has very little activity on mouse cells whereas rat SCF is active on human cells.

Sources

Bone marrow stromal cells, brain, liver, kidney, lung, placenta, fibroblasts, oocytes, testis.

Bioassays

Synergy with CSFs in progenitor bone marrow colony assay. Proliferation of MO7e cell line.

Physicochemical properties of SCF

Property	Human	Mouse
Amino acids – precursor	273	273
– mature[a]	248/220	248/220
M_r (K) – predicted[b]	27.9/18.5	27.7/18.3
– expressed[c]	36	28–36
Potential *N*-linked glycosylation sites[d]	5	4
Disulfide bonds	2	2

[a] Long and short membrane-bound forms after removal of predicted signal peptide.
[b] Long membrane form and mature soluble form.
[c] Mature soluble form.
[d] One site is lost in short form. Also evidence for *O*-linked glycosylation. Nonglycosylated SCF is biologically active.

3-D structure

Noncovalently linked homodimer. Contains extensive α-helix and β-pleated sheets[5].

Gene structure[6]

Scale

Exons 50 aa

Translated

500 bp

Untranslated

Introns 1Kb

Chromosome

5 38 21 57 52 28 37 35

hSCF

12q22-q24

mouse 10

The gene for SCF is located on human chromosome 12q22–q24, spanning 10 exons, 8 of which are coding exons of 5, 38, 21, 57, 52, 28, 37 and 35 amino acids. The mouse gene is located on chromosome 10.

Amino acid sequence for human SCF[6]

Accession code: SwissProt P21583

-25 MKKTQTWILT CIYLQLLLFN PLVKT
1 EGICRNRVTN NVKDVTKLVA NLPKDYMITL KYVPGMDVLP SHCWISEMVV
51 QLSDSLTDLL DKFS**N**ISEGL S**N**YSIIDKLV NIVDDLVECV KE**N**SSKDLKK
101 SFKSPEPRLF TPEEFFRIF**N** RSIDAFKDFV VASETSDCVV SSTLSPEKDS
151 RVSVTKPFML PPVAASSLR**N** DSSSSNRKAK NPPGDSSLHW AAMALPALFS
201 LIIGFAFGAL YWKKRQPSLT RAVENIQINE EDNEISMLQE KEREFQEV

Disulfide bonds are formed between Cys4–89 and 43–138. Alternative splicing gives rise to two membrane-bound forms[4]. The longer form contains a cleavage site between Ala164 and Ala165 or Ala165 and Ser166, yielding soluble SCF. The shorter form does not have amino acids 150–177, which contains the cleavage site, and is predominantly membrane bound.

Amino acid sequence for mouse SCF[7–9]

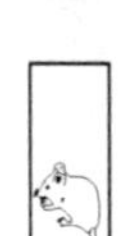

Accession code: SwissProt P20826

-25 MKKTQTWIIT CIYLQLLLFN PLVKT
1 KEICGNPVTD NVKDITKLVA NLPNDYMITL NYVAGMDVLP SHCWLRDMVI
51 QLSLSLTTLL DKFS**N**ISEGL S**N**YSIIDKLG KIVDDLVLCM EENAPKNIKE
101 SPKRPETRSF TPEEFFSIF**N** RSIDAFKDFM VASDTSDCVL SSTLGPEKDS
151 RVSVTKPFML PPVAASSLR**N** DSSSSNRKAA KAPEDSGLQW TAMALPALIS
201 LVIGFAFGAL YWKKKQSSLT RAVENIQINE EDNEISMLQQ KEREFQEV

Disulfide bonds are formed between Cys4–89 and 43–138. Alternative splicing gives rise to two membrane bound forms[3]. The longer form is shown here. By analogy with rat SCF[10], The longer form contains a cleavage site between Ala164 and Ala165 or

Ala165 and Ser166, yielding soluble SCF. The shorter form does not have amino acids 150–177, which contains the cleavage site, and is predominantly membrane-bound. Conflicting sequence A→S at position 182[9].

THE SCF RECEPTOR

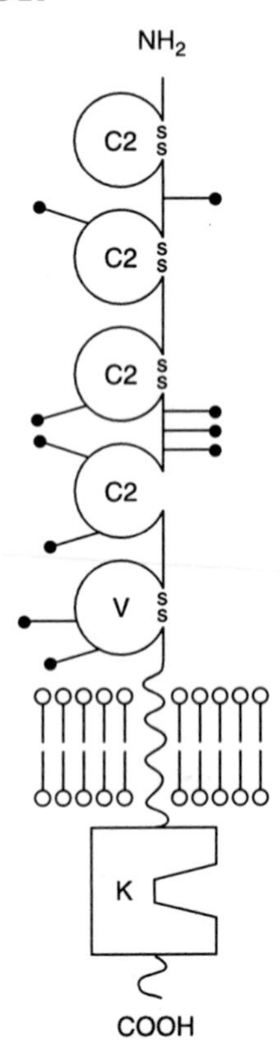

The c-*kit* proto-oncogene is the receptor for SCF (CD117). It is a single transmembrane glycoprotein with five extracellular Ig-SF domains and an intracellular tyrosine kinase domain split by a unique insertion sequence of 77 amino acids which is highly conserved between species. c-*kit* is structurally closely related to the CSF-1 receptor (c-*fms*) and the PDGF receptor. The functional receptor is probably a homodimer induced by binding of SCF. In mice, mutations of the W locus, which encodes c-*kit*, lead to changes in coat colour, anaemia and defective gonad development. In humans, a mutation resulting in a Gly→Arg substitution at position 642 has been identified in piebaldism.

Distribution

Almost all haematopoietic cell progenitors except B-lineage precursors which form colonies in response to IL-7. Also expressed on mast cells, melanocytes, spermatagonia and oocytes[2].

Physicochemical properties of the SCF receptor (c-*kit*)

Property	Human	Mouse
Amino acids – precursor	976	975
– mature[a]	954	953
M_r (K) – predicted	107.4	106.6
– expressed	145	145–150
Potential *N*-linked glycosylation sites	10	9
Affinity K_d (M)	?	?

[a] After removal of predicted signal peptide.

Signal transduction

The SCF receptor (c-*kit*) belongs to the split tyrosine kinase receptor family which includes the PDGF and FGF receptors. The interkinase domain of c-*kit* contains the binding site for PI-3′ kinase[11]. Ligand binding to the SCF receptor results in autotyrosine phosphorylation, and tyrosine phosphorylation of MAP kinase, GAP and PLCγ, as well as serine phosphorylation of Raf-1[12].

Chromosomal location

Human chromosome 4q12 and mouse chromosome 5.

Amino acid sequence for human SCF receptor (c-*kit*)[13]

Accession code: SwissProt P10721

```
-22  MRGARGAWDF LCVLLLLLRV QT
  1  GSSQPSVSPG EPSPPSIHPG KSDLIVRVGD EIRLLCTDPG FVKWTFEILD
 51  ETNENKQNEW ITEKAEATNT GKYTCTNKHG LSNSIYVFVR DPAKLFLVDR
101  SLYGKEDNDT LVRCPLTDPE VTNYSLKGCQ GKPLPKDLRF IPDPKAGIMI
151  KSVKRAYHRL CLHCSVDQEG KSVLSEKFIL KVRPAFKAVP VVSVSKASYL
201  LREGEEFTVT CTIKDVSSSV YSTWKRENSQ TKLQEKYNSW HHGDFNYERQ
251  ATLTISSARV NDSGVFMCYA NNTFGSANVT TTLEVVDKGF INIFPMINTT
301  VFVNDGENVD LIVEYEAFPK PEHQQWIYMN RTFTDKWEDY PKSENESNIR
351  YVSELHLTRL KGTEGGTYTF LVSNSDVNAA IAFNVYVNTK PEILTYDRLV
401  NGMLQCVAAG FPEPTIDWYF CPGTEQRCSA SVLPVDVQTL NSSGPPFGKL
451  VVQSSIDSSA FKHNGTVECK AYNDVGKTSA YFNFAFKGNN KEQIHPHTLF
501  TPLLIGFVIV AGMMCIIVMI LTYKYLQKPM YEVQWKVVEE INGNNYVYID
551  PTQLPYDHKW EFPRNRLSFG KTLGAGAFGK VVEATAYGLI KSDAAMTVAV
601  KMLKPSAHLT EREALMSELK VLSYLGNHMN IVNLLGACTI GGPTLVITEY
651  CCYGDLLNFL RRKRDSFICS KQEDHAEAAL YKNLLHSKES SCSDSTNEYM
701  DMKPGVSYVV PTKADKRRSV RIGSYIERDV TPAIMEDDEL ALDLEDLLSF
751  SYQVAKGMAF LASKNCIHRD LAARNILLTH GRITKICDFG LARDIKNDSN
801  YVVKGNARLP VKWMAPESIF NCVYTFESDV WSYGIFLWEL FSLGSSPYPG
851  MPVDSKFYKM IKEGFRMLSP EHAPAEMYDI MKTCWDADPL KRPTFKQIVQ
901  LIEKQISEST NHIYSNLANC SPNRQKPVVD HSVRINSVGS TASSSQPLLV
951  HDDV
```

Tyr801 is site of autophosphorylation.

Amino acid sequence for mouse SCF receptor (c-*kit*)[14]

Accession code: SwissProt P05532

-22 MRGARGAWDL LCVLLVLLRG QT
1 ATSQPSASPG EPSPPSIHPA QSELIVEAGD TLSLTCIDPD FVRWTFKTYF
51 NEMVENKKNE WIQEKAEATR TGTYTCSNSN GLTSSIYVFV RDPAKLFLVG
101 LPLFGKEDSD ALVRCPLTDP QVS**N**YSLIEC DGKSLPTDLT FVPNPKAGIT
151 IKNVKRAYHR LCVRCAAQRD GTWLHSDKFT LKVREAIKAI PVVSVPETSH
201 LLKKGDTFTV VCTIKDVSTS VNSMWLKMNP QPQHIAQVKH NSWHRGDFNY
251 ERQETLTISS ARVDDSGVFM CYA**N**NTFGSA **N**VTTTLKVVE KGFI**N**ISPVK
301 **N**TTVFVTDGE NVDLVVEYEA YPKPEHQQWI YMNRTSANKG KDYVKSDNKS
351 NIRYVNQLRL TRLKGTEGGT YTFLVSNSDA SASVTFNVYV NTKPEILTYD
401 RLINGMLQCV AEGFPEPTID WYFCTGAEQR CTTPVSPVDV QVQ**N**VSVSPF
451 GKLVVQSSID SSVFRH**N**GTV ECKASNDVGK SSAFFNFAFK EQIQAHTLFT
501 PLLIGFVVAA GAMGIIVMVL TYKYLQKPMY EVQWKVVEEI NGNNYVYIDP
551 TQLPYDHKWE FPRNRLSFGK TLGAGAFGKV VEATAYGLIK SDAAMTVAVK
601 MLKPSAHLTE REALMSELKV LSYLGNHMNI VNLLGACTVG GPTLVITEYC
651 CYGDLLNFLR RKRDSFIFSK QEEQAEAALY KNLLHSTEPS CDSSNEYMDM
701 KPGVSYVVPT KTDKRRSARI DSYIERDVTP AIMEDDELAL DLDDLLSFSY
751 QVAKAMAFLA SKNCIHRDLA ARNILLTHGR ITKICDFGLA RDIRNDSNYV
801 VKGNARLPVK WMAPESIFSC VYTFESDVWS YGIFLWELFS LGSSPYPGMP
851 VDSKFYKMIK EGFRMVSPEH APAEMYDVMK TCWDADPLKR PTFKQVVQLI
901 EKQISDSTKH IYSNLANCNP NPENPVVVDH SVRVNSVGSS ASSTQPLLVH
951 EDA

Tyr799 is site of autophosphorylation.

References

[1] Witte, O.N. (1990) Cell 63, 5–6.
[2] Morrison-Graham, K. and Takahashi, Y. (1993) BioEssays 15, 77–83.
[3] Flanagan, J.G. et al. (1991) Cell 64, 1025–1035.
[4] Anderson, D.M. et al. (1991) Cell Growth Differ. 2, 373–378.
[5] Arakawa, T. et al. (1991) J. Biol. Chem. 266, 18942–18948.
[6] Martin, F.H. et al. (1990) Cell 63, 203–211.
[7] Anderson, D.M. et al. (1990) Cell 63, 235–243.
[8] Zsebo, K.M. et al. (1990) Cell 63, 213–224.
[9] Huang, E. et al. (1990) Cell 63, 225–233.
[10] Lu, H.S. et al. (1991) J. Biol. Chem. 266, 8102–8107.
[11] Lev, S. et al. (1992) Proc. Natl Acad. Sci. USA 89, 678–682.
[12] Miyazawa, K. et al. (1991) Exp. Haematol. 19, 1110–1123.
[13] Yarden, Y. et al. (1987) EMBO J. 6, 3341–3351.
[14] Qiu, F. et al. (1988) EMBO J. 7, 1003–1011.

Other names

Pre-B cell growth-stimulating factor[1,2].

THE MOLECULE

Stromal cell-derived factor 1 (SDF-1) is a chemoattractant for pre-B cells, T cells, monocytes, microglia, megakaryocytes, endothelial cells, astrocytes and CD34+ progenitor cells[3–6]. It also acts as a growth factor for pre-B cells and can inhibit the proliferation of haematopoietic progenitors cells[7]. SDF-1 knockout mice exhibit defects in B lymphopoiesis, myelopoiesis and heart development[8].

Crossreactivity

Mouse and human SDF-1 differ only by a single amino acid, and act across species[5].

Sources

Constitutive expression in multiple tissues including bone marrow stroma, heart, liver, kidney, lung, brain, muscle and spleen[1,2].

Bioassays

Costimulation of pre-B cell proliferation with IL-7, chemotaxis of T cells.

Physicochemical properties of SDF-1

Property	Human	Mouse
Amino acids – precursor	93	89
– mature[a]	68	68
M_r (K) – predicted	8.0	8.3
– expressed	?	?
Potential *N*-linked glycosylation sites	0	0
Disulfide bonds	2	2

[a] SDF-1 can be inactivated by proteolytic removal of the N-terminal dipeptide by CD26 (dipeptidylpeptidase IV)[9].

3-D structure

Typical chemokine β–β–β–α structure but with a different packing of the α-helix against the β-sheet[10]; N-terminal 8 residues are involved in receptor activation[11].

Gene structure

The gene for human SDF-1 is on chromosome 10q[12].

Amino acid sequence for human SDF-1

Accession code: SwissProt P48061

```
-18  MNAKVVVVLV LVLTALCL
  1  SDGKPVSLSY RCPCRFFESH VARANVKHLK ILNTPNCALQ IVARLKNNNR
 51  QVCIDPKLKW IQEYLEKALN KRFKM
```

Amino acid sequence for mouse SDF-1[12]

Accession code: SwissProt P40224

```
-18  MDAKVVAVLA LVLAALCI
  1  SDGKPVSLSY RCPCRFFESH IARANVKHLK ILNTPNCALQ IVARLKNNNR
 51  QVCIDPKLKW IQEYLEKALN K
```

A variant sequence, SDF-1B, which has a C-terminal extension of RLKM has been reported[12] .

THE SDF-1 RECEPTOR, CXCR4

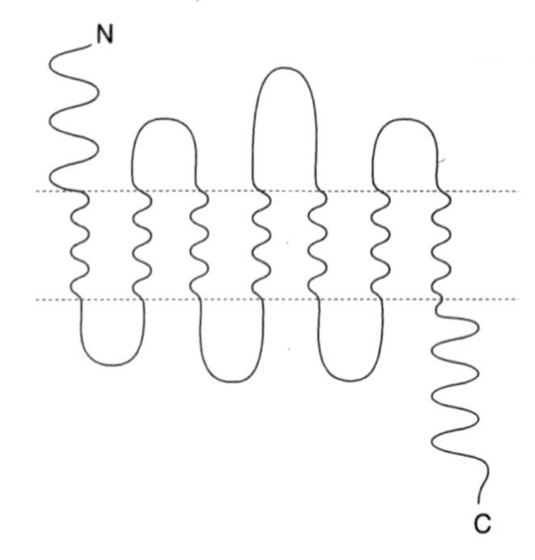

The SDF-1 receptor, CXCR4, is also known as LESTR or fusin and also functions as a co-receptor for lymphotrophic strains of HIV-1 and HIV-2[3,13,14]. SDF-1 is the only known chemokine ligand for CXCR4 and acts as an antagonist for HIV infection.

Distribution

CXCR4 mRNA is found in a wide range of tissues, with high level receptor expression on haematopoietic, glial and dendritic cells[3,4,6,15–17]. CXCR4 knockout mice are embryonically lethal, and exhibit defects in haematopoiesis, cardiogenesis and vasculogenesis[18].

Physicochemical properties of the SDF-1 receptor

Property	Human	Mouse
Amino acids – precursor	352	359
– mature	?	?
M_r (K) – predicted	39.7	40.4
– expressed	44.0	?
Potential *N*-linked glycosylation sites	1	2
Affinity K_d (M)	5×10^{-9}	1.9×10^{-10}

Signal transduction

Binding of SDF-1 to CXCR4-bearing cells causes a rapid mobilization of calcium, G protein activation, inositol phosphate generation, phosphorylation of focal adhesion complexes, activation of p44/42 MAP kinase and NFκB, followed by receptor internalization[3,5,6,13,19,20]. Astrocytes and mature B cells mobilize calcium but have been reported to be refractory to stimulation with SDF-1 despite expressing CXCR4 on their cell surface[5,6]. Recent reports, however, have shown that B cells exhibit polarized migration in response to SDF-1[21].

Chromosomal location

The human gene is located on chromosome 2q21[22].

Amino acid sequence for human SDF-1 receptor CXCR4

Accession code: SwissProt P30991

1 MEGISIYTSD **NYT**EEMGSGD YDSMKEPCFR EENANFNKIF LPTIYSIIFL
51 TGIVGNGLVI LVMGYQKKLR SMTDKYRLHL SVADLLFVIT LPFWAVDAVA
101 NWYFGNFLCK AVHVIYTVNL YSSVLILAFI SLDRYLAIVH ATNSQRPRKL
151 LAEKVVYVGV WIPALLLTIP DFIFA**NVS**EA DDRYICDRFY PNDLWVVVFQ
201 FQHIMVGLIL PGIVILSCYC IIISKLSHSK GHQKRKALKT TVILILAFFA
251 CWLPYYIGIS IDSFILLEII KQGCEFENTV HKWISITEAL AFFHCCLNPI
301 LYAFLGAKFK TSAQHALTSV SRGSSLKILS KGKRGGHSSV STESESSSFH
351 SS

Amino acid sequence for mouse SDF-1 receptor CXCR4

Accession code: SwissProt P70658, P70346, O09062, O09059

1 MEPISVSIYT SD**NYS**EEVGS GDYDSNKEPC FRDENVHFNR IFLPTIYFII
51 FLTGIVGNGL VILVMGYQKK LRSMTDKYRL HLSVADLLFV ITLPFWAVDA
101 MADWYFGKFL CKAVHIIYTV NLYSSVLILA FISLDRYLAI VHATNSQRPR
151 KLLAEKAVYV GVWIPALLLT IPDFIFADVS QGDISQGDDR YICDRLYPDS
201 LWMVVFQFQH IMVGLILPGI VILSCYCIII SKLSHSKGHQ KRKALKTTVI
251 LILAFFACWL PYYVGISIDS FILLGVIKQG CDFESIVHKW ISITEALAFF
301 HCCLNPILYA FLGAKFKSSA QHALNSMSRG SSLKILSKGK RGGHSSVSTE
351 SESSSFHSS

An I to V variant of CXCR4 at residue 216 has been reported.

References

[1] Tashiro, K. et al. (1993) Science 261, 600–603.
[2] Nagasawa, T. et al. (1994) Proc. Natl Acad. Sci. USA 91, 2305–2309.
[3] Oberlin, E. et al. (1996) Nature 382, 833–835.
[4] Aiuti, A. et al. (1997) J. Exp. Med. 185, 111–120.
[5] D'Apuzzo, M. et al. (1997) Eur. J. Immunol. 27, 1788–1793.
[6] Tanabe, S. et al. (1997) J. Immunol. 159, 905–911.
[7] Sanchez, X. et al. (1997) J. Biol. Chem. 272, 27529–27531.
[8] Nagasawa, T. et al. (1996) Nature 382, 635–638.
[9] Shioda, T. et al. (1998) Proc. Natl Acad. Sci. USA 95, 6331–6336.
[10] Dealwis, C. et al. (1998) Proc. Natl Acad. Sci.USA 95, 6941–6946.
[11] Crump, M.P. et al. (1998) EMBO J. 16, 6996–7007.
[12] Shirozu, M. et al. (1995) Genomics 28, 495–500.
[13] Bleul, C.C. et al. (1996) Nature 382, 829–833.
[14] Feng, Y. et al. (1996) Science 272, 872–877.
[15] Sozzani, S. et al. (1997) J. Immunol. 159, 1993–2000.
[16] Nagasawa, T. et al. (1996) Proc. Natl Acad. Sci. USA 93, 14726–14729.
[17] Heesen, M. et al. (1996) J. Immunol. 157, 5455–5460.
[18] Tachibana, K. et al. (1998) Nature 393, 591–594.
[19] Haribabu, B. et al. (1997) J. Biol. Chem. 272, 28726–28731.
[20] Ganju, R.K. et al. (1998) J. Biol. Chem. 273, 23169–23175.
[21] Vincentemanzanares, M. et al. (1998) Eur. J. Immunol. 28, 2197–2207.
[22] Baggioloini, M. et al. (1997) Annu. Rev. Immunol. 15, 675–705.

Other names
SCYA17 or A-152E5.3.

THE MOLECULE

Thymus and activation regulated chemokine (TARC) is a chemoattractant for T cells. It is produced by T cells and monocytes and is expressed predominantly by stromal cells in the thymus[1]. TARC production can be stimulated by Th2 cytokines and it acts on Th2 lymphocytes which selectively express the TARC/MDC receptor CCR4[1,3]. It may therefore play a role in thymic function and in Th2 immune responses.

Crossreactivity
No information.

Sources
TARC is produced by PHA-activated peripheral blood leukocytes, Th2 cytokine-stimulated monocytes and thymic stromal cells.

Bioassays
Chemotaxis of Th2 lymphocytes.

Physicochemical properties of TARC

Property	Human	Mouse
Amino acids		
– precursor	94	?
– mature	71	?
M_r (K) – predicted	10.5	?
– expressed	?	
Potential *N*-linked glycosylation sites	0	?
Disulfide bonds	2	?

3-D structure
No information.

Gene structure

The gene for human TARC is on chromosome 16q13[2].

Amino acid sequence for human TARC

Accession code: SwissProt Q92583

```
-23  MAPLKMLALV TLLLGASLQH IHA
  1  ARGTNVGREC CLEYFKGAIP LRKLKTWYQT SEDCSRDAIV FVTVQGRAIC
 51  SDPNNKRVKN AVKYLQSLER S
```

THE RECEPTORS, CCR4 and CCR8

TARC binds to both CCR4 and CCR8, mediating chemotaxis through both receptors in transfected cells[5,8]. For description of the CCR4 receptor, see MDC entry (page **373**). CCR8 (also known as TER-1, ChemR1, GPRCY6 or CKRL1) is expressed on T cells and NK cells, and also binds to I-309 and MIP-1β[6,7–10].

Distribution

CCR4 is predominantly expressed on Th2 lymphocytes, but can also be found on basophils, megakaryocytes and platelets. mRNA expression is particularly strong in thymus[4,5]. CCR8 is expressed on some Th2 cells and NK cell lines, weakly on IL-2-activated peripheral blood leukocytes, and in thymus and spleen[6,7,9,10].

Physicochemical properties of the CCR8 TARC receptor

Property	Human	Mouse
Amino acids – precursor	355	353
– mature		
M_r (K) – predicted	40.8	40.6
– expressed	?	?
Potential *N*-linked glycosylation sites	1	2
Affinity K_d (M)	?	?

Signal transduction

TARC binds specifically to CCR4 with high affinity (K_d 0.5–2.1 nM) and can mediate a Ca flux with EC_{50} of 8 nM in CCR4-transfected K562 cells. However, it has not yet been shown to mediate a Ca flux in normal human T cells, despite promoting chemotaxis at concentrations between 10 and 1000 nM[1,5]. The chemotactic responses were inhibited by pertussis toxin. TARC promotes maximal chemotaxis of CCR8-transfected Jurkat cells at 0.1 ng/ml, but did not mediate a Ca flux[8,10]. TARC did promote a Ca flux in murine Th2 cells at 100 ng/ml[10].

Chromosomal location

The human CCR4 and CCR8 genes are on chromosome 3p21–24[6].

Amino acid sequence for human CCR8 receptor

Accession code: SwissProt P51685

```
  1  MDYTLDLSVT TVTDYYYPDI FSSPCDAELI QTNGKLLLAV FYCLLFVFSL
 51  LGNSLVILVL VVCKKLRSIT DVYLLNLALS DLLFVFSFPF QTYYLLDQWV
101  FGTVMCKVVS GFYYIGFYSS MFFITLMSVD RYLAVVHAVY ALKVRTIRMG
151  TTLCLAVWLT AIMATIPLLV FYQVASEDGV LQCYSFYNQQ TLKWKIFTNF
201  KMNILGLLIP FTIFMFCYIK ILHQLKRCQN HNKTKAIRLV LIVVIASLLF
251  WVPFNVVLFL TSLHSMHILD GCSISQQLTY ATHVTEIISF THCCVNPVIY
301  AFVGEKFKKH LSEIFQKSCS QIFNYLGRQM PRESCEKSSS CQQHSSRSSS
351  VDYIL
```

Disulfide bond links Cys106–183.

Amino acid sequence for mouse CCR8 receptor

Accession code: SwissProt P56484

```
  1  MDYTMEPNVT MTDYYPDFFT APCDAEFLLR GSMLYLAILY CVLFVLGLLG
 51  NSLVILVLVG CKKLRSITDI YLLNLAASDL LFVLSIPFQT HNLLDQWVFG
101  TAMCKVVSGL YYIGFFSSMF FITLMSVDRY LAIVHAVYAI KVRTASVGTA
151  LSLTVWLAAV TATIPLMVFY QVASEDGMLQ CFQFYEEQSL RWKLFTHFEI
201  NALGLLLPFA ILLFCYVRIL QQLRGCLNHN RTRAIKLVLT VVIVSLLFWV
251  PFNVALFLTS LHDLHILDGC ATRQRLALAI HVTEVISFTH CCVNPVIYAF
301  IGEKFKKHLM DVFQKSCSHI FLYLGRQMPV GALERQLSSN QRSSHSSTLD
351  DIL
```

Disulfide bond links Cys104–181.

References

[1] Imai, T. et al. (1996) J. Biol. Chem. 271, 21514–21521.
[2] Nomiyama, H. et al. (1998) Cytogen. Cell Genet. 81, 10–11.
[3] Andrew, D.P. et al. (1998) J. Immunol. 161, 5027–5038.
[4] Imai, T. et al. (1998) J. Biol. Chem. 273, 1764–1768.
[5] Imai, T. et al. (1997) J. Biol. Chem. 272, 15036–15402.
[6] Napolitano, M. et al. (1996) J. Immunol. 157, 2759–2763.
[7] Tiffany, H.L. et al. (1997) J. Exp. Med. 186, 165–170.
[8] Bernardini, G. et al. (1998) Eur. J. Immunol. 28, 582–588.
[9] Stuber-Roos, R. et al. (1997) J. Biol. Chem. 272, 17251–17254.
[10] Zingoni, A. et al. (1998) J. Immunol. 161, 547–551.

Other names
SCYA-25.

THE MOLECULE

Thymus expressed chemokine (TECK) is a chemoattractant for macrophages, dendritic cells and thymic T cells. It is produced in thymus and gut by dendritic and epithelial cells[1,2]. It may therefore play a role in thymic function and in gut homing of T cells.

Crossreactivity
Human and mouse TECK act across species[1,3].

Sources
TECK is produced by thymic dendritic and epithelial cells, and by intestinal epithelial cells.

Bioassays
Chemotaxis of T cells.

Physicochemical properties of TECK

Property	Human	Mouse
Amino acids		
– precursor	173	147
– mature	150	124
M_r (K) – predicted	16.6	16.7
– expressed	?	?
Potential *N*-linked glycosylation sites	1	1
Disulfide bonds	2	2

3-D structure
No information.

Gene structure

The gene for human TECK is on chromosome 19p13.2 and mouse TECK is on chromosome 8[1,4]. This is distinct from the other CC chemokines which cluster on chromosomes 17 and 11 in human and mouse, respectively.

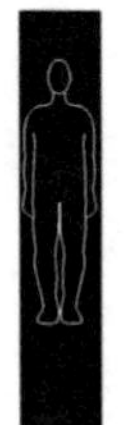

Amino acid sequence for human TECK

Accession code: SwissProt O15444

```
-23  MNLWLLACLV AGFLGAWAPA VHT
  1  QGVFEDC CLAYHYPIGW AVLRRAWTYR IQEVSGSCNL PAAIFYLPKR
 51  HRKVCGNPKS REVQRAMKLL DARNKVFAKL HHNMQTFQAG PHAVKKLSSG
101  NSKLSSSKFS NPISSSKRNV SLLISANSGL
```

Amino acid sequence for mouse TECK

Accession code: SwissProt O35903

```
-23  MKLWLFACLV ACFVGAWMPV VHA
  1  QGAFEDC CLGYQHRIKW NVLRHARNYH QQEVSGSCNL RAVRFYFRQK
 51  VVCGNPEDMN VKRAIRILTA RKRLVHWKSA SDSQTERKKS NHMKSKVENP
101  NSTSVRSATL GHPRMVMMPR KTNN
```

THE TECK RECEPTOR, CCR9

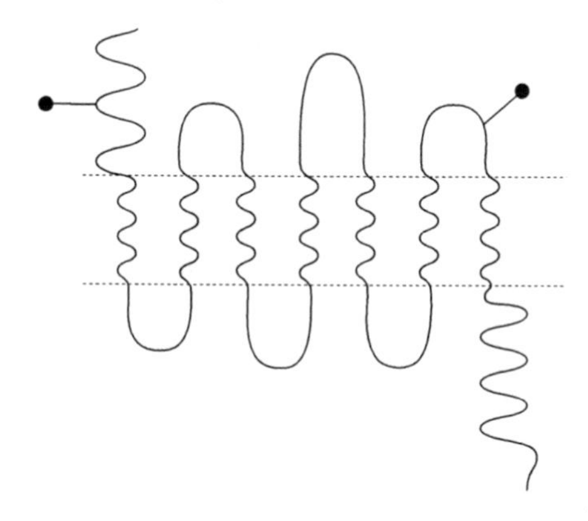

TECK binds to CCR9, also known as GPR-9-6, mediating a calcium flux and chemotaxis of T cells[3,5]. The human CCR9 exists as two splice variants, the more prevalent form, CCR9A, is more sensitive to TECK and has a 12-amino-acid N-terminal extension that is lacking in CCR9B[6].

Distribution

CCR9 is predominantly expressed on thymocytes and intestinal T cells.

Physicochemical properties of the CCR9 TECK receptor

Property	Human	Mouse
Amino acids – precursor	357	369
– mature	?	?
M_r (K) – predicted	40.7	41.9
– expressed	?	?
Potential *N*-linked glycosylation sites	2	2
Affinity K_d (M)	?	?

Signal transduction

TECK binding to CCR9 can mediate a Ca flux at concentrations as low as 6 nM with maximal responses at 100 nM. The calcium responses were only partially inhibited by pertussis toxin[3]. TECK promotes maximal chemotaxis of transfected cells at 100–200 nM.

Chromosomal location

The human CCR9 genes are on chromosome 3p21.3[7] and the mouse gene is on chromosome 9F1–F4[2].

Amino acid sequence for human CCR9A receptor

Accession code: SPTREMBLE Q9UQQ6

```
  1  MPTPDFTSPI PNMADDYGSE STSSMEDYVN FNFTDFYCEK NNVRQFASHF
 51  LPPLYWLVFI VGALGNSLVI LVYWYCTRVK TMTDMFLLNL AIADLLFLVT
101  PFWAIAAAD QWKFQTFMCK VVNSMYKMNF YSCVLLIMCI SVDRYIAIAQ
151  AMRAHTWREK RLLYSKMVCF TIWVLAAALC IPEILYSQIK EESGIAICTM
201  VYPSDESTKL KSAVLTLKVI LGFFLPFVVM ACCYTIIIHT LIQAKKSSKH
251  KALKVTITVL TVFVLSQFPY NCILLVQTID AYAMFISNCA VSTNIDICFQ
301  VTQTIAFFHS CLNPVLYVFV GERFRRDLVK TLKNLGCISQ AQWVSFTRRE
351  GSLKLSSMLL ETTSGALSL
```

Amino acid sequence for human CCR9B receptor

Accession code: SwissProt P51686

```
  1  MADDYGSEST SSMEDYVNFN FTDFYCEKNN VRQFASHFLP PLYWLVFIVG
 51  ALGNSLVILV YWYCTRVKTM TDMFLLNLAI ADLLFLVTLP FWAIAAADQW
101  KFQTFMCKVV NSMYKMNFYS CVLLIMCISV DRYIAIAQAM RAHTWREKRL
151  LYSKMVCFTI WVLAAALCIP EILYSQIKEE SGIAICTMVY PSDESTKLKS
201  AVLTLKVILG FFLPFVVMAC CYTIIIHTLI QAKKSSKHKA LKVTITVLTV
251  FVLSQFPYNC ILLVQTIDAY AMFISNCAVS TNIDICFQVT QTIAFFHSCL
301  NPVLYVFVGE RFRRDLVKTL KNLGCISQAQ WVSFTRREGS LKLSSMLLET
351  TSGALSL
```

CCR9B is a splice variant of CCR9A lacking the first 12 amino acids. Disulfide bond links Cys107–186.

Amino acid sequence for mouse CCR9 receptor

Accession code: SwissProt Q9WUT7

```
  1 MMPTELTSLI PGMFDDFSYD STASTDDYMN LNFSSFFCKK NNVRQFASHF
 51 LPPLYWLVFI VGTLGNSLVI LVYWYCTRVK TMTDMFLLNL AIADLLFLAT
101 LPFWAIAAAG QWMFQTFMCK VVNSMYKMNF YSCVLLIMCI SVDRYIAIVQ
151 AMKAQVWRQK RLLYSKMVCI TIWVMAAVLC TPEILYSQVS GESGIATCTM
201 VYPKDKNAKL KSAVLILKVT LGFFLPFMVM AFCYTIIIHT LVQAKKSSKH
251 KALKVTITVL TVFIMSQFPY NSILVVQAVD AYAMFISNCT ISTNIDICFQ
301 VTQTIAFFHS CLNPVLYVFV GERFRRDLVK TLKNLGCISQ AQWVSFTRRE
351 GSLKLSSMLL ETTSGALSL
```

Disulfide bond links Cys119–198.

References

[1] Vicari, A.P. et al. (1997) Immunity 7, 291–301.
[2] Wurbel, M.A. et al. (2000) Eur. J. Immunol. 30, 262–271.
[3] Zaballos, A. et al. (1999) J. Immunol. 162, 5671–5675.
[4] Nomiyama, H. et al. (1998) Genomics 51, 311–312.
[5] Youn, B.S. et al. (1999) Blood 94, 2533–2536.
[6] Yu, C.R. et al. (2000) J. Immunol. 164, 1293–1305.
[7] Maho, A. et al. (1999) Cytogenet. Cell Genet. 87, 265–268.

Other names
Sarcoma growth factor.

THE MOLECULE

Transforming growth factor α is a small integral membrane protein which shares biological and structural properties with EGF. The mature 50 amino acid cytokine is released by proteolytic cleavage[1–4]. TGFα can act as an autoinductive growth factor[5].

Crossreactivity
TGFα is active across species. TGFα is closely structurally related to EGF and to vaccinia growth factor, which all bind to the EGF receptor[6,7].

Sources
TGFα is made by monocytes, keratinocytes and many tissues and tumours.

Bioassays
Proliferation of the A431 carcinoma line.

Physicochemical properties of TGFα

Property	Human	Rat
Amino acids – precursor	160	159
– mature[a]	50	50
M_r (K) – predicted	6	6
– expressed	6	6
Potential *N*-linked glycosylation sites[b]	0	0
Disulfide bonds	3	3

[a] The C-terminal valine in the cytoplasmic tail of pro-TGFα is required for cleavage to mature TGFα[8].
[b] In mature 50 amino acid TGFα, there is one *N*-linked glycosylation site in the propeptide.

3-D structure
Similar to EGF[9].

Gene structure
Human TGFα is on chromosome 2.

Amino acid sequence for human TGFα[3]

Accession code: SwissProt P01135

```
-23  MVPSAGQLAL FALGIVLAAC QAL
  1  ENSTSPLSAD PPVAAAVVSH FNDCPDSHTQ FCFHGTCRFL VQEDKPACVC
 51  HSGYVGARCE HADLLAVVAA SQKKQAITAL VVVSIVALAV LIITCVLIHC
101  CQVRKHCEWC RALICRHEKP SALLKGRTAC CHSETVV
```

N-terminal and C-terminal sequences in italics are removed during processing to release the active molecule (amino acids 17–66). Disulfide bonds between Cys24–37, 32–48 and 50–59.

Amino acid sequence for rat TGFα[10]

Accession code: SwissProt P01134

```
-23  MVPAAGQLAL LALGILVAVC QAL
  1  ENSTPPLSDS PVAAAVVSHF NKCPDSHTQY CFHGTCRFLV QEEKPACVCH
 51  SGYVGVRCEH ADLLAVVAAS QKKQAITALV VVSIVALAVL IITCVLIHCC
101  QVRKHCEWCR ALVCRHEKPS ALLKGRTACC HSETVV
```

N-terminal and C-terminal sequences in italics are removed during processing to release the active molecule (amino acids 16–65). Disulfide bonds between Cys23–36, 31–47 and 49–58.

THE TGFα RECEPTOR

The TGF receptor (also known as c-*erbB*) is a class I receptor tyrosine kinase[11,12]. The receptor is also shared with epidermal growth factor (EGF), and with vaccinia virus growth factor. A viral oncogene v-*erbB* encodes a truncated EGF receptor lacking most of the extracellular domains. See EGF entry (page **203**).

Distribution

See EGF entry (page **203**).

Amino acid sequence for human TGF

Accession code: SwissProt XXXX

See EGF entry (page **203**).

Amino acid sequence for rat TGF

Accession code: SwissProt XXXX

See EGF entry (page **203**).

References

[1] Burgess, A.W. (1989) In British Medical Bulletin 45, Growth Factors, Waterfeld, M.D. ed., Churchill Livingstone, London, pp. 401–424.
[2] DeLarco, J.E. and Todaro, G. (1978) Proc. Natl Acad. Sci. USA 75, 4001–4005.
[3] Dernyck, R. et al. (1984) Cell 38, 287–297.
[4] Texido, J. et al. (1987) Nature 326, 883–855.
[5] Coffey, R.J. et al. (1987) Nature 328, 817–820.
[6] Montelione, G.T. et al. (1986) Proc. Natl Acad. Sci. USA. 83, 8594–8598.
[7] Stroobant, P. et al. (1985) Cell 383–393.
[8] Bosenberg, M.W. et al. (1992) Cell 71, 1157–1165.
[9] Campbell, I.D. et al. (1989) Prog. Growth Factor Res. 1, 13–22.
[10] Lee, D.C. et al. (1985) Nature 313, 489–491.
[11] Ullrich, A. et al. Nature 309, 418–425.
[12] Ullrich, A. and Schlessinger, J. (1990) Cell 61, 203–212.

Other names

Human TGFβ1 has been known as differentiation inhibiting factor and cartilage-inducing factor. Human TGFβ2 has been known as glioblastoma-derived T cell suppressor factor.

THE MOLECULE

Transforming growth factor β (TGFβ) is a pleiotropic cytokine involved in tissue remodelling, wound repair, development and haematopoiesis[1]. Its predominant action is to inhibit cell growth. TGFβ is also a switch factor for IgA. TGFβ is comprised of three related dimeric proteins, TGFβ1, 2 and 3, all of which are members of a superfamily including the activins, inhibins and bone morphogenic proteins. The expressed proteins are biologically inactive disulfide-linked dimers which are cleaved to active dimers of 112-amino-acid disulfide-linked peptides[2]. Platelet-derived TGFβ1 is covalently associated with an M_r 125 000–160 000 binding protein composed mainly of 16 EGF domain repeats[3] (human sequence in Genbank M34057, rat sequence in Genbank M55431). A similar protein is found in glioma and fibroblasts. TGFβ binds to proteoglycans such as decorin in the extracellular matrix and α_2-macroglobulin in blood[4].

Crossreactivity

There is greater than 98% homology between the functional regions of human and mouse TGFβ species; only the human sequences are given[5–7].

Sources

Platelets contain TGFβ1 and β2. Most nucleated cell types and many tumours also express TGFβ1, β2, β3 or combinations of the three forms.

Bioassays

Inhibition of growth of mink lung cell line MV-1-Lu.

Physicochemical properties of TGFβ1, 2 and 3

Property	Human		
	TGFβ1	TGFβ2	TGFβ3
Amino acids – precursor	390	414	412
– mature[a]	112	112	112
M_r (K) – predicted	44.3	47.8	47.3
– expressed	25	25	25
Potential *N*-linked glycosylation sites[b]	0	0	0
Disulfide bonds	1	1	1

[a] Functional TGFβ is a disulfide-linked dimer which must be cleaved from the inactive propeptide. Cleavage of the TGFβ1 propeptide at cell surfaces involves binding of the propeptide to the IGF type II receptor, and requires plasminogen activator and plasmin[8].
[b] There are two sites in TGFβ1 propeptide, three sites in TGFβ2 propeptide and four sites in TGFβ3 propeptide.

3-D structure

The crystal structure of TGFβ2 has been solved and shown to contain an unusual elongated nonglobular fold[9]. The structure can be used to model the other TGFβs.

Gene structure[10]

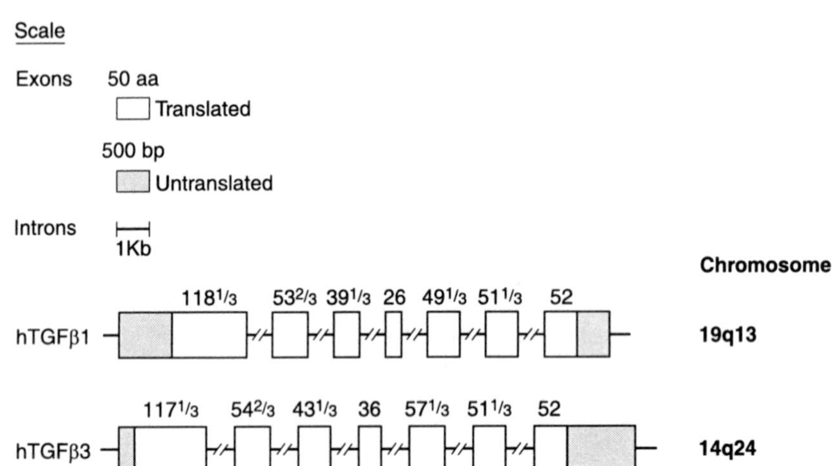

The gene for human TGFβ1 is located on chromosome19q13, TGFβ2 on 1q41 and TGFβ3 on 14q24. The gene for murine TGFβ1 is located on chromosome 7, β2 on 1 and β3 on 12. The gene for TGFβ1 and TGFβ3 contain seven coding exons.

Amino acid sequence for human TGFβ1[5]

Accession code: SwissProt P01137

```
-29  MPPSGLRLLL LLLPLLWLLV LTPGRPAAG
  1  LSTCKTIDME LVKRKRIEAI RGQILSKLRL ASPPSQGEVP PGPLPEAVLA
 51  LYNSTRDRVA GESAEPEPEP EADYYAKEVT RVLMVETHNE IYDKFKQSTH
101  SIYMFFNTSE LREAVPEPVL LSRAELRLLR LKLKVEQHVE LYQKYSNNSW
151  RYLSNRLLAP SDSPEWLSFD VTGVVRQWLS RGGEIEGFRL SAHCSCDSRD
201  NTLQVDINGF TTGRRGDLAT IHGMNRPFLL LMATPLERAQ HLQSSRHRRA
251  LDTNYCFSST EKNCCVRQLY IDFRKDLGWK WIHEPKGYHA NFCLGPCPYI
301  WSLDTQYSKV LALYNQHNPG ASAAPCCVPQ ALEPLPIVYY VGRKPKVEQL
351  SNMIVRSCKC S
```

Amino acid sequence for human TGFβ2[6]

Accession code: SwissProt P08112

```
-20  MHYCVLSAFL ILHLVTVALS
  1  LSTCSTLDMD QFMRKRIEAI RGQILSKLKL TSPPEDYPEP EEVPPEVISI
 51  YNSTRDLLQE KASRRAAACE RERSDEEYYA KEVYKIDMPP FFPSENAIPP
101  TFYRPYFRIV RFDVSAMEKN ASNLVKAEFR VFRLQNPKAR VPEQRIELYQ
151  ILKSKDLTSP TQRYIDSKVV KTRAEGEWLS FDVTDAVHEW LHHKDRNLGF
201  KISLHCPCCT FVPSNNYIIP NKSEELEARF AGIDGTSTYT SGDQKTIKST
251  RKKNSGKTPH LLLMLLPSYR LESQQTNRRK KRALDAAYCF RNVQDNCCLR
301  PLYIDFKRDL GWKWIHEPKG YNANFCAGAC PYLWSSDTQH SRVLSLYNTI
351  NPEASASPCC VSQDLEPLTI LYYIGKTPKI EQLSNMIVKS CKCS
```

Amino acid sequence for human TGFβ3[7]

Accession code: SwissProt P10600

```
-23  MKMHLQRALV VLALLNFATV SLS
  1  LSTCTTLDFG HIKKKRVEAI RGQILSKLRL TSPPEPTVMT HVPYQVLALY
 51  NSTRELLEEM HGEREEGCTQ ENTESEYYAK EIHKFDMIQG LAEHNELAVC
101  PKGITSKVFR FNVSSVEKNR TNLFRAEFRV LRVPNPSSKR NEQRIELFQI
151  LRPDEHIAKQ RYIGGKNLPT RGTAEWLSFD VTDTVREWLL RRESNLGLEI
201  SIHCPCHTFQ PNGDILENIH EVMEIKFKGV DNEDDHGRGD LGRLKKQKDH
251  HNPHLILMMI PPHRLDNPGQ GGQRKKRALD TNYCFRNLEE NCCVRPLYID
301  FRQDLGWKWV HEPKGYYANF CSGPCPYLRS ADTTHSTVLG LYNTLNPEAS
351  ASPCCVPQDL EPLTILYYVG RTPKVEQLSN MVVKSCKCS
```

THE TGFβ RECEPTORS

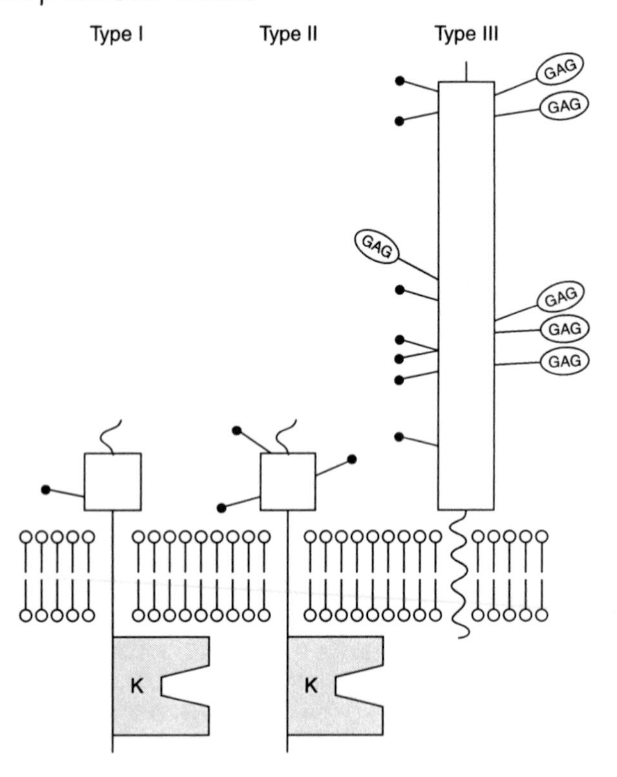

There are three types of TGFβ receptor[11]: high-affinity types I (M_r 55 000)[2,12] and II (M_r 80 000)[13], and low-affinity type III (M_r 250 000–350 000)[14,15]. The human receptors bind all three TGFβ isoforms. The type I receptor binds β1 = β2 > β3, type II β1 > β2 > β3 and type III β1 = β2 = β3. The human and pig type II receptor and the rat type III receptors have been cloned. The human and murine type I receptors have also been cloned[12,16]. The type I and II receptors are serine/threonine kinases related to the activin receptor. The type III receptors include β-glycan, an integral membrane protein modified by attachment of glycosaminoglycans, and endoglin (CD105) which is a homodimer of two M_r 95 000 disulfide-linked subunits related to β-glycan. Endoglin does not bind TGFβ2[17]. The type I and II receptors are thought to associate to mediate signal transduction events probably by serine/threonine phosphorylation. It is unclear if the receptors autophosphorylate. There appears to be an emerging family of receptors related to the type I receptor whose role in TGFβ signalling is not yet clear[12,18]. The type III receptors have not been shown to transduce signals, but may function to concentrate TGFβ on the cell surface and present the ligand to its other receptors. Coexpression of type II and III receptors increases the ability of the type II receptor to bind TGFβ. The type III receptor can be released from the cell surface by proteolysis. Two potential cleavage sites exist in the transmembrane region: KK and LAVV. The LAVV sequence is also used to release TGFβ. The glycosaminoglycan on the type III receptor is not involved in binding TGFβ.

Distribution

Most cell types. The type II receptor is lacking in retinoblastoma cells.

Physicochemical properties of the TGFβ receptors

Property	Type I Human	Type I Mouse	Type II Human	Type III Rat
Amino acids – precursor	503	509	565	853
– mature	479	492	542	829
M_r (K) – predicted	51	53	60	91.6
– expressed	53	68	65	250–350
Potential *N*-linked glycosylation sites	1	1	3	7
Affinity K_d (M)	$5–30 \times 10^{-12}$	$3–30 \times 10^{-11}$		

Signal transduction

Requires the formation of a heteromeric complex of type I and type II receptors (Ser/Thr kinases) for effects, possibly also homodimers[19]. Purified type II receptor can autophosphorylate on Ser and Thr.

Chromosomal location

Not known yet.

Amino acid sequence for human type I TGFβ receptor ALK-1[12]

Accession code: Genbank L11695

```
-24  MEAAVAAPRP RLLLLVLAAA AAAA
  1  AALLPGATAL QCFCHLCTKD NFTCVTDGLC FVSVTETTDK VIHNSMCIAE
 51  IDLIPRDRPF VCAPSSKTGS VTTTYCCNQD HCNKIELPTT VKSSPGLGPV
101  ELAAVIAGPV CFVCISLMLM VYICHNRIVI HHRVPNEEDP SLDRPFISEG
151  TTLKDLIYDM TTSGSGSGLP LLVQRTIART IVLQESIGKG RFGEVWRGKW
201  RGEEVAVKIF SSREERSWFR EAERYQTVML RHENILGFIA ADNKDNGTWT
251  QLWLVSDYHE HGSLFDYLNR YTVTVEGMIK LALSTASGLA HLHMEIVGTG
301  GKPAIAHRDL KSKNILVKKN GTCCIADLGL AVRHDSATDT IDIAPNHRVG
351  TKRYMAPEVL DDSINMKHFE SFKRADIYAM GLVFWEIARR CSIGGIAGDY
401  QLPYYDLVPS DPSVEEMRKV VCEQKLRPNI PNRWQSCEAL RVMAKIMREC
451  WYANGAARLT ALRIKKTLSQ LSQQEGIKM
```

Amino acid sequence for human type II TGFβ receptor[13]

Accession code: Genbank M85079

-23 MGRGLLRGLW PLHIVLWTRI AST
1 IPPHVQKSVN NDMIVTDNNG AVKFPQLCKF CDVRFSTCDN QKSCMS**N**CSI
51 TSICEKPQEV CVAVWRKNDE **N**ITLETVCHD PKLPYHDFIL EDAASPKCIM
101 KEKKKPGETF FMCSCSSDEC NDNIIFSEEY **N**TSNPDLLLV IFQVTGISLL
151 PPLGVAISVI IIFYCYRVNR QQKLSSTWET GKTRKLMEFS EHCAIILEDD
201 RSDISSTCAN NINHNTELLP IELDTLVGKG RFAEVYKAKL KQNTSEQFET
251 VAVKIFPYEE YASWKDRKDI FSDINLKHEN ILQFLTAEER KTELGKQYWL
301 ITAFHAKGNL QEYLTRHVIS WEDLRNVGSS LARGLSHLHS DHTPCGRPKM
351 PIVHRDLKSS NILVKNDLTC CLCDFGLSLR LGPYSSVDDL ANSGQVGTAR
401 YMAPEVLESR MNLENAESFK QTDVYSMALV LWEMTSRCNA VGEVKDYEPP
451 FGSKVRDPVV ESMKDNVLRD RGTRNSSFWL NHQGIQMVCE TLTECWDHDP
501 EARLTAQCVA ERFSELEHLD RLSGRSCSEE KIPEDGSLNT TK

Potential phosphorylation sites at Ser263, 378, 418, 462 and Thr180, 183, 398, 435 and 540.

Amino acid sequence for mouse type I TGFβ receptor[16]

Accession code: Genbank L15436

-17 MVDGVMILPV LMMMAFP
1 SPSVEDEKPK VNQKLYMCVC EGLSCGNEDH CEGQQCFSSL SIYDGFHVYQ
51 KGCFQVYEQG KMTCKTPPSP GQAVECCQGD WCNR**N**ITAQL PTKGKSFPGT
101 QNFHLEVGLI ILSVVFAVCL LACILGVALR KFKRRNQERL NPRDVEYGTI
151 EGLITTNVGD STLAELLDHS CTSGSGSGLP FLVQRTVARQ ITLLECVGKG
201 RYGEVWRGSW QGENVAVKIF SSRDEKSWFR ETELYNTVML RHENILGFIA
251 SDMTSRHSST QLWLITHYHE MGSLYDYLQL TTLDTVSCLR IVLSIASGLA
301 HLHIEIFGTQ GKSAIAHRDL KSKNILVKKN GQCCIADLGL AVMHSQSTNQ
351 LDVGNNPRVG TKRYMAPEVL DETIQVDCFD SYKRVDIWAF GLVLWEVARR
401 MVSNGIVEDY KPPFYDVVPN DPSFEDMRKV VCVDQQRPNI PNRWFSDPTL
451 TSLAKLMKEC WYQNPSARLT ALRIKKTLTK IDNSLDKLKT DC

Potential phosphorylation sites at Ser222 and 255.

Amino acid sequence for rat type III TGFβ receptor (β-glycan)[14,15]

Accession code: Genbank M80784 and M77809

-24 MAVTSHHMIP VMVVLMSACL ATAG
1 PEPSTRCELS PI**N**ASHPVQA LMESFTVLSG CASRGTTGLP REVHVLNLBS
51 TDQGPGQRQR EVTLHLNPIA SVHTHHKPIV FLLNSPQPLV WHLKTERLAA
101 GVPBLFLVSE GSVVQFPSG**N** FSLTAETEER NFPQENEHLL RWAQKEYGAV
151 TSFTELKIAR NIYIKVGEDQ VFPPTCNIGK NFLSLNYLAE YLQPKAAEGC

201 VLPSQPHEKE VHIIELITPS SNPYSAFQVD IIVDIRPAQE DPEVVKNLVL
251 ILKCKKSVNW VIKSFDVKGN LKVIAPNSIG FGKESERSMT MTKLVRDDIP
301 STQENLMKWA LDNGYRPVTS YTMAPVANRF HLRLENNEEM RDEEVHTIPP
351 ELRILLDPHD PPALDNPLFP GEGSPNGGLP FPFPDIPRRG WKEGEDRIPR
401 PKQPIVPSVQ LLPDHREPEE VQGGVDIALS VKCDHEKMVV AVDKDSFQTN
451 GYSGMELTLL DPSCKAKM**N**G THFVLESPLN GCGTRHRRST PDGVVYYNSI
501 VVQAPSPGDS SGWPDGYEDL ESGDNGFPGD GDEGETAPLS RAGVVVF**N**CS
551 LRQLR**N**PSGF QGQLDG**N**ATF NMELYNTDLF LVPSPGVFSV AENEHVYVEV
601 SVTKADQDLG FAIQTCFLSP YSNPDRMSDY TIIENICPKD DSVKFYSSKR
651 VHFPIPHAEV DKKRFSFLFK SVF**N**TSLLFL HCELTLCSRK KGSLKLPRCV
701 TPDDACTSLD ATMIWTMMQM KKTFTKPLAV VLQVDYKENV PSTKDSSPIP
751 PPPPQIFHGL DTLTVMGIAF AAFVIGALLT GALWYIYSHT GETARRQQVP
801 TSPPASENSS AAHSIGSTQS TPCSSSSTA

Potential glycosaminoglycan sites at Ser29, 118, 511, 522, 558.

References

[1] Roberts, A.B. and Sporn, M.B. (1990) In Handbook of Experimental Pharmacology, Vol. 65, Sporn, M.B. and Roberts, A.B. eds, Springer-Verlag, Heidelberg, pp. 419–472.
[2] Brown, P.D. et al. (1990) Growth Factors 3, 35–43.
[3] Kanzaki, T. et al. (1990) Cell 61, 1051–1061.
[4] Yamaguchi, Y. et al. (1990) Nature 346, 281–284.
[5] Derynck, R. et al. (1985) Nature 316, 701–705.
[6] De Martin, R. et al. (1987) EMBO J. 6, 3673–3677.
[7] Derynck, R. et al. (1988) EMBO J. 7, 3737–3743.
[8] Dennis, P.A. and Rifkin, D.B. (1991) Proc. Natl Acad. Sci. USA 88, 580–584.
[9] Daopin, S. et al. (1992) Science 257, 369–373.
[10] Derynck, R. et al. (1987) Nucleic Acids Res. 15, 3188–3189.
[11] Massague, J. (1992) Cell 69, 1067–1070.
[12] Franzen, P. et al. (1993) Cell 75, 681–692.
[13] Lin, H.Y. et al. (1992) Cell 68, 775–785.
[14] Lopez-Casillas, F. et al. (1991) Cell 67, 785–795.
[15] Wang, X.-F. et al. (1991) Cell 67, 797–805.
[16] Ebner, R. et al. (1993) Science 260, 1344–1348.
[17] Cheifetz, S. et al. (1992) J. Biol. Chem. 267, 19027–19030.
[18] Attisano, L. et al. (1993) Cell 75, 671–680.
[19] Wrana, J.L. et al. (1993) Cell 71, 1003–1014.

Other names
Tumour necrosis factor (TNF), cachectin, macrophage cytotoxin, necrosin, cytotoxin, haemorrhagic factor, macrophage cytotoxic factor, differentiation-inducing factor.

THE MOLECULE

Tumour necrosis factor α (TNFα) is a potent paracrine and endocrine mediator of inflammatory and immune functions. It is also known to regulate growth and differentiation of a wide variety of cells types. TNFα is selectively cytotoxic for many transformed cells, especially in combination with IFNα. *In vivo*, it leads to necrosis of methylcholanthrene-induced murine sarcomas. Many of the actions of TNFα occur in combination with other cytokines as part of the cytokine network [1–3]. TNFα is expressed as a type II membrane protein attached by a signal anchor transmembrane domain in the propeptide, and is processed by a matrix metalloproteinase, termed TNFα-converting enzyme (TACE)[4].

Crossreactivity
There is 79% homology between human and mouse TNFα and significant cross-species reactivity. Human TNF binds to mouse p55 receptor but not to mouse p75 receptor. Mouse TNF binds to both human receptors.

Sources
TNFα is secreted by activated monocytes and macrophages, and many other cells, including B cells, T cells and fibroblasts.

Bioassays
Cytotoxicity on murine fibroblast lines L929 or L-M. Assay is faster and more sensitive in the presence of 0.1 μg/ml of actinomycin D. Specific neutralizing antibodies can be used to distinguish between TNFα and TNFβ.

Physicochemical properties of TNFα

Property	Human	Mouse
pI	5.6	5.6
Amino acids – precursor	233	235
– mature[a]	157	156
M_r (K) – predicted	17.4	17.3
– expressed[b]	52	18–150[c]
Potential *N*-linked glycosylation sites	0	1
Disulfide bonds[d]	1	1

[a] Processing is by proteolytic cleavage of an atypical signal/propeptide of 76 residues in human TNFα and 79 residues in mouse TNFα. The unprocessed pro-form of TNF-α is expressed as a type II membrane protein by a signal anchor domain in the propeptide.
[b] TNFα is normally secreted as a homotrimer. Monomeric TNF is not biologically active.
[c] Differential processing of the murine propeptide and glycosylation results in several higher molecular weight isoforms.
[d] The disulfide bond is not required for biological activity.

3-D structure

TNFα exists as a homotrimer characterized by edge-to-face association of the antiparallel sandwich structure of the wedge-shaped monomers. The tertiary structure is very similar to the so-called 'jelly roll' motif of some plant and animal virus capsids[5].

Gene structure[6,7]

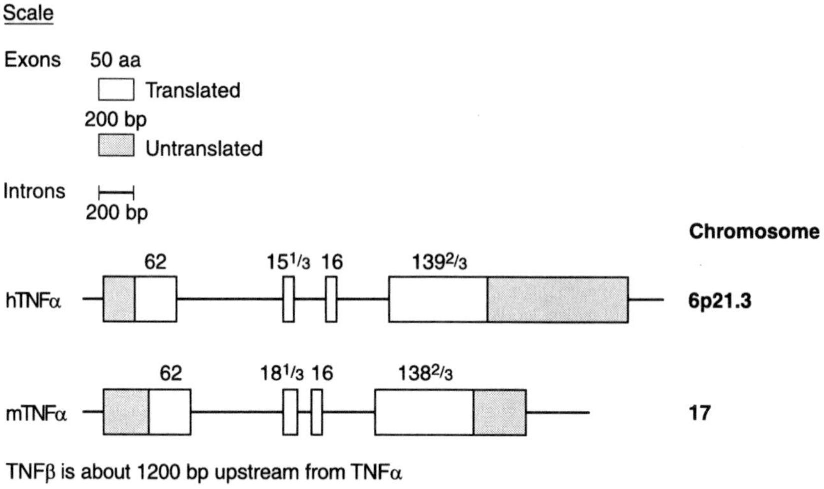

Scale

Exons 50 aa

Translated

200 bp

Untranslated

Introns

200 bp

Chromosome

hTNFα — 62 — $15^{1/3}$ — 16 — $139^{2/3}$ — 6p21.3

mTNFα — 62 — $18^{1/3}$ — 16 — $138^{2/3}$ — 17

TNFβ is about 1200 bp upstream from TNFα

The gene for TNFα is located on human chromosome 6p21.3 and contains four coding exons of 62, $15^{1/3}$, 16 and $139^{2/3}$ amino acids. The mouse gene also has four exons of 62, $18^{1/3}$, 16 and $138^{2/3}$ amino acids and is located on chromosome 17. TNFβ is ~1200 bp upstream of TNFα.

Amino acid sequence for human TNFα[8]

Accession code: SwissProt P01375

```
-76  MSTESMIRDV ELAEEALPKK TGGPQGSRRC LFLSLFSFLI VAGATTLFCL
-26  LHFGVIGPQR EEFPRDLSLI SPLAQA
  1  VRSSSRTPSD KPVAHVVANP QAEGQLQWLN RRANALLANG VELRDNQLVV
 51  PSEGLYLIYS QVLFKGQGCP STHVLLTHTI SRIAVSYQTK VNLLSAIKSP
101  CQRETPEGAE AKPWYEPIYL GGVFQLEKGD RLSAEINRPD YLDFAESGQV
151  YFGIIAL
```

Conflicting sequence F→S at position 14. Disulfide bond between Cys69–101. Signal anchor sequence 41–21 (underlined). Myristylation on Lys58/57.

Amino acid sequence for mouse TNFα[9]

Accession code: SwissProt P06804

```
-79  MSTESMIRDV ELAEEALPQK MGGFQNSRRC LCLSLFSFLL VAGATTLFCL
-29  LNFGVIGPQR DEKFPNGLPL ISSMAQTLT
  1  LRSSSQNSSD KPVAHVVANH QVEEQLEWLS QRANALLANG MDLKDNQLVV
 51  PADGLYLVYS QVLFKGQGCP DYVLLTHTVS RFAISYQEKV NLLSAVKSPC
101  PKDTPEGAEL KPWYEPIYLG GVFQLEKGDQ LSAEVNLPKY LDFAESGQVY
151  FGVIAL
```

Conflicting sequence G→R at position 152. Disulfide bond between Cys69–100 (by similarity). Signal membrane anchor sequence 44–24 (underlined). Alternative terminus at L −10 reported.

THE TNF RECEPTORS

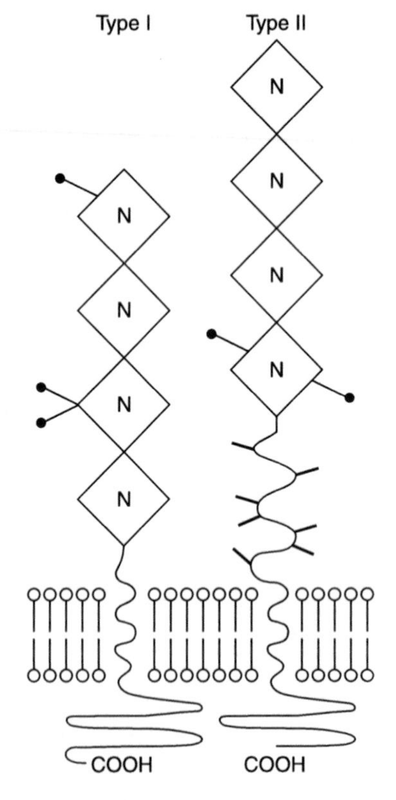

There are two receptors for TNF. The type I receptor (CD120a) has an M_r of 55 000 and the type II receptor (CD120b) has an M_r of 75 000. Both receptors bind TNFα and TNFβ (lymphotoxin). The mouse p75 receptor does not bind human TNFα which may explain some cases of nonspecies crossreactivity. Both receptors are members of the TNFR superfamily with four Cys-rich repeats in the extracellular domain (see Chapter 4). The two receptors are < 25% identical and have no more homology to each other than to other members of the superfamily. There is no significant homology between the intracellular domains of the two TNF receptors, indicating different signalling mechanisms[10]. Soluble forms of both the human p55 and p75 receptors have been found in serum of cancer patients and in urine. The soluble receptors are derived from the extracellular domain of each receptor as indicated on the sequence and are thought to act as inhibitors of TNF action[11]. Many viruses encode soluble protein related to the TNF receptor[12] (see Chapter 3).

Distribution

TNF receptors are present on nearly all cell types with few exceptions such as erythrocytes and resting T cells. The type I p55 receptor is found on most cell types

	p55 (type I)		p75 (type II)	
Property	Human	Mouse	Human	Mouse
Amino acids – precursor	455	454	461	474
– mature[a]	426	425	439	452
M_r (K) – predicted	48.3	48	46.1	47.9
– expressed	55	55	75–80	75
Potential *N*-linked glycosylation sites[b]	3	3	2	2
Affinity K_d (M)	5×10^{-10}	2×10^{-10}	10^{-10}	5×10^{-11}

[a] After removal of predicted signal peptide.
[b] The p75 receptor is also *O*-glycosylated.

whereas the type II p75 receptor seems more restricted to haematopoietic cells.

Physicochemical properties of the TNF receptors

Signal transduction

Receptor crosslinking by the TNF trimer is important for signal transduction. Signal transduction by the type I TNFR leads in some transformed cell types to cell death (by apoptosis and/or necrosis), but in most cell types TNF increases the expression of a large number of proinflammatory genes. Signalling occurs by a series of protein–protein interactions that lead to activation of caspase-8 or caspase-2 in the case of the apoptotic pathway, or protein kinases in the case of inflammatory gene expression. The death domain of the type I TNFR mediates signalling and recruits the adapter protein TNF receptor-associated death domain (TRADD). For apoptosis signalling, TRADD recruits Fas-associated death domain (FADD), which via its death effector domain recruits and activates caspase-8[13]. This triggers an apoptotic cascade. TRADD can also recruit a protein kinase termed RIP (receptor interacting protein), which in turn recruits another adapter termed RAIDD[14]. This has a death effector domain which recuits caspase-2, again promoting apoptosis. RIP can also couple via an as yet unknown kinase to the I-κB kinase (IKK) complex, leading to the activation of the central proinflammatory transcription factor NFκB[14]. This process can also be activated by another adapter, Traf-2, which activates the IKK complex, although again the linking kinase is not known[15]. Traf-2 and RIP are also implicated in the activation of Jun N-terminal kinase and p38 MAP kinase by TNF[16]. These kinases acts to stabilize mRNAs induced by TNF which contain AU repeats in their 3′ untranslated regions[17]. The role of the type II TNFR in signalling is not fully understood. Because it lacks a death domain its signalling will differ from the p55 TNFR[10]. Mice with the type I TNFR gene deleted are resistant to TNF-mediated toxicity and susceptible to infection by *Listeria monocytogenes*[18].

Chromosomal location

The human type I receptor is on chromosome 12p13 and the type II receptor is on chromosome 1p36-p32.

Amino acid sequence for human TNF type I (p55) receptor[19,20]

Accession code: SwissProt P19438

```
-21  MGLSTVPDLL LPLVLLELLV G
  1  IYPSGVIGLV PHLGDREKRD SVCPQGKYIH PQNNSICCTK CHKGTYLYND
 51  CPGPGQDTDC RECESGSFTA SENHLRHCLS CSKCRKEMGQ VEISSCTVDR
101  DTVCGCRKNQ YRHYWSENLF QCFNCSLCLN GTVHLSCQEK QNTVCTCHAG
151  FFLRENECVS CSNCKKSLEC TKLCLPQIEN VKGTEDSGTT VLLPLVIFFG
201  LCLLSLLFIG LMYRYQRWKS KLYSIVCGKS TPEKEGELEG TTTKPLAPNP
251  SFSPTPGFTP TLGFSPVPSS TFTSSSTYTP GDCPNFAAPR REVAPPYQGA
301  DPILATALAS DPIPNPLQKW EDSAHKPQSL DTDDPATLYA VVENVPPLRW
351  KEFVRRLGLS DHEIDRLELQ NGRCLREAQY SMLATWRRRT PRREATLELL
401  GRVLRDMDLL GCLEDIEEAL CGPAALPPAP SLLR
```

Conflicting sequence Pro391 missing, GPAA→APP at positions 422–425. Soluble receptor amino acids 20–180.

Amino acid sequence for human TNF type II (p75) receptor[21]

Accession code: SwissProt P20333

```
-22  MAPVAVWAAL AVGLELWAAA HA
  1  LPAQVAFTPY APEPGSTCRL REYYDQTAQM CCSKCSPGQH AKVFCTKTSD
 51  TVCDSCEDST YTQLWNWVPE CLSCGSRCSS DQVETQACTR EQNRICTCRP
101  GWYCALSKQE GCRLCAPLRK CRPGFGVARP GTETSDVVCK PCAPGTFSNT
151  TSSTDICRPH QICNVVAIPG NASMDAVCTS TSPTRSMAPG AVHLPQPVST
201  RSQHTQPTPE PSTAPSTSFL LPMGPSPPAE GSTGDFALPV GLIVGVTALG
251  LLIIGVVNCV IMTQVKKKPL CLQREAKVPH LPADKARGTQ GPEQQHLLIT
301  APSSSSSSLE SSASALDRRA PTRNQPQAPG VEASGAGEAR ASTGSSDSSP
351  GGHGTQVNVT CIVNVCSSSD HSSQCSSQAS STMGDTDSSP SESPKDEQVP
401  FSKEECAFRS QLETPETLLG STEEKPLPLG VPDAGMKPS
```

Amino acid sequence for mouse TNF type I (p55) receptor[22]

Accession code: SwissProt P25118

```
-21  MGLPTVPGLL LSLVLLALLM G
  1  IHPSGVTGLV PSLGDREKRD SLCPQGKYVH SKNNSICCTK CHKGTYLVSD
 51  CPSPGRDTVC RECEKGTFTA SQNYLRQCLS CKTCRKEMSQ VEISPCQADK
101  DTVCGCKENQ FQRYLSETHF QCVDCSPCFN GTVTIPCKET QNTVCNCHAG
151  FFLRESECVP CSHCKKNEEC MKLCLPPPLA NVTNPQDSGT AVLLPLVILL
201  GLCLLSFIFI SLMCRYPRWR PEVYSIICRD PVPVKEEKAG KPLTPAPSPA
251  FSPTSGFNPT LGFSTPGFSS PVSSTPISPI FGPSNWHFMP PVSEVVPTQG
301  ADPLLYESLC SVPAPTSVQK WEDSAHPQRP DNADLAILYA VVDGVPPARW
351  KEFMRFMGLS EHEIERLEMQ NGRCLREAQY SMLEAWRRRT PRHEDTLEVV
401  GLVLSKMNLA GCLENILEAL RNPAPSSTTR LPR
```

Amino acid sequence for mouse TNF type II (p75) receptor[22]

Accession code: SwissProt P25119

```
-22  MAPAALWVAL VFELQLWATG HT
  1  VPAQVVLTPY KPEPGYECQI SQEYYDRKAQ MCCAKCPPGQ YVKHFCNKTS
 51  DTVCADCEAS MYTQVWNQFR TCLSCSSSCT TDQVEIRACT KQQNRVCACE
101  AGRYCALKTH SGSCRQCMRL SKCGPGFGVA SSRAPNGNVL CKACAPGTFS
151  DTTSSTDVCR PHRICSILAI PGNASTDAVC APESPTLSAI PRTLYVSQPE
201  PTRSQPLDQE PGPSQTPSIL TSLGSTPIIE QSTKGGISLP IGLIVGVTSL
251  GLLMLGLVNC IILVQRKKKP SCLQRDAKVP HVPDEKSQDA VGLEQQHLLT
301  TAPSSSSSSL ESSASAGDRR APPGGHPQAR VMAEAQGFQE ARASSRISDS
351  SHGSHGTHVN VTCIVNVCSS SDHSSQCSSQ ASATVGDPDA KPSASPKDEQ
401  VPFSQEECPS QSPCETTETL QSHEKPLPLG VPDMGMKPSQ AGWFDQIAVK
451  VA
```

References

1. Manogue, K.R. et al. (1991) In The Cytokine Handbook, Thomson, A.W. ed., Academic Press, London, pp. 241–256.
2. Fiers, W. (1991) FEBS 285, 199–212.
3. Ruddle, N.H. (1992) Curr. Opin. Immunol. 4, 327–332.
4. Gearing, A.J.H. et al. (1994) Nature 370, 555–557.
5. Jones, E.Y. et al. (1989) Nature 338, 225–228.
6. Nedwin, G.E. et al. (1985) Nucleic Acids Res. 13, 6361–6373.
7. Semon, D. et al. (1987) Nucleic Acids Res. 15, 9083–9084.
8. Pennica, D. et al. (1984) Nature 312, 724–729.
9. Pennica, D. et al. (1985) Proc. Natl Acad. Sci. USA 82, 6060–6064.
10. Tartaglia, L.A. and Goeddel, D.V. (1992) Immunol. Today 13, 151–153.
11. Nophar, Y. et al. (1990) EMBO J. 9, 3269–3278.
12. Upton, C. et al. (1991) Virology 184, 370–382.
13. Hsu, H. et al. (1996) Cell 84, 299–308.
14. Stanger, B.Z. et al. (1995) Cell 81, 513–523.
15. Rothe, M. et al. (1995) Science 269, 1424–1427.
16. Wallach, D. et al. (1999) Annu. Rev. Immunol. 17, 331–367.

[17] Ridley, S.H. et al. (1997) J. Immunol. 158, 3165–3173.
[18] Rothe, J. et al. (1993) Nature 364, 798–801.
[19] Loetscher, H. et al. (1990) Cell 61, 351–359.
[20] Schall, T.J. et al. (1990) Cell 61, 361–370.
[21] Smith, C.A. et al. (1990) Science 248, 1019–1023.
[22] Lewis, M. et al. (1991) Proc. Natl Acad. Sci USA 88, 2830–2834.

Other names
Megakaryocyte colony-stimulating factor, c-MPL ligand.

THE MOLECULE

Thrombopoietin (Tpo) is a megakaryocytic lineage-specific growth and differentiation factor[1–3]. It acts in an analogous fashion to erythropoietin (Epo), functioning as a circulating regulator of platelet numbers[1,5].

Crossreactivity
Human Tpo is active on murine cells and vice versa[2–4]. There is 23% identity between the first 153 amino acids of Tpo and erythropoietin.

Sources
Serum from aplastic individuals, liver, kidney and skeletal muscle.

Bioassays
Generation of megakaryocytes/megakaryocytic colonies from bone marrow cultures[2–5].

Physicochemical properties of thrombopoietin

Property	Human	Mouse
Amino acids – precursor	353	356
– mature[a]	332	335
M_r (K) – predicted	38	35
– expressed[b]	60	?
Potential *N*-linked glycosylation sites	6	7
Disulfide bonds[c]	2	2

[a] The mature protein may be proteolytically processed to give a minimal functional unit corresponding to the N-terminal, Epo-like domain. Dibasic (RR) cleavage sites are present in the C-terminal domain. Both mature protein and the Epo-like domain are biologically active.
[b] The M_r 60 000 form represents the fully glycosylated mature protein. Other species of human Tpo of M_r 18 000, 28 000 and 30 000 have been described.
[c] By homology to erythropoietin, the first and last cysteines may form a critical disulfide bond.

3-D structure
The structure of the N-terminal 156 amino acids could be modelled on erythropoietin.

Gene structure
No information is available.

Amino acid sequence for human thrombopoietin[2]

Accession code: SwissProt P40225

```
-21  MQLTQLLLVV MLLLTARLTL S
  1  SPAPPACDLR VLSKLLRDSH VLHSRLSQCP EVHPLPTPVL LPAVDFLGES
 51  WKTQMEETKA QDILGAVTLL LEGVMAARGQ LGPTCLSSLL GQLSGQVRLL
101  LGALQSLLGT QLPPQGRTTA HKDPMAIFLS FQHLLRGKVR FLMLVGGSTL
151  CVRRAPPTTA VPSRTSLVLT LNELPNRTSG LLETNFTASA RTTGSGLLKW
201  QQGFRAKIPG LLNQTSRSLD QIPGYLNRIH ELLNGTRGLF PGPSRRTLGA
251  PDISSGTSDT GSLPPNLQPG YSPSPTHPPT GQYTLFPLPP TLPTPVVQLH
301  PLLPDPSAPT PTPTSPLLNT SYTHSQNLSQ EG
```

Epo-like domain amino acids 1–156.

Amino acid sequence for mouse thrombopoietin[3]

```
-21  MELTDLLLAA MLLAVARLTL S
  1  SPVAPACDPR LLNKLLRDSH LLHSRLSQCP DVDPLSIPVL LPAVDFSLGE
 51  WKTQTEQSKA QDILGAVSLL LEGVMAARGQ LEPSCLSSLL GQLSGQVRLL
101  LGALQGLLGT QLPLQGRTTA HKDPNALFLS LQQLLRGKVR FLLLVEGPTL
151  CVRRTLPTTA VPSSTSQLLT LNKFPNRTSG LLETNFSVTA RTAGPGLLSR
201  LQGFRVKITP GQLNQTSRSP VQISGYLNRT HGPVNGTHGL FAGTSLQTLE
251  ASDISPGAFN KGSLAFNLQG GLPPSPSLAP DGHTPFPPSP ALPTTHGSPP
301  QLHPLFPDPS TTMPNSTAPH PVTMYPHPRN LSQET
```

Epo-like domain amino acids 1–157.

THE THROMBOPOIETIN RECEPTOR (c-*mpl*)

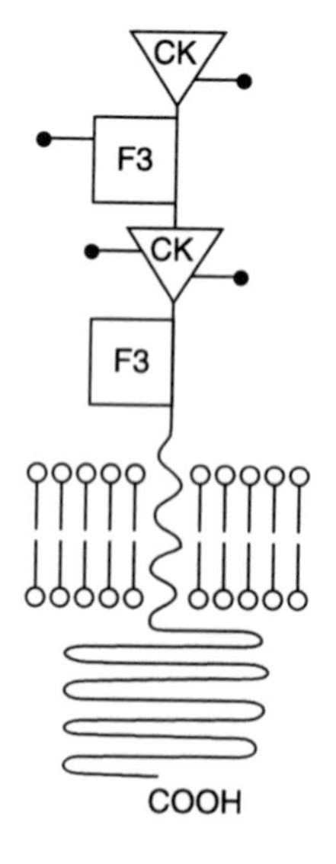

The thrombopoietin receptor (also known as c-*mpl*) is a member of the cytokine receptor superfamily with two extracellular segments each with a CKR and a FNIII domain containing a WSXWS (or WGXWS) motif similar to the LIFR and the common β-subunit of the IL-3, IL-5 and GM-CSF receptors. A soluble form of the mouse receptor has been identified[6]. A portion of the c-*mpl* (thrombopoietin receptor) gene has been found fused to viral sequences encoding the envelope protein of a mutant strain of Friend leukaemia virus called myeloproliferative leukaemia virus (MPLV)[7]. The viral oncogene v-*mpl* includes the entire cytoplasmic and transmembrane domains of the c-*mpl* gene and 40 amino acids including the WSXWS motif of the extracellular domain. The remainder of the extracellular domain is replaced by viral envelope sequences.

Distribution

Megakaryocytes and megakaryocyte precursors, platelets. c-*mpl* mRNA is found in BAF3 and HEL cell lines.

Physicochemical properties of the thrombopoietin receptor (c-*mpl*)

Property	Human	Mouse
Amino acids – precursor	635 (579)[a]	625 (457)[a]
– mature[b]	610 (554)	600 (432)
M_r (K) – predicted	68.6 (62.8)	67.1 (48.5)
– expressed	?	78 (55)
Potential *N*-linked glycosylation sites	4	4

[a] Truncated form in parenthesis.
[b] After removal of predicted signal peptide.

Signal transduction

The mechanism of signal transduction is not known.

Chromosomal location

The human thrombopoietin receptor is on chromosome 1p34, and the mouse receptor is on the D band of chromosome 4.

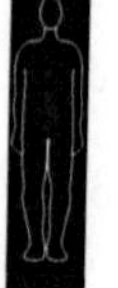

Amino acid sequence for human thrombopoietin receptor (c-*mpl*)[8]

Accession code: GenEMBL M90102

```
-25  MPSWALFMVT SCLLLAPQNL AQVSS
  1  QDVSLLASDS EPLKCFSRTF EDLTCFWDEE EAAPSGTYQL LYAYPREKPR
 51  ACPLSSQSMP HFGTRYVCQF PDQEEVRLFF PLHLWVKNVF LNQTRTQRVL
101  FVDSVGLPAP PSIIKAMGGS QPGELQISWE EPAPEISDFL RYELRYGPRD
151  PKNSTGPTVI QLIATETCCP ALQRPHSASA LDQSPCAQPT MPWQDGPKQT
201  SPSREASALT AEGGSCLISG LQPGNSYWLQ LRSEPDGISL GGSWGSWSLP
251  VTVDLPGDAV ALGLQCFTLD LKNVTCQWQQ QDHASSQGFF YHSRARCCPR
301  DRYPIWENCE EEEKTNPGLQ TPQFSRCHFK SRNDSIIHIL VEVTTAPGTV
351  HSYLGSPFWI HQAVRLPTPN LHWREISSGH LELEWQHPSS WAAQETCYQL
401  RYTGEGHQDW KVLEPPLGAR GGTLELRPRS RYRLQLRARL NGPTYQGPWS
451  SWSDPTRVET ATETAWISLV TALHLVLGLS AVLGLLLLRW QFPAHYRRLR
501  HALWPSLPDL HRVLGQYLRD TAALSPPKAT VSDTCEEVEP SLLEILPKSS
551  ERTPLPLCSS QAQMDYRRLQ PSCLGTMPLS VCPPMAESGS CCTTHIANHS
601  YLPLSYWQQP
```

A second mRNA species GenBank M90103 predicts a truncated form with a cytoplasmic domain of 66 amino acids compared with 122 amino acids for the full-length form. The truncated form is identical to the full-length form in the extracellular and transmembrane domains and the first nine amino acids of the cytoplasmic domain.

Amino acid sequence for mouse thrombopoietin receptor (c-*mpl*)[6,9]

Accession code: GenEMBL Z22649

```
-25  MPSWALFMVT SCLLLALPNQ AQVTS
  1  QDVFLLALGT EPLNCFSQTF EDLTCFWDEE EAAPSGTYQL LYAYRGEKPR
 51  ACPLYSQSVP TFGTRYVCQF PAQDEVRLFF PLHLWVKNVS LNQTLIQRVL
101  FVDSVGLPAP PRVIKARGGS QPGELQIHWE APAPEISDFL RHELRYGPTD
151  SSNATAPSVI QLLSTETCCP TLWMPNPVPV LDQPPCVHPT ASQPHGPAPF
201  LTVKGGSCLV SGLQASKSYW LQLRSQPDGV SLRGSWGPWS FPVTVDLPGD
251  AVTIGLQCFT LDLKMVTCQW QQQDRTSSQG FFRHSRTRCC PTDRDPTWEK
301  CEEEEPRPGS QPALVSRCHF KSRNDSVIHI LVEVTTAQGA VHSYLGSPFW
351  IHQAVLLPTP SLHWREVSSG RLELEWQHQS SWAAQETCYQ LRYTGEGRED
401  WKVLEPSLGA RGGTLELRPR ARYSLQLRAR LNGPTYQGPW SAWSPPARVS
451  TGSETAWITL VTALLLVLSL SALLGLLLLK WQFPAHYRRL RHALWPSLPD
501  LHRVLGQYLR DTAALSPSKA TVTDSCEEVE PSLLEILPKS SESTPLPLCP
551  SQPQMDYRGL QPCLRTMPLS VCPPMAETGS CCTTHIANHS YLPLSYWQQP
```

A truncated form of murine thrombopoietin with a 257-bp deletion coding for 55 amino acids of the extracellular domain including the WSXWS motif, the transmembrane domain and the first eight amino acids of the intracellular domain has also been identified (amino acids 402–488). The deletion generates a frameshift and terminates after a further 30 amino acids.

References

[1] McDonald, T.P. (1988) Exp. Hematol. 16, 201–205.
[2] De Sauvage, F.J. et al. (1994) Nature 369, 533–538.
[3] Lok, S. et al. (1994) Nature 369, 565–568.
[4] Kaushanksky, K. et al. (1994) Nature 369, 568–571.
[5] Wendling, F. et al. (1994) Nature 369, 571–574.
[6] Skoda, R.C. et al. (1993) EMBO. J. 12, 2645–2653.
[7] Souyri, M. et al. (1990) Cell 63, 1137–1147.
[8] Vignon, I. et al. (1992) Proc. Natl Acad. Sci. USA 89, 5640–5644.
[9] Vignon, I. et al. (1993) Oncogene 8, 2607–2615.

Other names

RANKL (receptor activator of NFκB ligand), OPGL (osteoprotegerin ligand), ODF (osteoclast differentiation factor) and TNFSF-11.

THE MOLECULE

TNF-related activation-induced cytokine (TRANCE) is a member of the TNF superfamily, which was independently cloned by three groups. Initially, TRANCE was cloned during a search for apoptosis-regulatory genes using a somatic cell approach in T-cell hybridomas[1]. Subsequently, RANKL and its cognate receptor RANK were cloned independently and shown to augment the ability of dendritic cells to stimulate naïve T-cell proliferation[2]. In 1998, RANK and RANKL were cloned using a different approached by a group aiming to characterize the protein responsible for osteoclast differentiation using osteoprotegerin (OPG) as a probe. This factor, termed OPG ligand or osteoclast differentiation factor (ODF), was subsequently shown to be identical to the previously characterized RANKL and TRANCE[3]. TRANCE is a type II transmembrane protein that is displayed on the cell surface. The TNFα convertase TACE has been implicated as a cell-associated protease capable of cleaving membrane-bound TRANCE[4]. TRANCE is found to be rapidly cleaved off the surface of activated T cells and in T-cell leukaemia[5]. TRANCE therefore plays a role in the interaction of T cells and antigen-presenting cells during the generation of immune responses and in regulating bone resorption processes[2,3,6].

Crossreactivity

Human and mouse TRANCE are 85% identical[2,7]. Both human and mouse TRANCE act on mouse cells and murine TRANCE can signal through human RANK.

Sources

Cellular sources of TRANCE include activated T cells, bone marrow precursor cells (myeloid), stromal fibroblasts and synovial cells and osteoblasts.

Bioassays

Stimulation of alloreactive T-cell proliferation using dendritic cells[1,2]. Induction of osteoclast differentiation[7].

Physicochemical properties of TRANCE

Property	Human	Mouse
pI[a]	7.25	7.76
Amino acids – precursor	317	316
– mature[b]	317/177	316/177
M_r (K) – predicted	35.47	35.4
– expressed	45	45
Potential *N*-linked glycosylation sites	2	2
Disulfide bonds	0	0

[a] Predicted.
[b] Human and mouse TRANCE lack signal peptides, however a 177-amino-acid form is shed.

3-D structure

The crystal structure for TRANCE is unknown, but is predicted to contain all 10 β-sheaths in its extracellular domain, which are folded into an antiparallel β-sandwich structure with a 'jelly roll' topology (three monomers associating about a 3-fold axis of symmetry forming a bell-shaped trimer), based on its homology with TNFα[8–10]. Probably expressed as homotrimer.

Gene structure

TRANCE is expressed on human chromosome 13q14[2,7].

Amino acid sequence for human TRANCE[2]

Accession code: SPTREMBLE O14788

1 MRRASRDYTK YLRGSEEMGG GPGAPHEGPL HAPPPPAPHQ PPAASRSMFV
51 ALLGLGLGQV VCSVALFFYF RAQMDPNRIS EDGTHCIYRI LRLHENADFQ
101 DTTLESQDTK LIPDSCRRIK QAFQGAVQKE LQHIVGSQHI RAEKAMVDGS
151 WLDLAKRSKL EAQPFAHLTI **N**ATDIPSGSH KVSLSSWYHD RGWAKIS**N**MT
201 FSNGKLIVNQ DGFYYLYANI CFRHHETSGD LATEYLQLMV YVTKTSIKIP
251 SSHTLMKGGS TKYWSGNSEF HFYSINVGGF FKLRSGEEIS IEVSNPSLLD
301 PDQDATYFGA FKVRDID

Conflicting sequence A→G at position 194[7].

Amino acid sequence for mouse TRANCE[11]

Accession code: SPTREMBLE O35235

1 MRRASRDYGK YLRSSEEMGS GPGVPHEGPL HPAPSAPAPA PPPAASRSMF
51 LALLGLGLGQ VVCSIALFLY FRAQMDPNRI SEDSTHCFYR ILRLHENAGL
101 QDSTLESEDT LPDSCRRMKQ AFQGAVQKEL QHIVGPQRFS GAPAMMEGSW
151 LDVAQRGKPE AQPFAHLTIN AASIPSGSHK VTLSSWYHDR GWAKIS**N**MTL
201 SNGKLRVNQD GFYYLYANIC FRHHETSGSV PTDYLQLMVY VVKTSIKIPS
251 SHNLMKGGST K**N**WSGNSEFH FYSINVGGFF KLRAGEEISI QVSNPSLLDP
301 DQDATYFGAF KVQDID

Conflicting sequence G→D at position 99[3].

THE TRANCE RECEPTOR, RANK

The receptor for TRANCE/RANKL was initially identified as an EST sequence obtained from a dendritic cell cDNA library[2], but which is also implicated in osteoclast differentiation and activation. RANK (also known as ODFR, ODAR (osteoclast differentiation and activation receptor), and TRANCE-R) encodes a novel TNFR-related protein which is a key regulator of bone and immune homeostasis. RANK is most closely related to CD40 and is about 40% similar in the ligand-binding domain. RANK is the largest member of the TNFR superfamily. RANK, as its name suggests, is capable of activating NFκB.

Distribution

RANK is widely expressed in intestinal epithelium, kidney, liver, bone and growth plate cartilage, lymph nodes, spleen, thymus and heart and at very high levels on osteoclast progenitor cells.

Physicochemical properties of the RANK

Property	Human	Mouse
Amino acids – precursor	616	?
– mature[a]	594	?
M_r (K) – predicted	66.03	?
– expressed	?	?
Potential *N*-linked glycosylation sites	2	
Affinity K_d (M)	3×10^{-9}	?

[a] After removal of predicted signal peptide.

Signal transduction

RANK functions as a type I transmembrane signalling molecule and associates with members of the Traf family (TNF receptor-associated factors 2, 5 and 6)[12] upon ligand binding, leading to the activation of NFκB and JNK/SAPK pathways[11]. RANK also initiates signalling pathways resulting in gene expression patterns associated with osteoclast lineages.

Chromosomal location

RANK is located on human chromosome 18q22.1.

Amino acid sequence for human RANK[2]

Accession code: SPTREMBLE Q9Y6Q6

```
-22  MAPRARRRRP LFALLLLCAL LA
  1  RLQVALQIAP PCTSEKHYEH LGRCCNKCEP GKYMSSKCTT TSDSVCLPCG
 51  PDEYLDSWNE EDKCLLHKVC DTGKALVAVV AGNSTTPRRC ACTAGYHWSQ
101  DCECCRRNTE CAPGLGAQHP LQLNKDTVCK PCLAGYFSDA FSSTDKCRPW
151  TNCTFLGKRV EHHGTEKSDA VCSSSLPARK PPNEPHVYLP GLIILLLFAS
201  VALVAAIIFG VCYRKKGKAL TANLWHWINE ACGRLSGDKE SSGDSCVSTH
251  TANFGQQGAC EGVLLLTLEE KTFPEDMCYP DQGGVCQGTC VGGGPYAQGE
301  DARMLSLVSK TEIEEDSFRQ MPTEDEYMDR PSQPTDQLLF LTEPGSKSTP
351  PFSEPLEVGE NDSLSQCFTG TQSTVGSESC NCTEPLCRTD WTPMSSENYL
401  QKEVDSGHCP HWAASPSPNW ADVCTGCRNP PGEDCEPLVG SPKRGPLPQC
451  AYGMGLPPEE EASRTEARDQ PEDGADGRLP SSARAGAGSG SSPGGQSPAS
501  GNVTGNSNST FISSGQVMNF KGDIIVVYVS QTSQEGAAAA AEPMGRPVQE
551  ETLARRDSFA GNGPRFPDPC GGPEGLREPE KASRPVQEQG GAKA
```

Potential disulfide bonds are predicted (based on similarity) between residues 34 and 46, 47 and 60, 50 and 68, 71 and 68, 71 and 86, 92 and 112, 114 and 124, 126 and 133, 127 and 151, 154 and 169, 175 and 194.

References

[1] Wong, B.R. et al. (1997) J. Exp. Med. 186, 2075–2080.
[2] Anderson, D.M. et al. (1997) Nature 390, 175–179.
[3] Yasuda, H. et al. (1998) Proc. Natl Acad. Sci. USA 95, 3597–3602.
[4] Lum, L. et al. (1999) J. Biol. Chem. 274, 13613–13618.
[5] Kong, Y.Y. et al. (1999) Nature 397, 315–323.
[6] Fuller, K. et al. (1998) J. Exp. Med. 188, 997–1001.
[7] Lacey, D.L. et al. (1998) Cell 93, 165–176.
[8] Eck, M.J. and Sprang, S.R. (1989) J. Biol. Chem. 264, 17595–17605.
[9] Eck, M.J. et al. (1992) J. Biol. Chem. 267, 2119–2122.
[10] Banner, D.W. et al. (1993) Cell 73, 431–445.
[11] Wong, B.R. et al. (1997) J. Biol. Chem. 272, 25190–25194.
[12] Arch, R.H. et al. (1998) Genes Dev. 12, 2821–2830.

Other names

None.

THE MOLECULE

Thymic stromal lymphopoietin (TSLP) was originally described as a factor derived from a novel thymic stromal cell line Z210R.1 that stimulated B-cell development from murine fetal liver[1–3]. It is functionally related to IL-7 and binds to a receptor complex that uses the IL-7R α-chain but it has no significant sequence homology with IL-7. TSLP supports B lymphopoiesis from early progenitor cells probably to a more mature stage than IL-7. It does not have growth-promoting activity on CD4–/CD8– thymocytes.

Crossreactivity

There is no crossreactivity between human and mouse TSLP.

Sources

Thymic stromal cells. TSLP mRNA is expressed in thymus, spleen, kidney, lung and bone marrow of normal mice.

Bioassays

Proliferation of the B lymphoid line NAG8/7.

Physicochemical properties of TSLP

Property	Mouse
PI (calculated)	5
Amino acids – precursor	140
– mature[a]	121
M_r (K) – predicted	14
– expressed[b]	~20 and ~24
Potential *N*-linked glycosylation sites	3
Disulfide bonds[c]	3

[a] After removal of predicted signal peptide.
[b] Two species expressed by Z210R.1 cells and transfected CV-1/EBNA due to differential glycosylation.
[c] Between Cys residues 6 and 79, 38 and 44, and 59 and 102.

3-D structure

Not known but probably similar to IL-7.

Gene structure

Not known. Mouse TSLP is on chromosome 18.

Amino acid sequence for human TSLP[4]

Has not yet been published, but a human cDNA clone resembling murine TSLP has been found[4].

Amino acid sequence for mouse TSLP[4]

Accession code: SwissProt Q9JIE6

```
-19   MVLLRSLFIL QVLVRMGLT
  1   YNFSNCNFTS ITKIYCNIIF HDLTGDLKGA KFEQIEDCES KPACLLKIEY
 51   YTLNPIPGCP SLPDKTFARR TREALNDHCP GYPETERNDG TQEMAQEVQN
101   ICLNQTSQIL RLWYSFMQSP E
```

THE TSLP RECEPTOR

TSLP binds to a receptor complex of the TLSR receptor and the IL-7R α-chain. The common γ-chain is not required for high-affinity binding of TSLP or signalling. The TLSR receptor is a type 1 transmembrane protein belonging to the haematopoietin receptor superfamily but it lacks the second of the four conserved cysteine residues present in other members of this family and instead of the characteristic WSXWS motif it has WTAVT[5,6]. It has significant homology with the common γ-chain[6].

Distribution

High-affinity TSLP binding has been found on B-cell, T-cell and macrophage cell lines. TSLP receptor mRNA is expressed widely in many different tissues.

Physicochemical properties of the TLSP receptor

Property	Mouse
Amino acids – precursor	359
– mature[a]	338
M_r (K) – predicted	35.7
– expressed	~50
Potential *N*-linked glycosylation sites	2
Affinity K_d (M)[b]	$\sim 10^{-9}$

[a] After removal of predicted signal peptide.
[b] Affinity for binding to cells expressing the TSLP receptor and the IL7Rα.

Signal transduction

Activates STAT5 but not Jak1, 2, 3 or 4[3,7].

Chromosomal location

Central region of mouse chromosome 5.

Amino acid sequence for mouse TSLP receptor[5,6]

Accession code: SwissProt Q9JIE7

```
-21  MAWALAVILL PRLLAAAAAA A
  1  AVTSRGDVTV VCHDLETVEV TWGSGPDHHG ANLSLEFRYG TGALQPCPRY
 51  FLSGAGVTSG CILPAGRAGL LELALRDGGG AMVFKARQRA SAWLKPRPPW
101  NVTLLWTPDG DVTVSWPAHS YLGLDYEVQH RESNDDEDAW QTTSGPCCDL
151  TVGGLDPVRC YDFRVRASPR AAHYGLEAQP SEWTAVTRLS GAASAASCTA
201  SPAPSPALAP PLLPLGCGLA ALLTLSLLLA ALRLRRVKDA LLPCVPDPSG
251  SFPGLFEKHH GNFQAWIADA QATAPPARTE EEDDLIHTKA KRVEPEDGTS
301  LCTVPRPPSF EPRGPGGGAM VSVGGATFMV GDSGYMTL
```

References

[1] Friend, S.L. (1994) Exp. Hematol. 22, 321–328.
[2] Ray, R.J. et al. (1996) Eur. J. Immunol. 26, 10–16.
[3] Levin, S.D. et al. (1999) J. Immunol. 162, 677–683.
[4] Sims, J.E. et al. (2000) J. Exp. Med. 192, 671–680.
[5] Park, L.S. et al. (2000) J. Exp. Med. 192, 659–670.
[6] Pandey, A. et al. (2000) Nature Immunol. 1, 59–64.
[7] Isaksen, D.E. et al. (1999) J. Immunol. 163, 5971–5977.

Other names

Apo3L, DR3-L (death receptor 3-L) and TNFSF-12.

THE MOLECULE

Mouse TNF-related ligand with weak ability to induce cell death (TWEAK) was cloned during the course of a study intended to clone an RNA that hybridized to an erythropoietin probe. It was named TWEAK based on its similarity to TNF and its weak apoptosis-inducing ability[1]. Human TWEAK was subsequently cloned and identified as a ligand for the death domain-containing receptor Apo3 (death receptor 3)[2]. TWEAK induces apoptosis in the presence of IFNγ[3] or very weakly in a limited set of cells[1]. It is synthesized as a membrane protein, then subsequently cleaved in the stalk region and readily secreted from cells[1]. Soluble TWEAK not only induces apoptosis but also NFκB activation in human cell lines[2] and chemokine secretion[1]. The observation that TWEAK-induced apoptosis occurs in such a limited number of cell types suggests that TWEAK-mediated cell death may occur via a weaker non-death domain-mediated pathway. In addition to Apo3 as a receptor for TWEAK, other non-death domain-containing receptors for TWEAK have also been postulated. Evidence has recently been presented suggesting that the apoptotic induction mediated by TWEAK is most likely indirect, mediated by TNF–TNFR interactions[4]. More recently, TWEAK has been shown to play a role in endothelial cell proliferation and angiogenesis, with potencies similar to that of VEGF and bFGF[5]. Evidence for the role of TWEAK in mediating CD4+ T-cell killing of antigen-presenting macrophages has also been demonstrated[6].

Crossreactivity

Mouse and human TWEAK are unusually conserved with 93% amino acid identity in the receptor-binding domain, greater than that observed for any of the other TNF superfamily members[1].

Sources

TWEAK transcripts are widely expressed and abundant in most tissues[1]. Among transformed cell lines, nonhaematopoietic cells exhibit higher level expression[1].

Bioassays

Induction of U937 cell apoptosis in presence of IFNγ[1].

Physicochemical properties of TWEAK

Property	Human	Mouse
pI[a]	9.5	7.85
Amino acids – precursor	249	225
– mature[a]	157	?
M_r (K) – predicted	27.216	24.781
– expressed	30–35	?
Potential *N*-linked glycosylation sites	1	1
Disulfide bonds	0	?

[a] During expression of recombinant TWEAK, cleavage occurs between Arg92 and Arg93, leading to a 17–18-kDa secreted form of TWEAK.

3-D structure

The crystal structure of TWEAK is unknown but is predicted to resemble that of TNF based on similarity.

Gene structure

Human and mouse TWEAK are located on chromosome 17p13 and 11 respectively. The genomic organization of TWEAK resembles that of APRIL and lies within 1 kb of the gene for APRIL, with six exons.

Amino acid sequence for human TWEAK[1]

Accession code: SPTREMBLE O43508

```
  1  MAARRSQRRR GRRGEPGTAL LVPLALGLGL ALACLGLLLA VVSLGSRASL
 51  SAQEPAQEEL VAEEDQDPSE LNPQTEESQD PAPFLNRLVR PRRSAPKGRK
101  TRARRAIAAH YEVHPRPGQD GAQAGVDGTV SGWEEARINS SSPLRYNRQI
151  GEFIVTRAGL YYLYCQVHFD EGKAVYLKLD LLVDGVLALR CLEEFSATAA
201  SSLGPQLRLC QVSGLLALRP GSSLRIRTLP WAHLKAAPFL TYFGLFQVH
```

The potential membrane anchor (transmembrane) sequence is underlined.

Amino acid sequence for mouse TWEAK[1]

Accession code: SPTREMBLE O54907

```
  1  VLSLGLALAC LGLLLVVVSL GSWATLSAQE PSQEELTAED RREPPELNPQ
 51  TEESQDVVPF LEQLVRPRRS APKGRKARPR RAIAAHYEVH PRPGQDGAQA
101  GVDGTVSGWE ETKINSSSPL RYDRQIGEFT VIRAGLYYLY CQVHFDEGKA
151  VYLKLDLLVN GVLALRCLEE FSATAASSPG PQLRLCQVSG LLPLRPGSSL
201  RIRTLPWAHL KAAPFLTYFG LFQVH
```

The potential membrane anchor (transmembrane) sequence is underlined.

THE TWEAK RECEPTOR

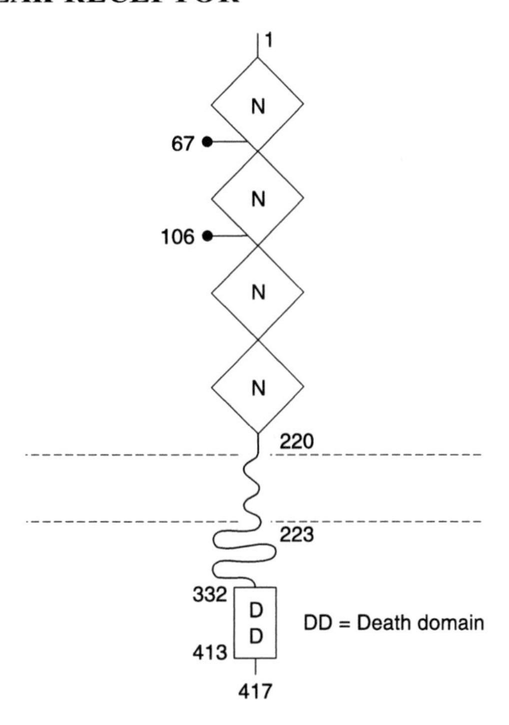

TWEAK has been shown to bind Apo3 (also known as DR3, WSL-1, TRAMP and LARD) and induce apoptosis[2], however more recent studies using both human and murine TWEAK ligand have shown that TWEAK and Apo3 do not interact in an *in vitro* binding assay and that TWEAK binds strongly to cells that do not express Apo3 on the cell surface. Biological activity of TWEAK was also observed in these cells. Finally, cells isolated from Apo3 knockout mice retain their ability to interact with TWEAK. Taken together, these results suggest that Apo3 is not the major receptor for TWEAK[7]. Additional TWEAK receptors are as yet unidentified.

Distribution

The human Apo3 gene is expressed primarily on the surface of thymocytes and lymphocytes[8]. Apo3 expression is detected in lymphocyte-rich tissues such as thymus, colon, intestine and spleen and is also found in the prostate.

Physicochemical properties of the TWEAK receptor

Property	Apo3 Human	Mouse
Amino acids – precursor	417	?
– mature[a]	393	?
M_r (K) – predicted	45.38	?
– expressed	?	?
Potential *N*-linked glycosylation sites	2	?
Affinity K_d (M)	?	?

[a] After removal of predicted signal peptide.

Signal transduction

Apo3 is capable of inducing both NFκB activation and apoptosis when over-expressed in mammalian cells. Apo3 has been shown to recruit TRADD to mediate interactions with downstream signalling pathways. Caspase inhibitors and a dominant-negative mutant of the cell death adapter protein Fas-associated death domain protein (FADD/MORT1), which is critical for apoptosis induction by TNF, have been shown to block apoptosis induction by TWEAK. Dominant-negative mutants of several factors that play a key role in NFκB induction by TNF also inhibit NFκB activation by TWEAK[2].

Chromosomal location

The Apo3 gene locus is tandemly duplicated on human chromosome band 1p36.2–p36.3[8].

Amino acid sequence for human Apo3

Accession code: SwissProt P78515

```
-24  MEQRPRGCAA VAAALLLVLL GARA
  1  QGGTRSPRCD CAGDFHKKIG LFCCRGCPAG HYLKAPCTEP CGNSTCLVCP
 51  QDTFLAWENH HNSECARCQA CDEQASQVAL ENCSAVADTR CGCKPGWFVE
101  CQVSQCVSSS PFYCQPCLDC GALHRHTRLL CSRRDTDCGT CLPGFYEHGD
151  GCVSCPTSTL GSCPERCAAV CGWRQMFWVQ VLLAGLVVPL LLGATLTYTY
201  RHCWPHKPLV TADEAGMEAL TPPPATHLSP LDSAHTLLAP PDSSEKICTV
251  QLVGNSWTPG YPETQEALCP QVTWSWDQLP SRALGPAAAP TLSPESPAGS
301  PAMMLQPGPQ LYDVMDAVPA RRWKEFVRTL GLREAEIEAV EVEIGRFRDQ
351  QYEMLKRWRQ QQPAGLGAVY AALERMGLDG CVEDLRSRLQ RGP
```

Mature Apo3 is formed after removal of the signal peptide. The putative transmembrane region is underlined. Three distinct isoforms of Apo3 can be generated as a result of alternative mRNA splicing. Conflicting sequence RPR→AAA at position 4–6, P→H at position 60, P→L at position 167, A→R at position 312, R→L at position 370 and R→H at position 381. The death domain is located at position 332–413 of precursor polypeptide.

References

[1] Chicheportiche, Y. et al. (1997) J. Biol. Chem. 272, 32401–32410.
[2] Marsters, S.A. et al. (1998) Curr. Biol. 8, 525–528.
[3] Nakayama, M. et al. (2000) J. Exp. Med. 192, 1373–1380.
[4] Schneider, P. et al. (1999) Eur. J. Immunol. 29, 1785–1792.
[5] Lynch, C.N. et al. (1999) J. Biol. Chem. 274, 8455–8459.
[6] Kaplan, M.J. et al. (2000) J. Immunol. 164, 2897–2904.
[7] Kaptein, A. et al. (2000) FEBS Lett. 485, 135–141.
[8] Grenet, J. et al. (1998) Genomics 49, 385–393.

Other names

Human VEGF has been known as vascular permeability factor VPF, folliculo-stellate cell-derived growth factor and glioma-derived vascular endothelial cell mitogen.

THE MOLECULE

Vascular endothelial growth factor (VEGF) is a heparin-binding, dimeric protein related to the PDGF/sis family of growth factors[1]. It is a mitogen for endothelial cells, activates and is chemoattractant for monocytes, enhances blood vessel permeability and is a procoagulant[2,3]. A homologue of VEGF is encoded by the orf virus[4].

Crossreactivity

There is about 88% homology between human and rat VEGF, 18% homology with PDGF-B and 15% homology with PDGF-A[1]. Rodent VEGF is active on human cells and vice versa[2].

Sources

Pituitary cells, monocyte/macrophages, smooth muscle, keratinocytes[2,3].

Bioassays

Proliferation of bovine endothelial cells[5].

Physicochemical properties of VEGF

Property	Human	Rat
pI	Basic	Basic
Amino acids – precursor	215, 191, 147	190
– mature	189, 165, 121[a]	164
M_r (K) – predicted	25	22.4
– expressed[b]	45	45
Potential *N*-linked glycosylation sites	1	1
Disulfide bonds[c]	8?	8?

[a] The 121- and 165-amino-acid forms are secreted, the 189-amino-acid form is cell associated[6].
[b] The protein forms dimers, it is not known whether these are hetero- or homodimers.
[c] There are 16 Cys residues with eight possible disulfide bonds.

3-D structure

No information, likely to be similar to PDGF.

Gene structure[7]

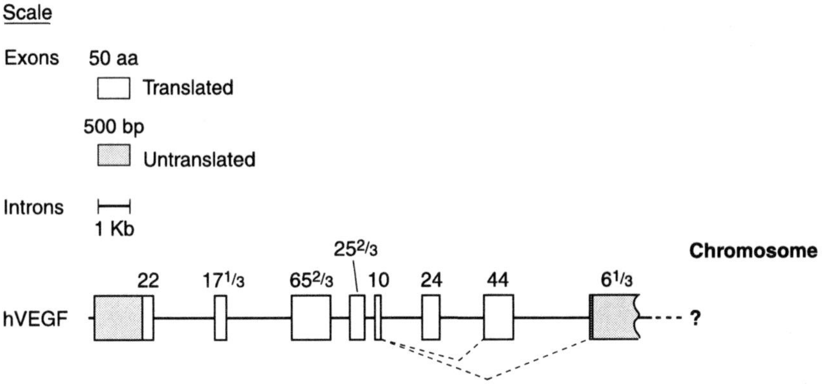
Scale
Exons 50 aa
Translated
500 bp
Untranslated
Introns
1 Kb
25 2/3
Chromosome
22
17 1/3
65 2/3
10
24
44
6 1/3
hVEGF
?

Shorter forms of VEGF are due to splicing out of exon 6 (191 aa) or exons 6 and 7 (147 aa)

The gene for human VEGF consists of 10 exons with 5′ and 3′ exons noncoding. The coding exons are of 22, $17^{1/3}$, $65^{2/3}$, $25^{2/3}$, 10, 24, 44 and $6^{1/3}$ amino acids. Shorter forms of VEGF are formed due to splicing out of exons 6 (191 amino acids) or exons 6 and 7 (147 amino acids).

Amino acid sequence for human VEGF[1,3]

Accession code: SwissProt P15692

-26 MNFLLSWVHW SLALLLYLHH AKWSQA
1 APMAEGGGQN HHEVVKFMDV YQRSYCHPIE TLVDIFQEYP DEIEYIFKPS
51 CVPLMRCGGC CNDEGLECVP TEES**N**ITMQI MRIKPHQGQH IGEMSFLQHN
101 KCECRPKKDR ARQEK*KSVRG KGKGQKRKRK KSRYKSWSVP CGPCSERRKH*
151 *LFVQDPQTCK CSCKNTDSRC KARQLELNER* TCRCDKPRR

Residues encoded by exons 6 and 7 (in italics) may be spliced out. Conflicting sequence K→N at position 116.

Amino acid sequence for rat VEGF[8]

Accession code: SwissProt P16612

```
-25  MNFLLSWVHW TLALLLYLHH AKWSQ
  1  AAPTTEGEQK AHEVVKFMDV YQRSYCRPIE TLVDIFQEYP DEIEYIFKPS
 51  CVPLMRCAGC CNDEALECVP TSESNVTMQI MRIKPHQSQH IGEMSFLQHS
101  RCECRPKKDR TKPENHCEPC SERRKHLFVQ DPQTCKCSCK NTDSRCKARQ
151  LELNERTCRC DKPRR
```

THE VEGF RECEPTORS

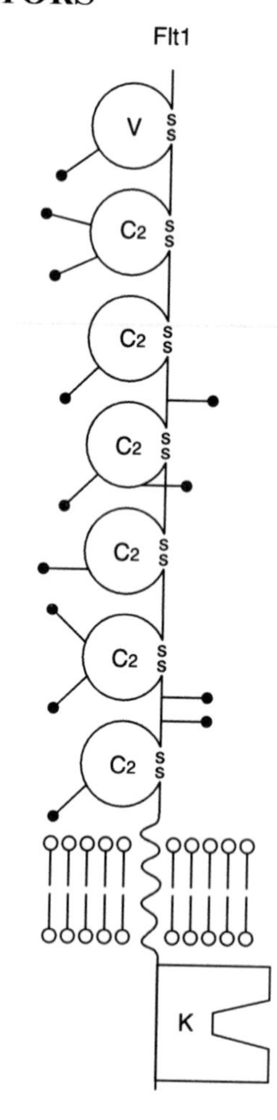

There are three published receptors for VEGF. One is the *fms*-like tyrosine kinase flt[9,10], the second receptor is the KDR gene product (the murine homologue of KDR is the flk-1 gene product)[11,12] and the third receptor is the flt4 gene product[13]. All are related to the PDGF receptor/M-CSF receptor/c-*kit* family of class III tyrosine kinases, but contain seven extracellular immunoglobulin domains, and an intracellular tyrosine kinase domain with a kinase insert. The receptors are thought to dimerize in the presence of ligand to transmit a signal. It is unknown if heterodimers of the different receptors can form.

Distribution

The receptors are found on endothelial cells, primitive haematopoietic stem cells, monocytes and erythroleukaemia cells[14].

Physicochemical properties of the VEGF receptors

	Human		
Property	flt1	flt4	KDR
Amino acids – precursor	1338	1298	1356
– mature	1316	1276	1336
M_r (K) – predicted	150	145	151
– expressed	160	?	195/235
Potential *N*-linked glycosylation sites	13	12	17
Affinity K_d (M)	0.2×10^{-10}	0.75×10^{-10}	

Signal transduction

Ligand binding causes receptor dimerization, tyrosine phosphorylation of a 200 000 M_r protein in endothelial cells, autophosphorylation and a calcium flux[12].

Chromosomal location

Flt1 is located on human chromosome 13q12[14], kdr is on human chromosome 4, and flt4 is found on chromosome 5q33–qter[13].

Amino acid sequence for human flt1 VEGF receptor[9]

Accession code: SwissProt P17948

```
  -22 MVSYWDTGVL LCALLSCLLL TG
    1 SSSGSKLKDP ELSLKGTQHI MQAGQTLHLQ CRGEAAHKWS LPEMVSKESE
   51 RLSITKSACG RNGKQFCSTL TLNTAQANHT GFYSCKYLAV PTSKKKETES
  101 AIYIFISDTG RPFVEMYSEI PEIIHMTEGR ELVIPCRVTS PNITVTLKKF
  151 PLDTLIPDGK RIIWDSRKGF IISNATYKEI GLLTCEATVN GHLYKTNYLT
  201 HRQTNTIIDV QISTPRPVKL LRGHTLVLNC TATTPLNTRV QMTWSYPDEK
  251 NKRASVRRRI DQSNSHANIF YSVLTIDKMQ NKDKGLYTCR VRSGPSFKSV
  301 NTSVHIYDKA FITVKHRKQQ VLETVAGKRS YRLSMKVKAF PSPEVVWLKD
  351 GLPATEKSAR YLTRGYSLII KDVTEEDAGN YTILLSIKQS NVFKNLTATL
  401 IVNVKPQIYE KAVSSFPDPA LYPLGSRQIL TCTAYGIPQP TIKWFWHPCN
  451 HNHSEARCDF CSNNEESFIL DADSNMGNRI ESITQRMAII EGKNKMASTL
  501 VVADSRISGI YICIASNKVG TVGRNISFYI TDVPNGFHVN LEKMPTEGED
  551 LKLSCTVNKF LYRDVTWILL RTVNNRTMHY SISKQKMAIT KEHSITLNLT
  601 IMNVSLQDSG TYACRARNVY TGEEILQKKE ITIRDQEAPY LLRNLSDHTV
  651 AISSSTTLDC HANGVPEPQI TWFKNNHKIQ QEPGIILGPG SSTLFIERVT
  701 EEDEGVYHCK ATNQKGSVES SAYLTVQGTS DKSNLELITL TCTCVAATLF
  751 WLLLTLLIRK MKRSSSEIKT DYLSIIMDPD EVPLDEQCER LPYDASKWEF
  801 ARERLKLGKS LGRGAFGKVV QASAFGIKKS PTCRTVAVKM LKEGATASEY
  851 KALMTELKIL THIGHHLNVV NLLGACTKQG GPLMVIVEYC KYGNLSNYLK
  901 SKRDLFFLNK DAALHMEPKK EKMEPGLEQG KKPRLDSVTS SESFASSGFQ
  951 EDKSLSDVEE EEDSDGFYKE PITMEDLISY SFQVARGMEF LSSRKCIHRD
 1001 LAARNILLSE NNVVKICDFG LARDIYKNPD YVRKGDTRLP LKWMAPESIF
 1051 DKIYSTKSDV WSYGVLLWEI FSLGGSPYPG VQMDEDFCSR LREGMRMRAP
 1101 EYSTPEIYQI MLDCWHRDPK ERPRFAELVE KLGDLLQANV QQDGKDYIPI
 1151 NAILTGNSGF TYSTPAFSED FFKESISAPK FNSGSSDDVR YVNAFKFMSL
 1201 ERIKTFEELL PNATSMFDDY QGDSSTLLAS PMLKRFTWTD SKPKASLKID
 1251 LRVTSKSKES GLSDVSRPSF CHSSCGHVSE GKRRFTYDHA ELERKIACCS
 1301 PPPDYNSVVL YSTPPI
```

An alternative terminus of GV at amino acid 1255 has been described. Tyr1031 is phosphorylated. Thr770 can be replaced with Phe.

Amino acid sequence for human flt4 VEGF receptor[3]

Accession code: GenEMBL X69878, X68203

```
  -22 MQRGAALCLR LWLCLGLLDG LV
    1 SDYSMTPPTL NITEESHVID TGDSLSISCR GQHPLEWAWP GAQEAPATGD
   51 KDSEDTGVVR DCEGTDARPY CKVLLLHEVH ANDTGSYVCY YKYIKARIEG
  101 TTAASSYVFV RDFEQPFINK PDTLLVNRKD AMWVPCLVSI PGLNVTLRSQ
  151 SSVLWPDGQE VVWDDRRGML VSTPLLHDAL YLQCETTWGD QDFLSNPFLV
  201 HITGNELYDI QLLPRKSLEL LVGEKLVLNC TVWAEFNSGV TFDWDYPGKQ
  251 AERGKWVPER RSQQTHTELS SILTIHNVSQ HDLGSYVCKA NNGIQRFRES
  301 TEVIVHENPF ISVEWLKGPI LEATAGDELV KLPVKLAAYP PPEFQWYKDG
  351 KALSGRHSPH ALVLKEVTEA STGTYTLALW NSAAGLRRNI SLELVVNVPP
```

401 QIHEKEASSP SIYSRHSRQA LTCTAYGVPL PLSIQWHWRP WTPCKMFAQR
451 SLRRRQQQDL MPQCRDWRAV TTQDAVNPIE SLDTWTEFVE GK**N**KTVSKLV
501 IQNA**N**VSAMY KCVVSNKVGQ DERLIYFYVT TIPDGFTIES KPSEELLEGQ
551 PVLLSCQADS YKYEHLRWYR L**N**LSTLHDAH GNPLLLDCKN VHLFATPLAA
601 SLEEVAPGAR HATLSLSIPR VAPEHEGHYV CEVQDRRSHD KHCHKKYLSV
651 QALEAPRLTQ **N**LTDLLV**N**VS DSLEMQCLVA GAHAPSIVWY KDERLLEEKS
701 GVDLADSNQK LSIQRVREED AGPYLCSVCR PKGCV**N**SSAS VAVEGSEDKG
751 SMEIVILVGT GVIAVFFWVL LLIFCNMRR PAHADIKTGY LSIIMDPGEV
801 PLEEQCEYLS YDASQWEFPR ERLHLGRVLG YGAFGKVVEA SAFGIHKGSS
851 CDTVAVKMLK EGATASEQRA LMSELKILIH IGNHLNVVNL LGACTKPQGP
901 LMVIVEFCKY GNLSNFLRAK RDAFSPCAEK SPEQRGRFRA MVELARLDRR
951 RPGSSDRVLF ARFSKTEGGA RRASPDQEAE DLWLSPLTME DLVCYSFQVA
1001 RGMEFLASRK CIHRDLAARN ILLSESDVVK ICDFGLARDI YKDPDYVRKG
1051 SARLPLKWMA PESIFDKVYT TQSDVWSFGV LLWEIFSLGA SPYPGVQINE
1101 EFCQRVRDGT RMRAPELATP AIRHIMLNCW SGDPKARPAF SDLVEILGDL
1151 LQGRGLQEEE EVCMAPRSSQ SSEEGSFSQV STMALHIAQA DAEDSPPSLQ
1201 RHSLAARYYN WVSFPGCLAR GAETRGSSRM KTFEEFPMTP TTYKGSVDNQ
1251 TDSGMVLASE EFEQIESRHR QESGFR

Amino acid sequence for human KDR VEGF receptor[12]

Accession code: GenEMBL X61656, L04947

-19 MSKVLLAVAL WLCVETRAA
1 SVGLPSVSLD LPRLSIQKDI LTIKANTTLQ ITCRGQRDLD WLWPNNQSGS
51 EQRVEVTECS DGLFCKTLTI PKVIG**N**DTGA YKCFYRETDL ASVIYVYVQD
101 YRSPFIASVS DQHGVVYITE NK**N**KTVVIPC LGSISNL**N**VS LCARYPEKRF
151 VPDGNRISWD SKKGFTIPSY MISYAGMVFC EAKINDESYQ SIMYIVVVVG
201 YRIYDVVLSP SHGIELSVGE KLVL**N**CTART ELNVGIDFNW EYPSSKHQHK
251 KLVNRDLKTQ SGSEMKKFLS TLTIDGVTRS DQGLYTCAAS SGLMTKK**N**ST
301 FVRVHEKPFV AFGSGMESLV EATVGERVRI PAKYLGYPPP EIKWYKNGIP
351 LES**N**HTIKAG HVLTIMEVSE RDTG**N**YTVIL TNPISKEKQS HVVSLVVYVP
401 PQIGEKSLIS PVDSYQYGTT QTLTCTVYAI PPPHHIHWYW QLEEECANEP
451 SQAVSVTNPY PCEEWRSVED FQGGNKIEVN KNQFALIEGK **N**KTVSTLVIQ
501 AA**N**VSALYKC EAVNKVGRGE RVISFHVTRG PEITLQPDMQ PTEQESVSLW
551 CTADRSTFE**N** LTWYKLGPQP LPIHVGELPT PVCKNLDTLW KL**N**ATMFS**N**S
601 TNDILIMELK **N**ASLQDQGDY VCLAQDRKTK KRHCVVRQLT VLERVAPTIT
651 GNLE**N**QTTSI GESIEVSCTA SGNPPPQIMW FKD**N**ETLVED SGIVLKDGNR
701 **N**LTIRRVRKE DEGLYTCQAC SVLGCAKVEA FFIIEGAQEK TNLEIIILVG
751 TTVIAMFFWL LLVIILGTVK RANGGELKTG YLSIVMDPDE LPLDEHCERL
801 PYDASKWEFP RDRLNLGKPL GRGAFGQEIE ADAFGIDKTA TCRTVAVKML
851 KEGATHSEHR ALMSELKILI HIGHHLNVVN LLGACTKPGG PLMVIVEFCK
901 FGNLSTYLRS KRNEFVPYKT KGARFRQGKD YVGAIPVDLK RRLDSITSSQ
951 SSASSGFVEE KSLSDVEEEE APEDLYKDFL TLEHLICYSF QVAKGMEFLA
1001 SRKCIHRDLA ARNILLSEKN VVKICDFGLA RDIYKDPDYV RKGDARLPLK
1051 WMAPETIFDR VYTIQSDVWS FGVLLWEIFS LGASPYPGVK IDEEFCRRLK

```
1101 EGTRMRAPDY TTPEMYQTML DCWHGEPSQR PTFSELVEHL GNLLQANAQQ
1151 DGKDYIVLPI SETLSMEEDS GLSLPTSPVS CMEEEEVCDP KFHYDNTAGI
1201 SQYLQNSKRK SRPVSVKTFE DIPLEEPEVK VIPDDNQTDS GMVLASEELK
1251 TLEDRTKLSP SFGGMVPSKS RESVASEGSN QTSGYQSGYH SDDTDTTVYS
1301 SEEAELLKLI EIGVQTGSTA QILQPDTGTT LSSPPV
```

References

1. Leung, D.W. et al. (1989) Science 246, 1306–1309.
2. Clauss, M. et al. (1990) J. Exp. Med. 172, 1535–1545.
3. Keck, P.J. et al. (1989) Science 246, 1309–1312.
4. Lythe, D.J. et al. (1994) J. Virol. 68, 84–92.
5. Connolly, D.T. et al. (1989) J. Clin. Invest. 84, 1470.
6. Houck, K.A. et al. (1992) J. Biol. Chem. 267, 26031–26037.
7. Tischer, E. et al. (1991) J. Biol. Chem. 266, 11947–11954.
8. Conn, G. et al. (1990) Proc. Natl Acad. Sci. USA 87, 20017–20024.
9. De Vries, C. et al. (1992) Science 255, 989–991.
10. Shibuya, M. et al. (1990) Oncogene 5, 519–524.
11. Millauer, B. et al. (1993) Cell 72, 835–846.
12. Terman, B.I. et al. (1992) Biochem. Biophys. Res. Commun. 187, 1579–1586.
13. Galland, F. et al. (1993) Oncogene 8, 1233–1240.
14. Jakeman, L.B. et al. (1992) J. Clin. Invest. 89, 244–253.
15. Myoken, Y. et al. (1991) Proc. Natl Acad. Sci. USA 88, 5819–5823.

Appendix I: Cytokine standards

The inherent variability of immunoassays, and particularly biological assays, means that cytokine assays should include a standard or reference preparation. Each laboratory should have an aliquoted frozen preparation of its own cytokine standards. Each assay should include at least one standard curve of the lab standard. The lab standard can be assigned an arbitrary potency of *x* units/ml which can then be compared with the activity of unknown samples by probit analysis. For most human cytokines, and some murine cytokines, an official WHO standard or reference reagent is available from NIBSC, Blanche Lane, South Mimms, Potters Bar, Herts, UK or in the USA from BRMP, National Cancer Institutes, Frederick, Maryland 21701. Some common ones are listed below. The official standards should be used to calibrate lab standards.

Preparation	Code
Interleukin 1α rDNA (International Standard)	86/632
Interleukin 1β rDNA (International Standard)	86/680
Interleukin 2 cell line-derived (International Standard)	86/504
Interleukin 2 rDNA	86/564
Interleukin 3 rDNA	88/780
Interleukin 4 rDNA	88/656
Interleukin 5 rDNA	90/586
Interleukin 6 rDNA (International Standard)	89/548
Interleukin 7 rDNA	90/530
Interleukin 8 rDNA	89/520
Interleukin 9 rDNA	91/678
Interleukin 10 rDNA	92/516
Interleukin 11 rDNA	72/788
Interleukin 12 rDNA	95/544
Interleukin 13 rDNA	94/622
Interleukin 15 rDNA	95/554
basic FGF rDNA	90/712
BDNF	96/534
BMP-2 rDNA	93/574
EGF rDNA	91/530
Flt3 ligand	96/532
G-CSF rDNA (International Standard)	88/502
GM-CSF rDNA (International Standard)	88/646
Growth hormone (International Standard)	80/505
HGF rDNA (International Standard)	96/564
IGF I (International Standard)	91/554
IGF II	96/538
IFNα human leukocyte (International Standard)	94/784
IFNω (International Standard)	94/754
Leukaemia inhibitory factor rDNA	91/602
M-CSF rDNA (International Standard)	89/512
MIP-1α rDNA	92/518
PF-4 purified human (International Standard)	83/505
RANTES rDNA	92/520
Stem cell factor/MGF rDNA	91/682
TGFβ1 rDNA	89/514

TGFβ1 (natural bovine)	89/516
TGFβ2 rDNA	90/696
TNFα rDNA (International Standard)	87/650
TNFβ rDNA	87/640

Appendix II: WWW Sites for DNA and Protein Databases

Entrez

http://www3.ncbi.nlm.nih.gov/Entrez/

This is an interlinked database of DNA sequences, protein sequences, protein structures and a subsection of MEDLINE containing references (including abstracts) for those references with sequence-related information.

European Bioinformatics Institute (EBI)

http://www.ebi.ac.uk

A side range of databases including the Protein Data Bank (PDB), SwissProt, EMBL and additional specialized databases.

GenBank

http://ncbi.nlm.nih.gov

The complete nucleotide database, including access to additional databases and BLAST search tools for homology-based searches.

PDB

http://www.rcsb.org

An archive of experimentally determined three-dimensional structures of biological macromolecules compiled at the Brookhaven National Laboratory.

SwissProt

http://www.expasy.ch/sprot

This database contains all SwissProt entries. The database has links to entries in PDB and references in Entrez.

TNF Ligand and Receptor Superfamily Database

http://www.gene.ucl.ac.uk/users/hester/tnftop.html

Contains information on the new TNF nomenclature scheme with information on ligands and receptors from this superfamily.

CD40L Defect Database

http://www.expasy.ch/cd40lbase

An archive of clinical and molecular data on CD40 ligands defects leading to X-linked hyperIgM syndrome.

Appendix III: Internet Resources for Transgenic and Targeted Mutation Research

TBASE (The Transgenic/Targeted Mutation Database)	http://tbase.jax.org/
IMR (Induced Mutant Resources of the Jackson Laboratory)	http://www.jax.org/resources/documents/imr/
Database of Gene Knockouts (full records are taken directly from TBASE)	http://www.bioscience.org/knockout/knochome.htm
BioMedNet Mouse Knockout Database	http://biomednet.com/db/mkmd
UCD Medpath Transgenic Mouse Searcher 2.0	http://www-mp.ucdavis.edu/personaltgmouse1.html
Nagy Laboratory: Cre Transgenic and Floxed Gene Skarnes Laboratory Resource of Gene trap Insertions	http://socrates.berkeley.edu/~skarnes/resource.html
MGD (The Mouse Genome Database)	http://www.informatics.jax.org/
The Portable Dictionary of the Mouse Genome	http://mickey.utmem.edu/front.html
Genetic and Physical Maps of the Mouse Genome (MIT)	http://www-genome.wi.mit.edu/cgi-bin/mouse/index
The European Collaborative Interspecific Mouse Backcross (EUCIB)	http://www.hgmp.mrc.ac.uk/MBx/MBxHomepage.html

Index

Lightning Source UK Ltd.
Milton Keynes UK
05 January 2010

148179UK00001B/9/P